Green Culture

Green Culture

An A-to-Z Guide

KEVIN WEHR, GENERAL EDITOR
California State University, Sacramento

PAUL ROBBINS, SERIES EDITOR
University of Arizona

Los Angeles | London | New Delhi
Singapore | Washington DC

Los Angeles | London | New Delhi
Singapore | Washington DC

FOR INFORMATION:

SAGE Publications, Inc.
2455 Teller Road
Thousand Oaks, California 91320
E-mail: order@sagepub.com

SAGE Publications Ltd.
1 Oliver's Yard
55 City Road
London EC1Y 1SP
United Kingdom

SAGE Publications India Pvt. Ltd.
B 1/I 1 Mohan Cooperative Industrial Area
Mathura Road, New Delhi 110 044
India

SAGE Publications Asia-Pacific Pte. Ltd.
33 Pekin Street #02-01
Far East Square
Singapore 048763

Publisher: Rolf A. Janke
Assistant to the Publisher: Michele Thompson
Senior Editor: Jim Brace-Thompson
Production Editors: Kate Schroeder, Tracy Buyan
Reference Systems Manager: Leticia Gutierrez
Reference Systems Coordinator: Laura Notton
Typesetter: C&M Digitals (P) Ltd.
Proofreader: Rae-Ann Goodwin
Indexer: Kathy Paparchontis
Cover Designer: Gail Buschman
Marketing Manager: Kristi Ward

Golson Media
President and Editor: J. Geoffrey Golson
Author Manager: Ellen Ingber
Editors: Mary Jo Scibetta, Kenneth Heller
Copy Editors: Tricia Lawrence, Holli Fort, Barbara Paris

Printed in the United States of America

Library of Congress Cataloging-in-Publication Data

Green culture : an A-to-Z guide / Kevin Wehr, editor.

p. cm. — (The Sage reference series on green society: toward a sustainable future)
Includes bibliographical references and index.

ISBN 978-1-4129-9693-8 (cloth) — ISBN 978-1-4129-7571-1 (ebk)

1. Sustainable living. 2. Environmentalism.
I. Wehr, Kevin, 1972-

GE196.G74 2011 304.2—dc22 2011007300

11 12 13 14 15 10 9 8 7 6 5 4 3 2 1

Contents

About the Editors

Green Series Editor: Paul Robbins

Paul Robbins is a professor and the director of the University of Arizona School of Geography and Development. He earned his Ph.D. in Geography in 1996 from Clark University. He is general editor of the *Encyclopedia of Environment and Society* (2007) and author of several books, including *Environment and Society: A Critical Introduction* (2010), *Lawn People: How Grasses, Weeds, and Chemicals Make Us Who We Are* (2007), and *Political Ecology: A Critical Introduction* (2004).

Robbins's research focuses on the relationships between individuals (homeowners, hunters, professional foresters), environmental actors (lawns, elk, mesquite trees), and the institutions that connect them. He and his students seek to explain human environmental practices and knowledge, the influence nonhumans have on human behavior and organization, and the implications these interactions hold for ecosystem health, the local community, and social justice. Past projects have examined chemical use in the suburban United States, elk management in Montana, forest product collection in New England, and wolf conservation in India.

Green Culture General Editor: Kevin Wehr

Kevin Wehr is an associate professor of sociology at the California State University at Sacramento. He has a B.A. in Sociology from the University of California, Santa Cruz (1994), and an M.S. (1998) and Ph.D. (2002) in Sociology from the University of Wisconsin–Madison. He specializes in environmental sociology, political sociology, social theory, criminology, and popular culture. His first book, *America's Fight Over Water* (2004), examined the social, political, and cultural changes that gave rise to the modern environmental movement against large dams. His second book, *Hermes on Two Wheels* (2009), is an ethnographic examination of the culture of bicycle messengers. He is currently working on a book analyzing the Do It Yourself (DIY) movement in the United States.

Introduction

Culture can be a rather slippery concept to define—something that seems to bleed into almost all aspects of our social lives. It can seem to be everywhere at once, and thus may be nowhere at all. Culture is a set of common human practices (like saying "bless you" after someone sneezes), but is also a set of dispositions or ways of seeing (or not seeing) the world around us. The need to acknowledge that someone has sneezed, and to lay upon him or her some sort of secularized blessing, is both a common ritual and a set of understandings about what is appropriate and polite. In some other cultures, such an acknowledgment might be embarrassing—culture is variable by geography, society, language, race, ethnicity, and many other social cleavages. In short, though, culture is a way of doing things and a way of understanding the world that is common to a specific group of people. One could attempt a definition thus: Culture is a durable system of meanings, symbols, signs, and understandings common to an identifiable social group.

Culture can be made visible in many ways, but even when it is relatively invisible to us, culture is salient. We might speak of "material culture" in the form of a comic book, a newspaper, or a novel. We see cultural expression in fashion, architecture, or public art. Popular culture is a form that is mass-produced explicitly for consumption by a large audience—movies, television, pop music, and video games are clear examples.

So what is "green culture"? There are multiple ways that culture affects the environment (and vice versa). The phrase *green culture* is a way of wrapping those many layers together into a coherent concept. Of course, there might be a culture of environmentalists (an identifiable group of people); in fact there are many cultures of different environmentalists (radical greens, deep greens, ecofeminists, and so forth). These groups of people may choose to live in sustainable ways, and their cultural understandings, rituals, and symbols may reflect their ecological worldview. Thus, there are cultural aspects to the way we live, and how we live can be more and less sustainable.

Similarly, there are cultural aspects to how we produce and consume goods in society. In the Western world, of course, it is primarily through the market that most people meet their daily needs. We shop in grocery stores or eat at restaurants for food, we buy clothes and necessary items at stores, and sometimes we treat ourselves with a luxury purchase. All of this implies a certain set of cultural understandings: many would be baffled by someone who grows his or her own food, makes his or her own clothes, and lives a life of voluntary simplicity.

But we don't buy everything we need from the market—we also pay taxes and have some services provided to us by local, regional, or federal government. This, too, has a cultural aspect to it. We have a set of understandings about the quality of the water that comes out of our faucet, about the cleanliness of the air we breathe, and about the regulation of traffic, commerce, and international relations. Some of these are sustainable practices, but many are not.

On top of this, there are sets of cultural understandings about the proper relationship between humans and the natural world—systems of understandings that are relatively durable, but change over longer periods of time. For instance, it once was considered appropriate to shoot buffalo from passenger trains. Now, we value the American bison as an endangered species. Our culture has changed, and with it the way we behave toward animals and nature.

The concept of sustainability is also a difficult idea to define simply, but for our purposes here, let's call it a set of human practices that allows for continued use of resources for all foreseeable future generations (and, one could argue, for future generations of other species as well). In an attempt to bring all this together, then, "green culture" is a way of understanding human practices common to identifiable groups, which has an ecological component that affects sustainability.

The volume is organized to reflect the many ways in which culture cross-cuts everyday ecological practices. The dominant form of culture, of course, is the media. The entries in this section focus on the ways in which environmentalism is portrayed in the media, and how ecological communication happens. Environmentalism is also nearly synonymous with activism. The section on "Actions and Activists" highlights both important individuals and social groups who engage in environmental activist, but also global, regional, national, and local actions and activities. These are often exceptional or spectacular moments, but how people live and work every day has clear cultural and environmental implications, so the section on "Living" is a central one. Where we live, how we get to work, and what we do on the weekends are all subject to strong cultural forces, which contribute to our level of sustainability. Perhaps the most important contributor to our ecological impact is what, how, and where we eat, so "Food" has its own section. Finally, a special section on "People" highlights the important contributions of selected individuals and important groups.

This volume provides an overview of the many elements of green culture and associated institutions, movements, organizations, and key actors and locations. The many entries are from diverse academic perspectives, and represent the latest thinking on the topics at hand. Some are about environmental "goods" and others about environmental "bads," but all represent important aspects of our shared human culture. These are key characteristics that we must consider as we contemplate moving to a more ecologically sustainable culture and society.

Kevin Wehr
General Editor

Reader's Guide

Actions, Activists, and Law

Antiglobalization Movement
Antiwar Actions/Movement
Art as Activism
Censorship of Climatologists and Environmental Scientists
Chipko Movement
Communication, Global and Regional
Communication, National and Local
Corporate Green Culture
Corporate Social Responsibility
Cyber Action
Demonstrations and Events
Earth First!
Earth Liberation Front (ELF)/Animal Liberation Front (ALF)
Ecocriticism
EcoPopulism
Environmental "Bads"
Environmental "Goods"
Environmental Justice Movements
Environmental Law, U.S.
Global Warming and Culture
Green Anarchism
Green Consumerism
Greenwashing
Hybrid Cars
Individual Action, Global and Regional
Individual Action, National and Local
Institutional/Organizational Action
International Law and Treaties
Law and Culture
Nongovernmental Organizations
Organic Clothing/Fabrics
Organic Foods
Political Persuasion
Rainforest Action Network
Social Action, Global and Regional
Social Action, National and Local
Tree Planting Movement
United Farm Workers (UFW) and Antipesticide Activities
Veganism/Vegetarianism as Social Action

Consumption

"Agri-Culture"
Air Pollution
Alternative/Sustainable Energy
Animals (Confinement/Cruelty)
Biodiversity Loss/Species Extinction
Body Burden
Cooking
Diet/Nutrition
Energy
Fish
Locavores
Obesity and Health
Organics
Population/Overpopulation
Soil Pollution
Space/Place/Geography and Materialism
Veganism/Vegetarianism
Water (Bottled/Tap)
Water Pollution

Daily Living

Alternative Communities
Artists' Materials
Auto-Philia/Auto-Nomy
Bartholomew I: The Green Patriarch

Media Culture

People and Places

List of Articles

List of Contributors

Adams, Jennifer
DePauw University

Adney, Karley
Independent Scholar

Amster, Randall
Prescott College

Antczak, Willa
Vermont Design Institute

Ballamingie, Patricia
Carleton University

Bardecki, Michal
Ryerson University

Barnhill, John H.
Independent Scholar

Boslaugh, Sarah
Washington University in St. Louis

Bridgeman, Bruce
University of California, Santa Cruz

Buettner, Angi
Victoria University of Wellington

Buhr, Susan M.
University of Colorado, Boulder

Cadzow, Daniel
State University of New York, Buffalo

Carveth, Rod
Fitchburg State University

Clark, Woodrow W., II
Clark Strategic Partners

Collins, Timothy
Western Illinois University

Connell, Robert
University of California, Berkeley

Davey, Gareth
Hong Kong Shue Yan University

Dougherty, Michael L.
University of Wisconsin–Madison

Ferber, Michael P.
King's University College

Gachechiladze-Bozhesku, Maia
Central European University, Hungary

Gardner, Robert Owen
Linfield College

Gayer, Dianne Elliott
Independent Scholar

Gonshorek, Daniel O.
Knox College

Good, Jennifer
Brock University

Goodsell, Eli
California State University, Chico

Green, Brandn Q.
Pennsylvania State University

Grosswiler, Paul
University of Maine

Gunter, Michael M., Jr.
Rollins College

Hanson, Lorelei
Athabasca University

Harris, Kristine
California State University, Sacramento

Heffner, Leanna R.
University of Rhode Island

Hein, James Everett
The Ohio State University

Helfer, Jason A.
Knox College

Hosansky, David
Independent Scholar

Ikeda, Kayo
Hiroshima University

Islam, Md Saidul
Nanyang Technological University

Jackson, Ellen M.
Knox College

Jain, Priyanka
University of Kentucky

Janos, Nik
University of California, Santa Cruz

Johnson, Erik W.
Washington State University

Jovanovic, Spoma
Independent Scholar

Juris, Jeffrey S.
Northeastern University

King, Katherine
University of Michigan

Knigge, LaDona
California State University, Chico

Kolodinsky, Jane
University of Vermont

Konieczka, Stephen
University of Colorado

Kozlowski, Anika
Ryerson University

Kremer, Joseph
Washington State University

Kte'pi, Bill
Independent Scholar

Lanfair, Jordan K.
Knox College

Lapp, Julia L.
Ithaca College

Lawrence, Kirk S.
University of California, Riverside

Lee, Megan
University of Georgia

Leonard, Liam
Independent Scholar

Lippert, Ingmar
University of Augsburg

Lubitow, Amy
Northeastern University

Mapp, Christopher
University of Louisiana at Monroe

McClellan, Sara
University of Colorado

McCreery, Anna C.
The Ohio State University

Munday, Pat
Montana Tech of the University of Montana

Nande, Kaustubh
Independent Scholar

Norville, Kylee M.
Knox College

Öztürk, Şule Yüksel
Anadolu University

Phelps, Jess
Independent Scholar

Pitts, Lewis
Independent Scholar

Podeschi, Christopher W.
Bloomsburg University of Pennsylvania

Pólvora, Alexandre
University Paris 1, Pantheon-Sorbonne

Purdy, Elizabeth Rholetter
Independent Scholar

Reed, Matt
Independent Scholar

Rodnitzky, Jerry
The University of Texas at Arlington

Roth-Johnson, Danielle
University of Nevada, Las Vegas

Salsedo, Carl A.
University of Connecticut

Santana, Mirna E.
Independent Scholar

Schewe, Rebecca L.
University of Wisconsin–Madison

Schroth, Stephen T.
Knox College

Schupp, Justin
The Ohio State University

Sheley, Loran E.
California State University, Sacramento

Smith, Dyanna Innes
Antioch University New England

Stough-Hunter, Anjel
The Ohio State University

Townsend, Patricia K.
Independent Scholar

Trevino, Marcella Bush
Barry University

Trumpy, Alexa J.
The Ohio State University

Tyman, Shannon K.
Independent Scholar

Uppal, Charu
Karlstad University

Walker, David M.
Ohio Wesleyan University

Weger, Krista
York University

Whalen, Ken
University of Brunei Darussalam

Willits, Jordan
Knox College

Willits, Logan
Knox College

Yao, Qingjiang
Fort Hays State University

York, Richard
University of Oregon

Young, Cory Lynn
Ithaca College

Young, Sebnem Yucel
Izmir Institute of Technology

Zehner, Ozzie
University of California, Berkeley

Green Culture Chronology

12,000–6,000 B.C.E.: During the Neolithic Revolution, early humans learn to domesticate plants and animals, developing agriculture and the beginnings of settlements in the Fertile Crescent. Previously gathered plants are sowed and harvested, while wild sheep, goats, pigs, and cattle are herded instead of hunted.

1306: England's King Edward I tries unsuccessfully to ban open coal fires in England, marking an early attempt at national environmental protection.

c. 1530: Commercial whaling begins as the Basques begin the pursuit of right whales in the North Atlantic, taking an estimated 25,000 to 40,000 whales over the next 80 years.

1690: Progressive Governor William Penn requires that one acre of forest be saved for every five that are cut down in the newly formed city of Philadelphia.

1854: Henry David Thoreau reflects upon living in nature in *Walden.*

1862: With much of the agricultural south not voting because of the Civil War, the United States creates the Department of Agriculture, which is charged with promoting agriculture production and the land grant university system.

1864: George Perkins Marsh publishes *Man and Nature*, a book that argues that many civilizations have fallen because of environmental degradation.

1872: President Ulysses Grant signs into law a bill designating the area of Yellowstone as the world's first national park.

1892: British reformer Henry S. Salt, a socialist, pacifist, and vegetarian, publishes a landmark work on animal welfare, *Animal Rights Considered in Relation to Social Progress.*

1892: The Sierra Club is founded in San Francisco by preservationist John Muir.

1900: Porsche develops the world's first hybrid-electric car.

1905: Upton Sinclair publishes his novel *The Jungle* in serial format in the socialist magazine *Appeal to Reason.* Public outcry over the filthy conditions of the meatpacking industry portrayed in this novel lead to passage of the Pure Food and Drug Act in 1906.

1905: The U.S. Forest Service is established.

1916: The National Park Organic Act establishes the National Park Service.

1918: The Save the Redwoods League forms in the United States for the purpose of purchasing the remaining redwood forests that have been extensively harvested for lumber.

1923: The dam at Hetch Hetchy, site of the infamous dispute between Gifford Pinchot and John Muir, is completed.

1942: Jerome Irving Rodale begins publication of *Organic Farming and Gardening*, popularizing the concept of organic food production as advocated by the British writers Sir Albert Howard and Lord Northbourne.

1946: The International Convention for the Regulation of Whaling holds its first meeting in Washington, D.C., and sets quotas for whaling that are intended to allow the whaling industry to continue at reduced levels so that whales are not hunted to extinction.

1949: The International Convention for the Regulation of Whaling establishes the International Whaling Commission, which is intended to regulate whaling but has been beset by conflicts between nations with traditional whaling industries (e.g., Japan, Iceland, and Norway) and those that wish to impose a moratorium on all whaling.

1949: Aldo Leopold publishes *A Sand County Almanac*, in which he states that humanity should adopt an "ethic dealing with man's relation to land and to the animals and plants which grow upon it."

1952: An atmospheric inversion in London, coupled with particulate matter in the air from motor vehicles and coal-burning stoves and factories, causes nearly 3,000 excessive deaths in a single week and highlights the importance of controlling man-made sources of air pollution.

1962: Rachel Carson's *Silent Spring* serialized in the *New Yorker.*

1962: *Silent Spring* calls attention to the harmful effects of human activity on the environment, including air pollution and the use of harmful chemicals in agriculture.

1964: U.S. President Lyndon B. Johnson signs the Wilderness Act into law. Over 9 million acres of land are closed to excavation.

1969: A major oil spill in Santa Barbara spills up to 100,000 barrels of crude, washing onto California beaches and sparking public outcry and engendering a period of environmental legislation.

1969: The National Environmental Policy Act (NEPA) is signed, creating a procedural requirement that federal agency actions be evaluated for environmental impacts.

1970: The first Earth Day is celebrated internationally, drawing attention to worldwide interest in environmental protection and reform.

1970: The Environmental Protection Agency (EPA) is created to enforce federal environmental regulations. The agency's mission is to regulate chemicals and protect human health by safeguarding air, land, and water.

1970: Dr. Norman Borlaug, father of the "green revolution," which is credited with substantially increasing crop yield in the third world, wins the Nobel Peace Prize. Although few question that the green revolution saved millions of people from starvation, many, particularly in more recent years, criticize Borlaug's reforms because they rely heavily on chemical fertilizers and irrigation and on seeds that must be purchased annually from multinational corporations, thus increasing corporate control of third world agriculture.

1971: Frances Moore Lappé publishes *Diet for a Small Planet*, introducing the concept of "complementary proteins" (now considered by some scientists to be fallacious). Lappé advocates for the adoption of a vegetarian diet, both for reasons of health and because of the much greater resources required to produce meat rather than vegetables and grains.

1973: Ernst Friedrich Schumacher's *Small Is Beautiful: Economics as if People Mattered* criticizes the assumption that economic development requires adoption of large-scale Western technologies and a lifestyle based on acquisition of consumer goods.

1973: The first action of the Chipko movement takes place in rural India, where largely women villagers embraced the trees in an act of civil disobedience against planned logging operations. Eventually, this sparked change in Indian forest policy.

1978: U.S. President Jimmy Carter declares Love Canal, a toxic waste dump in Moagara Falls, New York, a national emergency due to chemical pollution. Over 1,000 families are evacuated at public expense.

1978: New York City begins Operation Green Thumb to encourage community gardening. The city allows residents to use vacant lots for gardens for the nominal fee of $1 per year. By 1991, the city reports that there are over 500 community gardens in the city.

1982: Over one million gather in New York City's Central Park to protest the nuclear arms race.

1983: The Environmental Protection Agency recommends that the residents of Times Beach, a small town in eastern Missouri, evacuate due to dioxin contamination because the chemical was a contaminant in oil spread on the roads to control dust in the 1970s.

1983: The World Commission on Environment and Development, commonly known as the Brundtland Commission, is held.

1985: A group of activists found the Rainforest Action Network in San Francisco, California, with the purpose of protecting the world's rainforests and the people who live in them from environmental destruction. Their first major action is a boycott of the American fast-food chain Burger King, which at that time imported much of its beef from Central and South America, where rainforest destruction was hastened by the economic incentive of clearing the forest and turning it into grazing land for cattle.

1987: Burger King announces that it is no longer importing beef from rainforest areas.

1987: American activist Dave Foreman publishes *Ecodefense: A Field Guide to Monkeywrenching*, which advocates sabotage to prevent environmentally destructive development and other commercial activities. Many of Foreman's suggested tactics are illegal, including driving metal spikes into trees to prevent their being logged, sabotaging earth-moving equipment such as bulldozers, removing surveyor's stakes, and pulling down power lines. The book's title refers to *The Monkey Wrench Gang*, a 1975 novel by Edward Abbey that calls for individuals to take direct action to halt the destruction of wilderness.

1987: The United Church of Christ Commission for Racial Justice issues a study demonstrating that the location of toxic waste sites is more closely related to the race of neighborhood residents than to either income or social class.

1989: The worst oil spill in American history to date occurs when the supertanker ship *Exxon Valdez* grounds on a reef and spills over 11 million gallons of oil into Prince William Sound near Valdez, Alaska. The resulting oil slick extends to 50 miles and is estimated to kill 10 percent of the region's bird population.

1990: Joseph Hazelwood, captain of the *Exxon Valdez*, is convicted on one misdemeanor count related to the spill.

1991: Exxon pleads guilty to four misdemeanor counts relating to its infamous oil spill and pays about $1 billion in fines and environmental damage payments.

1992: The Earth Summit held in Rio de Janeiro results in the document "Agenda 21," which calls for national governments to adapt strategies for sustainable development and to cooperate with nongovernmental organizations and other countries in implementing them.

1992, 1995: Veganism captures worldwide publicity when American vegan chefs Ken Bergeron and Brother Ron Pickarski win gold medals at the International Culinary Olympics in Berlin.

1993: The U.S. Green Building Council is founded as a nonprofit trade organization that promotes self-sustaining building design, construction, and operation. The council develops the Leadership in Energy and Environmental Design (LEED) rating system and organizes Greenbuild, a conference promoting environmentally responsible materials and sustainable architecture techniques.

1994: U.S. President Bill Clinton signs Executive Order 12898, requiring federal agencies to determine the impact that environmental degradation has on low-income communities.

1996: Monsanto plants the first commercial fields with the Roundup Ready soybean, the first commercially genetically modified crop in the United States. The beans are engineered to resist the common herbicide glyphosate, which can therefore be sprayed on the fields without damaging the soybean crop. By 2002, about 70 percent of soybeans grown in the United States are engineered to be resistant to herbicides.

1996: Adam Werbach becomes the youngest president of the Sierra Club.

1997: The Fair-Trade Labeling Organization (FLO) International is founded in Germany with the goals of bringing together disparate fair-trade organizations and harmonizing standards for fair-trade certification.

1997: The Kyoto Protocol, an international agreement linked to the United Nations' Framework Convention on Climate Change that aims to reduce or prevent global warming, is adopted. Under the protocol, which goes in effect in 2005, most industrialized countries agree to reduce their emissions of greenhouse gases: some were set specific targets, some were given the goal of reducing their emissions to 1990 levels, while others were allowed to reduce their levels. The protocol also allows countries to trade carbon emissions in order to meet their goals.

1999: Mass protest by environmental and labor activists at the World Trade Organization's Ministerial Conference in Seattle is widely covered in the media, raising groups' concerns about the impacts of globalization. Before the event, it was known as N30, and since then, it has been dubbed the "Battle of Seattle."

2002: William McDonough and Michael Braungart popularize the term "cradle to cradle," which was introduced by Walter Stahel in the 1970s. "Cradle to cradle" refers to the principle that companies should be responsible for recycling the materials from their products after they are discarded.

2003: Widespread global protests against the planned war in Iraq occur.

2004: Wangari Muta Maathai receives the Nobel Peace Prize for her work on sustainable forestry and women's rights through the Greenbelt Movement in Kenya.

2005: Hurricane Katrina makes landfall in New Orleans and the Mississippi Gulf Coast, killing thousands and causing scores of billions of dollars in damages.

2006: California's Global Warming Solutions Act of 2006 is signed by Governor Arnold Schwarzenegger and is widely heralded as the nation's leading climate bill.

2006: Walmart launches an initiative requiring its suppliers to reduce packaging to the lowest possible levels.

2006: The documentary *Who Killed the Electric Car?* investigates the history of General Motors EV-1, from its deployment to its recall and destruction.

2007: *An Inconvenient Truth* wins the Academy Award for best documentary feature.

2008: Presidential candidates Barack Obama, Hillary Clinton, and John McCain all tout a "green jobs" agenda during the U.S. presidential election.

2010: The largest offshore oil spill in U.S. history occurs as an explosion rocks the Deepwater Horizon oil rig, spilling approximately 35,000 to 60,000 barrels of oil per day into the Gulf of Mexico and severely harming the region's environment.

Dustin Mulvaney
University of California, Berkeley

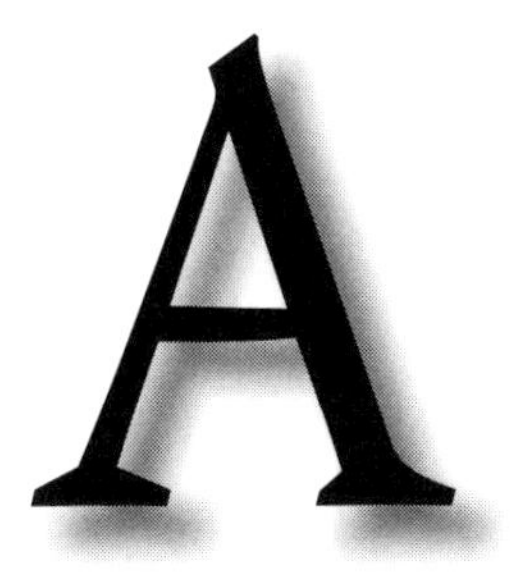

ADVERTISING

As ecological concerns have evolved and developed throughout society, advertisers have become interested in developing green messages and in targeting green customer segments. Historically, environmental friendliness was regarded as an extra feature for a product, while today it has become an ever-increasing selling point. In this regard, advertising is increasingly utilized to effectively present the environmental practices in the production processes of products or businesses. Businesses often produce ads that inform consumers of the environmental perspective of companies. Today, a debate exists on whether the environmental consciousnesses created in the advertisements of environmentally friendly products are artificial or actually stimulate consumers to action. Advertisements that raise environmental sensibility in society may bring prestige to a brand that could then gain wider acceptance and increased market share. Advertisements with an environmental theme often emphasize a product's sensitivity for environment. In these ads, such features as "ozone friendly" and "recyclable" are highlighted.

The advertising community is proactive and, as such, has created a standard lexicon for many overused terms such as "environmentally friendly," "recyclable," "safe," and "natural." Various legal regulations have been passed to determine whether companies that try to differentiate their products and brand name with environmental ads are in earnest. For a product to carry marks such as "ozone friendly" or "recyclable," it must be certified. Only after the product is certified can companies use these marks in their advertisements. A frequent problematic situation in environmental advertising is seen in ads that include ambiguous and vague expressions such as "environmentally friendly" and "environmentally responsible." Legal restrictions and regulations may be necessary since advertising accuracy is uncertain and could cause misunderstandings.

Strategies Used in Environmental Ads

Advertising strategies that take place in many fields of advertising are also applied to environmental advertising. Advertising is based on a certain dramatic code. Advertisers aim to influence their target audience with symbolism, and they transmit collective unconscious images through a combination of discourse and action. This discourse is utilized effectively in environmental ads. Individuals are encouraged to want to save the environment and are cast as playing a divine role in this effort. Scare tactics are often used in environmental advertisements, with the potential consumer first being scared into wanting to help the environment and then presented with the purchase and consumption of a "green" product as a means of doing so. Consumers choose to adopt the role cast in such advertisements and, ironically, consume these products as a way to rid themselves of their consumption guilt.

Advertising is inherently influenced by culture. Ads collectively present images that are created inside and outside of a culture and transmitted from history, literature, art, music, cinema, and television. These are the expression of efforts for understanding the importance of humans, objects, life, the environment, and experiences. Advertisers use these as intellectual shortcuts for introducing a product and evoking reactions. It is possible to express these narrations in a series of comparative features. Each of them takes its correlative definitions from vertical groups and its rival definitions from horizontal differences. These vertical and horizontal features are frequently used in ads on the environment. Comparative characteristics, which advertisers use in the advertisements about the environment, are given as a list in Figure 1. Advertisers employ these features to evoke environmental sensibility. In these features, sometimes nature and tradition, and sometimes modernity and nature, are used jointly. For the ads in which nature and tradition are used commonly, nostalgia is emphasized along with pastoral images. Ads in which modernity and nature are used jointly are generally negative ones. Depending on the quality of the product or the effect that advertiser wants to create, nature or technology is disparaged in this type of advertisement.

	Nature	**Tradition**	**Modernity**
Location	Wilderness	Nation	City
Time	Eternal	Past/Now	Now/Future
Scale	Individual	Small	Mass
Information	Traditional	Crafting	Technology
Qualities	Beauty Challenge	Simplicity Balance	Complexity Change
Promise	Freedom	Commonalty	Progression
Ambiance	Fear and Silence	Nostalgia	Optimism
Person	Adorer Challenger	Inheritor	Doer Victim

Figure 1 Comparative Characteristics of Advertising

When ads are examined in the light of these features, it can be seen that ads with nature scenes present themselves in four categories. The first category is ads that contain panoramic views, vast fields, wild nature, and, even if minimally, snowy mountains. The second category is ads that display sun, sand, beaches, and seas, reminiscent of holidays and entertainment. The third category is ads that show a rural environment in which plants or lakes may be presented in close-up. The last category is ads in which birds and wild animals are the focus. The least common are ads that depict nature in a negative light. Such advertisements are generally seen in the presentation of modern products like allergy medications and mosquito repellents. Vast tropical or unoccupied rural areas are displayed as an asset, and they might also be a place of escape, peace, and regeneration. Wild forests are a representation of purity, immaculateness, and genuineness. Domestic animals have been regarded as creatures exhibiting some qualities valued by humans. The most frequently highlighted meaning of nature is based on supremacy and wondrousness, with nature being depicted as the epitome of beauty and worthy of worship and protection.

Long before environmentalism found a prominent niche in the public mind, it had been seen in the ads of a Canadian logging company. In the first of the three different ads, the protective attitude of the company toward nature was emphasized, showing loving woodsmen performing their duties. The second ad contained the works by famous environmentalist Tommy Tompkins where he familiarized children living in the city with nature. The last ad showcased a bird sanctuary established by the company. The beauty and splendor of swans—among other birds—was presented in extraordinary views. In these ads, the message was "Humans and nature can cohabit." As green philosophy has gained prominence, greater sophistication can be seen in the environmental features of advertisements. Incredible natural beauty has been presented in ads, and the reaction of viewers has generally been that of admiration. An insinuation, however, underlies these actions. In the 1980s, ads glorifying technology were given prominence. In the 1990s, a decline occurred due to the variety of enthusiasms of the 1980s being replaced with a variety of concerns. This was caused by the new enthusiasm and environmentalism that emerged for the marketing of the green.

Green Advertising Forms

Easwar Iyer, Subhabrata Banerjhee, and Charles Gulas studied green ads in three different forms. The first type establishes a positive relationship between product and environment including ads emphasizing an environmentally friendly product. An example of this is hard disks advertised by the Western Digital Company with the slogan "nature friendly." The second type of ad emphasizes a product or service that encourages an environmentalist lifestyle. Many companies use this second type. These ads are a part of the consumption culture in that they offer a lifestyle. The third type proposes cooperation in the responsibility for environmental protection.

Les Carlson, Stephen Grove, and Norman Kangun, on the other hand, studied environmental ads in five categories. These categories contain five environmental theme orientations: product orientation, process orientation, image orientation, environmental situation orientation, and any combination of these four. Product orientation depicts environmentally friendly attitudes of products. In this kind of ad, explanations such as "environmentally friendly," "nonharmful to the environment," or "recyclable" are present. In the process-oriented ads, it is explained that the organization, which produces a product or provides a service, pays regard to the environment in the technology that it utilizes in the manufacture, development, production, and marketing of its products or services. Statements such as "For the production of this product, 25 percent recycled raw material was used," or "No creature was harmed during the tests of this product" are found in this kind of advertisement. In image-oriented ads, companies will showcase their cooperation with environmental organizations supported by a large number of people, or make a stand or promise that will likely be supported by the mass community. Statements such as "We are protecting our forests," or "Five percent of the income from each product you buy goes to the Society for the Protection of Nature" are used in this kind of ad. Advertisers use environmental-situation-oriented ads for drawing attention to an ecological subject that does not include a product or a service. In these ads, expressions like "Agricultural lands of the nation vanish" are used. Finally, in combination, the advertiser places ads that include two or more of these orientations.

In the past several years, green ads have especially focused on the subject of global warming. Ads related to global warming typically show themselves in three forms. The first is ads given by companies that are generally held responsible for global warming. Fossil fuels are one of the biggest causes of global warming. Automobile companies, for example, produce ads explaining that their responsibility is minimal and that they have developed solutions on the matter. Ford Motor Company's advertising campaign "Global Warming. There. We Said It." is a prime example of this form. The second type is ads by companies emphasizing that their products are natural. These ads invite people to stop global warming and appeal to social responsibility and individual conscience. The Body Shop's advertising campaign "Sign This Petition to Stop Global Warming Now!" is an example of this second type. In the third type, companies highlight fears of global warming and claim that consumers can solve the problem by using their products. Ben & Jerry's advertising campaign "Help Put the Freeze on Global Warming" is an example of this third type.

A Critical View of Environmental Ads

By their very nature, ads aim to direct people to more consumption. Environmental ads, however, have a cultural and social dimension as well, instead of a purely materialist approach. An ad is a cultural product, and cultural changes are also reflected in advertising. The thesis of advertising as cultural text is supported by the fact that changes occurring in cultural values as a result of social change are reflected in ads. Gillian Dyer, as cited by Dağtaş, explains it: "An ad, on the one hand, provides a material and materialist motivation by offering perpetual consumption, while, on the other hand, modern advertising paradoxically tells us that the material world is not sufficient. And ads get in touch with culture and values when they try to overtop." The concept of *green consumer* emerges at this juncture. Green advertising is a strategy that is developed for consumers who are sensitive to environmental problems. Green advertising strategies try to create a green consumer culture and identity.

Studies conducted on the subject of advertising and consumer culture have examined relations between behaviors and styles of consumption and consumer identities. As a result of these studies, it has generally been understood that the relationship among people, objects, and consumption necessitates acquisition of products. This is generally seen as the foundation of consumer identity in capitalist societies. The context needed to form consumer identity is constituted by the discourse of ads. Critical theoretician Raymond Williams describes this discourse as "official art of the modern capitalist society." The discourse of advertisement, according to Williams, is not only a materialist discourse—even if it might seem so at first sight—but also a magical discourse. For products and services to be sold, advertisements have to form and teach social and individual values. To do so, a cultural structure must exist where objects are not sufficient, but they also have to be valued. To achieve this, directly accessible social and individual meanings must be attributed to products or services in a different cultural pattern, even if by fantasy.

Within this scope, the ever-increasing environmental problems, social ecological movements toward environmental protection, societal sensitivities on the environment, and attention among world governments all bring environmental sensitivity to the forefront. The explicit global influences of ecological problems also put environmentalism forth as a culture. Strategies are found in green ads toward spreading this culture. A review of environmental ads shows that environmentalism is presented as fashionable, and that companies wishing to profit from this fashion display environmental sensibility. Green ads also play an important role for consumers. Consumers are often highly concerned about the quality of products they use, and of the possible harms of using the products they bought. Consumers do not wish to be encumbered by the consequences of their purchases. However, looking at the other side of the problem, İrfan Erdoğan and Nazmiye Ejder underline that environmentalism is a social phenomenon, and they stress the inaccuracy of reducing consumers to individuals: "Other than a few exceptions like nuclear energy and radiation, environmental pollution is generally attributed to the consumption by masses and thus masses are indicated as the source of the problem. [The] forming of environmental problems is a social phenomenon and it occurs at every stage of social activities from mining of raw materials, to processing, production, distribution and consumption. Therefore, when it [is] reduced to a single field, especially to daily usage and waste by consumers, this comes to mean that a significant part of the problem is ignored."

Capitalist economies emphasize that it is the consumer who decides what will be produced, how it will be produced, packaged, and presented in the mass production. Erdoğan and Ejder illustrate the fallacy of this emphasis by the comparison between usage of net bags and nylon bags: "In the mass production system, an individual is the user of waste-producing substances rather than the decision-maker on the forming of home-scope wastes. The best example of explaining this is the vanishing of the use of net bags in shopping and its being replaced by nylon bags as a mass production tool. Nylon bag is a usage infused by a certain industry. Mass production cannot survive without mass consumption and, therefore, consumption may not be left to chance and to classical market conditions: Consumers and necessities have to be created. In this market of created necessities are impulses towards instant-use and instant-disposal, the choice of consumers is limited with what is presented to them." Advertisement is a tool used in the creation of needs, and it makes the consumer feel those needs.

Reshaping, coloring, presenting in different forms and packages, and creating and supporting unnecessary consumption habits are characteristics of the mass consumption industry to stimulate consumption. The use of green ads is a method of selling products by putting them in an environmentalist package. Consumers sensitive toward the environment are often influenced by green ads that impose a philosophy of "find yourself, understand yourself, and gain value by consuming." Through media, ads can support thoughtless splurge consumption, unnecessary consumption, and the use of production methods and behaviors that contribute to environmental pollution. Many sectors from white goods to food and cleaning products do not acknowledge that they pollute. Even those industries that have been identified as a major cause of the current state of nature have begun to announce themselves nature friendly. This race of "we are more environmentally friendly" takes place through ads, as well as other forms of media.

It is understood that the global warming problem has been felt in daily life with water cuts, decreases of dam water levels, and the loss of agricultural production due to drought. Images of cracking soil becoming desert and emotional glacier scenes with polar bears and penguins are used to express the problem. These themes are used in ads to advertise products. A feeling is aroused in viewers that they can help starving polar bears unable to hibernate by "consuming more, buying more" and thus preventing global warming. These ads with environmental themes can sometimes prevent the stimulation of people truly acting to correct environmental problems and they suggest, instead, that the only way for people to express their sensitivity for these problems is through more consumption. In a capitalist system, it is obvious that companies, whose aim is to make profit, will not consider the environment as a field of responsibility. The understanding of "first economy, then ecology" as remarked by former president of the United States George W. Bush forms the basis of green advertising. Advertising is essentially an economic phenomenon; however, it may seem to project much social responsibility.

See Also: Environmental Media Association; Green Consumerism; Media Greenwashing; Print Media, Advertising; Television, Advertising.

Further Readings

Bannerjee, Subhabrata, Charles Gulas, and Easwar Iyer. "Shades of Green: A Multidimensional Analysis of Environmental Advertising." *Journal of Advertising,* 24/2 (1995).

Carlson, Les, Stephen Grove, and Norman Kangun. "A Content Analysis of Environmental Advertising Claim: A Matrix Method Approach." *Journal of Advertising,* 22/3 (1993).

Cohen, David. "The Regulation of Green Advertising: The State, the Market and the Environmental Goods." *The University of British Columbia Law Review* (1991).

Linder, Stephen. "*Cashing-In on Risk Claims: On the For-Profit Inversion of Signifiers for Global Warming.*" *Social Semiotics,* 16/1 (2006).

Peattie, Ken. "Towards Sustainability: The Third Age of Green Marketing." *The Marketing Review,* 2/2 (2001).

Şule Yüksel Öztürk
Anadolu University

"Agri-Culture"

Many environmentalists have pointed to the link between the environment and the culture of farming, in no small way because this is a fundamental link between society and the planet's ecosystems. It is possible to see the question of farm culture as being about the attitudes and behaviors of those who directly manage the land and raise crops but also about the wider relationship between society and agriculture. Environmental writer and thinker Jules Pretty in his book *Agri-Culture* (2002) observes that the Roman thinkers wrote of *agri* (the fields) and *cultura* (the culture)—that the farmed environment and culture were held as interwoven. For Pretty, after farming for 600 generations, the links between food and farming and our shared cultures are profound, with only recently that relationship between conducted them through commoditized goods. Pretty argues that by this hollowing out of agriculture we are losing a huge amount of our traditional cultures, as a narrow sense of economic and scientific rationalism overrides more meaningful relationships with nature.

The desire for organic food has gained prominence in the past few decades as a more sustainable alternative to technological and industrial farming. This field of bell pepper, sweet corn, and cucumber is in its final stage of the three-year transition to gaining USDA organic certification.

Source: U.S. Agricultural Research Service

Pretty's global survey of the links between farm and food culture with the environment sets the scene for a series of debates that beset the topic. It is important to note that although Pretty represents a stream of thinking about the importance of agriculture and its role in culture, for many years conservation thinking and farming were not seen as compatible. For many conservationists the only way of preserving an ecosystem or species was to separate it from farming altogether through reserves or parks. This argument is harder to maintain in densely populated areas where wildlife and agriculture have co-evolved for several thousands of years. In these instances, some forms of habitat can only be maintained through particular agricultural practices.

One of the reasons why conservationists tended to consider it important that areas were not farmed was not just a lack of appreciation of the importance of the farmed environment but of the difficulty of engaging with farmers. Although accounts such as Pretty's tend to group farmers together, there are substantial differences between how nations and communities relate to those who own and manage the land. In the United Kingdom, the residual influence of aristocratic land ownership emphasizes some of the class differences apparent in ownership of farmland, while in other nations such as South Africa and Zimbabwe, social differences over land ownership reflect colonial and ethnic histories that shape societal ideas about agriculture. In North America, agriculture is often seen as the harbor of republican values as land ownership has been more open to new entrants and less reflective of class differences.

One area in which *agri* meets *cultura* in contemporary debates is around the question of technology in farming. This debate started with the use of the steel plow but during the 20th century in the tensions it revolved around the intensification of agriculture in the green revolution and has continued into debates around genetic engineering/modification. In part, this reflects the reduction, by some powerful groups, of agriculture to be considered solely in terms of the extent of its production of commodities rather than of a broader culture. This productivism is often related to profit-centered businesses, but to its early proponents, it was an idealistic enterprise.

The Ford Foundation sponsored the development of varieties of maize during the 1940s in Mexico that responded well to irrigation and fertilizers. Among those working on these crops was Norman Borlaug, who, with his colleagues, saw these new technologies as warding off the mass hunger and outright starvation that was seen during the postwar years. Authors such as Fairfield Osborn in *Our Plundered Planet* (1948) warned of the ecological pressures on agriculture faced by a rapidly expanding population and the risk of starvation. The green revolution package of technologies marrying crops with fertilizers, pesticides, and machinery was seen by its pioneers such as Borlaug as an egalitarian measure only opposed by those who did not know hunger and discounted that of other people.

Agriculture and Unintended Consequences

It was the critique of authors such as Rachel Carson (1962), who exposed the unintended consequences of promiscuous use of pesticides in particular on wild ecosystems and human health, that began to question the technologies of the new agriculture. Later writers such as Vandana Shiva (1988) argued that the social tensions introduced by these new technologies were being exacerbated to the point of causing violence between the winners and losers of this new arrangement. In Europe there was considerable concern that these technologies, in combination with subsidy schemes designed to protect the incomes of European farmers, had resulted in huge agricultural surpluses, mountains of butter and lakes of wine. Many in the developing nations argued that European and North American surpluses were destroying their agricultural industry as cheap surpluses were dumped on local and global markets, undercutting local producers. However, the Soviet nations did not complain because after the early 1970s they became net importers of these surpluses, even during the Cold War period. These surpluses were in part the bedrock on which the rapid urbanization of many developing nations was achieved, as Mike Davis in *Planet of Slums* (2006) discusses, and without any capacity to produce food, the residents of these huge cities are reliant on the global surplus of grains especially. When global prices spiked in 2006–08, this unleashed a wave of rioting and hunger as the vulnerability of these populations was viscerally demonstrated.

For many interested in boosting the productivity of agriculture, the next stage was to improve the breeding of plants and animals through the use of the new genetic technologies and then augment those outcomes with new traits. The novel possibilities offered by these techniques included the introduction of toxins to insects borrowed from bacillus and resistance to particular pesticides. Other conjectures were made of the types of augmentations including drought resistance and improved nutrition, but those modifications that were most profitable to the companies developing the crops were first to the marketplace. In North America, these technologies were viewed by the Food and Drug Administration as "substantially equivalent" to nonmodified crops, and they were introduced to the food chain with little controversy. Their reception in Europe was markedly different.

For many Europeans, traditional forms of agriculture are viewed as expressions of a link to their culture and food habits in a way that informs the broader culture of their nature. Among those who were arrested after destroying genetically modified (GM) crops and leading protestors in dismantling a McDonald's was José Bové. He was happy to accept a jail term for what he saw as protest against the influence of the United States over European agriculture and traditions. Direct attacks against test crops were pioneered in the United Kingdom (UK), where consumers were struggling with the aftermath of a food contamination incident that had potentially left most of the population of the UK infected. During the 1980s a new disease began attacking cattle, causing them to stagger and fall, then die in confusion. This brain disease, popularly known as "mad cow disease," or formally as bovine spongiform encephalopathy (BSE), was always fatal. After several government investigations assured consumers that it was not possible for the disease to jump between species, in 1996, it was reported in the UK parliament that it had. The cause of the disease appeared to be the recycling of proteins in cattle feed, proteins that had been derived from cows—a technological mediated cannibalism. As part of the reaction to this disease outbreak, the British population blamed farmers, the food industry, and the government and were certainly not going to tolerate any new agriculture technologies. In a context of technologically sophisticated supply chains between the field and the consumer's plate, the chance of errors boomeranging back appeared to be amplified. Although no longer close to the consumer, the culture of farming retained a link to public health. European supply chains quickly moved to remove GM products, cleaving the increasingly global supply of food into two wealthy and influential spheres.

The Organic Movement

The most concerted critique of conventional agriculture has been through the organic farming and food movement that began in Europe in the 1920s. Since that time, the organic movement has developed both a practical alternative to mainstream agriculture and also a rebuttal of the dominance of a range of agricultural technologies. This started with the introduction of "artificial" nitrogen-based fertilizers, which the movement argued destroyed the ecosystem of the soil and threatened the future of global agriculture—pointing to the experience of the Dust Bowl. In the postwar period, they highlighted the ecological impacts of pesticides in the environment, including not only the effects on human health but also the reduction in the complexity of farmed ecosystems. During the 1960s in the UK, they also began a critique of the intensive production of animals on welfare grounds and also on the possibilities for the spread of antibiotic-resistant pathogens to humans. This critique continued to encompass GM crops that they saw as allowing the intensification of the use of chemical fertilizers and pesticides, as a threat to genetic diversity through genetic drift between wild and domesticated varieties as well as the potential for unknown health impacts. Although politically marginal for many years and dismissed as cranks when represented by figures such as Lady Eve Balfour and J. I. Rodale, as organic products became available during the 1970s and 1980s, the influence of the movement became more prominent.

The organic production system eschews technologies such as most contemporary fertilizers and pesticides, seeking to replace them with more sustainable alternatives. Fertility is produced through composting of waste materials, the planting of nitrogen-fixing plants, and the use of farmyard manure from animals. Pesticides used range from basic chemical mixes to soaps, plant-based sprays, and, in some instances, biological controls. The aim is to produce a farming system that is closer to natural systems and, whenever possible, augmenting and building their capacity. At a farm level, organic farmers are governed by a set of agreed standards that are inspected by third parties and allow the use of a logo designating a legally enforceable status of "organic." What the farmers actually do on their farms is defined by the technologies they are not allowed to use, leaving the individual farmer to determine what he or she wants to do on his or her own land.

The challenge from organic farming is useful in revealing the attitudes that surround debates about agriculture. Early criticisms accused the organic movement of being "vitalists," seeing in nature properties that could not be discerned by science. Certainly some proponents did, and still do, see nature in quasi-magical terms; others point out that the sum of the effects of an ecosystem still escape reductionist science. Later, they were accused of elitism in producing food in smaller quantities and at a greater price than other farming systems. The response to the question of productivity is a central axis of many debates about the nature of contemporary agriculture and, to a degree, society as well. How yield from a farm is measured can be conducted in a number of ways: the number of people used, the amount of land, the energy expended, or the resources used. The green revolution sought to boost productivity on the land available through the intensive use of energy in fossil fuels and machinery; other metrics of measurement might create a more favorable comparison between organic and nonorganic farming. Again, the question of the price of food relates to the end of farming and broader social questions. Food in many Western nations became very cheap, leading to its becoming an ingredient in highly processed foods to be wasted in consumers' homes as they bought more than they could eat and to the steady growth in the size of many consumers as they tended toward obesity. The organic riposte has been that farmers ought to grow, and sell, more basic foodstuffs that focus on creating health. In this debate, the question is often posed in terms of absolute shortage, while the social reality is about the virtues of relative abundance.

The challenges stemming from human-induced climate change have affected agriculture in new ways: First to consider is the role of agriculture in creating greenhouse gases (GHG) that contribute to the warming of the planetary climate system. Although many of these gases are attributed to the distribution of food through a global system of delivery, others are closely associated with agricultural production. Methane, a particularly long-lived GHG, can come from the paddy cultivation of rice and the production of ruminant animals—mostly cattle. A considerable effort has been put into attempting to calculate the carbon footprint of farming, with carbon serving as the proxy for all of the GHGs. In this way, the technologies used in production are being reconsidered, and the balance between calories used to produce and calories to consume is recalibrated. Others point to the role of farmlands in capturing and storing carbon, so that agriculture can mitigate climate change. In this way, the role of agriculture has to be reconsidered in combination with the other discussions outlined above.

An unlikely collision between communist politics and gastronomy in Italy in the 1970s gave rise to a social movement that brings the relationship between the field and culture a new twist. The slow food movement founded by Carlo Petrini and colleagues gained public prominence after a dispute about a proposed McDonald's at the Spanish Steps in Rome. With a manifesto of "Good, clean, fair?" the movement looked to preserve the traditions of an interlinkage between food and conviviality that they saw being threatened by the homogenous diet espoused by the golden arches. By "good," the movement means that the food should represent the locality and traditions of an area; through this the ecosystem is preserved and the communities that are embedded in that area continue. This is a focus on small producers, growers, and farmers who express a local culture through their agriculture. "Clean" for the movement means that food is produced with no or minimal intervention from industrial techniques. Crops are grown with minimal or no chemical aids, processing relies on smoking and older ways of preserving, with no GM products, irradiated or food additives being present. "Fair" means that the grower, farmer, and fisher are rewarded adequately for these efforts in a way that allows them to continue to follow their craft and often in a way that forms connections between producers and consumers. The slow food movement has become a global one, promoting the preservation of edible biodiversity, cultural traditions, and the pleasures of dining.

Agriculture has often been linked with living the "good life"; many urban and industrial cultures look on being in contact with nature through agriculture as the highest cultural goal. Those who protest against the dominant forms of agriculture often serve to expose the cultural ideas hidden in discussion of farm yields, crop protection, and animal rearing. As the crossroads through which human society connects to nature, agriculture has and will always have a central place in human society. Unlike many other ways in which people connect to nature, food and its production cannot be substituted.

See Also: Chávez, César (and United Farm Workers); Diet/Nutrition; Huerta, Dolores; Locavores; Organic Foods; Veganism/Vegetarianism; Veganism/Vegetarianism as Social Action.

Further Readings

Davis, Mike. *Planet of Slums*. London: Verso, 2006.
Osborn, Fairfield. *Our Plundered Planet*. Boston: Little Brown & Company, 1948.
Pretty, Jules. *Agri-Culture: Reconnecting People, Land, and Nature*. London: Earthscan, 2002.

Matt Reed
Independent Scholar

Air Pollution

From the beginning of the Industrial Revolution until the current debate about global warming, air pollution has been a political and cultural concern. In short, air pollution is the release of chemicals, particulate matter, or other airborne substances into the atmosphere in large enough quantities that they are harmful to humans, animals, or natural ecosystems. It includes a variety of different chemicals with different effects on human health and the ecosystem, and the predominant types of air pollution have varied significantly over time. However, the scientific definition of air pollution does not capture the effects it has had on public opinion or culture. The cultural understanding of air pollution includes the deadly smog events of the 19th and 20th centuries, and it has recently come to include a wider variety of pollutants. As scientific understanding about air pollution has grown, so have legal efforts to control it and define it in new and broader ways.

Smog: An Undefined Problem

Prior to the Industrial Revolution of the 18th and 19th centuries, air pollution was a small-scale problem in a few large cities around the world. Air pollution did not become a problem until machine manufacturing led to the widespread use of coal in industrializing nations. In the late 19th and early 20th centuries, air pollution was known as "smog" or "smoke" and, although it was believed to be harmful, its composition and specific health effects were unknown. Concerns about air pollution grew as the problem worsened, and local air pollution control efforts were passed in cities beginning at the start of the 20th century. However, national legal controls were still decades away.

During World War II and the years immediately following it, smog became an increasingly severe problem in the United States and elsewhere. Smog incidents in Los Angeles County led to the creation of the nation's first unified air pollution control district in 1947, and in 1948, smog in Donora, Pennsylvania, killed 20 people and hospitalized over 600 in a nationally publicized smog event. The deadliest smog event occurred across the Atlantic in London, England, where the Great Smog of 1952 lasted four days, killing over 4,000 people and causing health problems that eventually lead to another 8,000 deaths in the following months.

London "pea-soupers" were well known in the United States and added to the rising concerns about air pollution. In addition to human deaths and health impacts, air pollution during this time also blackened buildings and reduced visibility. For example, the population of gray peppered moths in London primarily consisted of lighter-colored individuals prior to the Industrial Revolution. After smog had darkened the buildings, however, the darker-colored moths were predominant because predators could not see them as easily on the dark buildings. These and other killer smog events produced a growing awareness of the danger of air pollution, while increased scientific understanding of the sources of the problem encouraged political action.

National Control Efforts: The Clean Air Act and Amendments

The first national effort to address air pollution was the Air Pollution Control Act, passed in 1955 to provide funding for research on air pollution. Specific chemicals were identified as components of air pollution, such as nitrogen oxides, volatile organic compounds, ground-level ozone, particulate matter, and sulfur dioxide. The term *air pollution* became more common with greater understanding of the phenomenon, and it replaced the older terms *smog* and *smoke*. The sources of air pollution were also studied; although factories had thus far been blamed as the primary culprit, automobiles, diesel trucks, and backyard incinerators were newly recognized as significant contributors of air pollution.

Although the Air Pollution Control Act only allocated money for research, the Clean Air Act of 1963 was the first national legislation regulating air pollution with a mechanism for enforcement. The Clean Air Act and its subsequent amendments have provided some of the primary legal tools for combating air pollution. The Clean Air Act of 1963 provided money for more research and enabled states to take federal legal action against air pollution through the U.S. Attorney General. Federal research money from this act and its precursors came to fruition in the 1960s, when the results of research confirming the impact of air pollution on public health became more widely known in the larger society.

Public opinion was firmly in favor of control, and many were concerned about the dangers of air pollution that research had demonstrated. The 1970 Clean Air Act Amendments greatly expanded the federal role in air pollution control, and this regulation fell under the purview of the newly created U.S. Environmental Protection Agency (EPA). Air quality standards established maximum allowable levels of six criteria air pollutants: carbon monoxide, sulfur dioxide, nitrogen dioxide, ground-level ozone, particulate matter smaller than 10 micrometers, and lead. Sanctions were then used to help improve air quality in localities that did not meet those standards. Additionally, the 1970 amendments regulated both stationary sources such as factories and power plants and mobile sources such as cars and light trucks. Although air pollution was still a significant problem in many parts of the United States, improvements were beginning to occur.

One improvement in air quality that can be traced to the 1970 Clean Air Act Amendments is the dramatic reduction in lead pollution in the air. Leaded gasoline had produced high levels of lead air pollution in U.S. cities, but the 1970 Amendments required the gradual phase-out of leaded gasoline. Lead was finally eliminated entirely from gasoline in 1995, and blood lead levels in adults and children decreased steadily throughout the 1970s and 1980s. Specifically, the percentage of U.S. children aged 1–5 with blood lead levels high enough to be dangerous dropped from over 88 percent in 1976 to under 5 percent in 1994. The reduction in lead air pollution thus provided significant improvements to public health.

Since 1970, the Clean Air Act has undergone two more updates—one in 1977 and one in 1990. Although the act was allowed to expire in 1981, Congress kept air pollution controls operating by appropriating funds until it was renewed in 1990. Despite this and other rollbacks, four of the six criteria air pollutants have decreased over time, and these improvements in air quality are at least partially due to the Clean Air Act and its amendments.

Redefining Air Pollution

Today, air pollution is considered to be broader than the six criteria air pollutants that have been regulated for decades. In 2007, the U.S. Supreme Court ruled that the EPA must regulate greenhouse gases or it would be in violation of the Clean Air Act. Greenhouse gases include carbon dioxide, methane, nitrous oxide, and others, all of which likely contribute to rising average temperatures in the global climate. These gases were ruled a threat to public health and welfare, both currently and in future generations, even though not all of them are directly considered to be pollutants. This ruling came at a time when debate about global warming had become both widespread and heated among the general public and political actors. Its implications are as yet unknown, but it nonetheless expands the legal definition of air pollution.

Additionally, there are other atmospheric chemicals recently added to the discussion about air pollution, specifically, substances that contribute to the depletion of the ozone layer surrounding Earth. Chlorofluorocarbons and related chemicals called halogenated hydrocarbons were used before the 1990s in air conditioning, refrigerators, and industrial applications, and their release into the air depleted the protective layer of ozone in the upper atmosphere. The 1987 Montreal Protocol phased out the use of some of these chemicals, and others are now considered a type of air pollution to be regulated by the EPA.

Public debate over the appropriateness of regulating greenhouse gases and halogenated hydrocarbons as air pollution is likely to continue. Scientific analysis of the effect of increased greenhouse gases in the atmosphere and the associated climate changes is ongoing. Historically, increased scientific understanding of other air pollutants has led to greater legal controls, and in recent years, Congress has discussed several bills to address greenhouse gas pollution. Although there is currently no comprehensive legislation ordering the regulation of greenhouse gases, it is a possibility for the future. The success of historical air pollution control efforts was enough that most cities in the United States rarely have significant problems. Except in a few places, air pollution has gone from a constant concern for city dwellers to an issue that is no longer on the minds of much of the public.

See Also: Energy; Environmental Law, U.S.; Environmental Protection Agency; Global Warming and Culture; Indoor Air Pollution.

Further Readings

Dunlap, Riley E. and William Michelson. *Handbook on Environmental Sociology*. Westport, CT: Greenwood Press, 2003.

McKitrick, Ross. "Why Did U.S. Air Pollution Decline After 1970?" *Empirical Economics*, 33 (2007).

Wark, Kenneth, Cecil F. Warner, and Wayne T. Davis. *Air Pollution: Its Origin and Control*, 3rd ed. Menlo Park, CA: Addison-Wesley, 1998.

Anna C. McCreery
The Ohio State University

Alternative Communities

Alternative communities are voluntary countercultural groups that provide their membership with nonroutine forms of collective activity. Emerging from the countercultural movements of the 1960s, alternative communities formed with a utopian vision of creating a better—more peaceful, just, and egalitarian—society. While traditionally centered on mistrust of government, rampant consumerism, urban growth and development, and sociocultural alienation, in recent decades, alternative communities have increasingly organized around notions of environmental sustainability. Ecocommunities, including self-contained ecovillages and cooperative housing models, have added an emphasis on environmental justice, which puts sustaining all forms of life on Earth at the forefront of their collective mission. Often referred to as utopian or intentional communities, the concept of alternative community provides a blanket term, which includes all of its varied forms. Alternative communities range widely in size, scope, and purpose from stable residential communities or cooperative housing arrangements representing three or four families to temporary mobile collectives like Burning Man or the Rainbow Family of Living Light, whose evolving membership represents tens of thousands of individuals.

Common Features

Despite their varied manifestations, "alternative communities" describes those collectives whose membership seeks relations of an intimate and communal nature outside mainstream social and cultural institutions. In fact, many alternative communities are united around a prefigurative cultural politics of community in which participants seek to challenge, alter, or re-create the rules and practices that organize our collective social experience. The goal of these communities is often to raise collective awareness by educating others about the social and environmental benefits of often unconventional ways of living, including (but not limited to) collective meals, shared parenting, organic or biodynamic gardening, permaculture, food and human waste composting, recycling and reuse of materials, participatory art, alternative energy, and direct political action.

One common feature of these communities is a decentralized and nonhierarchal organizational structure that places a strong emphasis on participatory action and collective decision making. Regular and open council meetings invite participation of all members to share their insights about the goals and activities of the collective. Key tasks are often broken down into committees or working groups in which a rotating point person organizes the key responsibilities central to the functioning of that particular part of the community. Adopting an ethic of experimentation and improvisation, alternative communities often take on an organic, evolving quality to protect the community from external or internal threats and to ensure the structure of the community fully reflects the needs and values of its membership.

Though membership in these communities may be transitory and fleeting, most consist of a core group of committed members who guide the mission and direction of the community. Peripheral members usually play a more marginal role in the overall mission of the organization, and their membership is often temporary and fleeting, making way for newcomers to join. Occasionally, schisms in the core membership over the structure and organization of the community lead to a splintering of these groups into smaller collectives representing the divergent ideologies or values of their members.

Examples

One of the first alternative communities, The Farm (www.thefarm.org; www.thefarm community.com), represented an original group of 300 individuals who left San Francisco in 1971 to build a utopian community in rural Tennessee around their shared countercultural ideals. This "back to the land" community was built intentionally around sustainable environmental themes of vegetarianism, self-sufficiency, local sourcing and recycling of building materials, midwifery, and organic food production. While the population ballooned to over 1,000 residents, the growing pains experienced by the community led to some internal strife, which ultimately led to the exodus of many residents and initiated a restructuring process of the organizational practices. Today, The Farm is home to about 200 permanent residents spanning four generations. Members have also broadened their outreach through the nonprofit Plenty International to build homes, schools, water systems, and municipal infrastructure in poor and disaster-ravaged communities.

In Portland, Oregon, Tryon Life Community Farm (www.tryonfarm.org) is home to a cooperative living community, which began when a dozen or so college students, artists, gardeners, and natural builders decided to live together in community in a defunct yoga retreat center. Until 2002, the community was housed on a seven-acre rental property abutting Tryon Creek State Park, a 650-acre urban preserve. When the rental property was to be sold by the owners to land developers in 2003, the group mobilized to raise $1.4 million to purchase the land. After a successful campaign to purchase the property, the residential farm community now hosts student groups, summer camps, and weekly workshops to experientially educate the public about environmental sustainability, alternative energy, green building, and organic farming practices. Through hands-on projects, the farm community attempts to demonstrate ways of living that retain a close, sustainable, communal, and spiritual relationship with land and other people.

Another recent movement toward sustainable forms of community includes the transition towns initiative (www.transitiontowns.org). Transition towns bring together small groups of motivated individuals who wish to prepare their communities for the challenges of peak oil and climate change. The focus of these groups is to educate and raise awareness about the changes in lifestyle and community design that must be undertaken to weather the impacts of depleting energy sources and potentially catastrophic climate change. Transition town participants do this by connecting with existing groups in the local community and building bridges with local government to launch a community action plan to address strategies for living more sustainability, including alternative forms of transportation, energy, housing, architecture, and food production.

Other examples of alternative communities include those that are temporary or transitory in nature and either move from place to place or converge in recurrent intervals on a weekly, monthly, or yearly basis. One example of these "portable" or recurrent, temporary communities is the countercultural Burning Man Festival (www.burningman.com), which converges around 50,000 hippies, artists, environmentalists, and voyeurs in the Black Rock Desert of Nevada. Every September, participants converge over a weeklong period to build a temporary residential community of creatively themed camps and larger-than-life art installations representing various aspects of alternative art, culture, and lifestyle. The event culminates in the burning of an enormous wooden effigy of the "Burning Man." In 2007, the event adopted a "Green Man" theme and now includes a "leave no trace" ethic of sustainability to minimize the impact of the event on the local landscape and surrounding community. Festival participants are increasingly organizing their art installations and camps around the use of alternative energy and reusable resources.

See Also: Alternative/Sustainable Energy; Art as Activism; Grassroots Organizations; Individual Action, National and Local; Permaculture.

Further Readings

Chen, Katherine. *Enabling Creative Chaos: The Organization Behind the Burning Man Event*. Chicago, IL: University of Chicago Press, 2009.

Fairfield, Richard and Timothy Miller. *The Modern Utopian: Alternative Communities Then and Now*. London: Process Press, 2009.

Fellowship for Intentional Community (FIC). *A Comprehensive Guide to Intentional Communities and Cooperative Living*. Rutledge, MO: FIC, 2007.

Gilmore, Lee, et al. *After Burn: Reflections on Burning Man*. Albuquerque: University of New Mexico Press, 2005.

Niman, Michael. *People of the Rainbow: A Nomadic Utopia*. Knoxville: University of Tennessee Press, 1997.

Robert Owen Gardner
Linfield College

Alternative/Sustainable Energy

Although the use of alternative sustainable energy and its related technologies is growing rapidly, fossil fuels dominate the energy economy today. Fossil fuels are the cornerstone of the Second Industrial Revolution (2IR) and remain the primary source of energy in the United States into the 21st century. Strong vested interests among oil and gas companies and their main clients have held the United States back from initiatives that the European Union (EU) and Asian countries began two decades ago. The result in the other parts of the developed world has been the Third Industrial Revolution (3IR), which has gone on for almost two decades now and which has taken the United States by surprise.

The 3IR probably began at the end of the 20th century, as Jeremy Rifkin, the environmental economist, explained in his 2004 book *The European Dream*. The Cold War had ended, and by the 1990s, economic adjustments in Europe and Asia to meet a new global economy had begun. These changes occurred first in the EU and Japan. Rebuilding, in the post–World War II period, provided an opportunity to rebuild and recreate businesses and industries as well as to commercialize new technologies. However, the historical roots for the 3IR go back decades, and even to the early part of the 20th century, as Herbert Blumer illustrated in his 1969 book *Symbolic Interaction*.

Nonetheless, the United States was the first to invent and commercialize many of the technologies developed into mass markets by the EU, Japan, South Korea, Taiwan, and now China. The Prius hybrid car from Toyota uses regenerative braking—a technology invented and patented from the U.S. Department of Energy labs in the 1990s. The United States failed in the past two decades to advance in the use of renewal energy and has failed to provide political leadership in the global community to recognize and develop policies to address global climate change.

The issue of climate change has been one of the key motivations for the development of alternative sustainable energy technologies, systems, programs, and businesses. However, the topic is not new. It has been of global concern for over three decades starting with the Brundtland Report for the United Nations (UN) in the late 1980s. This led to the Rio Summit in 1992, which resulted in UN support for its Intergovernmental Panel on Climate Change (UN IPCC) in the early 1990s, an agency that subsequently shared the Nobel Peace Prize in 2007 with former U.S. Vice President Al Gore and his film on the issues, *An Inconvenient Truth,* along with other UN committees and groups. The Brundtland Report defined sustainability as "at a minimum, sustainable development must not endanger the natural systems that support life on Earth: the atmosphere, the waters, the soils and the living beings." With that definition in mind, a number of communities have sought to become sustainable over the past 20 years.

The end of the Cold War in the early 1990s spurred the technological transformation that had already begun ushering in the 3IR. Germany, Japan, South Korea, and Taiwan jumped into the lead producing cars, electronics, and other high-tech products for an expanding global market. Frequently, technologies inspired, created, and developed in America are manufactured in other countries, such as Japan or China.

Europe, instead of the United States, leads the world in solar technology businesses that were originally developed in the 1970s in the United States. By the second decade of the 21st century, other countries like Germany took the lead in the development, manufacture, and installation of solar energy systems. For example, 6.4 GW of solar photovoltaics were installed in the EU in 2009, where Germany was the largest market. Italy ranked second in 2010, and had installed more solar panels in one quarter of 2009 than all of the solar panels installed in California in that same year.

Europe and Asia aggressively pursued policies of sustainable alternative energy, including technologies and programs that protected the environment, energy conservation, and the development of new technologies and market subsidies for clean, alternative, sustainable energy technologies. Sustainability policies were created to provide financial support and programs from their governments, which benefited from their new economic strategies, growth, and business leadership. The basic results are measurable and have placed the United States in a "catch-up" economic mode. For example, since the early 1990s, Germany has had an aggressive economic program called Feed-In Tariffs (FiT) that by the end of the first decade of the 21st century has made it the leading nation in solar manufacturing and installation. A feed-in tariff provides a guaranteed price for kilowatts produced by renewable sources. Over 250,000 jobs have been created as well, and this job growth has been replicated in other countries that embraced the 3IR.

There is ample evidence that FiT and similar kinds of "green" business development create jobs. This evidence suggests that the key to growth is government leadership in both programs and financing, much like the 3IR has done in the EU and Asia. As some scholars point out, that is exactly what the 2IR did in economics. The generation of power throughout the 2IR has been with a central plant providing power primarily by hydroelectric energy generation and fossil fuels. The problem is that these plants need long, costly, and dangerous transmission lines. There is now a shift to "on-site" power generation by family homes, office buildings, workplaces, factories, college campuses, and shopping malls.

The use of renewable energy is a major paradigm change and significant for these on-site power systems and is growing through the use of creative long-term financing such as FiT. Homes, buildings, and complexes can now have solar systems for power as well as for vehicles. The use of storage devices for the renewable energy means that the sun and wind do not always have to be available. The power grid can also act as a "storage device" or battery. So the energy systems are flexible, meaning that both central grid and on-site power can be used. Moving from the conventional central grid power system to an on-site power system that uses renewable energy and integrated technologies means that alternative sustainable energy will be in high demand.

Alternative Sustainable Energy Systems

Wind Generation

Wind generation has been used as power sources for hundreds of years. A large propeller is placed in the path of the wind, the force of the wind turns it, a gear coupling interacts with a turbine, and electricity is generated and captured. While ancient in form, there have been significant technological advances. The new generation of wind turbines are stronger, more efficient, quieter, and cheaper. Today, wind turbines are being installed in communities, and even smaller systems are now located on rooftops as part of the natural flow of air over buildings.

Solar Generation Systems

Solar generation systems capture sunlight including ultraviolet radiation via solar cells (silicon). This process of passing sunlight through silicon creates a chemical reaction that generates a small amount of electricity. This process is described as a photovoltaic or PV reaction and is at the core of solar panel systems. A second process uses sunlight to heat liquid (oil or water), which is then converted to electricity. A number of communities are now looking into solar "concentrated" systems in which the sun is captured in heat tubes and used for heating and cooling. This renewable technology can be used in water systems and buildings that have swimming pools.

Biomass

Biomass is a remarkable chemical process that converts plant sugars into gases (ethanol or methane), which are then burned or used to generate electricity. The process is referred to as "digestive" and is not unlike an animal's digestive system. The ever-appealing feature of this process is that it can use abundant and seemingly unusable plant debris—such as rye grass, wood chips, weeds, grape sludge, or almond hulls.

Geothermal

Geothermal power is extracted from heat stored in the Earth that originates from the formation of the planet, from radioactive decay of minerals, and from solar energy absorbed at the surface. It has been used for space heating and bathing since ancient Roman times, but is now better known for generating electricity. Worldwide, geothermal plants have the capacity to generate about 10 GW as of 2007, and in practice generate 0.3 percent of global electricity demand. In the last few years, engineers have developed several remarkable devices called geothermal heat pumps, ground source heat pumps, and geo-exchangers that gather ground heat to provide heating or cooling for buildings. Many communities with concentrations of buildings, like colleges, government centers, and shopping malls, are turning to geothermal systems.

Ocean and Tidal Waves

Ocean and tidal waves generate power, the collection of which was pioneered by France and Ireland with their revolutionary SeaGen tidal power system. France has been generating power from the tides since 1966, and now Électricité de France has announced a large commercial-scale tidal power system that will be able to generate 10 megawatts per year. The United States is equally capable of producing massive amounts of energy with the right technology. Ocean power technologies vary, but the primary types are wave power conversion devices that bob up and down with passing swells, tidal power devices that use strong tidal variations to produce power, ocean current devices that are similar to wind turbines but are placed below the water surface to capture the power of ocean currents, and ocean thermal energy conversion devices that extract energy from the differences in temperature between the ocean's shallow and deep waters.

Energy Storage Technologies

Energy storage technologies, such as fuel cells, are electrochemical cells that convert a source fuel into an electrical current. Energy storage technologies generate electricity inside a cell through reactions between a fuel and an oxidant, triggered in the presence of an electrolyte. The reactants flow into the cell, and the reaction products flow out of it, while the electrolyte remains within it. Fuel cells can operate continuously as long as the necessary reactant and oxidant flows are maintained. Fuel cells are different from conventional electrochemical cell batteries in that they consume reactant from an external source, which must be replaced. Many combinations of fuels and oxidants are possible. A hydrogen fuel cell uses hydrogen as its fuel and oxygen as its oxidant. Other fuels include hydrocarbons and alcohols. Other oxidants include chlorine and chlorine dioxide.

Bacterial and Microbial Fuel Cells

Bacterial, or microbial, fuel cell energy generation is becoming more and more a focus of research; for example, BP granted $500 million over 10 years for studies at the University of California, Berkeley, and the University of Illinois, Urbana. The process uses living, nonhazardous microbial fuel cells bacteria to generate electricity. The researchers envision small household power generators that look like aquariums but are filled with water and microscopic bacteria. When the bacteria inside are fed, the power generator—referred to as a biogenerator—would produce electricity.

National Plans

It is clear that every nation needs to have a comprehensive sustainable energy plan. The plan should address how the government, business leaders, and the public will address basic infrastructure needs such as energy, water, waste, telecommunications, and transportation. Clear policies, plans, and funds for these efforts must be developed so that business leaders receive consistent price signals.

One goal of a comprehensive sustainable energy plan must be to reduce the dependence of local communities on central grid connected energy, since most of these power generation sources come from fossil fuels that contribute to climate change. Local on-site power can be more efficiently used and based on the region's renewable energy resources such as wind, solar, geothermal, and biomass, among renewable resources. Such a model is now in effect in Denmark, where many communities are generating power with wind and biomass energy to provide enough base load to reduce dependence on central resources. Denmark is currently on track to meet its goal of generating 50 percent of its energy from renewable resources (primarily from on-site and local resources) by 2015.

The entire EU has committed billions of euros over the next few years to developing sources of renewable energy and the development of smart communities. The same is being done in Japan, Taiwan, and South Korea. Even stronger public national policies and more funds are being allocated in China.

However, by the 1990s, the EU nations, especially Germany and the Nordic countries, became aware that there were serious social, political, and environmental issues connected to fossil fuel dependency. Japan had its own epiphany. For example, in the mid-1990s, Toyota licensed "regenerative braking technologies" from the U.S. Department of Energy National Laboratories, after all three of the American auto companies had rejected the technology.

Germany, as noted above, jumped out in the lead with Feed-In Tariffs (FiT) in 1990, which is an incentive structure to encourage the adoption of renewable energy through government legislation. The policy obligates regional or national electricity utilities to buy renewable electricity at above-market rates from wherever it is produced, whether it is a homeowner or private utility company. Germany is now the world's leading producer of solar systems, and it has more solar systems installed than any other nation despite the fact that the climate in Germany is not ideal for solar energy. Japan has also implemented a similar aggressive FiT system in order to stimulate its renewable energy sector and regain the market leadership in solar that it held in the early part of the 21st century.

Germany, Finland, France, the UK, Luxembourg, Norway, Denmark, and Sweden are on track to achieve and exceed their renewable energy generation goals of 20 percent by 2020, with Denmark being the most aggressive. However, other EU countries are lagging behind, especially in central and eastern Europe. There is a need for these countries to implement programs like those in western Europe to become independent from importing oil and gas from North Africa, the Middle East, and Russia.

As Europe gained clarity of political purpose, it nurtured the creation of the 3IR. All the necessary ingredients for a game-changing megatrend were sprouting in Europe during the latter 20th century. Regional cooperation, the shock of spiking oil prices, resource depletion, and environmental degradation were the drivers of 3IR. Germany eventually passed Renewable Energy Laws (EEG), and major corporations were investing and innovating, happy to make a profit on new demand. Today, major European companies like Holland's Vesta, and German companies like BESCO and SOLON, and France's Somfy are international in scope, positioned for decades of promise and profits.

Some Asian countries, like China, are trying to skip over the 2IR and move directly into the 3IR. China led the United States and the other developed nations in 2009 for annual "clean energy investments and finance," according to a new study by the Pew Charitable

Trusts. South Korea's central government established a Green Growth Task Force in 2009. The country has targeted 79 percent of its stimulus money for "green" technologies, the highest percentage of any nation, making South Korea the seventh largest green technology investment sector in the world. In Japan, the future of sustainable residential living can already be seen firsthand at the Zero Emission House, a model house built by the Ministry of Economy, Trade and Industry (METI) to advertise Japan's outstanding energy-efficient and environmentally friendly technologies.

The Grid of the Future

Distributed renewable sustainable energy communities are agile or flexible in their use of new technologies. The ability of local communities to generate surplus problems requires countries to address the reality of an aging grid system and the insufficient capacity of current facilities to meet growing energy demands. Originally, when nations electrified their cities and built large-scale electrical grids, the systems were designed to transmit from a few large-scale power plants. However, these systems are inefficient for smaller-scale distributed power from renewable sources. Although some systems will allow for individual households to either buy power or sell power back to the grid, the redistribution of power from numerous small-scale sources is not yet handled well.

The grid of the future has to be "smart" and flexible and based on the principles of sustainable development. Three basic concerns of sustainable, agile, and smart communities need to be addressed. The word *agile* is used to mean flexibility—the ability of a local home or office building to generate power for itself and also sell it to the power grid in exchange for power when the sun or wind or other renewable source of power is not available. First, there needs to be political agreement on policies and goals for the entire community. Second, issues pertaining to the siting of buildings and overall facility master planning must be addressed from the perspective of "green" (meaning renewable energy as well as recycling and conservation), energy, and efficient orientation. These structures need to be designed for multiple uses by the community. Developing dense, compact, walkable areas that enable a range of transportation choices leads to reduced energy consumption and environmental protection. Third, a sustainable smart community is part of a living network and draws residents from a broad region. This is true for an office cluster of buildings, residential housing developments, government offices, and shopping malls. These buildings and areas require communication and transportation along with water, waste systems, and energy. Each of these construction and building infrastructures, in any community, requires recycling or reuse of waste such as industrial symbiosis that also acts as a vibrant, "experiential" applied educational model that catalyzes creative learning among community members and broader regional and national areas.

Sustainable Communities

Communities must build green, energy-efficient, multipurpose buildings that are shared. In short, buildings are not alone, but comprise a community or cluster of structures that need and share common infrastructures. Such a sustainable perspective includes developing dense, compact, walkable communities that enable a range of transportation choices and lead to reduced energy consumption. A sustainable, agile, and smart community is a vibrant, "experiential" applied model that should catalyze and stimulate entrepreneurial activities, education, and creative learning along with research, commercialization, and new businesses.

Many sustainable communities, cities, and other organizations, such as academic institutions and private sector business, recognize the need for policies that direct their facilities and infrastructures to be "green" based upon some criteria, such as the U.S. Green Building Council (USGBC) certification for achieving LEED (Leadership in Energy, Environment, and Design) from basic criteria to higher standards. Individual buildings are to have "net-zero" carbon emissions. Many organizations are seeking to make their entire facilities "energy independent and carbon neutral." More recently, the USGBC has created "community" or LEED neighborhood standards. This set of criteria reflects the broader concerns for clusters of buildings with designs integrated with basic infrastructure needs.

Sustainable agile communities must include the facilities, land, and infrastructure sectors that intersect and support them, such as energy, water, IT, waste, and environment. The public sector needs to lead and then provide vision, standards, codes, measurements, and evaluations with oversight to monitor these goals, policies, and programs, along with finance. Public policies must be matched with funding and investment, not just with tax credits.

See Also: Environmental "Goods"; Environmental Protection Agency; Green Consumerism; Water (Bottled/Tap).

Further Readings

Blumer, Herbert. *Symbolic Interaction: Perspective and Method*. Englewood Cliffs, NJ: Prentice-Hall, 1969.

Clark, Woodrow W., II, ed. *Sustainable Communities*. New York: Springer, December 2009.

Clark, Woodrow W., II and Ted Bradshaw. *Agile Energy Systems: Global Lessons From the California Energy Crisis*. London: Elsevier Press, 2004.

Clark, Woodrow W., II and Grant Cooke. *The Third Industrial Revolution*. Greenwood, CT: Praeger, forthcoming 2011.

Clark, Woodrow W., II and Michael Fast. *Qualitative Economics: Toward a Science of Economics*. London: Coxmoor Press, 2008.

Clark, Woodrow W., II and Michael Intriligator. "Global Case Studies in Energy, Environment, and Climate Change: Some Challenges for the Field of Economics." *Contemporary Economic Policy Journal*, Special Issue (Winter 2011).

Clark, Woodrow W., II and Xing Li. "Social Capitalism: Transfer of Technology for Developing Nations." *International Journal of Technology Transfer*, 2003.

Clark, Woodrow W., II and Henrik Lund, eds. "Special Issue on Sustainable Energy and Transportation." *Utility Policy Journal*, 2008.

Funaki, Kenataro and Lucas Adams. "Japanese Experience With Efforts at the Community Level Toward a Sustainable Economy: Accelerating Collaboration Between Local and Central Governments." In *Sustainable Communities*, Woodrow W. Clark II, ed. New York: Springer, 2009.

Next 10. http://nextten.org (Accessed December 2010).

Østergaard, Poul and Henrik Lund. "Sustainable Towns: The Case of Frederikshavn – 100 Percent Renewable Energy." In *Sustainable Communities*, Woodrow W. Clark II, ed. New York: Springer, 2009.

Rifkin, Jeremy. *The European Dream: How Europe's Vision of the Future Is Quietly Eclipsing the American Dream*. New York: Tarcher/Penguin, 2004.
Yang, Robert. "Business Opportunities: The Key to Any Nation's Green Energy Industry Strategy." Keynote Presentation. Stanford University, CA: Smart Green Cities Summit, Stanford Program on Regions of Innovation and Entrepreneurship (SPRIE), 2010.

Woodrow W. Clark II
Clark Strategic Partners

Animals (Confinement/Cruelty)

Ruth Harrison's *Animal Machines* exposed the conditions animals were subjected to when they were considered merely food production objects. These conditions included castration, use of antibiotics, beak trimming, and overcrowding, like these young turkeys.

Source: iStockphoto

Humans use animals for a wide range of purposes including consumption, cosmetic testing, entertainment, farming, medical and scientific research, and as pets. The use of animals in society is a controversial issue because they are often confined to man-made environments—such as farms, laboratory cages, and zoo enclosures—that are very different from natural habitats. This issue has been approached from different perspectives that can be broadly divided into two opposing camps: animal liberation and animal welfare. The animal liberation or rights perspective opposes the human exploitation of animals, based on the viewpoint that animals should be afforded moral and legal consideration. It regards the use and confinement of animals as cruel and a form of imprisonment and punishment. Conversely, the animal welfare position posits that there is nothing inherently wrong with the use of animals as resources and needs, as long as there is no unnecessary animal suffering. Animal welfare tends to be separated from moral considerations about the use of animals and instead focuses on how to define and measure suffering and characteristics of welfare, particularly in terms of physiological and psychological health. Key aspects of good animal welfare include responsible care, health, housing, and other management practices.

Debates about the ethical use of animals can be traced back to ancient times. More recently, organized efforts to abolish animal cruelty in the late 18th and early 19th centuries led to initiatives such as the creation of the Society for the Prevention of Cruelty to Animals (1824) in the United Kingdom (UK) and the American Society for the Prevention of Cruelty to Animals (1866) in the United States. Early publications about animal rights included *Moral Inquiries on the Situation of Man and of Brutes* (1824) by Lewis Gompertz and *Animals' Rights* (1892) by Henry S. Salt.

Ethical debates concerning society's use of animals gained momentum in the 1960s and 1970s following the publication of several landmark books that revealed pervasive animal cruelty. In 1964, Ruth Harrison, in her book *Animal Machines,* critiqued the inhumane treatment of animals in the factory farming of chickens, calves, and pigs, in which animals were regarded by humans as mere production objects. The book provided a detailed account of the rearing conditions of animals in intensive farming systems, including inadequate housing, overcrowding, castration, beak trimming, use of antibiotics, etc. In the United Kingdom there was a strong public reaction to the book. The UK's Ministry of Agriculture ordered an investigation by the Brambell Committee that led to the formation of ethical principles for animal husbandry and the country's first farm animal welfare legislation, the Agriculture (Miscellaneous Provisions) Act in 1968. It also inspired the European Convention for the Protection of Animals Kept for Farming Purposes.

In 1971, Stanley and Rosalind Godlovitch and John Harris at the University of Oxford edited *Animals, Men and Morals: An Inquiry Into the Maltreatment of Non-Humans*, which was a collection of essays about the moral treatment of animals. Contributors to the book included Ruth Harrison, who wrote the opening chapter, and Richard Ryder, who is credited with coining the term *speciesism* to describe widely held beliefs that humans are inherently superior to other species. In 1973, Australian philosopher Peter Singer wrote a review of the book in the *New York Review of Books* in which he introduced the concept of "animal liberation."

Several years later, in 1975, Singer wrote the classic book *Animal Liberation*, which is now widely regarded as the foundation of the modern animal rights movement, and was one of the most influential books of the 20th century. Singer put forward a philosophical basis for animal rights, followed by a detailed examination of society's treatment of animals. In the opening pages, Singer popularized the term *speciesism*, explaining in detail why he believed animals deserve equal consideration to humans, and disagreed with the viewpoint that animals have no rights based on human privilege. He argued that the consideration of animals should be based on sentience—that is, their ability to feel pain and suffering—rather than other criteria such as intelligence. The book is centered on utilitarian principles in philosophy, which Singer argued can be applied to understanding and minimizing animal suffering. The book remains controversial and has undergone several revisions, with each edition chronicling the progress of the animal liberation movement.

Another influential book during this period was *The Case for Animal Rights* (1983) by Tom Regan, which argued that some animals should have the same moral rights as humans, extending the principles of deontological ethics.

These classic texts and other works lit the fuse for an explosion of interest in human-animal associations. The 1970s and the late 1980s was a time when the general public was unaware of animal cruelty, and the animal rights concept was unknown and marginalized in political and social activism. Increased public awareness led to a cultural shift in people's attitudes and behavior toward the treatment of animals: a voluminous literature was published on the topic; governments created animal welfare committees, legislation, regulations, standards, etc.; and many companies banned practices such as animal testing and now take animal welfare more seriously. As exemplified in the entries published in this volume, *Green Culture*, there is increasing public concern for animals and the natural environment. Nowadays, people's habits and lifestyles are increasingly based on ethical and moral considerations. The topic of animals in society has become a subject of study in university departments, and attracts interest from a wide variety of academics and professionals in a wide range of fields, such as lawyers, physicians, psychologists, and veterinarians.

The animal rights movement emerged in the 1970s; this worldwide social movement aims to eliminate the use of animals as human property. It includes philosophical debate as well as direct action. Animal rights advocacy is wide ranging, including animal rights organizations and activists. People for the Ethical Treatment of Animals (PETA) is the largest animal rights group in the world. Most groups engage in peaceful education and research, but a minority use direct action such as violence and criminal activities. Some have been classified as domestic terrorist organizations in both the UK and the United States.

There have been advances in the animal welfare perspective. Animal welfare science is the study of animal consciousness and emotions and scientific approaches to assessing and improving welfare. Tremendous progress has been made in the past 30 years to understand the cause and prevention of animal suffering. Although still a relatively young discipline, animal welfare science has multidisciplinary links with biology and a related fields, such as animal behavior, behavioral ecology, evolution, genetics, physiology, neuroscience, cognitive science, etc., and draws upon the latest techniques and methods. Animal welfare in practice includes animal welfare organizations such as the Humane Society International, Royal Society for the Prevention of Cruelty to Animals, and the World Society for the Protection of Animals.

See Also: "Agri-Culture"; Biodiversity Loss/Species Extinction; Fish; Organic Foods; Permaculture.

Further Readings

Bekoff, Marc. *Minding Animals: Awareness, Emotion and Heart.* Oxford, UK: Oxford University Press, 2002.

Dawkins, Marion. *Animal Suffering: The Science of Animal Welfare.* London: Chapman & Hall, 1980.

Singer, Peter. *Animal Liberation.* New York: HarperCollins, 2001.

Singer, Peter, ed. *In Defense of Animals: The Second Wave.* Oxford, UK: Wiley-Blackwell, 2005.

Gareth Davey
Hong Kong Shue Yan University

Antiglobalization Movement

The anticorporate globalization or global justice movement represents one of the most significant expressions of grassroots mobilization and popular dissent of the past two decades. In response to growing corporate power and the free market, or neoliberal policies of governments and global institutions such as the World Bank, the International Monetary Fund (IMF), and the World Trade Organization (WTO), workers, students, environmentalists, youths, peasants, indigenous peoples, and the urban poor have come together across the north–south divide to strive for social, economic, and ecological justice and democratic control over their daily lives. The movement addresses the harmful consequences of corporate or neoliberal globalization, including poverty, inequality, social dislocation, hunger, and ecological destruction. Global justice activists come from diverse spheres, including environmental and human rights organizations, political parties, trade unions, grassroots struggles, and informal collectives, and have combined multiple forms of action, including direct action, marches and rallies, public education, and lobbying.

Corporate Globalization and the Environment

Since the 1980s, governments and multilateral institutions have implemented free market policies such as privatization, trade liberalization, deregulation, export-oriented production, and cuts in social spending and basic subsidies. Although some populations have benefited, the results have been disastrous for many others, particularly in the global south. During the 1990s, the number of people living in poverty around the globe increased by 100 million, even as world income grew 2.5 percent per year, and more than 80 countries had per capita incomes lower than the decade before.

Environmentalists have denounced the ecologically destructive impacts of globalization. Some have emphasized the specific neoliberal policies of governments and global institutions that promote export production, free trade, and urban growth, threaten ecosystems, deplete natural resources such as forests and water, and increase pollution and the burning of fossil fuels that cause global warming. The WTO, in particular, has been singled out for its decisions that favor free trade and corporate profit over national environmental laws that protect wildlife, clean air and water, and public health. Others have voiced a more radical critique of globalization and its unsustainable model of capitalist growth that is provoking a rapid loss of biodiversity, fast exhausting the planet's resources, and generating a global climate catastrophe. Whereas moderates seek a global regime of environmental laws and regulations, including multilateral treaties that would curb the emissions of fossil fuels, many radicals question the entire global system, calling for a return to small-scale, locally sustainable economies, communities, and ecologies.

Green Roots of the Global Justice Movement

Given the grave ecological consequences of corporate globalization, environmentalists, indigenous peoples, and farmers have led a massive wave of resistance over the past few decades. Indigenous communities and peasant farmers, largely in the south, have struggled against the decimation of their land, communities, and environment due to free trade agreements and corporate exploitation of their territories through mining, oil, and other resource extraction projects. Meanwhile, northern environmentalists have organized against neoliberal policies that threaten local and global ecosystems and have worked to preserve green spaces, wilderness, and wildlife in their own communities. Some of the earliest critiques of corporate globalization came from radical environmental organizations and networks such as Earth First!, Reclaim the Streets, the Rainforest Action Network, Greenpeace, and the Ecologist, while environmental direct action organizers played a key role in the anti-WTO protests in Seattle on November 30, 1999.

During the years leading up to the Seattle protests, environmentalists began building alliances and networks with other sectors, including labor movements, human rights and economic justice organizations, anti-free trade and anti-debt campaigners, Zapatista solidarity activists, squatters, and antisweatshop activists, as well as southern peasant and indigenous movements. Coming on the heels of the Zapatista Gatherings against neoliberalism and for Humanity in Chiapas and Spain in 1996–97, activists founded the Peoples' Global Action Network in 1998, which inspired the first global days of action against capitalism, including the high-profile Seattle protest. Counter-summit actions soon spread around the world, including blockades against the World Bank/IMF meetings in Prague in September 2000 and the Free Trade Area of the Americas Summit in Quebec City in April 2001. Protests reached an explosive crescendo with the violent clashes in Gothenburg, Barcelona, and Genoa in summer 2001. Since then, movement focus has shifted to world and regional social forums, where thousands have gathered in cities such as Porto Alegre, Mumbai, Dakar, Caracas, London, and Atlanta to discuss alternatives to corporate globalization.

Decentralized Networking

In addition to emphasizing ecological issues, radical environmentalists have also brought to the movement their political culture revolving around a critique of hierarchy, an affinity for direct democracy, and a commitment to decentralized network forms. The consensus decision making and horizontal networking that characterize the wider movement, and which have been reinforced by the use of digital technologies and the underlying logic of late capitalism, have been practiced for many years by radical direct action activists within networks such as Earth First! and Reclaim the Streets, and the U.S.-based antinuclear power movement.

The global justice movement today is primarily organized around flexible, decentralized networks, such as the former Direct Action Network in North America, the Movement for Global Resistance in Barcelona, or Peoples' Global Action at the transnational scale. Although global justice activists are critical of hierarchy and centralization, they are not opposed to organization, leadership, or strategy per se. Instead, they are trying to build participatory structures that reflect their directly democratic ideals. At the same time, given the new technologies at their disposal, antiglobalization activists have been able to build networks beyond the local scale. In contrast to traditional parties and unions, global justice movements involve broad umbrella spaces, where diverse organizations, networks, and collectives converge around common hallmarks while preserving their autonomy and specificity. Such grassroots forms of political participation are widely seen as an alternative mode of democratic practice. Global justice movements thus promote global democracy, even as they emphasize autonomy and local self-management.

Direct Action Tactics

Radical environmentalists have also contributed their knowledge and experience with direct action. The mass action strategy, involving horizontal coordination among autonomous affinity groups and consensus decision making, comes out of the U.S.-based direct action wings of antinuclear, ecology, peace, gay rights, and Central America solidarity movements beginning in the 1970s. Many specific tactics, including lockdown, road occupation, and banner hang techniques, were first developed in the radical ecology movement, particularly Earth First! in the United States and the United Kingdom. The latter also gave rise to Reclaim the Streets and its mobile street parties. Finally, militant direct action, including White Overall and confrontational Black Bloc tactics, are rooted in Italian and German autonomous movements, which also included radical ecologists.

Despite arising in distinct contexts, the tactics employed by antiglobalization activists all produce theatrical images for mass media consumption, building on the tradition of organizing direct actions as "image events" developed by environmental organizations such as Greenpeace. Beyond their utilitarian goals—shutting down summit meetings—mass actions are complex cultural performances that allow participants to communicate symbolic messages while providing a forum for producing and experiencing symbolic meaning through ritual interaction. The theatrical performances staged by activists from diverse global justice networks, including giant puppets and street theater, mobile carnivals (Reclaim the Streets); spectacular protests involving white outfits, protective shields, and padding (White Overalls); and militant attacks against the symbols of corporate capitalism (Black Bloc), are designed to capture mass media attention while also embodying and expressing alternative political identities.

Conclusion and Future Directions

Despite their differences, antiglobalization activists from diverse sectors are all struggling for social and economic justice and ecological sustainability. What makes the movement unique is its capacity for coordinating across vast distances and high levels of diversity and difference, overcoming many political and geographic obstacles that have plagued past mass movements. Environmental activists and issues have long been an important focus of global justice activism, but with the recent direct action protests against the Global Climate Summit in Copenhagen in December 2009, climate change and environmental justice have been thrust once again to the forefront of the movement's agenda. It remains to be seen, however, whether the movement will be able to translate its innovative tactics and forms into concrete policy victories.

See Also: Antiwar Actions/Movement; Benjamin, Medea; Chipko Movement; Cyber Action; Demonstrations and Events; Environmental Justice Movements.

Further Readings

DeLuca, Kevin Michael. *Image Politics: The New Rhetoric of Environmental Activism.* New York: Guilford Press, 1999.

Epstein, Barbara. *Political Protest and Cultural Revolution.* Berkeley: University of California Press, 1991.

Juris, Jeffrey S. *Networking Futures: The Movements Against Corporate Globalization.* Durham, NC: Duke University Press, 2008.

Katsifiacas, George. *The Subversion of Politics.* Atlantic Highlands, NJ: Humanities Press, 1997.

Jeffrey S. Juris
Northeastern University

Antiwar Actions/Movement

Ecological concerns were included in protests of the Vietnam War because of America's use of dangerous chemicals, such as this napalm bomb, that obliterated forest areas.

Source: U.S. Air Force via Wikimedia Commons

Although the United States has had small pacifist groups such as Quakers from colonial times on, the first really mass antiwar movements did not come until the 1960s, and they were surprisingly based on college campuses. The 1960s protestors targeted U.S. involvement in all foreign wars starting with the Vietnam War. Although the United States had earlier foreign wars such as the Spanish–American War in 1898, World War I in 1917, World War II in 1941, and the Korean War in 1950, these had not generated mass protests for various reasons. For example, in World War II, the United States had been attacked, and the Korean War was seen as part of the early Cold War with the Soviet Union and China. Protests against the Vietnam War set the pattern for future protests from the Gulf War in the 1990s to the present-day Iraq and Afghanistan Wars, as well as "green" ecology protests. The rallies often seemed similar, yet there were important differences between the approaches later protestors used.

As Cold War fears faded in the 1960s, Vietnam was not seen as particularly dangerous by many Americans—especially draft-age youth attending college. The anti–Vietnam War movement on college campuses was predated by campus civil rights activism in the late 1950s. These were inspired in part by Martin Luther King Jr. and black Southern college students. King himself would soon renounce the Vietnam War. Thus, for the first time in American history, college campuses became the center of national protest. Student antiwar activists were also commonly involved simultaneously in other protest movements such as women's liberation and ecology.

Although almost all the antiwar protestors supported environmental reform, ecology issues were almost never brought up at antiwar protests for fear of diluting the message. The one issue that brought ecology directly into the antiwar protests was America's use of napalm and dangerous chemicals to obliterate Vietnam forest areas the enemy used for cover. These chemicals were later found to cause disease in Vietnam veterans years after their exposure. The other area where antiwar and ecology issues mixed was concerts by folk-protest singers such as Bob Dylan, Pete Seeger, and Joan Baez. These singers mixed protest songs about civil rights and ecology with antiwar songs.

College students and professors had long been seen as dangerous by dictatorial governments in many other countries. For example, South American dictators often closed down college campuses at the first sign of mass protest. These dictators knew students and professors spelled trouble for them. Most students were clearly Leftist; however, they were young and often not yet well educated. There were self-proclaimed Marxist students who had never read Marx. At rallies, students often relied on seasoned speakers and used protest songs and chants to suggest mass feelings. For example, a young, radical, supposedly pacifist speaker might ask a crowd to chant: "Two, four, six eight, organize and smash the state." Right-wing critics sometimes suggested that the antiwar demonstrations were just entertainment for many bored students and were substitutes for the water fights and panty raids on college campuses in the 1950s.

Professors involved in antiwar movements, especially historians, were heavily influenced by the so-called New Left Historians. In the 1960s, younger history professors such as Howard Zinn and Barton Bernstein published "New Left" histories and interpretations of America's 20th-century wars. They argued that Americans should never fight in foreign wars unless the nation had been attacked. Only World War II met their criteria as a necessary war.

Many younger protest singers against the Vietnam War such as Joan Baez and Phil Ochs identified with pacifism because of their involvement with the nonviolent civil rights movement. Also, singing protest marchers reflected a communal pacifist spirit. Yet some pacifist songs of the later 1960s had double-edged lyrics that mirrored the ambivalent nonviolence common to the youthful activists. While these songs spoke eloquently for nonviolence, they also warned of "the fire next time." Behind the New Testament gentleness was a violent Old Testament anger. Similarly, many student protestors claimed to be pacifists, yet supported military dictators in other countries. Thus, many protestors adopted the inconsistent position of antimilitarism at home, coupled with a call for revolutionary guerrilla warfare in underdeveloped countries.

Of course there were also many adult antiwar demonstrators and later even some Vietnam veterans who protested, mainly in big city rallies. Here, young college protestors mixed with antiwar adults waving signs such as "Send LBJ to Vietnam." Yet in the 1960s, there was an obvious gulf between old and young. Students blamed adults for electing first Lyndon Johnson and then Richard Nixon as pro-war presidents. Student signs and shirts that read "Don't Trust Anyone Over 30" and "Make Love Not War" were clearly not aimed at the older generation.

Protesters against the Gulf War and Iraq and Afghanistan Wars were and are often the older, seasoned Vietnam protesters, while the percentage of campus antiwar demonstrators had greatly diminished. Most college students from the 1980s on just wanted the economy to remain strong enough for them to cash in their degrees. However, if the U.S. government had not learned anything from the Vietnam fiasco, protesters against the Iraq War had. Iraq War protestors did not disrespect the troops in any way, but centered on President George Bush and his son President George W. Bush. The 1960s protesters had often considered the Vietnam troops as not bright enough to see this was a stupid war. Many may have done this out of personal guilt that while they hid behind their student deferments, someone working at a gas station had to fight in their place. They asked why the draftees didn't at least question conscription. The unspoken assumption was often that the conscripts were not particularly patriotic, just not very intelligent.

The difference in the two foreign wars can be pinpointed with one event—9/11. The Afghanistan War clearly involved American security, as no aspect of the Vietnam War ever did. And clearly, the Afghan government protected those who attacked the United States. Also, there was no draft by this time, the casualties were much lower, and high war costs were being paid for by the deficit. Meanwhile, income taxes had actually been cut.

The Vietnam War protests found many people and conditions to blame for the war and thus were more complicated. Protesters blamed generals who advised sending troops and the silent majority of citizens as well as presidents. The Iraq War protests usually blamed only a managed president, his powerful vice president, and political advisers. The Vietnam generation often took personal responsibility. Vietnam protesters more often said, look what we are doing. In a democracy, citizens as well as leaders are responsible. And Vietnam protest songs such as Tom Paxton's "We Didn't Know," which compared U.S. voters to Germans who ignored Nazism, made this precise point. It is a striking generational difference. Many 1960s activists felt they were superior to older generations, both in their understanding of new issues such as ecology and their willingness to respond. The present generation does not generally feel superiority or a special responsibility for change. Most college students oppose present wars and environmental damage, but do not feel a personal responsibility to stop them.

See Also: Antiglobalization Movement; Demonstrations and Events; Environmental Justice Movements; Rainforest Action Network; Social Action, National and Local.

Further Readings

Harris, David. *Goliath*. New York: Sidereal Press, 1970.
Philbin, Marianne, ed. *Give Peace a Chance*. Chicago, IL: Chicago Review Press, 1983.
Zinn, Howard. *The Politics of History*. Urbana: University of Illinois Press, 1983.

Jerry Rodnitzky
The University of Texas at Arlington

Art as Activism

"Art" involves diverse pursuits in which creativity and imagination are used to produce visual media (e.g., drawings, paintings, sculptures, photographs, mixed media), dance, drama, music, and literature. "Activism" typically involves direct action (e.g., public protests, demonstrations, blockades, boycotts, hunger strikes) to advance a particular perspective on a controversial issue. Thus, "art as activism" employs creative pursuits to make visible (and thus raise awareness about) a social, economic, political, and/or environmental issue, with a view to advancing a particular agenda. This article explores discrete examples of art as activism in order to illustrate the broader principles at work in this social phenomenon.

Photography as Activism

A number of photographers can be understood not just as artists but also as activists. For example, internationally acclaimed photographer Chris Jordan uncovers the material detritus of our consumer culture through large-format, oversized images. His mission

Artistic expression can attempt to affect social change as well, like the Bread and Puppet Theatre, which uses gigantic puppets with political messages. Pictured is their annual domestic resurrection circus in Glover, Vermont, in the mid-1980s.

Source: Walter S. Wantman and Samuel Wantman/ Wikimedia

is to make visible the true scale of American mass consumption and the unintended consequences of largely unconscious collective choices. To this end, in 2009, Jordan led a team of artists to the Midway Atoll to expose the devastating impacts of the North Pacific Gyre (also known as the Great Pacific Garbage Patch). By bearing witness to the tragedy of countless albatross chicks dying unnecessarily from ingesting bits of plastic, Jordan used his photography strategically to broadcast this truth and (hopefully) effect positive change. According to Jordan, art connects people to their feelings (and thus, to nature and each other).

Canadian Edward Burtynsky is yet another internationally renowned photographer whose work has an implicit activist element. Burtynsky systematically documents landscapes radically transformed through industrialization, offering an implicit critique of modernity and lament for the loss of nature. Much like Jordan, he makes visible that which would otherwise remain hidden to the masses; his work on oil, mines, and quarries is particularly provocative in this regard. His photographs portray an industrial landscape of exquisite (often described as "sublime") beauty. Burtynsky donates his photographs to help raise awareness about WorldChanging (a nonprofit organization for which he is one of three serving on the board of directors) and provides links to other grassroots environmental nongovernmental organizations on his website.

Street Art as Activism

Street art as activism entails predominantly unsanctioned political art in public spaces (distinct from government-sponsored town murals, territorial graffiti, vandalism, and/or corporate messaging). Urban space becomes understood as an unexploited canvas and an effective means through which to reach people vis-à-vis street installations, theatrical performances, video projections, yarn bombs, guerrilla gardens, stencils, and sticker art. The global Reclaim the Streets (RTS) movement is noteworthy in this regard.

Theater as Activism

Theater as activism engages its audience in a dialogical critique of the social dynamics portrayed. In particular, Augusto Boal (1979) bridges conceptually the long-standing gap between theater and politics in *Theatre of the Oppressed.* Work such as Bread and Puppet—a political theater in Vermont that uses gigantic puppets—attempts to affect social change through the revolutionary potential of theater.

Music as Activism

Music as activism began perhaps with the 1969 Woodstock Music and Arts Festival. Though ostensibly about the "hippie" counterculture (with its distinctive clothing, characteristic hairstyles and jewelry, illicit drug use, open sexuality, and rock and roll music), Woodstock included heavy political undertones (namely, opposition to the Vietnam War and general support for social justice). At the time, it was the largest festival of its kind to date, though many others have followed in its wake (notably, Bob Geldof's antipoverty and Ethiopian famine relief efforts vis-à-vis his 1984 Band Aid, 1985 Live Aid, 1986 Sport Aid, and 2005 Live 8 concerts).

Broader Context

The examples presented above are far from exhaustive. Documentary filmmaking, strategic radio programming (e.g., Farm Worker Justice uses radio to advocate pesticide safety, *human immunodeficiency virus* [HIV] prevention, and to raise awareness about immigration and labor laws), and culture jamming (such as the "subvertisements" presented in *Adbusters*), would all fall under the rubric of "art as activism" and thus warrant further investigation.

Deborah Barndt explores the tensions within and between the worlds of art, activism, and academia and challenges the divides between these modes of engagement. More specifically, she examines art as both an individual and collective practice and argues that understanding art as activism challenges the elite world of conventional art (making art more accessible while simultaneously challenging its commoditization). Without a doubt, what constitutes art remains highly subjective, and ultimately, controversial. And if art writ large has an activist element yet receives public funding, then messages critical of hegemonic powers may themselves become subject to critique. However, much like peaceful protest and civil disobedience, art as activism can be understood as a fundamental democratic urge and a legitimate form of civic expression. This article has sought to break apart the generally unchallenged dichotomy between art and activism and, in so doing, shed light on some of the ways artists seek to sway public opinion through their creative endeavors.

See Also: Demonstrations and Events; Film, Documentaries; Nongovernmental Organizations; Theater.

Further Readings

Adbusters Media Foundation. *Adbusters: Journal of the Mental Environment.* https://www.adbusters.org (Accessed June 2010).

Barndt, Deborah. *Wild Fire: Art as Activism.* Toronto: Sumach Press, 2006.

Boal, Augusto. *Theatre of the Oppressed.* Toronto: Hushion House, 1979.

Burtynsky, Edward. "Edward Burtynsky Photographic Works." http://www.edwardburtynsky.com (Accessed June 2010).

Jordan, Chris. "Chris Jordan: Photographic Arts." http://www.chrisjordan.com (Accessed June 2010).

Patricia Ballamingie
Carleton University

Artists' Materials

With the growing popularity of the sustainability movement, increasing attention has been paid to the use of green artists' materials. Green or sustainable artistic materials are those utensils and supplies recognized and chosen by artists for their lack or reduction of permanent negative environmental impact or toxicity. Such materials span many media and movements and are largely incorporated into the production aspect of the interrelated genres of environmental art, green art, sustainable art, earth art, land art, art in nature, ecological art, bio-art, and crop art, across the media of painting, printmaking, sculpture, site-specific performance, glass art, metalwork, and mixed-media artwork. While many attempts to make artists' materials greener involve artists' making choices regarding selection of utensils and supplies, technological advances in the manufacture of these also have positively affected the sustainability of the process.

Oil Paints

Oil paints have been used since at least 650 C.E. and became commonly used throughout Europe in the following centuries. Jan van Eyck, a Flemish painter who worked in the 15th century, was one of the first artists to use a siccative oil mixture that was used to combine mineral pigments. Later artists, including Antonello da Messina and Leonardo da Vinci, added other substances such as litharge (lead oxide) and beeswax, which improved oil paints' consistency and drying speed. Modern oil paints are composed of color pigments suspended in linseed and other oils. Traditional substances, such as lead oxide, have been replaced with more environmentally friendly alternatives such as zinc and titanium. Certain pigments commonly used in the past, such as copper acetoarsenite and arsenic sulfide, have also fallen out of use, although other pigments, including cadmium and mercuric sulfide, remain in use although toxic to a certain extent. While these substances alone may not contribute greatly to environmental degradation, the toxicity of chemicals used as vehicles for emulsion, binders, retarders, and sealers do have ramifications for the environment, as do the surfaces or materials upon which these chemicals are applied.

Artists using oil paints frequently use mineral spirits or turpentine to thin the paint or to assist in the cleanup process. Turpentine, which is a distillation of resins obtained chiefly from pine trees, produces vapors that can irritate the eyes and skin. Turpentine vapors can damage the lungs and respiratory system and the central nervous system when inhaled and can cause renal failure when ingested. Since turpentine is also flammable, extreme care must be taken in its usage, handling, storage, and disposal. Artists interested in the environmental impact of their work thus tend to use oil paints and other substances carefully and consider the impact of these choices.

Acrylic Paints

Many artistic innovations since the latter half of the 20th century are attributable to technological and scientific breakthroughs, especially in the field of chemistry. Acrylic paints, for example, were first made commercially available in the 1950s. Acrylics are fast-drying paints that contain pigment suspended in an acrylic polymer emulsion, which becomes water fast when dry. Acrylic paints use water as the vehicle for emulsion, unlike oil paints, which commonly use linseed oil, which can cause spontaneous combustion when used improperly. Artists working with acrylics sometimes use a retarder, an agent that is used to slow the drying time of the paints, giving more time for blending or layering highlights. Retarders usually contain glycol, such as propylene glycol or glycerin-based additives. While the acute toxicity of propylene glycol is quite low, in large quantities it may cause appreciable health damage to humans.

The scientific developments that led to the availability of acrylic paints have enabled contemporary artists to select from a vast array of bright and once-rare color pigments. Acrylic paints are less toxic than oil paints, and they dry quickly. Although works produced with oil paints can last for centuries, many darken over time and need restoration. As a result, many artists have become interested in working with acrylic paints because these problems have been greatly diminished. The technological advancement in color production has thus also aided in the development of more permanent media of artistic production that is less subject to degradation over time.

Other Media

Certain artistic media such as watercolor are popular because of their relative lack of expense and environmentally friendly composition. Watercolors consist of pigments that are suspended in a water-soluble vehicle. Although more sustainable than many other options, watercolors have been criticized for their lack of longevity. The widespread prominence associated with the environmentalist movement and sustainable artistry, however, has led some artists to question the environmental implications of such permanence.

Metalwork is another popular medium, used to produce sculpture, jewelry, and other decorative objects. The processes used to extract ores from the Earth and to manufacture various metals are energy intensive and often result in a variety of actions that potentially damage the environment, such as subsidence, erosion, and pollution. Metals commonly used by artists include bronze, brass, stainless steel, pewter, and copper. Processes used to create sculptures such as lost-wax casting, the piece mold process, welding, or other means often require the use of equipment that produces toxic emissions or uses lead or other hazardous materials. Despite these risks, metalwork remains prevalent because of the long-lasting works it produces as well as its aesthetic effect.

See Also: Art as Activism; Corporate Green Culture; Sculpture; Sustainable Art.

Further Readings

Art & Creative Materials Institute. http://www.acminet.org (Accessed October 2010).

California Environmental Protection Agency, Office of Environmental Health Hazard Assessment. "Guidelines for the Safe Use of Art and Craft Materials." (October 2009). http://oehha.ca.gov/education/art/artguide.html (Accessed October 2010).

Michel, K. *Green Guide for Artists: Nontoxic Recipes, Green Art Ideas and Resources for the Eco-Conscious Artist*. Beverly, MA: Quarry Books, 2009.

Stephen T. Schroth
Daniel O. Gonshorek
Knox College

Auto-Philia/Auto-Nomy

Automobiles have transformed society as fundamentally as did the factories that produced them. The nature of human life was altered by the invention of the automobile, with cities reshaped and suburbs invented in response to its ubiquity; entire towns owe their existence to the ability of their residents to drive elsewhere to work and shop. The institution of auto loans added to the debt load of the average American family, while credit cards were introduced in no small part because in the changing landscape of a mobile United States, Americans were more and more likely to make purchases at stores where no one knew them well enough to let them run a tab. Shopping malls, outlet stores, "big box" stores, and chain restaurants are all categories of businesses that developed in response to the convenience of the driving American. Major economic issues like the competition with the Japanese in the 1970s and 1980s, the relocation of American-owned factories to Mexico in the 1980s and 1990s, and the government bailout of General Motors and Chrysler in the 21st century have revolved around the importance of the automobile industry to the manufacturing sector. Getting one's driver's license has become a significant coming-of-age ritual, and automobile advertisements play up associations with freedom, luxury, mobility, the ability to provide for one's family and to juggle one's many obligations, and the American Dream.

The United States relies much more heavily on car travel than most of the industrialized world. Car ownership is nearly universal, not just among families but individuals, with most middle-class families having more than one car. Only in a very few cities does mass transit offer extensive-enough coverage and frequent-enough service to replace driving for most residents. In fact, one out of every three mass-transit passengers lives in New York City or its suburbs, as do two out of three rail passengers. In most cities, mass transit has very limited coverage, often centered on the travel patterns of a particular demographic: bus routes from retirement communities to malls and supermarkets, or from college campuses to downtown areas. In some areas, transit is geared in large part toward tourists and business visitors and is focused on airports, hotels, and shopping districts, with little to no coverage in residential areas. Meanwhile, the average price of a new car was approximately $28,000 in 2010, about $3,000 more than the median annual personal income for an American over 18. The price necessitates, for nearly all Americans, either financing their car purchase or relying on used cars, if not both; auto loans, in turn, form the cash flows that are used as the underlying assets in structured investment vehicles called asset-based securities.

Almost half of Americans breathe unhealthy air. Transportation is a leading contributor to air pollution in the industrialized world, and the average car in the United States every year emits 5 tons of carbon dioxide, 77 pounds of hydrocarbons, and consumes 581 gallons of gasoline; the figures are about one-third to one-half higher for light trucks. In 2010, more than 35 years after the fuel crises of the 1970s that were expected to catalyze a series of fuel efficiency reforms and a move away from fossil fuel dependence, federal fuel efficiency standards remained barely altered, still at the 1985 levels for automobiles (27.5 miles per gallon).

Electric cars have been talked about for decades but only two mass-produced models (the Tesla and the Nissan Leaf) have now reached the market. Hybrid cars, which still have tremendous environmental impact—as some environmentalists put it, there are no green cars, all you can hope for is to do less harm—sell for twice the price of standard cars and are less available on the secondary market. Electric cars using batteries are in development, but there are strong suggestions that they will be smaller and less roomy, an unlikely replacement vehicle for most U.S. households. In their present form, biofuels are not a viable long-term replacement for fossil fuels because of the environmental cost of large-scale adoption. Hydrogen-powered and fuel cell cars may be a possibility in the future, as may cars powered by compressed air or liquid nitrogen, but these are far off on the horizon.

Ultimately, the problem is as much with dependence on cars as with the specifics of the cars themselves. Present fuel economy would be sufficient if cars were used less often, reserved for purposes for which public transportation is impractical or impossible, like Christmas trips to relatives or picking up supplies at home improvement stores. The problem with the American automobile is its overuse. One solution may lie in car-sharing companies like Massachusetts-based Zipcar, which was founded in 2000 and now operates in 67 metropolitan areas. Car-sharing companies are an alternative to car ownership. Part of the trouble with American driving habits is that once a driver owns a car, the cost of that car, its finance charges, and its insurance are such that he needs to use the car regularly in order to justify the expense. Entire industries, after all, from drive-in movies to drive-through restaurants, exist not because they are appealing enough that people will buy cars in order to patronize these businesses, but because once a driver is in the car for one purpose, he or she is easily diverted to other purposes. Car-sharing services could be compared to the car phones of the 20th century rather than the cell phones of the 21st: they provide members with access to a technology when it is needed without making it ubiquitous in its convenience and availability such that its use becomes part of one's daily habits. If Americans drove only when driving was the best or only option available, they would drive a good deal less—provided the mass transit infrastructure is put in place to offer them alternatives to the automobile.

See Also: Air Pollution; Begley, Jr., Ed; Green Consumerism; Hybrid Cars; Social Action, National and Local.

Further Readings

Halberstam, David. *The Reckoning*. New York: Morrow, 1986.
Kay, Jane Holtz. *Asphalt Nation: How the Automobile Took Over America and How We Can Take It Back*. New York: Crown, 1997.
Williams, Heathcote. *Autogeddon*. New York: Arcade, 1991.

Bill Kte'pi
Independent Scholar

B

BARTHOLOMEW I: THE GREEN PATRIARCH

As the spiritual leader of Orthodox Christians, Bartholomew I (right) has emphasized environmental awareness by refining environmental theology and coordinating work with scientists and political leaders. Here, he meets with President Barack Obama.

Source: Pete Souza/White House, via Wikimedia Commons

The spiritual leader of Orthodox Christians was born Demetrios Archontonis on February 29, 1940, in a small village on the Turkish island of Imvros. Bartholomew I was elected to be Ecumenical Patriarch and Archbishop of Constantinople and New Rome in October 1991. At this time, he became "first among equals" in the Orthodox churches and assumed the responsibility of initiating and coordinating the actions among the national Orthodox churches, which contain a membership of nearly 300 million people worldwide. He has emphasized environmental awareness by refining environmental theology and coordinating work with scientists and political leaders. By declaring the manner in which humanity interacted with the natural world as sin in a 1997 talk, the Ecumenical Patriarch thrust environmental degradation squarely into the theological traditions of Orthodox Christianity and Orthodox Christianity squarely into the environmental movement.

Patriarch Bartholomew is not the first Orthodox patriarch to articulate the instrumental role religions play in the solutions to environmental problems. This mantle belongs to his predecessor Patriarch Demetrios I, who gave the first Patriarchal Message on the Environment in 1989. This message, and the leadership given by Patriarch Demetrios at the 1988 conference Revelation and the Future of Humanity, led to the creation of an annual encyclical on the environment. This encyclical, of which Patriarch Bartholomew has issued 18, is an annual call on September 1 for Orthodox churches to pray for creation. These messages offer the most complete collection of Patriarch Bartholomew's theological and ecclesial reflections on environmental issues, but they are only one part of the many efforts he makes as the Green Patriarch.

As Ecumenical Patriarch, Bartholomew is tasked with convening Church-wide councils and meetings, facilitating interfaith and interchurch dialogues, and identifying issues that require the attention of the entire Church. His role is one of spiritual leadership, and he does not hold unrestricted administrative power, nor is he seen as an infallible judge in matters of faith. The organizational structure of Orthodox Christianity is different from Catholicism. Orthodox structure allows each ecclesial province to have full autonomy and inclusion in full communion with the worldwide Church. This type of structure is called autocephalous. The Ecumenical Patriarch is the president of his Synod, Constantinople and New Rome, and because of this position he is the Ecumenical and Spiritual Leader of the entire Orthodox Church. His calls for environmental action are best understood as being part of his responsibility for identifying social problems and theological topics that demand the attention of the entire Church, but the autocephalous churches have freedom to reciprocate or ignore these calls for action.

In addition to his work within the Orthodox Church, Patriarch Bartholomew has established the Religious and Scientific Committee. This organization has hosted eight international, interdisciplinary, and interreligious forums and conferences to examine the ecological futures of waterways. Specifically, these symposia have examined the Aegean Sea, the Black Sea, the Danube River, the Adriatic Sea, the Baltic Sea, the Amazon River, the Arctic, and the Mississippi River. These conferences have brought together a diverse range of scientists and religious leaders at the Ecumenical Patriarch Bartholomew's request and in conjunction with political leaders. Furthermore, he has initiated work toward the creation of the Center for Environment and Peace on one of the Princess Islands near Istanbul. This center will focus on climate change and the ways humans need to change their behaviors.

The ethical calls informing the work on environmental degradation by Patriarch Bartholomew echo traditional concepts within the Orthodox faith. In his numerous encyclicals he draws upon four main themes to make his arguments for increased environmental activism. The self-emptying (*kenosis*) of ministry is a concept that points toward the need for humility in human life. Although scientific progress, Patriarch Bartholomew argues, has provided us with ample goods, we must be careful and temper our technological advancements with humility. The ministry (*diakonia*) of the community calls us toward concern for the ways our actions negatively impact other beings. Witness (*matryria*) is the idea that we may need to be willing to suffer—go without—for the betterment of ourselves and others. And finally, the concept of thanksgiving (*eucharistia*), a theological concept used by Patriarch Bartholomew, explains how we as humans must receive gifts and life from the world with thanksgiving and not avarice. By connecting his ecological awareness with the 2,000-year-old tradition of Orthodox Christianity, Patriarch Bartholomew is providing all people with insights into the spiritual and ethical dimensions of global environmental problems.

The works and efforts of the Green Patriarch have generated a remarkable amount of prestige, as reflected in numerous honorary doctorate degrees from theological and secular universities. Additionally, he was awarded the Sophia Prize, the Visionary Award for Environmental Achievement, and a Congressional Gold Medal by the U.S. Congress.

Most recently, he was named one of first group of seven world leaders to be recognized by the United Nations Environment Programme with the Champion of the Earth Award.

See Also: Biblical Basis for Sustainable Living; Environmental Justice Movements; Religion; Religious Partnership for the Environment.

Further Readings

Chryssavgis, John, ed. *Cosmic Grace, Humble Prayer: The Ecological Vision of the Green Patriarch Bartholomew*. Grand Rapids, MI: Wm. B. Eerdmans Publishing, 2009.

Ecumenical Patriarchate of Constantinople. "The Green Patriarch." http://www.patriarchate.org (Accessed June 2010).

Patriarch Bartholomew. *Encountering the Mystery: Understanding Orthodox Christianity Today*. New York: Doubleday, 2008.

"Religion, Science and the Environment." http://www.rsesymposia.org (Accessed June 2010).

Brandn Q. Green
Pennsylvania State University

Begley, Jr., Ed

Ed Begley, Jr., is an American actor and environmentalist. Since 2007, Begley and his wife, actress Rachelle Carson, have starred in their own reality television series, *Living With Ed*, on HGTV and Discovery's Planet Green channel. He is also the author of *Living Like Ed: A Guide to the Eco-Friendly Life* (2008) and *Ed Begley, Jr.'s Guide to Sustainable Living: Learning to Conserve Resources and Manage an Eco-Conscious Life* (2009).

Born September 16, 1949, in Los Angeles to Allene Jeanne Sanders and Ed Begley, Sr., an Academy Award–winning actor, Edward James Begley, Jr., spent the first 13 years of his life in Buffalo, New York. In 1962, his family moved backed to California, where he graduated from Notre Dame High School and attended Los Angeles Valley College in North Hollywood. In 1976, he married his first wife, Ingrid Taylor; together they had a daughter, Amanda Begley, in 1977, and a son, Nicholas Taylor Begley, in 1979, before deciding to divorce in 1989. Approximately one decade later, he married Rachelle Carson, and they had a daughter together whom they named Hayden.

Best known for his role as Dr. Victor Ehrlich (Dr. Mark Craig's intern) on the television series *St. Elsewhere* (for which he received six Emmy nominations), Begley says that he was inspired by his father to become an actor. Since that time, he has appeared in a multitude of movies, television shows, and theater projects. Begley's films include the Christopher Guest films *A Mighty Wind, Best in Show*, and *For Your Consideration.* On television, he has played recurring roles in the series *Six Feet Under*, *Arrested Development*, and *Boston Legal* and has directed several episodes of *NYPD Blue*. In the theater, he has appeared in several plays by David Mamet. In 2003, he wrote and directed a play titled *César and Ruben*, a musical about the late labor leader César Chávez and Ruben Salazar, the Mexican American reporter for the *Los Angeles Times* who covered union activities for the newspaper and who was slain by Los Angeles Sheriff's Deputy Tom Wilson in 1970. For this play, Begley received a Nosotros Award and four Valley Theater League Awards. The work was revived in 2007.

Environmental Activism

Beginning with the purchase of his first electric vehicle, a Taylor-Dunn golf cart, in 1970, Begley has consistently shown himself to be a committed environmental leader for 40 years. Often turning up at Hollywood events on his bicycle, he has served as chair of the Environmental Media Association and the Santa Monica Mountains Conservancy. He has also served on the boards of the Thoreau Institute, the Earth Communications Office, Tree People, and Friends of the Earth, among many others. A promoter of ecofriendly products such as the Toyota Prius, he is also the spokesperson for a line of environmentally friendly household cleaning products, Begley's Best.

In his own life, Begley attempts to practice what he preaches through practices such as recycling and by becoming a vegan. The Begley family home, which is prominently featured in the reality series *Living With Ed*, is a modest 1,585-square-foot (147.3 square-meter) structure that utilizes solar power and wind power to supply the family's energy needs. Arguing that suburban lawns are environmentally unsustainable, he has converted his own yard to a drought-tolerant garden composed of native California plants.

In November 2008, Begley traveled through five cities in five days as part of a campaign to generate awareness of simple ways to save on energy bills. In this campaign, he promoted the use of simple appliances such as ceiling fans and programmable thermostats to save money on energy bills.

Begley's work in the environmental community has earned him a number of awards from some of the most well-known environmental groups in the United States, including the California League of Conservation Voters, the Natural Resources Defense Council, the Coalition for Clean Air, Heal the Bay, and the Santa Monica Baykeeper.

See Also: Alternative/Sustainable Energy; Chávez, César (and United Farm Workers); Communication, Global and Regional; Communication, National and Local; *e² design*, TV Series; Energy; Environmental Communication, Public Participation; Environmental Media Association; Global Warming and Culture; Green Consumerism; Huerta, Dolores; Hybrid Cars; Individual Action, Global and Regional; Individual Action, National and Local; Leisure; Popular Green Culture; Technology and Daily Living; Television, Cable Networks; Television Programming.

Further Readings

Begley, Jr., Ed. *Ed Begley, Jr.'s Guide to Sustainable Living: Learning to Conserve Resources and Manage an Eco-Conscious Life*. New York: Clarkson Potter, 2009.

Living With Ed. http://planetgreen.discovery.com/tv/living-with-ed (Accessed September 2010).

Who Killed the Electric Car? Video recording. Culver City, CA: Sony Pictures Home Entertainment, 2006.

Danielle Roth-Johnson
University of Nevada, Las Vegas

Benjamin, Medea

Medea Benjamin is an American political activist best known for founding the human rights organization Global Exchange, an advocacy group that supports Fair Trade alternatives to corporate globalization (i.e., where environmental issues and fair wages for the goods produced are given greater priority over corporate profits). She is also one of the cofounders of Code Pink: Women for Peace, an organization devoted to the end of the war in Iraq and to a reorientation of U.S. budget priorities to healthcare, education, and housing.

Born Susan Benjamin on September 10, 1952, Benjamin (who changed her first name to Medea as a young adult) received a master's degree in Public Health from Columbia University and a master's degree in Economics from the New School for Social Research. A decade before founding Global Exchange in 1988, she worked as an economist and nutritionist in Latin America and Africa for the United Nations Food and Agriculture Organization, the World Health Organization, the Swedish International Development Agency, and the Institute for Food and Development Policy.

Since 1988, she has devoted much of her energy to improving the labor and environmental practices of U.S. multinational corporations. She has also been openly critical of the policies of international institutions such as the World Trade Organization (WTO), the International Monetary Fund (IMF), and the World Bank. During the WTO meeting in Seattle in December 1999, for example, Benjamin's organization, Global Exchange, helped focus world attention on the need to prioritize labor and environmental concerns over corporate profits. Critical of unfair global trade policies, Benjamin has promoted Fair Trade alternatives beneficial to both producer and consumer, helping form a national network of retailers and wholesalers in support of Fair Trade. She was also instrumental in convincing coffee retailers such as Starbucks to begin carrying Fair Trade coffee.

A key figure in the anti-sweatshop movement, Benjamin has led campaigns against Nike and clothing companies such as the GAP. In 1999, for example, she helped expose the problem of indentured servitude among garment workers in the U.S. territory of Saipan (the Marianas Islands), an action that led to a billion-dollar lawsuit against 17 American retailers.

After several fact-finding visits to China, she cosponsored, with the International Labor Rights Fund, an initiative to improve the labor and environmental practices of U.S. multinationals in China. The Human Rights Principles for American Businesses in China that ensued from this action have subsequently been endorsed by major companies such as Cisco, Intel, Reebok, Levi Strauss, and Mattel.

In 2000, she ran for the U.S. Senate in California on the Green Party ticket with a focus on living wages, schools-not-prisons, and universal healthcare. Since that time, she has remained active in the Green Party and has also supported efforts by the Progressive Democrats of America.

In 2001, she devoted the majority of her efforts to California's energy crisis. Heading a powerful coalition of consumer, environmental, union, and business leaders working for clean and affordable power under public control, she worked to end the manipulation of the market by big energy companies and the ensuing rate hikes that caused hardship for low-income consumers and small businesses. Two years later, in September 2003, she traveled to Cancun, Mexico, to challenge the policies of the WTO. In November of that same year, she also appeared in Miami to protest the proposed Free Trade Area of the Americas (FTAA).

Benjamin is the author of eight books, among them *Bridging the Global Gap, The Peace Corps and More*, and the award-winning *Don't Be Afraid, Gringo: A Honduran*

Woman Speaks From the Heart. She has also written extensively about Cuba in *No Free Lunch: Food and Revolution in Cuba Today* (1989) and *The Greening of the Revolution: Cuba's Experiment With Organic Agriculture* (1994). More recently, with Jodie Evans, she coedited *Stop the Next War Now: Effective Responses to Violence and Terrorism*, a varied collection of essays from writers such as Barbara Ehrenreich, Eve Ensler, Arianna Huffington, Alice Walker, Helen Thomas, Camilo Mejia, and Jody Williams. She has also assisted in producing various television documentaries such as the anti-sweatshop video *Sweating for a T-Shirt*.

In 2005, she was nominated for the project "1000 Women for the Nobel Peace Prize 2005," an undertaking that selected 1,000 women from around the globe for collective nomination for the Nobel Peace Prize to represent the multitude of anonymous women who work for peace, justice, human rights, security, and education worldwide.

See Also: Antiglobalization Movement; Antiwar Actions/Movement; Corporate Social Responsibility; Environmental Justice Movements; Fair Trade.

Further Readings

Benjamin, Medea, Joseph Collins, and Michael Scott. *No Free Lunch: Food and Revolution in Cuba Today*, 3rd ed. Princeton, NJ: Princeton University Press, 1989.

Evans, Jodie and Medea Benjamin, eds. *How to Stop the Next War Now: Effective Responses to Violence and Terrorism*. Novato, CA: New World Library, 2005.

Kiefer, Chris and Medea Benjamin. "Solidarity With the Third World: Building an Environmental-Justice Movement." In *Toxic Struggles: The Theory and Practice of Environmental Justice*, Richard Hofrichter, ed. Salt Lake City: University of Utah Press, 2002.

Rosset, Peter and Medea Benjamin, eds. *The Greening of the Revolution: Cuba's Experiment With Organic Agriculture*. San Francisco, CA: Global Exchange, 1994.

Danielle Roth-Johnson
University of Nevada, Las Vegas

Biblical Basis for Sustainable Living

The Brundtland Report, also known as *Our Common Future*, describes three foundations of sustainability: social, economic, and environmental. According to this fundamental document, sustainable development is development that meets the needs of the present without compromising the ability of future generations to meet their own needs. The Brundtland Report centers upon the connections between environmental health, economic development, and social justice and particularly emphasizes the connection between environmental deterioration and social development. Long before the Brundtland Report was published, these themes were discussed in detail in the Bible.

Social sustainability and economic sustainability are commonly recognized themes of Scripture. Some of the most frequently cited social sustainability verses include the golden rule of doing unto others as you would have them do unto you and loving your neighbor as yourself. Economic themes are also well known and include, among other precepts, the tithing of income for the poor. Environmental sustainability is a less-recognized Biblical theme, yet Scripture is pervasive with passages concerning sustainable living.

Examples of Environmental Sustainability in Scripture

Passages concerning environmental sustainability can be found from Genesis to Revelations in the Bible. The earliest interactions between God and humanity described in the first book of the Bible directly relate to environmental stewardship. Genesis 2:15 depicts God placing the first human into a garden in order that he may work it and take care of it. In the rest of the narrative of Genesis, God's concern for other creatures beyond humankind is apparent, as demonstrated in the covenant God made with Noah in Genesis 9 whereby God pledged never again to destroy the world by flood. This covenant was made not only with humans but also with the birds, the livestock, the wild animals, and every living creature on Earth.

The rest of the Pentateuch is also rife with commands to live in an environmentally sustainable manner. In the law of Moses cited in Exodus 23, God commands his people to leave the ground fallow every seventh year in order for the poor and the wild animals to eat. This command is reiterated in Leviticus 25, and in Leviticus 26 it is reinforced with the promise that, if the land is not left fallow in the seventh year, God will remove the people from the land in order that it may enjoy the Sabbaths it did not have while it was being exploited.

Some of the most poetic passages from antiquity regarding the environment originate in the wisdom literature of the Bible. Job 39 to 41 is especially poignant as it describes the detailed knowledge God possesses of the creation in comparison to the naïveté of Job's limited ecological understanding. Themes include how the universe was created and what its dimensions entail, how the oceans and atmosphere were created and how they function, the complexities of light and weather, the workings of the human mind, the intricacies of complex ecosystems, and the splendor of many particular species such as lions, goats, ravens, bears, ostriches, horses, hawks, and the behemoth. Psalms also contains numerous passages emphasizing the grandeur of creation and God's expectation that humankind respect its limits. Psalm 8 exemplifies the theme of creation's majesty as the psalmist reflects upon the works of God in the heavens and upon the Earth and recognizes humankind's responsibility of stewarding the works of God's hands. Psalm 19 directly links the majesty of creation in the beginning of the psalm with the laws and statues of Scripture in the middle portion, ending with a plea for God to help the psalmist identify the hidden sins of the heart in order that the author may be blameless. Other key psalms include 24, which asserts that the Earth and everything in it is the Lord's, and 115, which denotes that though the Earth is the Lord's, it has been given to humankind to be stewarded.

Additionally, the books of the prophets depict the consequences of human-induced environmental degradation. Isaiah 5:8 provides a warning to those who add house to house and field to field, leaving no space in the land. Jeremiah 12 describes how the land will be made a wasteland after its vineyards are trampled and the pleasant fields destroyed because there is no one who cares.

Finally, the New Testament continues the theme of environmental sustainability by linking stewardship to Jesus Christ. Romans 1 associates the invisible qualities of God to the splendor of creation. Philippians 2 depicts the act of kenosis, in which Jesus emptied himself of divinity in order to participate in and redeem creation. Colossians 1 describes Jesus as both the source of creation and the source of the reconciliation of creation to the creator. The final book of the Bible, Revelation, contains the frightening prophecy in 11:18 whereby God states that the time will come to destroy those who have destroyed the Earth. So, from Genesis to Revelation, sustainable living is a key theme of Scripture, which discusses both sustainable practices and the consequences of environmental degradation.

Sustainability and the Christian Worldview

Despite the vast number of passages in the Bible describing sustainable living, some scholars have indicated that Christianity is responsible for the current scale of global environmental degradation. Historian Lynn White initiated this dialogue in 1966 through a talk at the American Association for the Advancement of Science when he asserted that Christianity bears a huge burden of the guilt for the contemporary ecological crisis and that the world will have a worsening ecologic crisis until we reject the Christian axiom that nature has no reason to exist save to serve man. White roots his theological argument in Genesis 1:28, asserting that most Christians believe that God issues dominion to humans to be fruitful and increase in number, fill the Earth and subdue it, and rule over the creation.

While many Christians do accept the idea of dominion, most scholars resist this interpretation of Genesis 1:28 and favor a response to environmental stewardship derived from multiple books of the Bible rather than from one isolated passage. For instance, Albert Wolters asserts that the core worldview of Christianity calls believers to participate in Christ's redemption of all of creation. J. Walsh and R. Middleton build upon this by identifying three gods of the present age: science, technology, and economic growth. Faith in these three gods is waning throughout Western cultures as science and technology have led to ecological degradation rather than to a healthier world, and economic growth appears to be reaching its limit in the finitude of creation. The biblical response to these crises involves renouncing idols, embracing stewardship, and engaging community.

Examples of the application of Bible-based sustainable living in contemporary culture are myriad. Many initiatives have been launched through the National Religious Partnership for the Environment (NRPE), an association of independent faith groups including the Evangelical Environmental Network (EEN), the U.S. Conference of Catholic Bishops, the National Council of Churches (NCC), and the Coalition on the Environment and Jewish Life. Examples of initiatives aimed at sustainable living are too numerous for this report. One example is the United Church of Christ Toxic Wastes and Race Report. Additionally, many denominational and parachurch organizations advocate for Bible-based sustainable living, including A Rocha, Au Sable, Blessed Earth, Catholic Conservation Center, Christian Environmental Association, Echo, Eco-Justice Ministries, Flourish, National Council of Churches Eco-Justice Working Group, National Religious Partnership for the Environment, North American Coalition for Christianity and Ecology, Plant with a Purpose (formerly Floresta), Quaker Earthcare Witness, Renewal: Students Caring for Creation, Restoring Eden, Target Earth, The Regeneration Project, and Web of Creation.

See Also: Religion; Religious Partnership for the Environment; United Church of Christ.

Further Readings

Carter, Jimmy. *Our Endangered Values: America's Moral Crisis*. New York: Simon & Schuster, 2006.
Walsh, B. J. and J. R. Middleton. *The Transforming Vision*. Downers Grove, IL: Intervarsity Press, 1984.

White, L. "The Historical Roots of Our Ecologic Crisis." *Science*, 155/3767 (1967).
Wolters, A. *Creation Regained.* Grand Rapids, MI: Eerdmans, 2005.
World Commission on Environment and Development. *Our Common Future.* Oxford, UK: Oxford University Press, 1987.

Michael P. Ferber
King's University College

BIODIVERSITY LOSS/SPECIES EXTINCTION

As an increasing number of species become endangered or extinct, this reduction in biodiversity is being fought against by reintroduction programs by the American Zoological Association, which successfully breed the animals in captivity and release them into the wild. These programs increased the population of California condors from its low of 23 to over 300 a decade later, including this chick as it prepares to take flight.

Source: Mike Wallace, Zoological Society of San Diego/ U.S. Fish and Wildlife Service

Biodiversity is the term used to describe the variety of living organisms on the planet. A contraction of "biological diversity," biodiversity is an indicator of planetary health. Biodiversity provides valuable ecosystem services that balance and support life on Earth, including clean air, pure water, stable climate, and pollination. Biodiversity also provides humans with inspiration, tranquility, and improved well-being. It is estimated that species diversity is experiencing a 30 percent decline, which is accelerating. Nearly one-quarter of the world's mammal species, one-third of amphibians, and one out of eight bird species are threatened with extinction or already extinct.

Humans are causing species loss through exploitation, introduction of invasive species, and the destruction and fragmentation of habitat. The planet has experienced five mass extinction events, with the last event occurring 65.5 million years ago. Scientists widely believe that the rate of biodiversity loss indicates we are heading into the next event, referred to as the sixth extinction. Species recovery, reintroduction, and other conservation strategies are showing positive impacts toward increasing some populations of threatened species. The discovery of new, unrecorded species and populations of species already thought to be extinct are also positively affecting biodiversity.

Areas of high levels of species richness are known as biodiversity "hotspots." These areas have increased numbers of species endemic, or specific only to that particular region. The Mediterranean Basin is an example of one of over 30 identified hotspots for biodiversity. It contains 22,500 endemic vascular plant species. The Mediterranean Basin is identified as a hotspot in part due to the level of threat to its biodiversity: only 4.5 percent of the area remains undisturbed. The basin is also home to the Iberian lynx, the most endangered cat in the world. The population of Iberian lynx fell from an estimated 4,000 in 1960 to as few as 100 in 2005. The lynx's decline is attributed to habitat loss, road casualties, poaching, poisoning, feral dogs, and disease affecting rabbits in Spain and Portugal, its primary prey.

The lynx is part of a global decline in biodiversity that some estimate to have dropped by 30 percent over the past 40 years. The status of most of the world's estimated 8–30 million species remains poorly known, with fewer than 2 million species identified and named. Fewer still have been recorded over time to assess changes in their population. Disagreements over taxonomic classification, bias toward terrestrial vertebrates, and the sheer volume of species to monitor make it difficult to quote definitive statistics on biodiversity.

Status of the World's Species

The International Union for Conservation of Nature (IUCN) Species Survival Commission maintains a collection of data that is used to assess the risk of extinction for 2.7 percent of the world's described species. The Red List Index compiles information on the population status of 44,838 species being monitored throughout the world. Of this group, 16,928 are threatened with extinction: 3,246 are critically endangered; 4,770 are endangered; and 8,912 are listed as vulnerable.

The 2008 Red List shows a total of 869 known extinctions, including 65 species listed as extinct in the wild. If those listed as "possibly extinct" are added, the total rises to 1,159 extinct species. This list provides a baseline for documenting trends in species populations, where data is available. Because of the high level of undocumented or data-deficient species, the true number of extinctions is thought to be significantly underrepresented. Threat is measured by reductions in population over the course of years or beyond three generations; the extent of occurrence over geographic range; and the number of individuals of mature, reproductive age.

There are a few taxonomic groups in which every species has been evaluated. These comprehensive assessments are used as indicators of overall trends in biodiversity. They include mammals, birds, cycads and conifers, warm-water reef-forming corals, freshwater crabs, and groupers.

Amphibians and cycads appear to have the highest number of threatened species among their groups. It is thought that their restricted mobility makes them more vulnerable than highly mobile groups like birds and dragonflies, which can relocate from distressed habitats more successfully.

Almost one-quarter of the world's mammal species are threatened or already extinct. Since 1500, 76 mammal species have been identified as extinct, plus 29 species listed as "possibly extinct." Countries with the highest diversity of mammal species are, in order, Indonesia, Brazil, China, and Mexico. The greatest threat to mammal populations is loss of habitat. The second-greatest threat is the use of mammals for food and medicine, particularly in Asia.

Nearly one-third of total amphibian species are threatened or extinct (with 25 percent data deficiency). South America, southeastern United States, and tropical West Africa have areas of high amphibian biodiversity. The status of amphibians in Colombia demonstrates how habitat loss and disease can impact species populations. Colombia has over 700 amphibian species, and is second only to Brazil in amphibian diversity. However, it has the highest number of threatened amphibian species, at 214. The intense topography of the Andes Mountains creates small, restricted populations that are vulnerable to even small changes in their habitat. Without surrounding healthy habitat to escape to, the loss of territory combined with exposure to the fungal disease *chytridiomycosis* can weaken or destroy entire populations rapidly. In general, amphibians are most dramatically affected by habitat loss, pollution, and disease.

In contrast to amphibians, birds are one of the best documented groups (only 1 percent data deficiency). The Red List shows one out of eight species of birds threatened or extinct. Over 27 percent of marine birds and nearly 12 percent of terrestrial birds are threatened. Since 1500, the planet has lost 134 species of birds, plus four species extinct in the wild. At least 15 species are labeled "possibly extinct." Most of these extinctions have occurred on islands, due to the introduction of invasive and nonnative predatory species such as cats or rats. Brazil and Indonesia currently have the highest number of birds that are threatened with extinction.

Over 85 percent of the threat to bird species, and the primary cause of their global population decline, is attributed to the loss of habitat from agriculture. Logging and invasive species have a negative impact on more than half of the birds under threat. Birds also suffer from the harmful effects of pollution and overexploitation. As with all species, it is predicted that human-induced climate change could pose the greatest risk to bird populations long term.

In general, freshwater species loss is blamed on pollution and invasive species spread through the highly connected aquatic systems. Development of water resources without regard to ecosystem health is another cause of biodiversity loss. Marine species are vulnerable to overfishing, climate change, coastal development, pollution, and invasive species. Reported statistics show that 17 percent of 1,045 shark and ray species are threatened with extinction, as well as 12 percent of groupers, and six of the seven species of marine turtles.

Out of intense concern over the health and status of the world's ocean reef systems, the Red List documented over 800 species of warm-water reef-building corals, finding 27 percent threatened and 17 percent data deficient.

Geographic patterns in species decline are evident. Rapid deforestation in south and southeast Asia since the 1990s has created sharp declines in bird, amphibian, and mammal species. Large mammals in this area have also been highly impacted by hunting. Invasive alien species are causing substantial bird declines in the island regions of the Pacific, including Micronesia, Fijian Islands, and most of Polynesia. Amphibians are most under threat in the neotropical zone of South and Central America, Mexico, Caribbean islands, and southern Florida.

Plants are as threatened as mammals, having one in five species currently threatened with extinction. Trends in the plants studied show that the major threats include logging, uncontrolled fire, grazing, conversion of habitat to agricultural land, the spread of human development, large-scale mining or hydroelectric projects, and harvest by collectors.

Assessing the world's plant populations is even more difficult. Much of the focus is put on studying currently threatened plant species, so there is a reporting bias toward those already in jeopardy. Cycads, a unique and ancient lineage of tropical and subtropical plants, have survived five periods of mass global extinction. Although common during the Jurassic period, cycads are today the most threatened plant group known. Confusion over classifications complicates population estimates, which range from 150 to 300 different species and quotes of between 23 percent and 82 percent threatened with extinction.

Human Impact on Biodiversity

Hunting and collecting are two types of human exploitation of biodiversity. Whales, large mammals, and game fish have suffered under the pressures of overhunting. Poaching and other illegal activities offering financial reward bring sharp declines to species of high interest, such as the gorilla and rhino. The ivory trade is infamous for its devastating effect on African and Asian elephant populations. Illegal plant collecting is equally devastating, since the value of a specimen increases with its rarity. Plant collectors have been responsible for decimating the remaining populations of many species known to have only a few surviving individuals left in the wild.

The introduction of foreign species into an ecosystem, whether intentional or not, can be devastating to an endemic population. Nonnative species take over by outcompeting their native counterparts for space, food, and other resources necessary for survival. A vivid example of this occurred in Africa's Lake Victoria when Nile perch were introduced in the 1950s to encourage commercial fishing. A top predator, the perch caused the loss of over 80 percent of the native biodiversity of the lake. Ironically, the Nile perch is now being overfished in Lake Victoria, causing its decline.

Habitat loss is largely the result of converting land to agricultural use, urbanization and development, and installation of technologies like hydroelectric. Native habitat is either destroyed completely or fragmented into patches. Species able to live on the edges of their habitats are surviving fragmentation; those requiring larger areas of internal habitat are not. Some species of neotropical migrant songbirds, for example, are in decline due to forest fragmentation. As they migrate from forested habitat in Central and South America where they overwinter, songbirds are experiencing fewer available forested areas for stopping points along their route and at their destination breeding grounds. As large tracts of forested lands continue to be carved up by human activity, birds have less nesting success, resulting in population decline.

Small, localized populations are also under threat from shrinking habitat. Their restricted ranges make it impossible for them to relocate when their habitat comes under distress. For these populations, habitat loss has created an island-like effect, eliminating corridors and other types of connectivity that an isolated population could use to move to a safer location. The numbers of individuals within the population decline, reducing genetic diversity within the species, and making the entire species unable to survive the impacts of disease, further loss of habitat, or other disturbances.

Mass Extinction

From a deep time perspective, the planet has undergone five mass extinction events in the past 540 million years, losing more than half of the animal species during each event. The most recent mass extinction occurred 65.5 million years ago. Referred to as the Cretaceous-Tertiary (or K-T) extinction event, it is connected with the loss of 75 percent of Earth's species at the time, including nearly all of the dominant animal species: the dinosaurs. Although there are conflicting scientific theories over the cause of the K-T event, it is most widely believed that this extinction was caused by the catastrophic events of volcanic activity, the impact of one or more massive asteroids, or a combination of both. This scenario, based on observations of fossil and geologic evidence, relies on a thin band of sediment called the K-T boundary, which is thought to document the event at locations around the world. Mammals and birds of the time appear to have survived the extinction, but mosasaurs and plesiosaurs (large aquatic reptiles), pterosaurs (winged reptiles, also called pterodactyls), and large numbers of plant and invertebrate species did not survive.

It is also theorized that atmospheric conditions surrounding the extinction event dramatically reduced the amount of sunlight available for photosynthesis, causing broad-scale ecological collapse. Photosynthesizing plants and phytoplankton provided the basic foundation of the food chain at the time of the K-T extinction event. The loss of plants led to the loss of the plant-eating herbivores, proceeding up the food chain to the loss of their predators, including the dominant *Tyrannosaurus rex*. The only species that survived the extinction were those based on a food chain of dead or decaying plant and animal matter, including snails, worms and insects, and the birds and mammals who ate them.

Today's ecology has the same basis in photosynthesis. It is widely debated whether the current trends in species loss on the planet are ushering in the next mass extinction event.

It is thought that trends in mass extinctions reveal long-term stress of a biotic system shocked into rapid decline by a sudden event. These mass extinctions result in the destruction and decline of the dominant species of the time. For the K-T event, that meant the transition from reptiles as dominant to mammals as dominant. Humans are clearly the dominant mammal of our time. Our use of resources combined with ever-increasing population growth has put Earth into a period of biotic stress. Comparison of these extinction trends is causing increasing speculation over the inevitability of what many scientists are calling the sixth extinction.

In 1998, timed with the unveiling of a new hall dedicated to biodiversity, the American Museum of Natural History released a survey stating that 7 out of 10 biologists were in agreement that the next mass extinction was already under way. This extinction is happening faster than any other in our history and is the result of human activity rather than natural phenomena. The effects of our dramatic loss of species include the loss of resilience and ability of Earth's systems to recover from natural and human disasters, loss of systems that purify our air and water, loss of species available for new medicines, and weakened social and political stability around the globe due to losses in the world's economies.

The Future of Biodiversity

Besides the intrinsic value of nature on the planet, global biodiversity contributes to our health and well-being through breathable air, clean water, food, fibers, energy, climate regulation, and aesthetic and spiritual values. Biodiversity provides a vital resource for tourism, fisheries, forestry, and other major economic activities. The economic impact of a massive loss of biodiversity could become catastrophic, destabilizing the world's food security, health systems (which for most of the world rely on the medicinal value of plants), and other biotic resources.

While many species are in decline, direct conservation interventions are having a positive impact on others. Species can recover through sustainable harvesting efforts. Commercial fishing programs that promote responsible harvesting are allowing some fish populations to recover. The lobster industry provides an apt example. Besides laws restricting legal catch size to protect juvenile lobsters and making it illegal to take egg-bearing females, a voluntary V-notch program has gained widespread participation. When a gravid lobster is caught in a trap, a V-shaped notch is cut into the tail before the egg-bearing lobster is released. When a lobster with a V-notch is caught in a trap, it is in turn released, protecting the population's brood stock.

Reintroduction programs are successfully restoring populations, some of which were previously extinct in the wild. Black-footed ferrets have been reintroduced in the western United States and in Mexico through a program of the U.S. Fish and Wildlife Service that operated from 1991 to 2008. The program started when the 18 individuals left in the wild were taken into care by partnering zoos of the American Zoological Association (AZA). Captive breeding programs of the AZA also increased the population of California condors from its low of 23 existing in the wild in 1982 to over 300 individuals a decade later, including 171 wild condors.

Species thought to be extinct and new species are being discovered, adding to the world's biodiversity. The saola, for example, is an antelope relative discovered in 1992. Its elusive nature makes its status difficult to report, but it is estimated that fewer than 500 exist in its habitat in the Annamite Mountains of Laos and Vietnam. The Laotian rock rat was discovered in 2005—one of a group of rodents thought to have gone extinct 11 million years ago. The coelacanth was thought to have gone extinct 65 million years ago until it was rediscovered in 1938. Javan elephants were rediscovered on Borneo in 2006, 800 miles from where they were thought to have been hunted to extinction in the 1800s.

See Also: Fish; Population/Overpopulation; Rainforest Action Network; Sprawl/Suburbs/Exurbs; Water Pollution.

Further Readings

MacLeod, N., et al. "The Cretaceous-Tertiary Biotic Transition." *Journal of the Geological Society*, 154/2 (1997).

Quammen, David. *The Song of the Dodo: Island Biogeography in an Age of Extinctions*. New York: Simon & Schuster, 1997.

Vie, J. C., et al. "Wildlife in a Changing World: An Analysis of the 2008 IUCN Red List of Threatened Species." Gland, Switzerland: World Conservation Union, 2009.

Wilson, Edward O. and Frances M. Peter. "Biodiversity." Paper from the First National Forum on Biodiversity, September 1986. Washington, DC: National Academy of Sciences, 1996.

Dyanna Innes Smith
Antioch University New England

Blogs

Blogs serve as a useful for way for citizens to share their opinions about political issues, and blogging, in turn, is considered a political act. For this reason, blogs relating to green culture are especially important since some politicians do not make the environment a primary concern on their agendas. Some activists offer incentives to green bloggers to endorse particular green movements; situations like this emphasize both the political power of the blogger and the potential political force amassed by a given activist through blogger support. Thousands of green blogs exist, although some are considered more reputable and have many dedicated followers committed to learning about green living. Blogging allows writers to reach an immediate worldwide audience while discussing all facets of green practices, ranging from green parenting to green fashion. The most respected and successful green blogs take an optimistic approach rather than instilling fear in readers about environmental peril and destruction.

Some green activists harness the energy, passion, and commitment of the worldwide blogging community by offering writers incentives for creating awareness about platforms the activist supports. One such example is the incentive offered by Brighter Planet's 350 Challenge. The challenge takes its name from the maximum number (350) of parts per million of carbon dioxide allowed in the air if the environment is to be protected. This site is affiliated with Bill McKibben and his fight to create awareness about the significance of the number 350 concerning green issues. Brighter Planet encourages bloggers to write about the effects of climate change, and if they address the topic on their blog, they can apply for a Brighter Planet badge to place on their blog's home page, which states "My blog fights climate change!" When a blogger posts the badge, Brighter Planet offsets 350 pounds of carbon monoxide in the blogger's name. For those who feel passionately about environmental justice, this incentive from Brighter Planet is worthwhile.

Several sites, like Green Blog and Best Green Blogs, both offer green tips to viewers and link to hundreds of green blogs. Green Blog provides readers with links to scores of other blogs with different emphases, including green interests, technology, food and health, shopping, gardening, politics, science, energy, activism, and the news and media. The site's home page also provides readers with coverage of issues pertaining to the environment and allows registered users to comment on the posted materials. This particular blog does indeed educate readers on green topics; the function of this blog is to primarily link readers to other blogs of interest. Green Blog serves as a network for those who want to explore other blogs that promote environmental health and consciousness. Similarly, the site Best Green Blogs features what editor Timothy Latz considers to be the best blogs about environmentally conscious practices. While the site highlights various blogs each week, Best Green Blogs also provides a blog directory for readers on the home page; the site prides itself on covering and linking to blogs around the world. Its focus remains worldwide, and one of the site's goals is to connect bloggers with others in their areas of interest to unite them in their passion and activism. Site visitors can select an area of emphasis (consumerism, energy efficiency, conservation, eco-art, energy audits, eco-psychology, endangered species, low-impact living, organic farming, and organic parenting are examples of only some categories); readers can also select blogs by country, a useful search tool if one wants to study a green issue relevant to a particular part of the world. Readers also have the ability to rate the quality of blogs they visit, and the blogs with the best reviews are often highlighted for other readers.

Two websites known for their goal to create general awareness about green living are TreeHugger and The Daily Green. Treehugger remains one of the most popular and reputable green blogs. Sponsored by the Discovery Company, TreeHugger aims to encourage conversation about a wealth of environmental topics and green practices. Besides providing readers with the latest news on green issues, the blog also shows subscribers how to become activists themselves and provides forums in which bloggers can interact with one another. Another blog, The Daily Green, gives subscribers tips for how to live a green life. The blog offers a newsletter, hosts videos, and includes an RSS (really simple syndication) feed. The site remains dedicated to supplying readers with the latest environmental news, but is also popular for its other features, including "Weird Weather Watch." This section of the blog is known for its photos, since the goal of "Weird Weather Watch" is to document climate change via photographs from around the world, submitted by both professional and amateur photographers. Another popular aspect of the blog is the section called "New Green Cuisine," in which bloggers find information about food labels, farmers markets, eating organic, and a community cookbook (that includes recipes from around the world). The Daily Green also hosts an ecopedia, which, as the website explains, supplies subscribers with information on "everything green."

Some green blogs are known for their devotion to creating awareness about a specific environmental concern. The Oil Drum, edited by Nate Hagens, hosts discussions about the Earth and energy practices. According to the website, the mission of the blog is fourfold: to "raise awareness," to "host a civil discussion," to "conduct original research in a transparent manner," and to "create a global community working toward a common goal." While some green bloggers attempt to persuade subscribers to adhere to environmentally responsible practices by pleading, The Oil Drum continuously presents its readers with hard evidence on which they can make decisions to become green citizens.

Many of today's most visited blogs concern families and how they can incorporate good green practices into their households. The Natural Nursery Blog provides readers with resources about green parenting. Readers provide reviews of and comments about products like baby slings, organic blankets, cloth diapers and wipes, and glass bottles; other posts concern breast-feeding and midwife-supported births. Similarly, The Green Parent advertises itself as a resource emphasizing green practices; the blog promotes green events around the country and world, but also focuses specifically on practices helpful to parents trying to encourage green household practices or who want to educate their children in environmentally friendly practices. In the blog's section titled "Green Your Kids," bloggers list ideas for ecofriendly crafts and have even posted quizzes parents can give their children to teach them about green living. Still other blogs like No Impact Man provide green education for families. This blog by Colin Beavan gives readers suggestions about how to combat environmental problems. When Beavan launched his blog in February 2007, he explained to subscribers that his goal was to see that he and his family (wife, child, and pet) made zero impact on the Earth while living in New York City. His strategy involved measuring the negative impact he and his family made on the planet and then creating a positive impact that would equalize or outweigh the negative impact. Beavan blogs about the steps he and his family took to become a "no impact" family like living without making garbage (e.g., packaging) and reducing their consumption.

Another popular breed of green blogs can be classified as ecofashion blogs. This form of blogging remains popular since, in part, bloggers and subscribers champion that being green is fashionable and part of a new cultural tradition that makes one appear more aware than citizens who are uneducated or not interested in green issues. The blog Style Saves the World promises to teach readers how to live green, but in a chic way; Little Green Stilettos advertises that it will show subscribers how to go green "beautifully." The blog Green Lashes and Fashion also guarantees to teach readers how to make green choices about fashion, beauty, and traveling, but with style. Similarly, several green activists devote blogs to discussing ecofriendly weddings. One such site, Ethical Weddings: Give Everyone Something to Celebrate, encourages readers to plan weddings with little waste or damage to the environment. Tips from the Ethical Weddings blog include urging brides to buy gowns made from organic materials or to refuse gowns made by sweatshop employees; other green suggestions include having future brides and grooms grow their own flowers to be used at the ceremony and reception.

A trope used by some green bloggers is to manipulate readers into action by making readers feel guilty for not living as green as possible, or to frighten them into being green by emphasizing Earth's impending danger. While these blogs receive little respect for their rhetorical tactics, blogs that use a more respectful approach and that even employ humor gain many followers. Sites that do not play on subscribers' guilt or fear but that take an optimistic approach to green education via blogging include Grist and Mother Nature Network. Grist: A Beacon in the Smog takes a humorous but didactic approach, with specific topics like climate and energy, food, living, placemaking, business, and politics. The blog also supports columns like "Farmer's Daughter," which lists organic Southern recipes. While the messages of bloggers on the site are both genuine and serious, the light-hearted way in which writers approach issues (consider, for instance, the popular article about the surprising success of carrots in vending machines or its motto "Gloom and doom with a sense of humor") appeals to readers in way that blogs focused only on dire consequences do not. The Mother Nature Network blog takes an approach similar to that of Grist. Mother Nature Network creates green awareness via optimistic pieces like "World's Oldest Man Offers Diet Advice" and "Six Things You Can Do to Plan Your Spring Garden." Instead of using a more forceful approach to teach readers about eating healthy and organic or giving guidelines for organic gardening, blog posts like the above encourage readers to learn and become more passionate about environmentally friendly practices implicitly.

Blogging remains a powerful political force for activists whose voices may otherwise be marginalized in mainstream media. Some bloggers specialize in a specific green issue while others work tirelessly to educate subscribers in all things green. Ultimately, regardless of the rhetoric used, these sites create awareness about green issues on local, national, and international levels.

See Also: Communication, Global and Regional; Communication, National and Local; Cyber Action; Environmental Communication, Public Participation; Fashion; Individual Action, Global and Regional; Individual Action, National and Local; Organic Clothing/Fabrics; Organic Foods; Popular Green Culture; Social Action, Global and Regional; Social Action, National and Local; Technology and Daily Living.

Further Readings

Best Green Blogs. http://www.bestgreenblogs.com (Accessed September 2010).

Brighter Planet's 350 Challenge. http://www.350.brighterplanet.com (Accessed September 2010).

The Daily Green: The Consumer's Guide to the Green Revolution. http://www.thedailygreen.com (Accessed September 2010).

Ethical Weddings: Give Everyone Something to Celebrate. http://www.ethicalweddings.com/blog (Accessed September 2010).

Green Blog. http://www.green-blog.org (Accessed September 2010).

Green Lashes and Fashion. http://www.greenlashesandfashion.blogspot.com (Accessed September 2010).

The Green Parent. http://www.thegreenparent.com (Accessed September 2010).

Grist: A Beacon in the Smog. http://www.grist.org (Accessed September 2010).

Haute Nature. http://www.hautenature.com (Accessed September 2010).

Little Green Stilettos: Beautiful Green Things and Thoughts for Girly Girls. http://www.littlegreenstilettos.com (Accessed September 2010).

Mother Nature Network. http://www.mnn.com (Accessed September 2010).

The Natural Nursery Blog: Helping You Grow Your Green Family. http://www.naturalnurseryblog.co.uk (Accessed September 2010).
No Impact Man. http://www.noimpactman.typepad.com (Accessed September 2010).
The Oil Drum. http://www.theoildrum.com (Accessed September 2010).
Style Saves the World. http://www.stylesavestheworld.blogspot.com (Accessed September 2010).
TreeHugger. http://www.treehugger.com (Accessed September 2010).

Karley Adney
Independent Scholar

Body Burden

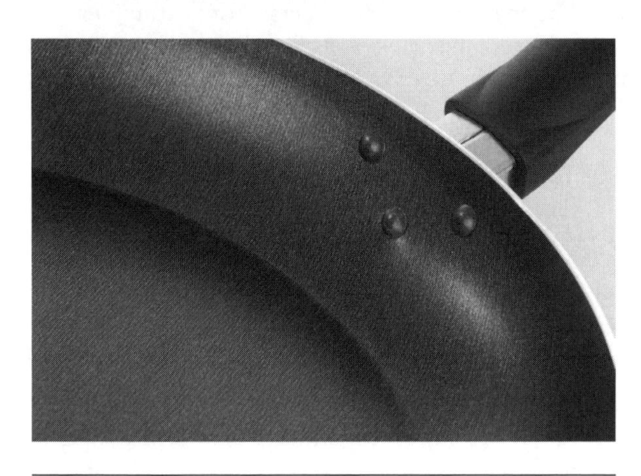

The more than 80,000 chemicals in commerce today can enter the body through a variety of household items; some nonstick pans, for example, contain the chemical PFOA, which was found in humans during a study by the CDC.

Source: iStockphoto

Body burden is a term used to describe the presence of chemicals in the human body that are not naturally present. This can include synthetic or man-made chemicals as well as naturally occurring substances, but refers to chemicals that one would not ordinarily find in the body. The concept has raised questions related to the ubiquity of chemicals in everyday life and has challenged previous notions regarding the relationship between the human body and the built environment. The term's prevalence within environmental health fields and in relation to calls for chemical regulatory reform makes it an important part of an understanding of green culture.

Biomonitoring

Biomonitoring is the scientific means of assessing an individual's body burden. Using samples of tissue or fluid (most commonly used are urine, blood, and breast milk, though hair, nail, or bodily tissues may also be used), biomonitoring studies can reveal the presence of chemicals in the body. Although biomonitoring cannot reveal how a chemical entered the body or what the health effects will be, the use of biomonitoring is a vital tool in developing information about human body burden. When used in tandem with laboratory studies about the effects of certain chemicals on animal or human bodies, biomonitoring can greatly inform our understanding of the relationship between our bodies and our environment.

Human Body Burden

Since 2001, the Centers for Disease Control and Prevention (CDC) has published four national reports investigating the body burden of the American population. Using a representative sample of more than 2,000 Americans each testing period, these studies have tested for the presence of chemicals in the human body. Although this research has revealed reductions in human exposures to lead and tobacco, synthetic chemicals, pesticides, and flame retardants have been consistently found in human bodies throughout the CDC testing process. The most recent version of the report (as of 2010) tested for the presence of 212 different chemicals. The tests found that nearly all participants' bodies contained polybrominated diphenylethers (PBDEs), which are flame retardants used in a variety of consumer products. Bisphenol-A (BPA) is another chemical used in manufacturing consumer goods and was present in 90 percent of persons tested, while perfluorooctanoic acid (PFOA), a chemical widely used to produce nonstick pans, was also found in measurable levels in most of the CDC study participants. Similar studies, though smaller in scale, have routinely been conducted by environmental health advocates across the United States. The "Is It in Us?" report (2007) tested 35 Americans across the country and found results similar to those of the CDC: all participants carried some chemical body burden, while nearly all were found to have measurable levels of BPA and PBDEs in their bodies. Another study, conducted in 2005, examined the level of chemicals present in umbilical cord blood. Prior to this study, it was often assumed that a placental barrier prevented toxic chemicals from being passed through the bloodstream. However, the Environmental Working Group's report "Body Burden: The Pollution in Newborns" demonstrated that chemicals are present in umbilical cord blood and are passed from the mother to the developing fetus. Tests revealed nearly 300 chemicals present in the cord blood of 10 U.S.-born babies, including PFOA, PBDEs, and PCBs (polychlorinated biphenyls).

Unequal Exposure to Toxins

It has been widely noted that the distribution of environmental hazards does not fall equally across a population, but in fact, exposures to toxic substances are often predicted along race and class lines, with low-income or minority communities experiencing the highest level of exposure to environmental contaminants. For example, Mexican Americans sampled by the CDC have consistently had higher levels of pesticide residues in their bodies than other Americans, mostly likely because large proportions of the Mexican American population live in states that engage in large-scale farming and/or may be employed as farm laborers. Children of all races are also found to have the highest body burdens of many chemicals for which they were tested. This is likely due to their higher consumption of food, air, and water pound for pound, compared to adults, but raises important questions about consistent and life-long exposure to certain chemicals, particularly when many chemicals (BPA, phthalates) are known to interfere with development processes.

Despite some evidence of unequal body burdens, many environmental justice advocates note that biomonitoring is, by itself, an insufficient means to truly assess environmental health hazards. In this light, biomonitoring is limited in its capacity to control environmental health risks because it measures exposure after its occurrence. Rather than simply allowing biomonitoring data to inform the concept of individual body burden, environmental justice advocates suggest that community-focused, participatory research and health data collection can remedy this shortcoming. This type of research allows community groups and individuals to participate in the data collection and dissemination processes, allowing communities to choose which chemicals or hazards to investigate, while also increasing transparency in the research process. This practice uses biomonitoring as an instrument that is both scientific and democratically useful.

Chemical Regulation

As biomonitoring advances have improved our understanding of individual body burden, concerns have routinely been voiced over the regulation of chemical substances. Given the fact that a much greater proportion of the chemicals in consumer and industrial products are entering our bodies than was ever believed possible, questions have arisen regarding the safety of the chemicals present in the average person's body. While biomonitoring cannot assess the relationship between a chemical's presence and a particular health outcome, mounting scientific evidence suggests that there are relationships between many chemical substances and certain negative health outcomes.

The current chemical regulatory mechanism in the United States, the Toxic Substances Control Act overseen by the U.S. Environmental Protection Agency (EPA), is tasked with regulating and conducting safety assessments on the more than 80,000 chemicals in commerce today. Although in existence since 1976, the EPA has been able to evaluate only 250 chemicals for safety, leaving unanswered questions about the safety of these chemicals that enter our bodies. As a result of this lack of information about many chemical substances, a variety of public, environmental, and health stakeholders have recently begun to press for legislative reforms that can work to better assess the safety of industrial chemicals. Until such a time when chemical safety is more assured, it is important to continue to measure and record body burden data.

See Also: Environmental "Bads"; Environmental Justice Movements; Environmental Protection Agency; United Church of Christ.

Further Readings

Bullard, Robert, Paul Mohai, Robin Saha, and Beverly Wright. *Toxic Wastes and Race at Twenty: 1987–2007*. Cleveland, OH: United Church of Christ, 2007.

Environmental Working Group. "Body Burden: The Pollution in Newborns." http://www.ewg.org/reports/bodyburden2 (Accessed August 2010).

"Is It in Us? Toxic Trespass, Regulatory Failures and Opportunities for Action." http://isitinus.org (Accessed August 2010).

Urban Habitat. "Burden of Proof: Using Research for Environmental Justice." *Race, Poverty and the Environment*, 11 (2004).

U.S. Centers for Disease Control. "Fourth National Report on Human Exposure to Environmental Chemicals." http://www.cdc.gov/exposurereport (Accessed August 2010).

Amy Lubitow
Northeastern University

Burials

A green burial, also known as a natural burial, is the practice of interring a human or animal corpse in the ground in ways that do not discourage the process of natural decomposition. Although the name *green burial* is a relatively new concept, natural burials were the norm prior to the development of the modern funeral industry, which some believe may encourage environmental waste, interfere with the natural process of decay, and introduce potentially harmful dangerous chemicals into the ground. Contemporary green burials impact all areas of human deathcare, including body preparation, receptacle selection, interment practices, gravesite memorials, and cemetery maintenance.

Individuals who choose to have a green burial typically reject the use of chemicals in processing the body for burial. Mainstream mortuary processing usually involves the practice of embalming, or preserving the body with the use of formaldehyde-based chemical solutions. While these solutions are biodegradable over time, some individuals are nonetheless concerned and claim these volatile chemicals endanger funeral workers and the ground in which embalmed bodies are buried. Green burials do not involve the use of any chemicals that preserve or sanitize the body prior to burial, ensuring that the remains are exposed to natural microbial decomposers that will allow them to quickly return to nature.

The selection of a receptacle in which to bury a body is also impacted by environmental concerns. The majority of coffins and caskets sold in the American market are composed primarily of pressed steel, which impedes natural decomposition. Other coffins are produced from exotic or slow-growing hardwoods that are used in unsustainable ways. In reaction to these problematic environmental choices, those who choose green burials prefer to be buried in a biodegradable shroud, blanket, box, or coffin. Shrouds and blankets can be made of any organic material and are simply wrapped around the body prior to interment. Boxes or coffins can be made of cardboard, wicker, pine, or other renewable and biodegradable material. Another choice is to select post-consumer recycled materials or locally grown materials for the shroud, blanket, box, or coffin.

A desire for green burial impacts the choices made regarding interment, or burial, of the body as well. Those buried in the American market most commonly have their coffins sealed in a metal-reinforced cement burial vault designed to protect the receptacle and to discourage shifting of the ground above as decay progresses, and many cemeteries and memorial parks require vaults. Even so, those wanting to reflect sustainable choices in their burials frequently reject burial vaults. These individuals frequently object to the energy and natural resources that are needlessly used in the production and the distribution shipping of burial vaults. Furthermore, burial vaults further impede the exposure of the body to microbial decomposers and delay the process of natural decay. Those who opt for a green interment also consider the ideal depth of burial to encourage decomposition to be somewhat shallower than the typical six-foot rule of thumb, requiring instead to be buried at a depth much closer to the surface to allow the same aerial circulation required for composting. Finally, green burial interments typically avoid the use of bulldozers or other machines in favor of digging by hand; often, family members and loved ones will participate in digging the burial hole.

While a tombstone seems the ubiquitous choice in memorializing gravesites in modern cemeteries, they are not necessarily chosen for green burials. While there is no mandate against tombstones, they are often made of limestone, marble, or granite, which are mined at great expense to the environment and usually transported great distances before use at their final destination. Environmentalists note that even the carving of words and etchings into stone using contemporary means involves the use of fuels. To avoid these environmental costs, those who choose green burial often opt for natural markers, such as planting a native tree, shrub, or perennial plant. Other options include having a local stone engraved with the name of the deceased, or leaving the spot completely free of monumentation.

Traditional cemetery planning and maintenance is another aspect of human deathcare that some environmentalists see as wasteful or otherwise problematic. For example, many cemeteries have nonnative plantings or grasses that require maintenance and watering with few or no trees, while others use pesticides and herbicides made of dangerous chemicals. Individuals who wish to avoid these practices can opt for a green cemetery. Green cemeteries generally strive to preserve the natural habit within their enclosures and eschew the use of pesticides, chemical fertilizers, and herbicides in the maintenance of their grounds. Unlike most traditional cemeteries, green cemeteries often encourage the planting of several native plantings on a cemetery plot in order to promote healthy bird, wildlife, and insect habitats. Some green cemeteries even offer a conservation easement with the purchase of a plot, so that the land is protected in perpetuity from improper use or maintenance. Although there are some cemeteries that are entirely green, many larger traditional cemeteries have dedicated acres as a green park where all the practices common to green cemeteries are practiced but often without conservation easement.

The environmental claims of green cemeteries remain unregulated by any federal agency, but legislation regulating the practices of these cemeteries is being drafted in a few states. Cemeteries can also apply to the independent nonprofit Green Burial Council for certification and verification of their green practices.

See Also: Environmental "Goods"; Gardening and Lawns; Green Consumerism; Space/Place/Geography and Materialism.

Further Readings

Bailey, Sue. *Grave Expectations: Planning the End Like There's No Tomorrow.* Kennebunkport, ME: Cider Mill Press, 2009.

Butz, Bob. *Going Out Green: One Man's Adventure Planning His Natural Burial.* Traverse City, MI: Spirituality and Health Books, 2008.

Harris, Mark. *Grave Matters: A Journey Through the Modern Funeral Industry to a Natural Way of Burial.* New York: Scribner, 2007.

Jennifer Adams
DePauw University

C

California AB 32

Greenhouse gas emissions, which can enter the atmosphere from a variety of natural and man-made sources, absorb and emit radiation within the thermal infrared range and greatly affect Earth's temperature. California aims to reduce greenhouse gas emissions, such as these aerosols, to pre-1990 levels.

Source: NASA Goddard's Scientific Visualization Studio

The Global Warming Solutions Act of 2006, also known as California AB 32, is an environmental law passed in California. California AB 32 establishes a timetable designed to bring California into compliance with most of the provisions of the Kyoto Protocol, an addendum to an international treaty controlling greenhouse gas emissions signed but not ratified by the United States. California AB 32 requires greenhouse gas emissions within the state to be reduced by 2020 to 1990 levels, an action that would require a reduction in such emissions of approximately 25 percent. While California AB 32 brought criticism from the business community, others have estimated that the legislation will ultimately create jobs for those employed in green industries. Similarly, cost estimates for California AB 32, and its effect on household incomes, have been widely divergent.

Background

In 1997, the Kyoto Protocol was adopted at a conference held under the auspices of the United Nations in Kyoto, Japan. The Kyoto Protocol set binding targets for 37 industrialized countries and the European Union to reduce greenhouse gas emissions. Although the United States signed the Kyoto Protocol, it was never ratified by the U.S. Senate, which was necessary for its implementation. Indeed, the administration of President George W. Bush affirmatively stated it was not seeking the Kyoto Protocol's ratification. The United States' failure to ratify the Kyoto Protocol subjected it to intense criticism, both domestically and abroad. California lawmakers, with a long history of progressive action with regard to environmental legislation, became interested in California AB 32 as a means of ensuring that state's compliance with the objectives of the Kyoto Protocol.

Greenhouse gases are emissions into the atmosphere from a variety of natural and man-made sources. Greenhouse gas emissions absorb and emit radiation within the thermal infrared range. Primary greenhouse gases include carbon dioxide, methane, nitrous oxide, hydrofluorocarbons, perfluorocarbons, and sulfur hexafluoride. The presence of greenhouse gases greatly affects the temperature on Earth. It has been estimated that without greenhouse gases, the average temperature of Earth would be approximately 59 degrees Fahernheit (33 degrees Celsius) cooler than currently is the case. The burning of fossil fuels, common since the beginning of the Industrial Revolution, has substantially increased the presence of carbon dioxide and other greenhouse gases in Earth's atmosphere. California AB 32 seeks to reduce these emissions in California to pre-1990 levels.

California AB 32 was coauthored by Assembly Speaker Fabian Nuñez and Assemblywoman Fran Pavley, both Democrats. After approval by both houses of the California legislature, California AB 32 was signed into law by Governor Arnold Schwarzenegger on September 27, 2006. Although California AB 32 required that greenhouse gas emissions be reduced to 1990 levels by 2020, it also required immediate action by California businesses and residents to begin complying with the statute's requirements. As with the Kyoto Protocol, California AB 32 was controversial and aroused great passions from various constituencies regarding the need for it and its potential ramifications for California's environment and business climate.

Issues Regarding the Statute

While proceeding through the legislature, business and energy interests vigorously opposed California AB 32. Criticism of the bill focused on concerns related to increased regulations, costs to be borne by small businesses and families, harm to the competitive business environment, and claims that the legislation was based on untried and untested science. Supporters of California AB 32 disputed these allegations. Small businesses were not regulated by California AB 32, and while there might be small increases in the cost of some items, these were anticipated to be minimal. Instead of increasing the expenses of small businesses and families, California AB 32 was predicted by its supporters to save consumers money in reduced energy expenditures and decreased utility bills. The statute was also trumpeted by supporters as a means of making California more competitive, asserting that its passage would lead to green jobs and draw investment and new businesses into the state. Finally, savings generated from past environmental initiatives were demonstrated as support for the passage of California AB 32.

California AB 32 essentially puts the state, Earth's 12th-largest generator of greenhouse emission gases, in compliance with the Kyoto Protocol, albeit with a later date of compliance than the agreement itself. In passing the legislation, California joined the states of Connecticut, Delaware, Maine, Maryland, Massachusetts, New Hampshire, New Jersey, New York, and Vermont in voluntarily attempting to comply with the Kyoto Protocol. These states combined account for approximately 20 percent of the population of the

United States. In addition, nearly 1,000 cities and municipalities passed similar statutes that were popular but largely had little effect.

Implementation

Although California AB 32 had a deadline for achieving its goals of 2020, it demanded regulatory changes almost immediately after it became law. By the end of 2006 (before California AB 32 went into effect), emitters were given an opportunity to join the California Climate Action Registry to be able to be grandfathered into the Air Resources Board's reporting and verification program. In November 2010, Californians voted against Proposition 23, which would have suspended California AB 32 until California's unemployment rate dropped to 5.5 percent or less for four consecutive quarters.

In 2007, the Air Resources Board published a set of discrete early action greenhouse gas reduction measures that could be implemented by emitters. Within a year of this, the Air Resources Board established California's 1990 baseline for greenhouse gas emissions and set the statewide 2020 goal limits, while also establishing the state's mandatory reporting and verification system. Early action emission reduction measures were identified by the Air Resources Board and have been implemented. Options that may be realized in the future include a market-based cap-and-trade system for greenhouse gas emissions, as well as more stringent emissions limits and emission reduction standards. The scope and stringency of California AB 32 has made it of interest to many concerned with environmental regulation.

See Also: Air Pollution; Clinton, William J. (Executive Order 12898); Environmental Law, U.S.; Kyoto Protocol; Social Action, Global and Regional.

Further Readings

Aldy, J. E. and R. N. Stavins, eds. *Architectures for Agreement: Addressing Global Climate Change in the Post-Kyoto World*. New York: Cambridge University Press, 2007.

Stewart, R. B. and J. B. Wiener. *Reconstructing Climate Policy: Beyond Kyoto*. La Vergne, TN: AEI Press, 2003.

Victor, D. G. *The Collapse of the Kyoto Protocol and the Struggle to Slow Global Warming*. Princeton, NJ: Princeton University Press, 2004.

Stephen T. Schroth
Ellen M. Jackson
Knox College

Censorship of Climatologists and Environmental Scientists

In December 2007, the U.S. House of Representatives Committee on Oversight and Government Reform issued a report on the "Political Interference With Climate Change Science Under the Bush Administration," the result of a 16-month-long investigation into allegations made by various government scientists and officials claiming they had been censored by the White House and government officials. Within the U.S. offices of the National Aeronautics and Space Administration (NASA), National Oceanic and Atmospheric Administration (NOAA), Environmental Protection Agency (EPA), and the U.S. Climate Change Science Program (U.S. CCSP), the report found many instances of alterations to climate change reports and testimonies to Congress, deletions and suppression of references to climate change facts and reports, and limitation of media access to scientists. The following synthesizes the findings of the report as well as information from news articles.

Censorship at the U.S. Climate Change Science Program

On June 8, 2005, the *New York Times* reported that several reports of climate research issued in 2003 and 2004 by government scientists were altered by Phillip A. Cooney, a White House official who at the time was the Chief of Staff for the Council on Environmental Quality (CEQ). Prior to his White House appointment, Cooney was a former lobbyist at the American Petroleum Institute. The "Strategic Plan of the Climate Change Science Program and Our Changing Planet," reports to Congress from the Climate Change Science Program, were written by the U.S. Global Change Research Program (now called the U.S. Climate Change Science Program) and included scientific findings regarding climate change and their implications. The alterations to several drafts of the reports included handwritten notes by Cooney, a lawyer with no scientific training—notes that played down the impacts of climate change, emphasized the uncertainty of scientific findings, and diminished the human role in climate change. In many cases, the changes made their way into final versions of the reports.

White House officials and other groups denied any wrongdoing, asserting that the edits made were part of the normal interagency review that takes place with all reports related to global environmental change. Critics, such as climate experts and environmental groups, countered this by saying that while government reports are routinely vetted, the scientific content should be reviewed by scientists. In the immediate days following the release of the *New York Times* article, Phil Cooney resigned from CEQ and immediately took a position with ExxonMobil.

Rick Piltz, a former senior associate of 10 years with the U.S. Climate Change Science Program (U.S. CCSP), first brought this story to the attention of the *New York Times*. Piltz resigned in 2005 in protest to the actions by the CEQ and Phil Cooney, which also included censoring and systematically deleting any references to the National Assessment of Climate Change Impacts, a previous climate report issued by the U.S. CCSP in 2001. The 2001 document, which had been commissioned by the Clinton administration, was a major report detailing the regional and national implications of climate change in the United States. Piltz asserted that beginning in 2002, the White House pressured the U.S. CCSP to delete any references to the National Assessment in any future plans or reports, including reports to Congress.

Censorship at the National Aeronautics and Space Administration

A *New York Times* article dated January 29, 2006, first broke a story from James Hansen, one of the world's leading climate experts and longtime director of the National Aeronautics and Space Administration (NASA) Goddard Institute for Space Studies, that he was being censored by NASA officials. Hansen stated that since he gave a lecture in

December 2005 calling for action to reduce greenhouse gas emissions, officials at NASA headquarters had ordered that his future lectures, papers, and postings on the Goddard website be reviewed by the public affairs staff prior to release and all requests from journalists for interviews be given prior approval. Among the restrictions was that during any news media interviews his supervisors could stand in for him.

Hanson also told the *New York Times* that a warning was relayed to him by headquarters officials and public affairs staff that he would face "dire consequences" if he continued making statements supporting the case for climate change. A representative from the public affairs administration, Dean Acosta, told the *New York Times* that there was no effort to censor Hansen and that the reviews applied to all NASA personnel.

In the days following the release of the *New York Times* article describing NASA's censorship, Representative Sherwood Boehlert (R-New York), the chairman of the House Science Committee, sharply criticized NASA. Other Republican lawmakers defended the agency and criticized Hansen. An internal investigation was requested by 14 senators and resulted in a 48-page report issued in 2008, which concluded that activities of censorship did occur and were inconsistent with the law that established NASA 50 years ago. The internal investigative office, however, stated that the activities appeared to be limited to the headquarters press office, with no evidence that officials higher at NASA or in the Bush administration were involved with interfering with the communication of climate information. The report also credited the agency's administrator, Michael Griffin, with quickly ordering a review and policy changes following the 2006 *New York Times* article.

Since 1988, Hansen had been issuing public warnings about the long-term threats due to greenhouse gas emissions from burning coal, oil, and other fossil fuels. In a 2007 interview with PBS, Hansen described censorship he had encountered from federal government officials dating back to the 1980s. Hansen described that beginning in 1989, he began to experience various instances of censorship. For example, in 1989, the Office of Management and Budget under the Reagan administration made alterations to testimony that Hansen gave before Congress.

Censorship at the Environmental Protection Agency

Several instances of censorship occurred at the Environmental Protection Agency (EPA). In 2002, edits made by Phil Cooney, with White House approval, resulted in the deletion of a chapter on climate change from EPA's "Air Pollution Trends," an annual report on air pollution. In 2003, Cooney and CEQ also extensively edited climate change material in EPA's "Report on the Environment," resulting in the deletion of the climate change chapter.

On July 8, 2008, Senator Barbara Boxer (D-CA), the chair of the Senate Environment and Public Works Committee, held a press conference denouncing the Bush administration, particularly Vice President Dick Cheney's office, for allegedly being involved in removing statements on the health risks posed by climate change. Jason Burnett, a former associate deputy administrator of EPA and chief adviser on climate to the EPA administrator, had stepped forward in a letter to Senator Boxer. The letter described that requests by the offices of Cheney and the CEQ had resulted in the removal of two sections of testimony to Congress by Julie Gerberding, the head of the Centers for Disease Control and Prevention. The two sections removed were "Climate Change Is a Public Health Concern" and "Climate Change Vulnerability."

White House officials rejected the criticism, standing behind previous statements made by White House Press Secretary Dana Perino claiming that changes in testimony were justified because the statements did not comport with the review of climate risks by the Intergovernmental Panel on Climate Change.

Censorship at the National Oceanic and Atmospheric Administration

At the National Oceanic and Atmospheric Administration (NOAA), restrictions regarding the communication between scientists and the media had been tightened during the Bush administration. According to the 2007 testimony to Congress by Kent Laborde, a career public affairs official at NOAA, climate scientists at the agency were only allowed to be interviewed by the press if approved by NOAA officials and the White House CEQ, and only if a public affairs officer was present or on the phone. Previously, the scientists had always been free to communicate with the press openly. Also, according to Laborde's testimony, following Hurricane Katrina there was a concerted effort by the White House and the Department of Commerce to specifically direct journalists to NOAA scientists who did not support the link between climate change and hurricane intensity.

In 2006, Thomas Karl, the director of NOAA's National Climatic Data Center, gave testimony to Congress regarding climate change science. Previously, his testimony had been extensively edited by the White House CEQ, political appointees at NOAA, and the Department of Congress to downplay the effects of climate change and human influence on climate.

General Censorship

In addition to NASA, EPA, NOAA, and the U.S. CCSP, the claims of censorship by the George W. Bush administration of government scientists have been widespread. The Union of Concerned Scientists (UCS) released a survey in 2007 reporting that 150 climate scientists from eight federal agencies personally experienced at least one incident of political interference during the five years prior, totaling at least 435 specific incidents.

See Also: Air Pollution; Environmental Protection Agency; Film, Documentaries; Global Warming and Culture; Gore, Jr., Al; Kyoto Protocol; Political Persuasion.

Further Readings

"Did White House Censor Science? Democrats and Republicans Spar Over Allegations on Global Warming." *ABC News* (December 10, 2007).

Hansen, James. Interview with PBS (January 10, 2007).

Piltz, Rick. Interview with PBS (November 13, 2006).

Revkin, Andrew C. "Bush Aide Softened Greenhouse Gas Links to Global Warming." *New York Times* (June 8, 2005).

Revkin, Andrew C. "Cheney's Office Accused of Editing Climate Change Testimony." *New York Times* (July 9, 2008).

Revkin, Andrew C. "Climate Expert Says NASA Tried to Silence Him." *New York Times* (January 29, 2006).

Revkin, Andrew C. "Investigators: NASA Officials 'Mischaracterized' and Limited Flow of Findings on Climate." *New York Times* (June 2, 2008).
Revkin, Andrew C. "Lawmakers Condemn NASA Over Scientist's Accusations of Censorship." *New York Times* (January 31, 2006).
U.S. House of Representatives Committee on Oversight and Government Reform. "Political Interference With Climate Change Science Under the Bush Administration." (December 2007).

Leanna R. Heffner
University of Rhode Island

Chávez, César (and United Farm Workers)

César Chávez was born in 1927 to a Chicano farming family near the Mexican–American border. When his family lost their ranch during the Great Depression, they traveled around California looking for work harvesting seasonal crops. Thus, from a young age, Chávez was inducted into the hardships of life as a migrant worker. In his 20s, after struggling to make a living for himself and his young family in the fields of California, Chávez was offered a position with the Los Angeles (L.A.)–based Community Service Organization (CSO), a group that formed in East L.A. to organize Mexican Americans in response to police brutality and other institutionalized injustices.

Having gained organizing experience and political contacts through his work at CSO, where he was promoted to executive director, César Chávez and a few of his colleagues started the National Farm Workers Association (NFWA) in 1962. The union was specifically for organizing field workers and indeed became a powerful voice for Hispanic farm workers. Philosophically, Chávez was influenced by the teachings of St. Francis of Assisi, Mohandas Gandhi, and Martin Luther King, Jr.; his belief in nonviolence would remain strong throughout his life and guided his organizing practices.

La Huelga

On September 8, 1965, the Great Delano Grape Strike and Boycott, also known as La Huelga, began when a group of Filipino workers under the Agricultural Workers Organizing Committee (AWOC) went on strike against nine vineyards. The demands included higher wages, better living conditions, and fair hiring practices. The NFWA joined the strike less than two weeks later, and eventually the two unions merged to become the United Farm Workers (UFW). As a united front, the UFW called for a nationwide boycott of all California table grapes, which was met with success, spreading to Canada and even Europe.

What had began as a fight for fair working conditions also became a fight against pesticide usage and chemical companies in 1968, when two separate incidences threatened the health of farm workers who were exposed to unsafe pesticides. Thus the UFW forged an important connection between the labor, social justice, and burgeoning environmental movements.

The strike officially ended on July 29, 1970, after Giumarra Vineyards Corporation signed the first union contracts for farm workers in the United States. After a five-year strike, the UFW finally drew up contracts with 150 grape growers, providing fair wages and safe working conditions for 30,000 workers. The contracts also eliminated the use of the most dangerous of the chemicals then in popular use and established recordkeeping requirements for all sprays conducted by the companies. This was by no means the end of the struggle against toxic poisons, but, for the first time, agricultural corporations were being held accountable for the impact of the chemicals they used on their fields.

Under the leadership of Chávez, the UFW led a strike against lettuce growers in 1970 and another strike against the grape industry in 1975. The latter led to the 1975 Agricultural Labor Relations Act, which provided a bill of rights for farm workers, protecting their rights to unionize, boycott, and vote by secret ballots in union elections.

Ending on August 21, 1988, César Chávez fasted for 36 days in a Fast for Life. He said of the fast, "During the past few years I have been studying the plague of pesticides on our land and our food. The evil is far greater than even I had thought it to be, it threatens to choke out the life of our people and also the life system that supports us all. This solution to this deadly crisis will not be found in the arrogance of the powerful, but in solidarity with the weak and helpless."

¡Sí se Puede! (Yes, It Can Be Done!)

La Huelga was more than just a labor fight; it also drew attention to the Latino presence in American agriculture and occurred within the context of a national labor-environment alliance in the late 1960s and early 1970s. Rachel Carson's now famously influential testimony against DDT, *Silent Spring*, was published in 1962, paving the way for general outrage at the impact of pesticides on human and environmental health. By 1969, DDT was banned in three states, and, in the midst of the UFW's struggle, the first Earth Day was held on April 22, 1970. That same year, though notoriously toothless, the Federal Environmental Pesticide Control Act (FEPCA) was passed.

The UFW's success in the fight against agribusiness has been attributed to a broad-based coalition of support, including middle-class consumers, organized labor, religious leaders, student activists, liberal Democrats, Latino activists, civil rights leaders, and environmentalists. Chávez recognized that developing a relationship with the environmental movement was crucial to the success of the UFW's battle against the California grape growers and the backstabbing tactics of the Teamsters, who took jobs from the UFW by signing contracts without pesticide accountability. As Chávez and the farm workers learned more about the impacts of pesticides, the link between the fight against hazardous working conditions and the fight for a healthy ecosystem seemed to strengthen both causes. When the UFW extended an invitation to environmental organizations to join in the fight against pesticides, however, they were met with mixed success. Smaller environmental organizations such as the Center for Science in the Public Interest, Environmental Action, Friends of the Earth, and the Environmental Policy Center accepted the invitation to join the struggle against California's grape and lettuce industry and lobbied the Environmental Protection Agency (EPA) to regulate pesticides. But more conservative environmental groups, including the Sierra Club, the National Wildlife Federation, and the Audubon Society, refused to support the boycott even while recognizing the dangers of pesticide use.

See Also: Huerta, Dolores; Organizations and Unions; United Farm Workers (UFW) and Antipesticide Activities.

Further Readings

Gordon, Robert. "Poisons in the Fields: The United Farm Workers, Pesticides, and Environmental Politics." *Pacific Historical Review*, 68/1 (February 1999).
Shaw, Randy. *Beyond the Fields: César Chávez, the UFW, and the Struggle for Justice in the 21st Century*. Berkeley: University of California Press, 2008.
United Farm Workers. http://www.ufw.org/_page.php?menu=research&inc=history/03.html (Accessed September 2010).

Shannon K. Tyman
Independent Scholar

China

China has long suffered from various environmental problems, which are the most severe among the major countries. Of 142 countries that have data, China's environmental sustainability ranked 129th.

China's economic boom is the major cause of the problem. China is the world's largest producer of steel, cement, agricultural food, and television sets, and the second-largest producer of electricity and chemical textiles. It is also the world's largest consumer of fertilizer and the second-largest producer and consumer of pesticides. Having increased the number of its cars six times from 1980 to 1994, China has made auto manufacture one of its four pillar industries and is becoming one of the top three manufacturers. Although in some areas where foreign competition and investment are intensive, such as automobile production, Chinese industries are as efficient as their foreign counterparts, much of China's economy, such as coal mining and chemical production, still uses inefficient and polluting technologies. A 1997 governmental survey found that 45 percent of the 7,555 chemical plants posed a severe environmental risk. China is the second-largest energy consumer, but its energy efficiency is only half of that in the developed countries. Meanwhile, from 1978 to 2002, China's per capita consumption of meat, milk, and eggs increased four-, four-, and eightfold, respectively. Increasing agricultural wastes, namely, animal and fish droppings, lead to terrestrial and aquatic pollution.

China's large population and rapid urbanization make things more challenging. Its one-child policy decreased the yearly population growth rate, but its average household size also decreased from 4.5 to 3.5 people, and the number of households increased dramatically. From 1952 to 2003, China's population has doubled, but its urban population has tripled—from 13 percent to 39 percent. The number of cities has increased fourfold to more than 660, and the cities themselves have also grown. Smaller households and urban dwellers consume more resources per person.

China's environmental deterioration is also influencing other countries. It is the largest contributor of sulfur oxides, carbon dioxide, and chlorofluorocarbons to the atmosphere and one of the two leading importers of tropical rainforest timber that drive tropical deforestation. It conducts 15 percent of the world fish catch and consumes 33 percent of the global fish and seafood. Its dust and air pollutants are transported eastward to adjacent countries and even to North America. Although China's environmental impact per capita is still quite small compared to developed countries, because of its largeness, it draws enormous environmental attention.

Current Environmental Situation

China's environmental problems include air pollution, disappearance of endangered species, cropland losses, depleted fisheries, etc. Two scholars, Jianguo Liu and Jared Diamond, summarize the problems within five categories:

Air

Industrial waste gases have made China's air quality worse. In 2000, China ranked first for SO_2 emissions and third for NO_x emissions among major countries. As of 2007, China is the world's leading emitter of carbon dioxide by total amount (the United States produces more on a per capita basis). A third of city residents breathe air below China's own quality standard. Of the 20 cities ranked most polluted by the Word Bank, 16 are in China, mostly because of air pollution. Polluted air is causing about 300,000 Chinese to die per year and has damaged 20–30 percent of Chinese men's fertility. The World Bank has estimated that China's health costs due to air pollution will be $98 billion by 2020.

Household indoor pollution created by cooking with coal and wood poses another major health threat. A 1997 World Bank study estimated that 111,000 premature deaths, 220,000 hospitalizations, and over 3 million cases with respiratory symptoms could be avoided if the indoor air quality were improved to China's class 2 standard.

Over 60 percent of China's energy comes from coal, particularly brown coal, which produce particulates and sulfur dioxide that cause acid rain. Acid rain severely influences 40 percent of China. In the 1990s, acid rain fell on a quarter of Chinese cities for more than 60 percent of rainy days per year. Fortunately, with governmental attention, the air quality in China is getting better.

Land

Nineteen percent of China's land suffers erosion. Seventy percent of the Yellow River has been eroded, and the erosion in the Yangtze River exceeds that in the Nile and the Amazon combined. Soil erosion and sediment shorted China's navigable river channels by 56 percent between 1949 and 1990. Fertilizer and pesticide usage and maul-irrigation has salinized 9 percent of China's land. Meanwhile, overgrazing and agricultural overdevelopment of the land has led a quarter of China's land to become desertified. The spread of the desert in northwest China was about 600 miles a year in the 1970s and 1,300 miles a year in the 1990s. Desert has reached less than 200 miles from Beijing and threatens its very existence. China's cropland is only 0.1 hectare (ha) per person, hardly half of the world average. Its forest is only 0.1 ha per person, compared with the world average of 0.6 ha per person. China's per capita grassland is only half of the world average, with 90 percent of its grassland degraded.

When land is eroded, dust storms, landslides, drought, and floods come more frequently. Since 1990, dust storms hit northwestern China almost every year, though they attacked the area on average only once every 31 years in the past thousands of years. A dust storm on May 5, 1993, even killed 100 people. Droughts damage about 160,000 square kilometers of cropland each year, double that in the 1950s. Deforestation generates floods. The 1996 flood created a loss of $27 billion. The 1998 flooding killed more than 3,000 people, destroyed 5 million homes, and inundated 52 million acres of land, creating a loss of at least $20 billion.

Freshwater

Wastewater discharges and fertilizer and pesticide runoffs polluted about 75 percent of China's lakes. In 1996, 50,000 Chinese people were affected by water pollution–related diseases. Average blood lead levels of Chinese urban residents are nearly double the level that may endanger children's mental development. China's per capita quantity of freshwater is only a quarter of the world average. In 20 of the years between 1972 and 1997, the lower Yellow River had flow stoppage, and the days of flow stoppage increased from 90 in the 1980s to 230 in 1997. Over 100 Chinese cities suffer from severe water shortage. A factory closure in Xi'an, due to water shortage, brought a loss of $250 million. Every day, 60 million people cannot obtain enough clean water for life needs. In rural areas, one in three people lacks access to safe drinking water.

Oceans

China has 3 million square kilometers of sea areas and has jurisdiction over 200 nautical miles off its coasts. Almost all of those seas are polluted, mainly by land pollutants, oil spills, and other marine activities.

Biodiversity

China is rich in world species. Unfortunately, about 20 percent of Chinese species are endangered. From 1983 to 2003, the Chinese government established almost 2,000 nature reserves and many zoos, museums, and botanical gardens. The area covered by these reserves is higher than the world average, but their management needs to be improved.

China's environmental degradation brings loss to not only its ecological system but also its people. Each year, China spends $72 million to curb the alligator weed. Sandstorm costs are about $540 million, and acid rain costs about $730 million. Desertification and alien species invasion each cost $7 billion, and water and air pollution together costs $54 billion. In each year of the past two decades, pollution and ecological damage have cost 7 percent to 20 percent of China's GDP.

Governmental Policies

China's political leadership once believed that humans should conquer nature and had little environmental awareness. During the Great Leap Forward (1958–60), the number of factories increased four times and tons of trees were felled to fuel the backyard steel production. The situation was changed in 1972, when China sent its first delegation to the

First United Nations Conference on the Human Environment. In 1973, China established its Leading Group for Environmental Protection, which evolved into the National Environmental Protection Agency in 1988 and into the State Environmental Protection Administration in 1998. China also set environmental protection as one of its basic national principles in 1983 and developed its first five-year environmental protection plan in 1996. China participates in most international treaties and conferences on the environment and has passed more than 100 environmental laws, policies, and regulations.

Although China's environmental law appears excellent on the paper, its implementation is far below expectation, due to the weak environmental regulation agencies and the low environmental awareness in the governmental system. Economic growth still takes priority. For example, in the national "Go West" campaign, it was the State Development and Reform Commission rather than the State Environmental Protection Administration that determined which environment-based projects should be proposed to the World Bank for funding.

China's nearly 25 million township and village enterprises hire more than 30 million workers, account for a third of Chinese production and half of its exports, but generate more than 70 percent of the rural pollution. The Chinese government's reliance on fines, though working for urban pollutants, is not effective among those rural industries too widely scattered to be monitored. Plus, as major local tax contributors, those township and village enterprises are usually warned of governmental inspections in advance and can escape from emission tests by stopping the polluting lines temporarily or flushing out waste pipes with clean water.

Courts at the local level usually do not support citizens in their claims against the polluters harming them because court officials, mainly dismissed military, are not capable of trying environmental cases and often are subject to pressure from local governments that support the pollutant enterprises for gross domestic product (GDP) growth. A professor from the China University of Politics and Law, Wang Canfa, has established the Center for Legal Assistance to Pollution Victims. Beyond his reach, however, people are still helplessly facing the environmentally ignorant courts.

Environmentalists

The first and the most influential Chinese environmental nongovernmental organization (NGO), Friends of Nature, was founded in 1994. Up to 2004, around 100 grassroots environmental NGOs were established, which stopped many antienvironment projects with help from the central government or international organizations such as the United Nations Educational, Scientific and Cultural Organization (UNESCO). Chinese environmentalists have also received much help from the media and journalists. Many journalists became environmentalists themselves. Because of limited financial resources and the restrictive political atmosphere, some environmentalism organizations, such as Greener Beijing and Green-Web, are merely based online, using the Internet to raise public awareness and engagement. The environmental government-organized NGOs, although often acting on behalf of the government, also have their own agenda.

College students are enthusiastic supporters of environmentalism. From 1990 to 1995, six university green associations were established in China. In 1996 and 1997, more college environmental organizations were established, which began to promote environmental awareness in local communities and attract media attention. In 1996, the cry of the student environmentalists and other activists led to the establishment of the habitat of the endangered golden monkey in Yunnan Province. Thereafter, the number of college student environmental organizations grew from 22 in 1997 to over 150 in 2002.

Elizabeth Economy, an influential writer on China's environment, believes that the younger generation of Chinese environmentalists is open and aggressive and may serve as the breaker of the political barriers, as happened in Western civil movements. Many Western scholars share the view. In China, some environmentalists think of their NGOs as laboratories of democratic participation. However, a 2002 survey of Beijing university students cast doubts on this projection because, although demonstrating a high environmental awareness, the students are reluctant to conflict with the government on any environmental issues and have a great deal of faith in growth's capability to solve those problems. Despite the awful present, they are quite optimistic about the future of China's environment.

See Also: Air Pollution; Biodiversity Loss/Species Extinction; Soil Pollution; Water Pollution.

Further Readings

Liu, Jianguo and Jared Diamond. "China's Environment in a Globalizing World." *Nature*, 435 (2005).

Stalley, Phillip and Dongning Yang. "An Emerging Environmental Movement in China?" *The China Quarterly*, 28 (2006).

Yao, Qingjiang. "Media Use, Postmaterialist Values, and Political Interest: The Making of Chinese Environmentalists and Their Views on Their Social Environment." *Asian Journal of Communication*, 18 (2008).

Qingjiang Yao
Fort Hays State University

Chipko Movement

Chipko is a grassroots, socioecological movement that originated in the 1970s in the districts of Garwal and Kumaon—in the Himalayan foothills in the erstwhile state of Uttar Pradesh, now Uttarakhand, in India. The Hindi word *chipko* literally means "embrace" or "to adhere." Influenced by the Gandhian principle of peaceful resistance to unjust policies, the Chipko activists hugged trees to prevent them from being felled.

Uttarakhand was abundant in natural forest and mineral resources and considered paramount for national economic development. After independence from British rule, the post-colonial Indian government continued to hold title to the forests via the Uttar Pradesh Forest Department (UPFD). Economic growth–oriented policies such as mining and extraction of forest resources affected the delicate balance between nature and the community. Conflict in Indo-China further worsened the scenario—a network of roads was built to improve accessibility and more resource-extraction industries followed. Further, the government transferred 10 percent of reserved forests to the Ministry of Defense, which drove away the poor who depended on the forest for fuel and fodder, as well as local small-scale extractors. Due to the conflict, trans-Himalayan trade carried out by the local

The Himalayan foothills of Uttar Pradesh produced the Chipko socioecological movement in the 1970s as a way to peacefully resist the destruction of lush forest areas like these in Gangotri, India.

Source: iStockphoto

community also suffered, and village men were forced to migrate to the city to earn wages while women were left behind to look after the old and children. Severe deforestation affected the quality of soil, and finally, the heavy monsoon of 1971–72 caused flooding, landslides, and damage to terrace cultivation. Severe economic backwardness, acute ecological crises, and neglect on the part of the government caused resentment among the locals and gave rise to Chipko.

Chipko-like activism is traced back to 1730, when 363 people from the Bishnois community of Rajasthan chose to be slaughtered by the axmen sent to get timber for the maharaja of Jodhpur while protecting their sacred trees by the act of hugging. However, the rise of Chipko in modern times is credited to two major events in the 1970s. The UPFD followed a contract system through which it auctioned the felling of trees to international or extra-regional corporations. The first instance occurred in April 1973 when Dasholi Gram Swarajya Sangh, a local cooperative based in Chamoli that promoted local community forest industries, was denied access to 12 trees for the needs of the local population, and it started organizing small mobilizations against the contract system. The second event occurred in 1974 in Reni village when, in the absence of men, women of the village decided to employ Chipko tactics to save the trees from the axes of lumbermen. Women's participation henceforth gained momentum and thousands of village women started employing Chipko. This particular instance marked the beginning of the rise of the ecofeminist perspective on deforestation in the Himalayas, where the women demanded the return of the forests to the local population but not at the expense of forest scarcity.

Although the rise of the Chipko movement relied on hundreds of acts of tree hugging by thousands of local people, two prominent figures who set the movement at the national and global stage are Chandi Prasad Bhatt, a prominent Gandhian activist from Chamoli district who aspired to reappropriation of forest resources for local use and development, and Sunderlal Bahuguna, who sought to reestablish the spiritual love for nature inherent in ancient Indian community. Bahuguna chanted the slogan "ecology is the permanent economy" and was able to influence India's then–prime minister, Indira Gandhi, to impose a 15-year ban on felling in the 1980s.

Increased participation of women in the Chipko movement led scholars like Vandana Shiva to believe that the movement is essentially a case of ecological-feminist activism against the scientific forestry approach. Scholars like Ramachandra Guha believe that

Chipko is a peasant grassroots movement that has the public face of an environment movement. Yet others categorize it as a Gandhian movement of nonviolence, paying attention to the deep relationship between Hinduism and ecology, where plants are worshipped.

At the local level, the Chipko movement succeeded in imposition of a 15-year ban on tree felling, which reduced environmental degradation related to acute deforestation. The success of Chipko affected women's workload in some areas as forest abundance reduced their time spent in collection of fuel and fodder. It established the tradition of local populism that helped in the constitution of the struggle for the much-required separate state of Uttarakhand that brought the regional issues of underdevelopment to the forefront of Indian politics. Nationally, the Chipko movement quickly spread to the states of Himachal Pradesh, Bihar, Rajasthan, Karnataka, and Vindhyas; and in the Western Ghats it was launched as "Appiko" where locals practiced tree hugging in protest against deforestation due to overexploitation by paper and pulp industries and a hydroelectric dam project. The powerful message of Chipko also energized other civil society movements that focused on the issues of tribal poor, women, and marginalized groups that had been displaced by big development projects. Hence Chipko is known for its critiques of the Western model of economic development. Chipko was the first organized environmental movement that originated in India, and it raised public awareness to environmental issues both nationally and globally. There have also been numerous Chipko-like protests in Switzerland, Japan, Malaysia, the Philippines, Indonesia, and Thailand.

The most important critique of the Chipko movement is put forth by Haripriya Rangan, which maintains that it is often over-romanticized, whereas the actual needs of the local community are overlooked amid passionate concerns for nature. The exalted interest in deforestation and ecology has suppressed the issue of sustaining livelihoods for the local poor community. For example, there has been an increase in the government control over local forests and, hence, decreased possibility of economic justice. Restrictions on forest use have resulted in the existence of a "timber mafia," abandonment of beneficial development projects for the local community, and a rise in governmental bureaucracy at different scales. The ban on green felling hinders routine maintenance activities such as lopping, thinning, and clearing, which are essential to improving the quality of timber. Numerous issues such as these make Rangan argue that the idea of the movement can be more powerful than the movement itself.

See Also: Environmental Justice Movements; Grassroots Organizations; Social Action, Global and Regional.

Further Readings

Chakraborty, S. *A Critique of Social Movements in India: Experiences of Chipko, Uttarakhand and Fishworkers Movement.* New Delhi, India: Indian Social Institute, 1999.

Guha, Ramachandra. *The Unquiet Woods: Ecological Change and Peasant Resistance.* Berkeley: University of California Press, 1989.

Rangan, Haripriya. *Of Myths and Movements: Rewriting Chipko Into Himalayan History.* London: Verso, 2000.

Priyanka Jain
University of Kentucky

Clinton, William J. (Executive Order 12898)

U.S. presidents obviously have great potential influence on environmental policy formation. They can call public attention to specific issues to set public agenda and shape public debates; lead the policy formulation by devoting presidential resources to a specific issue; shape the policy implementation by appointing federal officials, proposing federal budgets, and issuing executive orders; and shape the resolution of the issues with presidential international influence. Since the environment became a modern political issue in the United States in the late 1960s, Bill Clinton has been considered to be the president most successful at using those policy tools in favor of the environment.

It is unclear how much Clinton's environmental policy was shaped by his predecessor, President George H. W. Bush, who promised to be an "environmental president" and helped pass a new Clean Air Act. President Bush later retreated to a harsh stance on the environment due to economic recession and business pressures. The environmental debate, nevertheless, became a core issue in the 1992 presidential election campaign. While the incumbent Bush, who was running for reelection, criticized environmentalists as extremists who impede the Americans' progress, Bill Clinton selected the leading environmentalist in the U.S. Congress, Al Gore, as his vice presidential running mate. During their campaign, Gore published his best-selling book *Earth in the Balance*. Clinton's campaign included many environmental promises: raise the standards of automobile fuel consumption; increase use of natural gas and decrease reliance on nuclear power; support alternative energy research and development; pass a new Clean Water Act; and make "no net loss" of wetlands a reality. Clinton and Gore also argued that environmental cleanup would not decrease jobs but increase jobs. Consequently, most of the environmentalists cast their votes for Clinton and Gore.

Entering the White House, President Clinton's appointments of key environmental officials, most of whom were Gore's former environmental assistants, earned applause from the environmental community. Gore shouldered the chief responsibility for formulating and coordinating environmental policies. Other appointees to the cabinet and executive offices were also mainly pro-environment. Clinton's government is thus called "the green administration." He disappointed the environmentalists during his first two years by not taking much visible action in improving the environment. All in all, however, his administration was very efficient in responding to environmentalists' appeals, even though he faced a Republican-controlled Congress during six of his eight years as president. Clinton reversed many of the Reagan and Bush policies that were widely criticized by environmentalists. He also favored increasing spending on environmental programs, research in alternative energy and conservation, and international population policy. He earned praise from environmentalists when he spoke out forcefully against the unpopular environmental policy decisions made by the Congress, supported the controversial new Environmental Protection Agency (EPA) clean air standards for ozone and fine particulates, and strongly backed international action on climate change. With regained environmental support in Congress, he was able to cooperate with Congress to pass two environmental bills in 1996: the Food Quality Act and the Safe Drinking Water Act Amendments.

Although the Clinton administration failed several times in its promotion of environmental protection policy, including failure to elevate the EPA to cabinet rank, it still must be given credit for calling the government's high attention to environmental issues. The Office of Environmental Policy was contacted by the vice president's office, cabinet secretaries, and other White House staff on a daily basis, and it was eventually folded into the Council for Environmental Quality in order for it to function more effectively. On September 30, 1993, the president issued Executive Order 12866, requiring the administrative staffs to select approaches "that maximize net benefits (including potential economic, environmental, public health and safety, and other advantages; distributive impacts; and equality)" instead of just focusing on the potential economic gains.

Another important environmental action taken by President Clinton was Executive Order 12898 issued on February 11, 1994, which exclusively addressed the issue of environmental justice. The order required each federal agency to make achieving environmental justice part of its mission and to develop its own strategy to protect minorities and those with low incomes from environmental and health hazards. The order also created an interagency working group in environmental justice to guide and supervise the development and implantation of all federal agencies' environmental justice strategies. It is one of a series of governmental responses to the environmental justice movement initiated in early 1980s. The order pushed federal, state, and local governments to create many environmental justice programs and has been cited by judges in making decisions in favor of environmental equity. A few existing studies on its impact, however, still show that it has not been well implemented and has little impact in changing the pattern of environmental inequity.

The 1994 midterm election put Republicans in control of both chambers of Congress. The executive branch could have little voice in law-making process. However, when the House passed a drastic revision of the Clean Water Act, which turned out to be unpopular in the public according to the opinion polls, the president delivered an aggressive speech on May 30, 1995, in Washington, D.C., vowing to veto what he called the "Dirty Water Act." In August, the president again made highly publicized speeches at Baltimore Harbor and in Yellowstone National Park castigating the Republican proposal. The proposal finally was dropped before the president vetoed it.

After his reelection in 1996, Clinton still faced a Republican-controlled Congress, which did not support his pro-environment programs. In his 1998 State of the Union Address, he proposed a new Clean Water Action Plan and a program to implement the Kyoto Protocol, a United Nations treaty on climate change. Congress funded the first program but not the second one. In January 1999, he announced a Better American Bonds program. He and Vice President Gore also endorsed a "smart growth" strategy to control development and reduce congestion around cities, which became a major theme in Gore's 2000 presidential campaign. As some scholars estimated, however, about 15,000 Floridians, who were nominally Democrats but still felt frustrated with the Clinton administration's languid environmentalism, cast their ballots for Ralph Nader and the Green Party, which is arguably a deciding factor in Gore's losing the election.

See Also: Environmental Justice Movements; Global Warming and Culture; Gore, Jr., Al; Kyoto Protocol; Public Opinion.

Further Readings

Bullard, Robert D. and Glenn S. Johnson. "Environmental Justice: Grassroots Activism and Its Impact on Public Policy Decision Making." *Journal of Social Issues*, 56 (2000).

Kraft, Michael E. and Norman J. Vig. "Environmental Policy From the 1970s to 2000: An Overview." In *Environmental Policy*, 4th ed., M. E. Kraft and N. J. Vig, eds. Washington, DC: CQ Press, 2000.

Melosi, Martin V. "Environmental Justices, Political Agenda Setting, and the Myths of History." In *Environmental Politics and Policy, 1960s–1990s*, O. L. Graham, ed. University Park, PA: Pennsylvania State University Press, 2000.
O'Neil, Sandra George. "Superfund: Evaluating the Impact of Executive Order 12898." *Environment Health Perspectives*, 115 (2007).
Vig, Norman J. "Presidential Leadership and the Environment: From Reagan to Clinton." In *Environmental Policy*, 4th ed., M. E. Kraft and N. J. Vig, eds. Washington, DC: CQ Press, 2000.

Qingjiang Yao
Fort Hays State University

Communication, Global and Regional

The communications industry itself, as well as other businesses and industries, has responded to increased public environmental consciousness and demand for green communication products. The green culture movement has also increased demand for environmentally friendly methods of disposal of the mass quantities of electronic communication devices discarded each year, known as electronic waste or e-waste. The communications industry also serves the larger green culture movement through the use of communication tools such as the mass media and the Internet to unite green culture members and spread their message, both regionally and globally. Part of the reason for the green movement's success in acquiring new members over the last few decades is because of communication. Communication, however, can also have negative ramifications for the green culture movement, such as information overload and competing claims for public loyalty among various environmental organizations with competing agendas.

Green Communication Issues and Developments

Environmental activists have recognized the importance of green issues and technological developments within the communication industry due to its role as one of the world's leading industries. Fiber-optic cables, satellites, the use of microwaves, and wireless telephony have helped fuel the industry's global emergence. New telecommunications products and services such as cell phones and the Internet have been among the fastest and most widely adopted new technologies in the late 20th and early 21st centuries.

The United Nations (UN) recognizes the importance of communication within the environmental movement and fosters its development in several ways. The UN agency known as the International Telecommunication Union (ITU), founded in 1865 and headquartered in Geneva, Switzerland, works with governments, industries, scientists, the public sector, and regulatory agencies to create international standards in the field. Its initiatives focus in part on environmental issues related to communication, including climate change and disaster mitigation.

The UN and other regional and global confederations or organizations have also emphasized the need to expand the reach of environmental communication to developing countries and remote areas. The ITU works to increase the telecommunication infrastructure in developing nations, thereby increasing their access to environmental as well as other kinds of information. Many European nations have implemented or are in the process of implementing the 1998 UN agreement known as the Aarhus Convention. The Aarhus

Convention was designed to ensure broad public access to information on the environment, environmental justice, and public participation in governmental decisions on environmental matters at the regional level.

Corporations within the communication industry, like other businesses, have responded to increased public demand for environmentally friendly products and services. The public has become increasingly aware of environmental problems and the need for change and often receives communications from green organizations and individuals explaining the benefits of environmentally conscious purchasing and other lifestyle decisions. For example, Apple Computers product users began a successful online campaign in 2007 under the name "Green My Apple," seeking to force the company to produce more environmentally friendly computers and electronic devices.

A variety of corporations and industries rely on communications such as mass media advertising and the Internet to inform existing and potential customers of their green products, services, manufacturing, packaging, and other sustainable initiatives in an attempt to gain customer loyalty among the increasing membership of the green culture movement. Businesses and marketers also utilize the same communication tools to research and test the appeal of so-called green products. Such green marketing campaigns have faced criticism from some corners of the green culture movement for the lack of regulation and potential for false advertising to capitalize on the popularity of environmental consciousness.

Another key area of concern in the field of communication is the disposal of the rapidly growing number of electronic communication devices that are discarded each year, in large part due to the rapid technological developments leading to the introduction and marketing of ever-newer models. Such waste is known as electronic waste or e-waste. These unwanted older electronic communication devices include computers and accompanying hardware and cell phones. Corporations produce the majority of e-waste, most without regard to ecofriendly recycling methods. Therefore, a large amount of toxic chemicals is released into the air, adding to air pollution.

Many of these devices are shipped internationally to countries that accept electronic waste, such as India and China, where they are stripped of their parts, recycled, or buried. For example, hundreds of thousands of tons of e-waste are recycled annually in India. The recycling process for electronic waste can expose recyclers to radioactive tubes, posing a health threat as well as an environmental threat. Attempts to lessen the amounts of electronic waste produced include programs to donate old computers, cell phones, and other electronic communication devices to schools or charitable organizations.

Some environmental nongovernmental organizations (NGOs) focus on the interaction between the environmental movement and the communication industry. One of the leaders in this field is the Environmental Communication Network (ECN). ECN is a group of scholars, professionals, and activists who hold conferences, publish newsletters and journals, and facilitate the exchange of information and ideas in the field of environmental communication. The ECN website is housed on the web servers of the State University of New York College of Environmental Science and Forestry, although the organization itself is independent of the college.

The Use of Communication in the Green Culture Movement

Democratic, grassroots participation is one of the hallmarks of the environmental movement, which makes communication an essential component of the movement's success. The green culture movement relies on communication to unite its members, gain new supporters, and raise public awareness of environmental issues. Early international statements recognizing the importance of communication to the environmental movement include those of the 1972 United Nations Conference on the Human Environment in Stockholm, Sweden. The conference noted that such communication on environmental ideas included the gathering, transmitting, and analysis of information, scientific exchange, educational programs, and the mass media.

Key forms of communication utilized by the modern green culture movement include the print and mass media, cell phones, and video and digital cameras. The rise of the Internet as an effective means of regional and global communication resulted in its rapid adoption by the green movement, including the use of e-mail, list serves, chat rooms and message boards, websites, and social media applications such as Facebook, YouTube, and Twitter. Most environmental organizations as well as numerous individuals interested in environmental preservation maintain websites dedicated to informing the general public on key issues and publishing tips, and advice on public participation in the green culture movement and how to maintain a sustainable lifestyle.

The print media and mass communication were among the first tools utilized by green culture activists. The use of green books, newspapers, magazines, radio, television, and other forms of mass communication helped spread the environmental movement beyond the local to the regional and global levels. Green newspapers and magazines expanded their global presence in the late 20th century, helping to spur green activism in countries like England, which had previously lacked large-scale environmental communication. By the 21st century, environmental periodicals numbered in the thousands and merited their own catalog, known as the Environmental Index.

Green culture activists often rely on grand communication gestures through the media to raise public awareness or issue public challenges for involvement with environmental issues. For example, Greenpeace activists created an effective statement encouraging President Barack Obama to become a spokesperson for the environmental issue of global warming by hanging a huge banner on the national landmark Mt. Rushmore in South Dakota, a gesture sure to garner national and global media attention. Similarly, green politicians and political parties use communication to spread their message, gain new members, and encourage the public to vote.

The rise of Internet technology and wireless devices in the late 20th century provided green culture activists with a new and more effective tool to unite members and to spread their message beyond the local level to the regional, national, and international levels. Environmental activists can spread news of issues or developments and share their thoughts and ideas far beyond their local reach to ever-expanding audiences through online articles, blogs, and websites. This technology also enables the democratic exchange of thoughts and ideas and collaboration on possible approaches and solutions. The Internet's global reach also allows environmental messages to be translated into different languages and to reach isolated areas. Photographs and web cams enhance the visual impact of the message.

Environmental organizations with a more activist approach have also found the Internet an important communication tool to unify members and to inspire action as well as to inform. So-called cyber-activists seek to use Internet communication technologies to ensure that corporations, organizations, and governments protect the environment, practice sustainability, and do not violate local, national, or international environmental laws, regulations, and treaties. Examples have included the use of Internet and Global Positioning System (GPS) technology to map the land of indigenous peoples in the Amazon to prevent logging within their territories.

The rise of the Internet, however, has also led to communication problems and information overload within the green culture movement and among the general public. The public and interested members now have a variety of green organizations to choose from, each with its own website offering sometimes-contradictory recommendations. The public may become overwhelmed at the amount or contradictory nature of environmental information, which often results in disinterest, confusion, and inaction, the opposite of the green culture movement's intended goals. Wider communication sometimes fosters hostile communication or the public airing of grievances that can compromise movement unification, cooperation within and among regional and global organizations, and the effective enactment of environmental agendas.

The limited reach of Internet access in remote and developing countries, often referred to as the digital divide, hinders the outreach of groups and programs in those areas, which often lack large or cohesive environmental movements. Such diversity can also hinder the unification of green culture activists and groups themselves. Other areas face communication difficulties of government control of and/or censorship of the Internet and other communication mediums. Government tactics have included Internet censorship, the jamming of satellite communications, and the hacking or shutting down of environmental websites.

On the other hand, the green culture movements in countries where activists have faced censorship, such as Iran, often go underground and rely on the Internet as a method of sending and retrieving information that may otherwise be suppressed and communicating with one another when face-to-face meetings may draw suspicion or arrest. They also utilize other electronic communication devices such as cell phones and video cameras to film environmental street protests and other measures as well as their suppression by the government. The release of these videos can gain international attention and support for such countries' environmental issues and movements. For example, the suppression of Iranian environmental activists attracted the attention of activists in other countries, such as the United States, who sought to share software and technology to give Iranian environmental groups free access to the Internet.

See Also: Communication, National and Local; Cyber Action; Environmental Communication, Public Participation; Internet, Advertising, and Marketing; Public Opinion.

Further Readings

Carr, Nicholas G. *The Big Switch: Rewiring the World, From Edison to Google.* New York: W. W. Norton, 2008.

Cox, Robert. *Environmental Communication and the Public Sphere.* Thousand Oaks, CA: Sage, 2010.

Depoe, Stephen P., ed. *The Environmental Communication Yearbook.* London: Lawrence Erlbaum, 2006.

Dertouzos, Michael L. *What Will Be: How the New World of Information Will Change Our Lives.* San Francisco, CA: HarperEdge, 1997.

Dodd, Annabel Z. *The Essential Guide to Telecommunications*, 3rd ed. Upper Saddle River, NJ: Prentice Hall, 2001.

Enayat, Mahmood. "Iran's Green Communications: Beyond Twitter to Small Media" (June 15, 2010). http://www.enduringamerica.com/june-2010/2010/6/15/irans-green-communications-beyond-twitter-to-small-media-ena.html (Accessed October 2010).

Environmental Communication Network. http://www.esf.edu/ecn/ecn.htm (Accessed October 2010).
Esfandiari, Golnaz. "Green Supporters Want West to Help Iranians Access Internet, Uncensored Information." Radio Free Europe. Radio Liberty (June 2, 2010). http://www.rferl.org/content/Green_Supporters_Want_West_To_Help_Iranians_Access_Internet_Uncensored_Information/2059734.html (Accessed October 2010).
Freeman, Roger L. *Telecommunication System Engineering*. Hoboken, NJ: Wiley-Interscience, 2004.
"Greenpeace and Peaceful Protest." http://www.greenpeace.org/usa/en/campaigns/Actions/Peaceful-Protest (Accessed October 2010).
Hukill, Mark A. and Ryota Ono. *Electronic Communication Convergence: Policy Challenges in Asia*. Thousand Oaks, CA: Sage, 2000.
International Telecommunication Union. http://www.itu.int/en/pages/default.aspx (Accessed August 2010).
Katz, James Everett. *Handbook of Mobile Communication Studies*. Cambridge, MA: MIT Press, 2008.
Kennedy, Paul M. *Preparing for the Twenty-First Century*. New York: Random House, 1993.
McLuhan, Marshall and Bruce R. Powers. *The Global Village: Transformations in World Life and Media in the 21st Century*. New York: Oxford University Press, 1989.
Mitchell, William J. *E-topia: "Urban Life, Jim—But Not as We Know It."* Cambridge, MA: MIT Press, 1999.
Nagel, Jean. *Living Green*. New York: Rosen Publishing Group, 2009.
Neuman, W. Russell, Lee W. McKnight, and Richard Jay Solomon. *The Gordian Knot: Political Gridlock on the Information Highway*. Cambridge, MA: MIT Press, 1997.
Newberg, Paula R. *New Directions in Telecommunications Policy*. Durham, NC: Duke University Press, 1989.
OurEarth. http://www.ourearth.org/about (Accessed October 2010).
Ryan, Daniel J. *Privatization and Competition in Telecommunications: International Developments*. Westport, CT: Praeger, 1997.
Sheppard, P. J. and G. R. Walker. *Telepresence*. Boston, MA: Kluwer Academic Publishers, 1999.
Union of Concerned Scientists: Citizens and Scientists for Environmental Solutions. http://www.ucsusa.org (Accessed October 2010).
Wall, Derek. *Earth First! and the Anti-Roads Movement: Radical Environmentalism and Comparative Social Movements*. London: Routledge, 1999.
Wheeler, James O. and Yuko Aoyama. *Cities in the Telecommunications Age: The Fracturing of Geographies*. New York: Routledge, 2000.
Zielke, Katrin. "The Green Curtain: East Germany's Environmentalists Helped End Stalinism, but Worry About Unification." *Mother Jones* (April 1990).

Marcella Bush Trevino
Barry University

Communication, National and Local

The GfK Roper Green Gauge Global report, which examines the green habits of 36,000 consumers in 25 countries worldwide, found that American consumers are skeptical about the cost and efficacy of green products and their impact on the environment. Approximately two in three Americans perceive green products to be too costly, and one-third believe they do not work as well as "regular" products. While the percentage of people willing to pay a premium for ecofriendly products will go up and down as times change—environmental disasters increase awareness and interest in green products, for instance, while prosperous economic times often dull interest—in general, the market for green products and services is growing every year as people become more aware of the impact of their consumer choices.

That said, environmental marketing is a tricky business, especially given the relative skepticism most people have toward green branding claims. Greenwashing—claims of environmental benefits that are overstated or misrepresented—abounds in the world of ecofriendly products and services, and with so many companies failing to live up to their green business promises, consumers are understandably wary of dubious branding.

At the beginning of 2009, green marketing was showing signs of waning interest from consumers after reaching a high point in 2008. As the economy worsened, consumers began losing interest in spending more for green products. With the election of Barack Obama as president, there has been renewed interest in sustainability. President Obama gives green marketers greater visibility—and the green movement a more visible ally—than they would have otherwise.

Recognizing the importance of a green economy, in his first 100 days, the new president took significant steps to change the United States' direction on energy and the environment. President Obama committed billions of dollars to new spending on clean energy. He has adopted positions on efficiency, renewables, and climate change, including a green "dream team" that includes not only the heads of the Department of Energy (DOE) and the Environmental Protection Agency (EPA), but less-known yet critical appointments, including climate expert John Holdren as science adviser.

Joe Romm, a leading commentator on energy and climate, proclaimed, "Obama is the first president in history to articulate both the why and how of the sustainable vision—and to actively, indeed aggressively, pursue its enactment. And that is why he is likely to be remembered as the green FDR."

On Earth Day, April 22, 2009, Obama delivered a speech in which he emphasized American ingenuity to overcome hardships, particularly in the energy area. He talked about green innovation as a source of prosperity and job creation. And he said, "I think the American people are ready to be part of a mission," while acknowledging that it would not be easy. President Obama has also increased funding for energy education and training through a program called RE-ENERGYSE (short for REgaining our ENERGY Science and Engineering Edge).

To communicate his green agenda, Obama has made green issues more personal, for himself and for the American people. For example, the photos of Michelle Obama planting an organic garden at the White House with local children generated not only a lot of media attention, but symbolically represented that personal connection for which Obama is striving. In addition, it signaled a values-based approach to humanizing green issues, such as the debate over climate change.

As a reflection of the new interest in the green economy, at the 2009 North American International Auto Show in Detroit, every manufacturer had a hybrid, plug-in, or electric battery car to show off. This occurred, in part, because the auto industry had come under fire from members of Congress who, in the wake of a $17 billion government bailout, said Detroit had not invested enough in hybrid technologies—and automakers were eager to prove them wrong. Ford showed off an expanded hybrid selection, General Motors unveiled its plans to compete in the electric battery market, and Toyota plans to sell 180,000 units of the third-generation Prius in 2010.

For drivers who want to be more ecofriendly but cannot afford to buy a new electric car, the automakers helped launch the website www.EcoDrivingUSA.com, which provides ecofriendly driving tips, such as maintaining a steady speed.

On another front, Walmart has encouraged its suppliers to create more sustainable offerings, ones that will resonate with their customers. The company has found that some green products have sold well (rubber mulch made from recovered tires, for example), while others have not (the classic milk jug). Walmart is currently focusing on developing new green products in four key areas: waste improvement and recycling, natural resources, energy, and social or community impact.

In July 2010, Walmart announced that it was partnering with Seventh Generation, the nation's leading brand of nontoxic and environmentally friendly household products, to offer Seventh Generation goods at more than 1,500 Walmart stores as well as through Walmart.com. The Seventh Generation products that Walmart carries include laundry detergent, dish soap, disinfecting wipes, and all-purpose sprays.

In addition to these corporate examples, the following are three exemplars of national, state, and local green communication efforts.

On the national level, the U.S. Green Building Council (USGBC) is a 501(c)(3) nonprofit community of leaders working to make green buildings available to everyone within a generation. This organization developed the Leadership in Energy and Environmental Design (LEED) building program for energy-efficient construction and renovation, which is now the leading standard for green buildings. It has literally set the green standard against which all buildings are now measured.

The annual Greenbuild International Conference and Expo was launched by USGBC in 2002 as the world's largest conference and exposition dedicated to green building. The conference brings together those involved in designing, building, living in, and working in green buildings. Greenbuild features three days of speakers, networking opportunities, industry showcases, LEED workshops, and tours of the host city's green buildings. This event sells out every year.

In addition to Greenbuild, the USGBC Federal Summit 2010, held in Washington, D.C., saw more than 750 professionals from the green building industry come together for two days of education, networking, and critical discussions on greening federal, state, and local governments. Attendees, including public sector employees at the local, state, and federal level as well as private sector employees from throughout the industry, engaged in conversations and made connections that will shape green building in the government sector in the years to come. During the summit, U.S. General Services Administrator Martha Johnson announced a plan for a zero-environmental-footprint target for the General Services Administration (GSA).

In terms of state efforts, Wisconsin's Focus on Energy program offers incentives for homeowners and businesses to install energy efficiency and renewable energy projects. The program has gained popularity since it was launched 10 years ago as a pilot program. When the state made the decision to harness wind power and install 100 wind turbines, Hoffman York, the marketing communications agency that works with the program, faced a challenge and engaged in marketing efforts to combat some residents' "not in my backyard" (NIMBY) sentiment about the wind turbines. The program has successfully demonstrated to people that energy efficiency and renewable energy products are not only good for the environment, but save money in the long run.

Focus' residential programs show Wisconsin residents how to reduce their carbon footprints and lower their costs of living by being more energy efficient. Whether they are switching to Energy Star–qualified compact fluorescent light bulbs or building a Wisconsin

Energy Star Home, in 2009 alone, Wisconsin residents saved approximately $10 million in energy costs by making simple lifestyle changes. Once residents take steps to improve the energy efficiency of their homes, Focus helps them explore the possibility of acquiring and installing renewable energy resources.

On the local level, since its founding in 1999, Solar San Antonio has had as its goal to generate awareness and to educate the public about solar energy. Because of a small marketing budget, Solar San Antonio has been a bit nontraditional in many ways, so it comes as no surprise that the organization never quite embraced traditional marketing but instead has found much success with modern social media marketing. The organization has a presence on Twitter (@solarsanantonio) and Facebook (http://www.facebook.com/SolarSanAntonio) as well as social media outlets MySpace, LinkedIn, and YouTube.

For most homeowners, the largest barrier to installing solar energy is the large up-front cost. Solar San Antonio has worked with local financial institutions to facilitate San Antonio's first-ever low-interest solar loan programs. With these options, solar energy is easier and more affordable for San Antonio homeowners to acquire. Net solar installation prices are expected to be between $20/month and $50/month.

In August 2010, Solar San Antonio announced its "Bring Solar Home" campaign with support from CPS Energy, the City of San Antonio, Bexar County, and Wells Fargo Bank. The campaign will include a variety of print, outdoor, radio, and television ads to encourage the community to contact Solar San Antonio at a dedicated campaign phone number or at the campaign website, www.bringsolarhome.com. Using these tools, Solar San Antonio staff plan to gather preliminary information to determine what solar product is of interest and possible for the homeowner. This information will then be given to three solar installation companies to follow up with bids. Campaign staff will monitor the entire process to ensure quality and a good experience for the homeowner, solar installer, CPS Energy, and financial institution.

Greenbuild, Wisconsin's Focus on Energy, and Solar San Antonio demonstrate that successfully marketing a business or product as green is related to how well the company manages with respect to the environment. Green marketing can take many forms and emphasize ecofriendly aspects such as the following:

- *Operational sustainability*: Service companies and manufacturers can improve operational sustainability by reducing everyday energy and water consumption, minimizing pollution, using greener materials and processes, and properly managing waste.
- *Green products*: Companies producing products can do many things to improve the green factor of their offerings, including choosing sustainable materials, designing products to save energy and water, and making products that are less toxic and more natural than competitors' goods. Greener products will also be packaged in an ecofriendly manner and will be made to be easily recycled or composted.
- *Sustainable marketing*: Here, companies want to ensure that the actual marketing systems used are green, which is really distinct from operational sustainability (which focuses on manufacturing and production). For instance, when printing marketing materials, companies should use 100 percent postconsumer recycled paper made without chlorine (called "processed chlorine free") and printed using plant-based dyes (like soy inks). Tools like green web hosting, carbon offsets for any marketing emissions, recycling any unused materials from billboards and signage, ecological packaging, and so on should be explored.
- *Environmental causes*: Any organization can choose to promote environmental causes. Nonprofits and nongovernmental organizations (NGOs) engage in green marketing to get the word out about their activities, but for-profits can also support environmental causes by making donations and advertising for these environmental do-gooders as part of a green marketing campaign.

Knowing the audience is also a necessity if a company wants to have an effective environmental marketing campaign. According to the Green Gauge Report, there are five different groupings of green consumers:

- *True-Blue Greens*: the most environmentally active segment of society
- *Greenback Greens*: those most willing to pay the highest premium for green products
- *Sprouts*: fence-sitters who have embraced environmentalism more slowly
- *Grousers*: uninvolved or disinterested in environmental issues, or feel the issues are too large for them to solve
- *Apathetics*: the least-engaged group, who believe that environmental indifference is mainstream (also known as "Basic Browns")

Regardless of target market, an ecomarketing campaign should have several important factors in order to ensure long-term sustainability in the green space:

- Green claims should be genuine and verifiable. Any environmental claims need to be transparent and explicit.
- Informed consumers are loyal consumers, so consumers must be educated about the benefits of a product or service for the environment.
- Making it appear that customers are giving back to the environment by choosing the company's service or product is a successful strategy to adopt.

Green marketing can refer to anything from greening product development to the actual advertising campaign itself. Going by alternative names such as sustainable marketing, environmental marketing, green advertising, ecomarketing, organic marketing—all of which point to similar concepts though perhaps in a more specific fashion—green marketing is essentially a way to brand a marketing message in order to capture more of the market by appealing to people's desire to choose products and services that are better for the environment.

There are many environmental issues impacted by the production of goods and rendering of services, and therefore, there are also many ways a company can market its eco-friendly offerings. Green marketing can appeal to a wide variety of these issues: an item can save water, reduce greenhouse gas emissions, cut toxic pollution, clean indoor air, and/or be easily recyclable. As the prior examples demonstrate, when put side by side with the competition, the more environmental marketing claims a product or service can make, the more likely it is the consumer will select it, provided the price point is not too much higher than the alternative.

See Also: Advertising; Green Consumerism; Greenwashing; Media Greenwashing.

Further Readings

Focus on Energy, http://www.focusonenergy.com (Accessed September 2010).

GfK Custom Research North America. http://www.gfkamerica.com/practice_areas/roper_consulting/roper_greengauge/index.en.html (Accessed September 2010).

Solar San Antonio. http://www.solarsanantonio.org (Accessed September 2010).

U.S. Green Building Council. http://www.usgbc.org (Accessed September 2010).

Rod Carveth
Fitchburg State University

Commuting

Like these people in the United Kingdom, many commuters prefer the environmentally friendly practice of utilizing train stations and bicycles (of which over 1,000,000 are sold each year) instead of driving to work.

Source: iStockphoto

By the last quarter of the 20th century, it had become abundantly clear that the vast number of carbon dioxide–producing vehicles on the road was rapidly increasing the rate of global warming around the world. City, county, and national governments joined businesses, nongovernmental organizations (NGOs), and private citizens in an effort to reduce the number of commuters on roads. Those who lived close enough to their jobs were encouraged to walk or bike to work. Others were enjoined to carpool or to use public transportation. Scores of companies permitted employees to telecommute, negating the need for commuting. Many cities converted to hybrid vehicles, and legislatures at all levels began funding improved mass transit systems. Both public and private sectors increased their efforts to engage in innovative methods of providing commuters with viable alternatives to using their own vehicles to travel between work and home.

Commuters who live close enough to their workplaces often choose to walk, but for others, walking is not an option. For those who are environmentally conscious, biking or a combination of biking and public transportation provides them with the joint benefits of becoming physically and mentally healthier and protecting the environment. Worldometers reports that more than 100 million bicycles are sold each year compared to 43 million automobiles. Approximately half of all bicycle production occurs in China, the country with the world's largest population. Other nations are also full of cyclists. In Canada, half of the population rides bicycles. Many employers are now supporting cycling as a method of commuting by building changing rooms with showers and setting aside parking areas exclusively for cyclists.

Over the last four decades, residents of many large cities have led the way in choosing alternate means to commute rather than contributing to additional increases in the levels of carbon dioxide in the atmosphere. New Yorkers lead the United States in this practice, with approximately 82 percent of the working population traveling to work by means of public transportation, biking, or walking. Since the passage of the Clean Air Taxis Act in 2007, the city has further slashed carbon dioxide levels by instituting the use of hybrid vehicles for the city fleet by 2012. Residents of Portland, Oregon, are also making considerable strides toward sustainability, and the use of public transportation has grown by 75 percent in recent years. In Chattanooga, Tennessee, the city funds free electric buses for use downtown, a move that has reduced carbon dioxide emissions by some 3.5 million pounds a year and reduced the release of particulate matter by 600 pounds during that same time period. Officials in Minneapolis, Minnesota, voted to subsidize mass transit and carpooling. The city now has 85 miles of bike lanes, and at least 60 percent of all commuters opt for alternate methods of transportation. Some 125,000 commuters in Salt Lake City, Utah, ride city buses or TRAX, a newly installed light rail system.

A number of countries have likewise adopted alternative methods to public taxis and private automobiles. The concept of car sharing for commuters was a major element at the World Expo held in Shanghai in 2010. Bremen, Germany, has proven to be an innovator in this field. Acknowledging that most commuters drive their cars for only an hour a day, the Cambio company devised a plan to reduce the number of cars on the roads by furnishing a fleet of cars, vans, and minibuses to members for a fee of about $3.75 a month plus mileage charges. In 2009, Cambio, which boasted 23,000 members in Germany, expanded its operations to 17 cities in Belgium. Reports from the Eugene Car Co-op in Oregon indicate that with one car meeting the needs of 10 members, carbon dioxide levels are considerably lower than when each member commuted to work in his/her own vehicle.

One innovative solution to cutting down on carbon dioxide levels is to reduce commuting time by transferring employees to locations nearer their homes. After conducting studies for a number of Seattle, Washington, companies that included Starbucks, Boeing, and the city fire department, software developer Gene Mullins discovered that some employers were traveling up to 290 miles a day. He found that the commuting time of Boeing employees was equal to 85 circumnavigations of Earth each day.

Some experts believe that the wave of the future is connected to providing commuters with technologically advanced methods of transportation. In Canberra, Australia, researchers involved in the Intelligent Transport Systems Project, a division of CSIRO Mathematical and Information Services, are currently working on developing responsive buses that can be summoned to a stop by passengers, identify detours to avoid traffic jams, and deliver commuters directly to their doors at night. Passengers on these buses will be able to plan journeys from their homes, offices, and via the Internet. These state-of-the-art buses are meant to supplement the current system rather than to replace it. Another innovation is the electric car, such as the one introduced by Ford around the turn of the 21st century. The TH!NK Neighbor vehicle is designed for short-term commuting. The vehicle, which operates on six 12-volt lead-acid batteries, produces zero emissions. Running at a top speed of 25 mph, the vehicle can travel up to 30 miles before it needs to be recharged, a process that takes four to eight hours. A two-passenger version is also available.

See Also: Auto-Philia/Auto-Nomy; Corporate Social Responsibility; EcoPopulism; Green Consumerism; Individual Action, National and Local; Social Action, National and Local.

Further Readings

Boelte, Kyle. "Look, Ma, No Car!" *Sierra*, 95/2 (2006).

Foster, John Bellamy. *The Ecological Revolution: Making Peace With the Planet*. New York: Monthly Review Press, 2009.

Goodspeed, Brianne. "It's Not Easy Being Greenest: 10 Cities to Watch." *E: The Environmental Magazine*, 17/4 (2006).

Grazi, Fabio, et al. "An Empirical Analysis of Urban Form Transport and Global Warming." *Energy Journal*, 29/4 (2008).

Kennedy, Mike. "10 Paths to Green." *American School and University*, 81/4 (2008).

Kofoed, Philip. "Smart Moves Planned for Clean Commuters." *Ecos*, 94 (January/March 1998).

MacGregor, Sherilyn. *Beyond Mothering Earth: Ecological Citizenship and the Politics of Care*. Vancouver: UBC Press, 2006.

Masters, Coco. "13 Let Employees Work Close to Home." *Time*, 169/15 (April 9, 2007).
"The New Cycling Revolution." *Natural Life*, 134 (July/August 2010).
Schildgen, Bob. "Carless Behavior." *Sierra*, 83/2 (1998).
Schor, Juliet B. and Betsy Taylor. *Sustainable Planet: Solutions for the Twenty-First Century.* Boston: Beacon Press, 2002.
Siuru, Bill. "TH!NK Electric." *Poptronics*, 2/4 (2001).
Steffen, Alex. *Worldchanging: A User's Guide for the 21st Century.* New York: Abrams, 2006.
Torgerson, Douglas. *The Promise of Green Politics: Environmentalism and the Public Sphere.* Durham, NC: Duke University Press, 1999.
Zuurbier, Moniek, et al. "Commuters' Exposure to Particulate Matter for Pollution Is Affected by Mode of Transport, Fuel Type, and Route." *Environmental Mental Health Perspectives*, 18/6 (2010).

Elizabeth Rholetter Purdy
Independent Scholar

Cooking

Many consumers invested in green issues are also home cooks. Going green often intersects with a "Do It Yourself" ethos—which may lead as easily to gardening, raising livestock, and various home projects—but with an expanded awareness of the impact of and concerns over the various industries responsible for feeding Americans. Many things that saved lives or improved the quality of life when they were introduced have become suspect now that they may no longer be necessary—an abundance of sugar, the practice of heavily salting foods or treating them with nitrites to preserve them, the use of refrigerated transportation to transport perishable food across the country, the popularization of ground beef as a by-product of the beef industry that surged as a result of such, the factory farms that made poultry and eggs more affordable than ever, the agricultural subsidies that kept farmers in business, the use of vegetable fats instead of animal fats, the use of pesticides and antibiotics in raising our food.

Not all of these issues can be remedied in the home kitchen, but a greater awareness of what is in one's food—or especially what might be in it—seems often to lead to a greater interest in preparing that food oneself instead of relegating the task to others. Furthermore, in the case of many environmentally sound options like locally produced food (including produce, meat from locally pastured livestock, and local dairy products), cooking it yourself may be the only option available. The chicken in most restaurants could come from anywhere; the ground beef is of unknown provenance. At home, more and more often, it is possible to cook with meats bought from a ranch you can visit yourself to see the conditions in which the animals are raised. Cooking at home means knowing exactly what is in your food and how safely it was handled.

But not all cooking has equal environmental impacts. Cooking consumes a great deal of energy, and household ovens are rarely as energy efficient as they could be, particularly given how long they are usually kept. Dishwashers consume a good deal of water, and the wastewater carries detergents with chemicals that need to be treated to keep them from impacting local soil and water. Appliances account for 30 percent of the average household's energy use, with the refrigerator as the biggest "energy hog": even most Energy

Star–rated refrigerators cannot be counted on for efficiency, since manufacturers are allowed to run their efficiency tests with the freezer turned off.

Even cutting boards can be made greener. While there are now some on the market made of recycled cardboard, plastic, and cork, bamboo cutting boards have been in use for hundreds of years and are a sustainable material. Their downside is the travel cost from China. Glass, though environmentally friendly, is bad for knives and will lead to needing to replace them more often. Knives should ideally be sharpened regularly and treated with a sharpening steel after each use: properly cared for, a forged steel knife will last for generations, while blades made of cheaper material will need replacing every few years.

Some green behaviors may require replacing household items: convection ovens cook with less energy than conventional, for instance (in that they cook faster and use the same amount of energy per minute); garbage disposals make the waste management process more resource intensive and represent greater energy and water use in the home and should be replaced by composting whenever possible, for which there are many safe indoor methods available. But other changes need only a change in habits. While dishwashers are an energy-consuming luxury, if one must run them, they can be set on a delay to run in the middle of the night at off-peak hours, when power plants are running more efficiently. Refrigerators will use less energy overall when they are full, because cold food warms up more slowly than cold air; when there is empty space in the refrigerator, store pitchers of drinks there. The freezer can be kept full by making fewer trips to the supermarket or other food stores, thus also conserving gas; further, cooks with a chest freezer can take advantage of buying a "share" of meat from a local ranch, which is butchered, packaged, and ready to freeze at a much lower per-serving cost than meat bought a cut at a time. The stove uses less energy than the oven, for most tasks, and should be relied on more often. Appliances should be unplugged when not in use to reduce energy consumption. Microwaves are not suitable, aesthetically, for every cooking purpose, but are more efficient than stoves or ovens for some tasks; this is especially true for small portions of certain foods. An oven takes the same amount of time to bake four potatoes as one, but a microwave will cook the single potato faster. On the other hand, ovens are not portion-blind with all foods, as they are sometimes mistakenly described: a large turkey clearly takes longer to cook than a small one, and the cooking times of meatloaves and lasagnas are a function of their bulk. But for reheating, for cooking food for one or two people, and things of that nature, microwaves are certainly far more efficient.

Food waste can be reduced significantly. Most obviously, leftovers need not be thrown away. Trimmings from most vegetables, and bones, leftover meat, and trimmed fat can be saved in freezer bags to make soup stock. Citrus rinds can be candied or dried and used as a seasoning. Watermelon rind can be pickled. Cucumber peels can be used as a facial treatment or added to a pitcher of water in the refrigerator for a summer beverage. Many older cookbooks and church cookbooks have recipes for jams, jellies, and punches made from apple peels and cores. Corncobs and corn silk can be used to make corn broth for chowder or added to vegetable stock.

More can be grown at home than most people think, even in apartment buildings. Windowsill herb gardens are much cheaper and more energy efficient than store-bought herbs, can be grown year round, and can be used to grow a number of other small vegetables, such as radishes, microgreens of many species, watercress, and sunflower shoots. Larger vegetables can often be grown indoors in the summer if there is sufficient sunlight, though they may not be as water efficient as buying from a local farm.

Because meat is more resource intensive to grow than vegetables and grains, it should be consumed less. This does not mean that green cooking must be vegetarian. Michael Pollan's maxim sums up the best and most responsible approach: "Eat food. Not too much. Mostly plants." "Eat food" means to eat real food, not processed items; "not too much" refers to portion control, to avoid waste; "mostly plants" is simply a healthy and environmentally sound formulation. Only in recent history has meat overtaken plants on the typical plate; this recent trend may become more than just a fad.

See Also: Green Consumerism; Home; Nature Experiences; Veganism/Vegetarianism.

Further Readings

Heyhoe, Kate. *Cooking Green: Reducing Your Carbon Footprint in the Kitchen*. Cambridge, MA: Da Capo, 2009.

Newgent, Jackie. *Big Green Cookbook*. Hoboken, NJ: Wiley, 2009.

Bill Kte'pi
Independent Scholar

Corporate Green Culture

In recent years, more and more U.S.-based and global corporations have embraced environmentally friendly business practices and committed themselves to positive environmental change. Corporations choose to go "green" for many reasons, ranging from cost-cutting strategies to pressure from consumers. The development of greener corporate culture can typically be understood as a result of a combination of factors, including stricter government regulations, an emphasis on social responsibility and environmental commitment, partnerships with environmental nongovernmental organizations (NGOs), the development of new technology that is both more environmentally friendly and cost efficient, and pressure from corporate stakeholders such as consumers and investors. Most major corporations today produce sustainability reports outlining their commitment to green practices and protecting the environment. Yet while some corporate green initiatives are lauded as having a significant positive impact on the environment, others are dismissed as "greenwashing," wherein corporations make environmentally friendly claims without significantly altering their actions to support these claims.

The Rise of Corporate Green Initiatives

When the modern environmental movement first became popular in the United States in the 1960s, most U.S.-based corporations were reluctant to change business practices to become more environmentally friendly. Companies typically only changed their practices when they were forced to do so by the Environmental Protection Agency (EPA) or other government agencies. These changes typically involved cleaning up waste after it was produced. Yet as corporate officers began to calculate the cost of this cleanup, and as consumers, shareholders, and environmental groups began to pressure corporations to adopt more environmentally friendly practices, many corporations began implementing pollution-prevention policies. By the beginning of the 21st century, the majority of the world's largest corporations voluntarily produced sustainability reports outlining their commitment to environmental projects and sustainability goals. For example, Unilever reports that it recycles over 17 tons of waste annually at its toothpaste factory, teaches palm oil producers in Ghana how to reuse plant waste, and provides financing for Brazilian tomato farmers to convert to ecofriendly drip irrigation. Similarly, Dow Chemical reports that it is developing ecofriendly Styrofoam to be used for walls in low-cost housing. Internet search engine Google has pledged to invest hundreds of millions of dollars in renewable energy, such as solar and wind power. And Nike has launched the Nike Environmental Action Team, which developed Reuse-A-Shoe, a program that recycles shoes and turns them into new products.

Other corporations have demonstrated their green commitment by embracing environmentally friendly technology. Many large corporations have committed to using green building materials for corporate headquarters or production centers. For example, the financial conglomerate Citigroup has redesigned many of its bank branches to include more natural lighting and recycled materials. Similarly, snack food conglomerate Kettle Foods has installed wind turbines in its manufacturing facility, and many other corporations have installed solar panels on roofs and in parking areas. Yet the reasons corporations choose to adopt more green policies and further their commitment to environmentalism are often difficult to accurately pinpoint. Environmental changes may be made as a result of a sense of moral obligation or because being green is cost effective. Changes are also often the result of external pressure.

External Pressure

While corporations adopt new green initiatives for numerous reasons, one key motivating factor is the desire to meet the demands of corporate stakeholders. Customers, shareholders, environmental groups, and employees may all influence a corporation's environmental practices and strategies. As both voters and consumers of corporate services and products, customers can have a strong impact on corporate environmental policies. Corporate boycotts (choosing not to by a corporation's product) or buycotts (buying and promoting a corporation's products) based on a firm's environmental reputation can have a strong impact on profits, which may in turn cause a corporation to reevaluate its environmental practices.

Because poor environmental performance often leads to cost increases, investors often monitor corporations' environmental performances and voice their concerns. Investors do so by communicating with corporate management, filing shareholder resolutions, voting against management, or withdrawing their investments. Today, many individual investors, as well as larger investors like pension funds, consider a company's environmental performance before they invest. To aid in this endeavor, media sources such as *BusinessWeek* and *Newsweek* have produced reports ranking the world's leading sustainable or green corporations. For example, investors can look at *Newsweek*'s "Green Rankings" to find information on green policies and environmental impact for the world's 500 largest companies, including Hewlett-Packard, Dell, and Johnson & Johnson, which were ranked as the first, second, and third most environmentally friendly large corporations. Reports such as these publicly recognize environmentally friendly corporations, and these corporations may deepen their commitment to environmentally friendly policies as a result of positive feedback. Corporations that are ranked lower may also choose to increase green initiatives in order to avoid negative rankings in the future.

Pressure for corporations to become greener also comes from environmental groups and other nongovernmental organizations. In the 1980s, as the membership levels and financial resources of many environmental groups increased, representatives of these groups began to target companies directly, prompting companies to actively create environmental rules to demonstrate social responsibility. As corporations began to view environmental groups more positively and value the research studies these groups conducted, alliances between environmental groups and corporations began to increase. Eventually, many firms began to see environmentalism as an aspect of the corporate environment.

As a result, corporations like Coca-Cola have partnered with a variety of environmental NGOs, including Greenpeace and the World Wildlife Fund (WWF), to change the corporation's refrigeration and bottling practices, to conserve water, and to reduce water pollution. Similarly, the Washington, D.C.–based Environmental Defense Fund is working with the world's largest retailer, Walmart, on a range of issues, including decreasing global warming, reducing packaging waste, and promoting alternative fuels. As a result of these partnerships, corporations are able to measure and report environmental improvements to the government, suppliers, shareholders, and consumers. NGOs, on the other hand, are able to further their environmental agendas and are granted access to valuable corporate resources that would otherwise be unavailable.

Green Commitment Versus Greenwashing

Many corporations have realized that marketing green products boosts their profits while also lowering their carbon footprint. Yet this increase in profits has caused critics to question many corporations' true environmental commitment. For example, hotel chains across the United States have placed placards in rooms urging guests to reuse towels and to keep the same bedsheets throughout their stay in order to conserve water. Many critics have pointed out that hotels may only be using environmental language to decrease costs; critics argue that hotels are reluctant to commit to other environmental changes that may negatively impact their profits.

Furthermore, many corporations have been accused of greenwashing, or misleading customers about their environmental practices or the environmental benefits of their products. As a result, organizations like the Advertising Standards Agency (ASA) have received increasing numbers of complaints about suspected false or misleading green claims in corporate advertisements. Companies were particularly likely to be challenged if they claimed their products were carbon neutral or carbon negative, or if they claimed their products were 100 percent recycled or wholly sustainable. The motoring and utilities industries were most likely to receive complaints. As a result of greenwashing concerns, the communications firm Futerra released a greenwashing guide to help both business and the public recognize dangerous and misleading greenwashing claims. The guide urges consumers to be wary of words with no clear or concrete meaning, such as "ecofriendly," and to ignore claims that cannot be supported with evidence.

While some corporations may make false environmental claims due to confusion or deception, it is undeniable that over the past 40 years, corporations have become more environmentally conscious. Most large corporations publish environmental or sustainability reports detailing their green commitments. Whether as a result of executive and employee initiatives, external pressure, a concern for their bottom line, or a combination of all of these factors, many corporations are committed to acting in environmentally friendly ways and being perceived as green.

See Also: Advertising; Corporate Social Responsibility; Green Consumerism; Greenwashing; Nongovernmental Organizations.

Further Readings

Engardio, Pete, with Kerry Capell, John Carey, and Kenji Hall. "Beyond the Green Corporation." *BusinessWeek* (January 29, 2007).
Esty, Daniel and Andrew Winston. *From Green to Gold: How Smart Companies Use Environmental Strategy to Innovate, Create Value, and Build Competitive Advantage.* New Haven, CT: Yale University Press, 2006.
Hoffman, Andrew J. *From Heresy to Dogma.* Palo Alto, CA: Stanford Business Books, 2001.
Kosova, Weston. "It Ain't Easy Being Green." *Newsweek* (September 21, 2009).

Alexa J. Trumpy
The Ohio State University

Corporate Social Responsibility

Corporate social responsibility (CSR) is a concept ripe for debate over its meaning, practice, and implementation. As corporations themselves have evolved, so too have ideas of social responsibility. The focus on this article is to explain the evolution of a corporation and describe its social responsibilities.

Corporations in the United States evolved out of the post-Revolution era. At that time, corporations were small and specialized, had a limited lifetime, and were able to make political contributions. During industrialization, corporations began to gain influence over the public and started acting out against the government. During the Civil War, corporations were able to gain substantial wealth from the government. In 1886, the Supreme Court decided one of the most influential cases pertaining to corporations—*Santa Clara County v. Southern Pacific Railroad*—declaring that corporations had all the rights of a person under the Fourteenth Amendment: the right to due process, free speech, and protection from discrimination.

Mergers and Unions

In the 1880s and 1890s, rapid economic expansion occurred through horizontal mergers—when two or more companies in the same market combine to create a larger company. A substantial outcome of the mergers was the completion of the transcontinental railroad. This enabled companies that were initially regional to communicate across the country.

In 1932, Franklin Roosevelt created the Securities Exchange Act to protect the country's stock and the National Recovery Administration to create minimum wages and labor standards for the country. This change was part of the New Deal, in which many different relief efforts such as the Social Security Act and the Works Progress Administration (WPA) were created. The WPA became one of the largest employers in the country during that time, building many public buildings, areas, and roads. A couple of decades later, in the 1950s, unions were created.

Legal Cases, the Rights of Corporations, and Deregulation

The next four decades (1960–2000) yielded significant rulings in legal cases that impacted the personhood of corporations. In the case *Fong Foo v. United States* (1962), corporations were deemed to be protected from double jeopardy by the federal government under the Fifth Amendment. The case *First National Bank of Boston v. Bellotti* (1978), according to www.ReclaimDemocracy.org, "Struck down a Massachusetts law that banned corporate spending to influence state ballot initiatives, even spending by corporate political action committees. Spending money to influence politics is now a corporate 'right.'" Furthermore, the ruling in *Ross v. Bernhard* granted corporations the right to a trial jury under the Seventh Amendment. The decades of the 1980s and 1990s brought about deregulation, both positive and negative manifestations. In a controversial 2010 decision (*Citizen's United v. Federal Election Commission*), the Supreme Court of the United States ruled that corporations had an unrestricted right to freedom of speech, including campaign donations.

Corporate Scandals and the Sarbanes-Oxley Act

The first two years of the new millennium brought to light the worst corporate corruption and scandals associated with fraudulent accounting practices and inflation of stock values. In the aftermath, the Sarbanes-Oxley Act set standards for all U.S. public companies to prevent further corporate corruption and deception of stake/shareholders. Implementing this act was necessary in order to restore public confidence in corporate America and to strengthen the nation's capital market. The Sarbanes-Oxley Act also put the burden of proof on organizations to demonstrate how they were going to become more socially responsible.

Corporate Social Responsibility

As the parameters of corporations were defined, scrutinized, and challenged, so, too, were business practices. For what is a corporation responsible when it is legally defined as a person who is protected under various amendments? Is a business responsible to itself to ensure its own survival? Or is a corporation responsible to customers, share/stakeholders, investors, and local and global communities? The answers to these questions are located within the approved draft (May 2010) of ISO 26000, the guidelines by the International Organization for Standardization recommended for corporate social responsibility.

Currently, organizations are being pressured by various groups and industries to strategically integrate and implement more socially responsible standards into their internal and external policies and practices: reporting financial transactions in a transparent and ethical manner; caring for the physical natural environment by offsetting carbon footprints and creating energy-efficient and sustainable products; improving the welfare of employees by offering healthcare benefits and living wages; stimulating local economies by buying locally produced food; and improving public health, among other initiatives.

See Also: Corporate Green Culture; Environmental "Goods"; Global Warming and Culture; International Law and Treaties.

Further Readings

Connor, Amy and Michelle Shumate. "We Believe: How Corporations Communicate Social Responsibility." Conference paper presented at the International Communication Association Annual Meeting, 2009.

Ihlen, Oyvind. "Talking Green: The Rhetoric of 'Good Corporate Environmental Citizens.'" Conference paper presented at the International Communication Association Annual Meeting, 2008.

Jones, Kevin and Jennifer Bartlett. "The Strategic Value of Corporate Social Responsibility: A Relationship Management Framework for Public Relations Practice." Conference paper presented at the International Communication Association Annual Meeting, 2008.

McWilliams, Abagail, Donald S. Siegel, and Patrick M. Wright. "Corporate Social Responsibility: Strategic Implications." Rensselaer Working Papers in Economics, May 2005.

Wilcox, Tracy. "Human Resource Development as an Element of Corporate Social Responsibility." *Asia Pacific Journal of Human Resources*, 44 (2006).

Cory Lynn Young
Ithaca College

Cyber Action

Cyber action or cyberactivism, the use of digital technologies and the Internet for social change, was one of the most important political innovations of the final decade of the 20th century and continues to play a key role in contemporary environmental and other movements. Cyberactivism enables activists to build global alliances and networks, linking diverse local struggles with their counterparts elsewhere. It facilitates the transnational flow of information and resources, helps activists organize local and global actions, and provides a medium for developing new protest tactics. Particular cyberactivist tools and practices include the use of electronic list serves and websites, local/global networking, electronic civil disobedience, culture jamming and guerrilla communication, temporary media centers and hacklabs, and web-based alternative media. Environmentalists have been at the forefront in the development and use of cyberactivism, which has achieved particular visibility with the rise of the global justice movement.

History and Context

Environmental and other social justice groups started using electronic newsgroups and e-mail in the late 1980s, taking to the World Wide Web in the early 1990s. More established organizations such as Friends of the Earth and Greenpeace primarily used the web for advertising and outreach, although more radical environmental organizations such as Earth First! and the UK-based Reclaim the Streets soon began using e-mail, list serves, and websites for networking and coordination of anti-road and other direct action protests. In the mid-1990s, the Zapatistas and their transnational network of supporters made high-profile use of the Internet to circulate information about their plight and the ongoing human rights situation in Chiapas, weaving together an "electronic fabric of struggle." Some of the earliest denial-of-service attacks against opponents' websites were employed against the Mexican government during this period, providing a model for subsequent electronic civil disobedience in conjunction with global justice protests.

Environmental and other anti–corporate globalization activists soon began using e-mail, listservs, and websites to plan and coordinate global days of action against institutions such as the G8, World Trade Organization (WTO), World Bank, and International Monetary Fund (IMF). The June 1999 Carnival against Capitalism, when activists targeted London and other global financial centers to protest the G8 meeting in Birmingham, and the anti-WTO protests in Seattle that November were important milestones in the use of the Internet to organize global protests. These mass global actions, which continued throughout the next decade in cities such as Quebec, Genoa, and Barcelona, and most recently in Copenhagen against the December 2009 Global Climate Summit, also provided platforms for developing and experimenting with other forms of cyberactivism: culture jamming against corporate targets, the creation of the Independent Media Center (IMC, or Indymedia), and the building of temporary media centers and hacklabs. Environmental and other social justice activists have also begun to use new social networking sites such as Facebook or Twitter to share information and coordinate and mobilize for actions.

Local/Global Networking

Cyberactivism has been perhaps used to greatest effect as a tool for sharing resources, communicating, and coordinating at multiple scales. The Internet has allowed small groups of activists with limited resources to reach out beyond their local struggles to build wider regional and global connections. This flexible, decentralized mode of organization resonates with the anarchist ideals of many radical environmental activists. Mass global justice actions have been organized through a global web of list serves and websites, while activists have also used e-mail to share logistical information about meetings, protests, and activities; hold strategic debates; and exchange information about mobilizations and events. In addition, activists have used interactive web pages during mobilizations to provide information, resources, and links; post documents and calls to action; and house real-time chats. Particular networks and more traditional organizations have their own permanent websites, which they use for publicity, outreach, and mobilization. For example, Greenpeace developed an online Cyber Center in order to forge a global community of resistance to environmental destruction, which has coordinated campaigns such as the Corporate 100 actions against global warming.

Electronic Civil Disobedience

Electronic civil disobedience (ECD) brings together activists and programmers in the development of new forms of direct action in cyberspace. Initially conceived by the Critical Art Ensemble collective as a tool to allow small groups of highly skilled digital artists, activists, and techies to achieve political goals by making innovative and strategic use of the Internet, ECD has since become more popular and mass based. Electronic Disturbance Theater first used ECD as a mass tactic in support of the Zapatistas in 1998, organizing virtual sit-ins against the Mexican government to protest the war on the Zapatistas. Floodnet software facilitates these actions by allowing participants to click on a link, sending them to a targeted website along with thousands of others, overloading the server. Collectives such as the electrohippies have since organized ECD actions in conjunction with numerous global justice and environmental protests.

In addition to virtual sit-ins, these mobilizations have also included bombarding target servers and websites with e-mails, leaving behind ironic messages or "electronic graffiti," and "hijacking," or redirecting, surfers to mock sites. For example, activists built the clone "World Trade Organization/GATT Home Page" during the anti-WTO protests in Seattle featuring mock quotes by then–WTO Director-General Mike Moore. Environmental organizations such as Greenpeace have made creative use of similar digital tactics. In June 2000, Greenpeace activists installed a webcam at the end of an underwater radioactive discharge pipe run by the French nuclear agency Cogema in order to provide live documentation. The images were broadcast on the web and at the Convention for the Protection of the Marine Environment of the North-East Atlantic in Copenhagen. Visitors to the Greenpeace website could also send real-time messages to conference delegates. As a result, the convention called on France and the United Kingdom to end their nuclear reprocessing activities.

Culture Jamming/Guerrilla Communication

Culture jamming refers to the playful parodying of corporate advertisements and logos to generate alternative messages that challenge corporate power. Pirating billboards, ironic graffiti, and altering websites are all specific instances of culture jamming. The Canadian-based *Adbusters* magazine and the associated Media Foundation provide online commentary and multimedia resources that local participants can download to facilitate their participation in anticorporate campaigns such as the international Buy Nothing Day. Guerrilla communication is a related practice involving the creative juxtaposition of incommensurate elements to generate subversive meanings. Guerrilla communication uses paradox to shatter tacitly accepted notions and to open a space for alternative formulations. Culture jamming and guerrilla communication make use of the old Situationist strategy of *détournement*, which takes well-known phrases, images, and ideas from mass culture out of context and gives them an unexpected twist, or detour, in order to create surprising, often playful combinations.

Environmental direct action organizations have employed culture jamming and guerrilla communication tactics as part of their anticorporate campaigns. For example, before the Sydney Olympics in 2000, Greenpeace developed the spoof website Coke Spotlight (Coca-Cola was an Olympic sponsor) to highlight the company's use of greenhouse gases in their refrigerants. One of the images on the site featured the traditional "Enjoy Coke" logo, which was modified to read "Enjoy Climate Change." A month after the site went live, the Coca-Cola Company committed to phasing out greenhouse gases. As with ECD, culture jamming and guerrilla communication tactics are largely symbolic, achieving victories by generating mass media attention. However, they also form part of a wider cyber-activist culture that involves the creative use of new digital technologies, decentralized coordination, the free and open exchange of ideas and information, and the combination and recombination of software and cultural codes.

Indymedia

Web-based independent media represents another key terrain of cyberactivist practice. Grassroots activists have long relied on small-scale alternative news sources and 'zines to get the word out about their organizations and activities, but the web greatly facilitates these efforts, providing a low-cost means for cheaply and quickly producing and distributing publications. Environmentalists were among the first to make use of new digital technologies in this way, as mainstream organizations developed electronic newsletters while more radical networks such as Earth First! created their own online 'zines. However, it was the emergence of Indymedia that revolutionized web-based alternative media. The first IMC was established during the anti-WTO protest in Seattle, where independent journalists reported directly from the streets, while activists in Seattle and beyond uploaded their own text and image files. The network quickly expanded into a global alternative communications network involving multimedia platforms composed of electronic print, video, audio, and photography. The global network now has more than 200 local sites and receives two million page views per day.

Local Indymedia collectives are organized along decentralized, nonhierarchical lines, involving consensus decision making, autonomous working groups, and horizontal coordination. The global Indymedia process is similarly managed through a series of transnational editorial, technical, and logistical working groups, which communicate through global e-mail lists and periodic web meetings via Internet Relay Chat technology. Central to Indymedia's organization and philosophy is open publishing software, an innovative technical system that allows activists to create and distribute their own news stories. However, as Indymedia expands—and along with it, the number and diversity of contributions, including increasing spam and racist, sexist, and/or irrelevant posts—many activists see a growing need to develop mechanisms for shaping and controlling content. Finding a balance between openness and quality remains one of the most significant challenges facing Indymedia.

Temporary Media Centers and Hacklabs

Environmental activists were also pioneers in establishing temporary media centers and computer labs to provide Internet access and digital networking tools during mass direct actions. UK-based activists created a mobile office to support anti-roads and quarry and tree protection protests during the late 1990s. At around the same time, tactical media activists and squatters began building temporary media and "hack" labs to provide physical spaces for digital networking, technological experimentation, and the free exchange of information. These experiences provided models for the later development of Independent Media Centers and temporary computer labs during global justice mobilizations.

For example, the alternative media zone established during the July 2002 Strasbourg No Border Camp, ironically called "Silicon Valley," housed an IMC, Internet café, radio tent, web-based news and radio, and a double-decker media bus called the Publix Theater Caravan, which itself offered video screening, Internet access and streaming, and a bar and lounge. The Internet café and the entire zone were outfitted with a WiFi connection, and the radio tent was equipped with a 50-watt transmitter that generated simultaneous netcasts. Cyberactivists have built similar media labs during regional and World Social Forums. During the November 2002 European Social Forum in Strasbourg, the "Euraction Hub Project" provided an open space for sharing ideas and experiences, experimenting with digital technologies, carrying out autonomous actions, and organizing in a horizontal and participatory fashion. In this sense, temporary media centers and hacklabs do not simply provide digital tools—they also reflect the participatory, antihierarchical ethic of many radical activists.

Conclusion and Future Directions

Cyberactivism has provided an important tool for environmental, global justice, and other activists to pursue their goals of a more just, equitable, and sustainable world. The Internet and related online digital technologies have allowed activists to build networks that are locally rooted, yet globally connected, to carry out novel forms of electronic civil disobedience, to develop and implement innovative culture jamming and guerrilla communication campaigns, to build more effective alternative media projects, and to provide technical support for mass gatherings and actions through temporary computer centers and hacklabs. At the same time, cyberactivism has also allowed activists to generate organizational practices and forms and to experiment with new technologies that reflect their broader political cultures and egalitarian ideals. Most recently, cyberactivists have taken to new digital media, including blogs, wikis, and social networking sites such as Facebook and Twitter. We still know little about these emerging practices and their impact, suggesting an important avenue for future research.

See Also: Blogs; Communication, Global and Regional; Communication, National and Local; Environmental Communication, Public Participation; Environmental Justice Movements.

Further Readings

Cleaver, Harry M. "*The Zapatistas and the Electronic Fabric of Struggle*" (1995). http://www.eco.utexas.edu/faculty/Cleaver/zaps.html (Accessed December 2009).

Juris, Jeffrey S. *Networking Futures: The Movements Against Corporate Globalization.* Durham, NC: Duke University Press, 2008.

Juris, Jeffrey S. "The New Digital Media and Activist Networking Within Anti-Corporate Globalization Movements." *Annals of the American Academy of Political and Social Sciences*, 597 (2005).

Lasn, Kalle. *Culture Jam.* New York: Quill, 2000.

Lovink, Geert. *Dark Fiber: Tracking Critical Internet Culture.* Cambridge, MA: MIT Press, 2002.

Meikle, Grahame. *Future Active: Media Activism and the Internet.* New York: Routledge, 2002.

Pickerill, Jenny. "Radical Politics on the Net." *Parliamentary Affairs*, 59/2 (2006).

Richardson, Joanne. *The Language of Tactical Media.* In *An@rchitexts,* Joanne Richardson, ed. Brooklyn, NY: Autonomedia, 2003.

Wray, Stefan. *On Electronic Civil Disobedience* (1998). http://cristine.org/borders/Wray_Essay.html (Accessed December 2009).

Jeffrey S. Juris
Northeastern University

D

Demonstrations and Events

Each year on April 22, citizens nationwide celebrate the green movement on Earth Day by holding festivals, staging protests, or cleaning public areas, as in this beach cleanup in San Juan National Historic Site in Puerto Rico.

Source: National Park Service

Since the inception of Earth Day in 1970, scores of events that focus on protecting and celebrating the environment have been founded and supported both across the country and around the world. Various festivals, weeklong celebrations, protests, demonstrations, specific days, and even single hours represent the passion and dedication of members in the green movement for educating citizens about their responsibilities toward the Earth and how to protect the planet. The most prominent green festivals include the Green Apple and Music Arts Festival and the Green Festival. The Green Apple and Music Arts Festival, sponsored in part by the Earth Day Network, was held annually between 2006 and 2009. The primary goal of the festival was to educate attendees about issues concerning the environment. The festival prided itself on being accessible to a large and varied audience, thus spreading knowledge about green practices to many people regardless of age, ethnicity, or socioeconomic status. Hundreds of musical artists performed at the festival over the last few years, and founder Peter Shapiro ensured the performances were green—products and practices that are environmentally friendly were implemented in both the advertising and performances of the shows. The first Green Apple Festival was held in New York. Each year, other cities around the United States also began holding simultaneous offshoots of the Green Apple Festival, including Chicago, Denver, and San Francisco, among others.

Like the Green Apple Festival, the Green Festival has been held in various cities, but on different dates. The Green Festival, according to its website, is the result of a partnership between Green America and Global Exchange. The Green Festival highlights green businesses that support Fair Trade practices, includes a series of speakers, and offers educational opportunities and a celebratory atmosphere. The festival appears in cities like San Francisco, Chicago, Seattle, and Washington, D.C.

As environmental concerns have moved to the forefront of the public consciousness, certain weeks have also been devoted to creating awareness about environmental issues, including Green Office Week, National Green Week, and National Environmental Education Week. National Green Week, sponsored by the Green Education Foundation, aims to educate students across the country about green practices. According to its website, schools wishing to participate in the National Green Week tradition select a week between February 7 and April 22 (example dates used during the 2010 academic year) to focus on discussing healthy environmental practices. Students across the nation undertake various measures to celebrate National Green Week, including using reusable containers for bringing snacks and meals to school and examining the efficiency with which their respective schools use resources like water. National Environmental Education Week, known also as EE Week, is held annually the week before Earth Day. Sponsored by the National Environmental Education Foundation, the goal of the week is to educate students ranging from preschool to high school about effective environmentally friendly practices. The foundation happily provides educators with materials and resources to use in their classrooms to promote and celebrate EE Week.

Another important weeklong celebration is Green Office Week, which began in 2009 in the United Kingdom. The focus of this week was encouraging employees to make green choices; employees who participated were given copies of *The Green Office for Dummies Guide* to help them make decisions benefiting the Earth. Though employees felt somewhat flustered after the first Green Office Week in 2009 due to a lack of resources to assist employees in being green, Green Office Week was celebrated again in 2010, this time with each day devoted to a specific theme. These themes included "Make a Start Monday" (when employees agreed to devote the week to being more environmentally conscious) and "Waste Not Wednesday," a day for which the theme is obvious. Green Office Week is championed for its efforts to bring more ecofriendly practices to the workplace; employers hope that the environmental awareness celebrated in the workplace also carries over to each employee's home.

Several days of demonstration relating to green culture in particular deserve analysis. The most noted and celebrated day devoted to the green movement is Earth Day, held each year on April 22. This tradition began in 1970, founded by Senator Gaylord Nelson, a Democrat from Wisconsin. Scholars attribute the first Earth Day as the inception for the environmental movement in the United States and around the world. People honor Earth Day in various ways, ranging from adults staging protests or cleaning rivers and lakes to schoolchildren who make arts and crafts celebrating the beauty of Earth.

Perhaps the other most significant day devoted to creating awareness about green issues and the Earth is World Environment Day, which takes place on June 5. Though the day is acknowledged and honored throughout the world, each year a different city hosts the primary celebrations. Past cities honored with sponsoring World Environment Day include Bangkok, Thailand (1987); Beijing, China (1993); Havana, Cuba (2001); Beirut, Lebanon

(2003); Mexico City, Mexico (2009); and Kigali, Rwanda (2010), among many others. Each year, organizers emphasize a different theme concerning the environment, like desertification, global warning, the status of drinking water, the melting of the polar ice caps, and the future of the environment, just to name a few.

Other significant days of action include World Day for Water (March 22); World Wetlands Day (February 2); United Nations World Environment Day (June 5); Arbor Day (varies by state); and days concerning transportation methods like Bike to Work Day and Car Free Day (both vary).

Demonstrations lasting a single hour also exist, and have made an impact on green awareness throughout the world. Earth Hour, a ceremony started in 2007, occurs late in the month of March (the date varies each year). The idea originated in Sydney, Australia, on March 31, 2007, when millions of Australians turned off their lights to make a statement about climate change and control. With each passing year, millions more people celebrate Earth Hour. Over 100 countries participated in Earth Hour 2010.

Some demonstrations are held specifically in response to practices that have negative environmental consequences. Protests in response to G-20 summits, the Kyoto Protocol, and the World Trade Organization Ministerial scheduled to meet in Seattle also mark significant demonstrations in an effort to protect the Earth. Protests occur regularly at G-20 summits. These summits gather the finance ministers, heads of state, and bank governors from the European Union and the 19 countries in the world whose economies account for the vast majority of trade in the world, and who represent the majority of the Earth's population. Since the primary goal of these summits is to settle on and employ economically beneficial trading practices, environmentalists commonly protest G-20 summits for the gathering's lack of concern for the planet and how trade practices, no matter how beneficial they are in a monetary sense, disregard the health of the Earth. The 2009 G-20 London summit led a protest march called "March for Jobs, Justice, and Climate." Environmental protesters argued that if the G-20 agreed to spend billions to bail out financial institutions, the G-20 should also be able to devote money to addressing various societal concerns, especially those affecting the environment. George Monbiot of the United Kingdom's *The Guardian* claimed that "the G20 leaders appear to have decided to deal with [environmental] problems only when they have to—in other words, when it's too late."

Other significant political protests include those in response to the Kyoto Protocol and the World Trade Organization Ministerial scheduled to meet in Seattle. Major protests occurred in reaction to the Kyoto Protocol, a treaty focused on combating the problem of global warming. According to John Vidal and Terry Macalister of *The Guardian*, after the treaty was enforced in 2005, a group of Greenpeace activists arrived at the London energy exchange carrying a banner stating "Climate change kills. Stop pushing oil." Oil trading proponents injured protesters by toppling a metal bookcase on top of them, while kicking and hitting others. The Greenpeace activists approached the situation peacefully but their protest was met with unhesitating violence. In 1999, the World Trade Organization Ministerial, scheduled to meet in Seattle, was met with protesters who criticized the WTO for abusing the environment in favor of free trade practices. The demonstrations began on November 29, 1999, when protesters packed the streets so tightly that members of the WTO were unable to leave their hotel rooms and make their way to meetings. The police used tear gas and rubber bullets, among other means, to attack protesters.

While the duration and the focus of the event may vary from a week to a day to a single hour, the past 40 years have witnessed the development of numerous demonstrations and celebrations, beginning with the inception of Earth Day in 1970. The sponsors and organizers of the events may differ, but the goal of these events remains shared: to educate citizens about environmentally friendly practices and to honor the Earth.

See Also: Environmental Justice Movements; Grassroots Organizations; Individual Action, Global and Regional; Individual Action, National and Local; Institutional/Organizational Action; Nongovernmental Organizations; Organizations and Unions; Social Action, Global and Regional; Social Action, National and Local.

Further Readings

Earth Day. http://www.earthday.org (Accessed September 2010).
Earth Hour. http://www.earthhour.org (Accessed September 2010).
Green Apple Music Festival. http://www.greenapplemusicfestival.com (Accessed September 2010).
"Green Festivals." Green America: Come Together. http://www.greenamericatoday.org/greenbusiness/greenfestivals.cfm (Accessed September 2010).
Green Office Week. http://www.greenofficeweek.eu (Accessed September 2010).
Monbiot, George. "G20 Forgets the Environment." *The Guardian*. http://www.guardian.co.uk/environment/georgemonbiot/2009/apr/02/1 (Accessed November 2010).
"National Environmental Education Week." National Environmental Education Foundation. http://www.eeweek.org (Accessed September 2010).
"National Green Week." Green Education Foundation. http://www.greeneducationfoundation.org (Accessed September 2010).
Vidal, John and Terry Macalister. "Kyoto Protests Disrupt Oil Trading." *The Guardian* (February 17, 2005). http://www.guardian.co.uk/environment/2005/feb/17/activists.climatechange (Accessed November 2010).
"World Environment Day." United Nations Environment Programme. http://www.unep.org/wed/2010/english/about.asp (Accessed September 2010).

Karley Adney
Independent Scholar

DiCaprio, Leonardo

Born in 1974 in California, Leonardo DiCaprio is an American actor and a film producer. DiCaprio made his breakthrough in films in 1992 in *This Boy's Life*. Since then, he has received multiple awards for his performances, including the Golden Globe Award for *The Aviator* (2004) and the Silver Berlin Bear for *Romeo and Juliet* (1996). The success of the film *Titanic* triggered a social phenomenon dubbed "Leo-Mania" among teenage girls and young women in general.

His contribution to environmental issues can be divided into two areas: awareness raising and support of politicians. Through the media and the Internet, he has been successful in raising awareness among individuals, communities, and business leaders as well as encouraging them to take actions. Meanwhile, he has been trying to achieve a sustainable society by supporting Democratic politicians.

Awareness-Raising Activities

In 1998, DiCaprio established the Leonardo DiCaprio Foundation to foster awareness of environmental issues. Collaborating with organizations such as the Natural Resources Defense Council, Global Green USA, and the International Fund for Animal Welfare, DiCaprio and his foundation have committed to a wide range of environmental issues, such as global warming, biodiversity, and nuclear weapons.

DiCaprio and his foundation have been active in reaching out to individuals, communities, and business leaders by various means. The foundation created a website in both English and Spanish to foster awareness of environmental issues. In 2003, the Natural Resources Defense Council opened a new green building, featuring the Leonardo DiCaprio e-Activism Computer Zone. At the event, hosted by the Natural Resources Defense Council and the Rolling Stones, he showed his commitment to the global warming issue by saying: "Thousands of climate scientists agree that global warming is not only the most threatening environmental problem, but one of the greatest challenges facing all of humanity."

He was one of the first film stars to drive a hybrid car. He drove a Toyota hybrid car to the Academy Awards ceremony in 2005, organized by Global Green USA, to highlight the urgency for personal and political actions to address the global warming issue.

In 2006, collaborating with Tree Media Group founders, DiCaprio wrote and produced a documentary film, *The 11th Hour*, concerning the environmental crises caused by human actions. In the film, he examines the issue of global warming and offers "visionary and practical solutions" for reversing the harmful effects of global warming and restoring many of Earth's fragile ecosystems. Not only to raise awareness about the issue but also to urge everyone to take actions for a sustainable environment, the film was followed by a blog to form a social action network among individuals and communities worldwide. Besides the online public forum, 11th Hour Action was also launched in 2008. It brings the film to campuses, aiming at engaging and activating students in taking leadership in a sustainability movement. These events have taken place on various campuses in the United States and Spain.

DiCaprio has also been trying to reach out to business communities. In 2010, he honored TAG Heuer, intending to create a business model to protect the environment so that other corporations would "follow suit."

DiCaprio and his foundation have received various awards for their initiatives—the Martin Litton Environmental Warrior Award by Environment Now in 2001 and the Environmental Leadership Award from Global Green USA in 2002, to name two. In 2010, he was recognized by VH1, a New York–based cable television network owned by the MTV network. VH1's Do Something Award honored DiCaprio for his activities to empower and to inspire young people. The DiCaprio Foundation joined the California Community Foundation (CCF) in early 2008 and is now known as the Leonardo DiCaprio Fund at CCF. It continues to support environmental causes through grants and active participation.

Support of Politicians

DiCaprio has lobbied extensively for reducing greenhouse gases. He is also a strong advocate of protecting wilderness areas. In his speech at the 2002 Environmental Leadership Award, he criticized the Department of the Interior for not protecting permanent wilderness, calling it "a reversal of four decades of wilderness policy." In the same speech, he also criticized the Senate Committee on Energy and Natural Resources for rejecting Senator Dianne Feinstein's proposal on a Corporate Average Fuel Economy (CAFE) amendment to apply federal energy-efficiency standards to trucks and SUVs.

In 2007, DiCaprio confirmed his commitment to the global warming issue, appearing with former Vice President Albert Gore, Jr., at the Oscars ceremony. They announced that the Academy Awards had incorporated environmentally intelligent practices throughout the planning and production process. Recently, he joined the Natural Resources Defense Council Action Fund's This Is Our Moment Campaign to lobby the Clean Energy Jobs and American Power Act through the Senate.

DiCaprio actively campaigned for Democratic candidates during the past two presidential elections: John Kerry in 2004 and Barack Obama in 2008. Both candidates committed to promoting renewable energy sources and increasing CAFE standards to reduce U.S. dependency on foreign oil. In both elections, he made significant contributions as an individual.

Besides being an active environmental advocate, DiCaprio is a renowned philanthropist addressing issues of poverty and hardship. Through Global Green USA, he urged U.S. political leaders to attend the Earth Summit in South Africa in 2002. During the filming of *Blood Diamond* (2006), DiCaprio spent time with orphaned children in Mozambique, and following the devastating 2010 earthquake in Haiti, he donated $1 million for relief.

See Also: Alternative/Sustainable Energy; Energy; Environmental Law, U.S.; Global Warming and Culture; Gore, Jr., Al; Hybrid Cars; Individual Action, Global and Regional; Individual Action, National and Local; Media Greenwashing; Popular Green Culture; Social Action, Global and Regional; Social Action, National and Local.

Further Readings

11th Hour Action. http://11thhouraction.com/node/1369 (Accessed September 2010).
Leonardo DiCaprio Eco-Site. http://www.leonardodicaprio.org (Accessed September 2010).
This Is Our Moment. http://www.nrdcactionfund.org/thisisourmoment (Accessed September 2010).

Kayo Ikeda
Hiroshima University

Diet/Nutrition

The "green diet" revolution of the 21st century is associated with issues surrounding food safety, sustainable agriculture, the "back to the land movement," obesity, and alternatives to technologies, including genetic modification of both seeds and animals. It is a response to the growth of conventional agriculture, the industrial approach that burgeoned after World War II. From a small niche market arising from a backlash to conventional agriculture that gained momentum from the environmentalist movement in the 1960s and 1970s to what has become mainstream to our understanding of food, green diets and nutrition have grown to a prominent place in our culture.

The introduction of processed foods in the mid-19th century, combined with overeating, caused higher obesity levels. To bring attention to the benefits of natural and organic foods, the first White House vegetable garden since World War II was planted by First Lady Michelle Obama, White House Chef Sam Kass, and students from Bancroft Elementary.

Source: White House photo

Baby boomers of the 21st century need only look back to Rachel Carson's 1962 *Silent Spring* as the catalyst for the modern green diet movement, and the teens of today to Michael Pollan's 2006 *Omnivore's Dilemma*. But until the Industrial Revolution, a green diet was the usual diet. As early as 1580, Thomas Tusser wrote *Five Hundred Points of Good Husbandry*, which included information about the organic farming principle of crop rotation to ensure fertility of the land.

Post–World War II, conventional agriculture became the norm in our culture. This type of agriculture can be described as large scale, monoculture plantings, capital intensive, and overusing of fertilizers, herbicides, and pesticides. During the growth of the industrial approach, there were outliers with a small following of devotees. In the 1940s, Sir Albert Howard and Jerome Rodale were strong critics of what has become "conventional." They wrote about and lived an organic agriculture lifestyle. And, as with any cultural movement, they were a catalyst for polarization of conventional versus alternative agriculture views. During this period, the conflict was not really about eating food, but about producing it. But by the 1950s, the organic movement seemed to be asleep.

Silent Spring (1962) awoke a new generation of antiestablishment activists desiring a greening of agriculture. The new conflict still was not about eating food; this time, it was about not harming the environment. The year 1970 brought us Earth Day and *Mother Earth News,* and people were looking toward alternatives to technology, self-sufficiency, alternative lifestyles, and growing their own food. Green diets and nutrition were almost a by-product. With the publication of Frances Moore Lappé's *Diet for a Small Planet* (1971) and Mollie Katzen's *Moosewood Cookbook* (1977), the vegetarian movement became a symbol of the greening of the American diet. Yet green diets remained an alternative niche.

Polarization between alternative and conventional agriculture and the foods that resulted grew even bigger in the 1970s, compared to the 1940s and 1960s. Earl Butz, U.S. Secretary of Agriculture (1971–76) pushed a "big business" agriculture agenda. With the support of national agricultural policy, the number of U.S. farms decreased and individual farm size increased until 1990, with a resulting increase in the production of major commodities, including corn and soybeans. At the same time, a quiet proliferation of publications about alternatives to the industrial approach to agriculture in the mid- to late 1970s from academic, popular, and government sources were being published.

The impact of the industrial approach to agriculture—based on technology and efficiency—on the American diet was an increase in processed food. The 1940s through the 1960s marked the age of technological innovation in processed foods, with such introductions as Wonder bread, Tang, Cool Whip, and Minute Rice—not a typical list of green foods. The 1980s brought a host of changes in ingredients and a proliferation of brand extensions. Between 1982 and 1991, manufacturers brought to market over 90,000 new food products. Condiments, candy, gum, snacks, bakery products, and soft drinks led the way.

The long list of new product introductions contained one or both of two ingredients: sugar and fat. The farm policies of the 1970s, combined with an increase in free trade, introduced cheap ingredients into food manufacturing. These ingredients found their way into almost every processed food product. By the early 1980s, for example, both Coca-Cola and Pepsi-Cola switched from a 50–50 blend of cane sugar and corn syrup to 100 percent corn syrup. Prices fell and demand for processed foods soared. The fast food industry responded with "super sizes," and portion sizes began to grow in the 1970s, continuing through the turn of the century. Between 1976 and 2000, the percent of overweight Americans grew from 45 to 65 percent. Not only were Americans eating too much, they were eating too many processed and restaurant foods. The foods purchased at restaurants are reported to be at least 1.4 times higher in calories compared to foods prepared at home; they are also higher in sodium and saturated fat.

During the past 10 years, a food revolution has been gaining momentum again. This time it is backed not only by organic production and environmentalists, but also by the health of our nation. One way to assess the steam of a revolution is to look at the heat of the arguments. The Agricultural Productivity Act of 1985 perhaps started the new movement by mandating the development of low-input sustainable agriculture (LISA). Big industry responded with alternative names to call LISA—LILO (low input low output) and FIDO (few inputs declining outputs—"a real dog"). A host of best-selling books and documentaries were introduced to the general consumer, including *Fast Food Nation*, *Super Size Me*, *Food Politics*, *Food Inc.*, and *King Corn*. Organic farmland acreage more than doubled from 1997 to 2005. The National Organic Standards were passed in 2002. By 2006, 46 percent of organic foods were being sold in conventional supermarkets compared to 7 percent in 1991. And compared with 3 percent in 1999, in 2008, 68 percent of consumers said they bought organic foods at least occasionally.

Consumption of organic foods is only one indication of a move to green diets and nutrition. There has been growth in community supported agriculture (CSA) farms. The U.S. Department of Agriculture/National Institute for Food and Agriculture (USDA/NIFA) reported 12,549 CSAs in 2007, compared with 60 in 1990. These farms sell shares to consumers who receive a bundle of unprocessed agricultural products ranging from fruits and vegetables, meat and eggs to minimally processed products, including bread and cheese. There is also a growing farm to school (FTS) movement. By June 2010, almost 9,000 schools were involved in an FTS program. These programs link schoolchildren to locally produced agricultural products through classroom, cafeteria, and community activities. Organizations like Slow Food, with a mission to "counteract fast food and fast life, the disappearance of local food traditions and people's dwindling interest in the food they eat, where it comes from," are growing. Furthermore, 2009 brought the planting of the first White House vegetable garden since World War II, an organic garden spearheaded by First Lady Michelle Obama. Locavore groups that support sustainable food production, processing, distribution, and consumption are forming all over the country. The *New York Times* has even asked the question, "Is this a food revolution?" Indeed, it just may be. After all, *locavore* was the Oxford Dictionary word of the year in 2008.

See Also: "Agri-Culture"; Film, Documentaries; Locavores; Obesity and Health; Organic Foods.

Further Readings

Beus, Curtis and Riley Dunlap. "Conventional Versus Alternative Agriculture." *Rural Sociology*, 55/4 (1990).
Carson, Rachel. *Silent Spring*. Boston, MA: Houghton Mifflin, 1962.
Coffin, Raymond and Mark Lipsey. "Moving Back to the Land." *Environment & Behavior*, 13/1 (1981).
Critser, Greg. *Fat Land*. Boston, MA: Houghton Mifflin, 2003.
Farm-to-School. http://www.farmtoschool.org (Accessed June 2010).
Feenstra, Gail. "Creating Space for Sustainable Food Systems." *Agriculture and Human Values*, 19/2 (2002).
Gallo, Anthony. "Record Number of New Products in 1991." *Food Review*, 15/2 (1992).
Hanover, L. Mark and John White. "Manufacturing, Composition, and Applications of Fructose." *American Journal of Clinical Nutrition*, 58/5 (1993).
Jacob, Jeffrey and Merlin Brinkerhoff. "Alternative Technology and Part-Time Semisubsistence Agriculture." *Rural Sociology*, 51/1 (1986).
Katzen, Mollie. *The Moosewood Cookbook*. Berkeley, CA: Ten Speed Press, 1977.
Kenner, Robert (Director). *Food, Inc.* T-Sign Studios, 2009.
Knorr, Dietrich and Tom Watkins, eds. *Alterations in Food Productions*. New York: Van Nostrand Reinhold, 1984.
Lappé, Frances Moore. *Diet for a Small Planet*. New York: Ballantine Books, 1971.
National Agricultural Library, U.S. Department of Agriculture. http://www.nal.usda.gov (Accessed June 2010).
National Agricultural Statistics Service, U.S. Department of Agriculture. http://www.nass.usda.gov (Accessed June 2010).
Nestle, Marion. *Food Politics*. Berkeley: University of California Press, 2002.
Pollan, Michael. *The Omnivore's Dilemma*. New York: Penguin Press, 2006.
Rolls, Barbara. "The Supersizing of America." *Nutrition Today*, 38/2 (2003).
Schlosser, Eric. *Fast Food Nation*. Boston, MA: Houghton Mifflin, 2001.
Slow Food. http://www.slowfood.com (Accessed June 2010).
Spurlock, Morgan (Director). *Super Size Me*. USA: Sony Pictures, 2004.
Tusser, Thomas. *Five Hundred Points of Good Husbandry*. London: H. Denham, 1580.
U.S. Department of Agriculture Economic Research Service. http://www.ers.usda.gov (Accessed June 2010).
Woolf, Aaron (Director). *King Corn*. USA: Balcony Releasing, 2007.
Young, Lisa and Marion Nestle. "The Contribution of Expanding Portion Sizes to the U.S. Obesity Epidemic." *American Journal of Public Health*, 92 (2002).

Jane Kolodinsky
University of Vermont

e^2 *DESIGN*, TV SERIES

The Public Broadcasting Service (PBS) e^2 *design* television series showcased building and design projects under the sustainable design paradigm. The purpose of the series was to show that sustainable architecture and design solutions could ensure planet livability for future generations. The episodes served as case studies of solution-based architecture and design completed by innovators and thought leaders in the sustainable design industry. The primary objective of the series was optimistic and informative documentary-style storytelling intended to alter the viewers' sustainability paradigm. Viewers were challenged to ask questions regarding the role of architecture and design in maintaining planet livability for future generations.

The Producers

The e^2 *design* series was made for PBS by the production company Kontentreal and was underwritten by Autodesk. The New York–based Kontentreal is a documentary film and strategic entertainment company started in 2003 by advertising veteran Karena Albers and filmmaker/cinematographer Tad Edward Fettig. Autodesk is a leader in global design software for architects, designers, and engineers that supported the e^2 series because of the company's commitment to sustainable design.

About the e^2 *design* Series

The e^2 *design* episodes are part of the e^2*: The Economies of Being Environmentally Conscious Series*. The 18 e^2 *design* episodes run 25 minutes and are narrated by actor and sustainable design advocate Brad Pitt. In addition to the e^2 *design* episodes, e^2 *energy* was produced in 2007, and e^2 *transport* was produced in 2008 and narrated by actor Morgan Freeman. In 2010, season four includes e^2 *Middle East,* and the episode(s) examines the global impact of the region.

Each episode in the series makes the assumption that humanity as a whole can improve the current social, cultural, economic, and environmental issues through the implementation of sustainable design solutions. The series objective is to inform and inspire individuals by showcasing innovative ideas, initiatives, and leaders of the sustainable design paradigm.

The documentary style of "solution-based storytelling" is used to foster problem awareness without inciting despair. Viewers leave each episode with examples of large and small successful implementations of human ingenuity.

e² design Season One (2006)

Season one episodes included "The Green Apple," "Green for All," "The Green Machine," "Gray to Green," "China: From Red to Green," and "Deeper Shades of Green." The six episodes in the first season introduced the viewer to the three main objectives of the *e² design* series: the first objective was to interview influential thinkers in the sustainable design movement. In the episode "Deeper Shades of Green," three leaders articulate their beliefs and practices surrounding sustainable design. Architect Ken Yeang explains that intelligent or smart design involves technologies never seen by the user, such as an escalator that stops when not in use. Architect William McDonough explains his "Cradle-to-Cradle" philosophy, which promotes an open life cycle for all products to either return to soil or return to industry. Architect Werner Sobek promotes flexible use of space and repurposed products. The second objective of the series was to showcase solutions from around the globe to demonstrate the variety of methods used to achieve the same goal of sustainable design. The episode "Green for All" examines cultural values in relation to sustainable design in rural communities in Mexico. The third and most prevailing objective of season one was to showcase innovative architecture and design solutions that have been implemented. For instance, the episode "Gray to Green" followed the challenges surrounding a house design and build that used repurposed materials from a demolition site.

e² design Season Two (2007)

Season two episodes included "The Druk White Lotus School," "Greening the Federal Government," "Bogotá: Building a Sustainable City," "Affordable Green Housing," "Adaptive Reuse in the Netherlands," and "Architecture 2030." Season two followed the same overall objectives as season one with the addition of showcasing large-scale urban planning initiatives. The episode "Architecture 2030" examined a call-to-action initiative envisioned by architect Ed Mazria. The innovative nonprofit, nonpartisan, and independent organization Architecture 2030 actively promotes that all building practices should be carbon neutral by 2030. The episode "Bogotá: Building a Sustainable City" examined Mayor Enrique Peñalosa's drastic urban planning initiative to sustainably revitalize the busy metropolis of Bogotá.

e² design Season Three (2008)

Season three episodes included "A Garden in Cairo," "The Village Architect," "Melbourne Reborn," "The Art and Science of Renzo Piano," "New Orleans," and "Super Use." Season three incorporates the same objectives as the two previous seasons plus the addition of sustainable design and building implemented by the leaders and citizens of a community. The "New Orleans" episode focused on the efforts of community leaders in New Orleans, Louisiana, to rebuild after the 2005 environmental disaster Hurricane Katrina.

Series Summary

The optimism and solution-based storytelling showcased in each *e² design* series episode allows viewers to believe that sustainable design is feasible and that it will make a significant impact on the health and well-being of people and the planet.

See Also: Film, Documentaries; Popular Green Culture; Television, Cable Networks; Television Programming.

Further Readings

Architecture 2030. http://www.architecture 2030.org (Accessed August 2010).
Kontentreal. http://www.kontentreal.com (Accessed August 2010).
PBS. "e²: The Economies of Being Environmentally Conscious." http://www.e2-series.com (Accessed August 2010).

Megan Lee
University of Georgia

EARTH FIRST!

Since its original founding in 1980, Earth First! has played a unique, albeit controversial, role in the U.S. and global environmental movement, highlighted by its distinctive philosophical orientation, organizational structure, and political strategies of direct action and civil disobedience. Earth First! has sought from the beginning to set aside ecological preserves according to the science of conservation biology, but did not push this agenda meekly and was defined particularly by monkey-wrench tactics and the Federal Bureau of Investigation (FBI) label as an ecoterrorist association in its initial decade of existence.

Earth First! became recognized by its unlawful protest tactics, including tree sits, chaining themselves to or sabotaging equipment, cutting down billboards, and tree spiking. Pictured are Earth First! activists in 1988.
Source: Caroline Evans/National Park Service (NPS.gov)

Key splits within Earth First! in the early 1990s precipitated a change of direction, leading to less radical practices. The group officially neither advocates nor condemns monkey-wrenching today, for example, and FBI attention has largely shifted to the more radical splinter group Earth Liberation Front. Yet the Earth First! philosophy continues to rest fundamentally upon the ideas of deep ecology. This approach, as articulated by those such as Bill Devall, Arne Naess, and George Sessions, argues for a biocentric instead of anthropocentric philosophy, thus positioning ecological interests as more important than any human concerns. Campaigns aim to preserve ancient forests, protect endangered species, prevent urban sprawl, and ensure sustainability of indigenous cultures. Logging as well as grazing and mining, particularly on public lands, draw the lion's share of their attention.

Despite adapting and evolving over the years, Earth First! retains an overt disdain for corporate organizational structure and boasts no hierarchical pattern of leadership. Indeed, it exists not as an organization per se but rather a loose, autonomous collection of several thousand activists. Groups exist in the United States, Canada, Mexico, and United Kingdom, and in at least a dozen other nations, ranging from Australia and the Philippines to Spain and Slovakia. Within the United States, group examples include Blue Ridge EF! in Virginia, Everglades EF! in Florida, Chicago EF!, and California-based associations such as Earth First! Humboldt.

Cofounded in 1980 by several mainstream, Washington, D.C.–based environmental group staffers, most notably David Foreman and Mike Roselle, Earth First! was a product of their personal frustration with what they saw as the willingness of groups like The Nature Conservancy and the Wilderness Society to compromise in order to influence policy. The specific spark that raised their ire was the 1979 U.S. Forest Service Roadless Area Review and Evaluation (RARE II) that opened 36 million acres to logging in national forests. Foreman and Roselle believed mainstream environmental groups sold out their ecological principles during the review and advocated a return to grassroots pressure to counterweight these top-heavy mainstream groups inside the Beltway. Inspired by novelist Edward Abbey, the group adopted his monkey-wrench tactics as a step beyond mere civil disobedience and trumpeted the motto "No compromise in defense of Mother Earth!"

In its early years, Earth First! offered a mix of wilderness proposals based on conservation biology with innovative publicity stunts. Its big "coming out" party, so to speak, was a 1981 Glen Canyon Dam operation where a giant roll of black polyurethane created the illusion that the dam was leaking and generated national attention. Other examples of unlawful tactics include staging tree sits, chaining themselves to earthmoving equipment, sabotaging bulldozers, pulling out survey stakes, cutting down billboards, and blocking roads for logging or other development operations. Tree spiking, or inserting either metal or ceramic spikes into trees (without harming the tree) to destroy mill saws, was another popular tactic, and quickly became a signature one after its 1984 introduction. But after a tragic mill accident nearly decapitated a logger in 1987, several Earth First! groups began to seriously reconsider the merits of this tactic, and in 1990, Northern California Earth First! renounced the practice altogether. While several groups followed suit, the practice did continue among a number of others.

Around 1987, increasing radicalization attracted a number of new members into Earth First!, including some with Leftist, even anarchist, backgrounds. This counterculture presence was highlighted by events such as flag burnings and puke-ins at shopping malls, events that Foreman later pointed to as instrumental in pushing him to exit the group. In fact, by 1990, two factions had emerged, one represented by Foreman and his emphasis on deep ecology as a tool to save as much biodiversity as possible and the other represented by Roselle and his emphasis on Earth First!ers as social activists. Finally, a West Coast faction led by Judi Bari and Roselle joined with the newer, more radical individuals within Earth First!, while much of the old guard sided with Foreman and ended their Earth First! ties. Foreman, in fact, moved on to cofound a new group, Wildlands Project (now Wildlands Network), in 1991. Part of what also drove this political realignment was the

1990 arrest of Foreman by the FBI. While Foreman's arrest was connected to an operation to down transmission power lines from a nuclear power station in Arizona, he was only charged with conspiracy in the operation and pled guilty to a misdemeanor for giving two signed copies of his book *Ecodefense* to an FBI informant. By 1992, pressure to mainstream Earth First! forced a second rupture, this one beginning in the United Kingdom, with the more radical elements branching off to form Earth Liberation Front (ELF).

Today, the overall effectiveness of Earth First! remains contested. Proponents point to specific success stories postponing logging of Oregon old-growth Douglas firs and California coastal redwoods as well as the basic fact that Earth First! tactics raise the costs developers face in logging, mining, and ranching. Advocates also assert Earth First! attracts previously detached cohorts of environmentalists and fosters informal networking with events such as the annual weeklong campout known as the Round River Rendezvous, which features workshops, campaign discussions, and speakers in addition to musical performers. Finally, perhaps strategically just as significant, Earth First!ers point out that their very existence makes other environmental associations appear all the more reasonable. On the other hand, critics emphasize the negative by-product of monkey-wrench tactics, in particular arguing Earth First! has alienated too many mainstream environmentalists and engendered backlash that otherwise never would have developed if softer, more compromising tactics had been employed.

See Also: Biodiversity Loss/Species Extinction; Demonstrations and Events; Earth Liberation Front (ELF)/Animal Liberation Front (ALF); Ecocriticism; Environmental Justice Movements; Forest Service; Green Anarchism; Sierra Club, Natural Resources Defense Council, and The Nature Conservancy.

Further Readings

Davis, John, ed. *The Earth First! Reader: Ten Years of Radical Environmentalism.* Layton, UT: Gibbs Smith, 1991.

Devall, Bill and George Sessions. *Deep Ecology: Living as if Nature Mattered.* Layton, UT: Gibbs Smith, 2001.

Earth First! http://www.earthfirst.org (Accessed August 2010).

Earth First! The Radical Environmental Journal. http://www.earthfirstjournal.org (Accessed August 2010).

Federal Bureau of Investigation. "The Threat of Eco-Terrorism." Testimony of James F. Jarboe, Domestic Terrorism Section Chief, Counterterrorism Division, FBI before the House Resources Committee, Subcommittee on Forests and Forest Health, February 12, 2002. http://www.fbi.gov/congress/congress02/jarboe021202.htm (Accessed August 2010).

Foreman, Dave. *Confessions of an Eco-Warrior.* New York: Harmony Books, 1991.

Foreman, Dave. *Ecodefense: A Field Guide to Monkeywrenching.* Tucson, AZ: Ned Ludd Books, 1985.

Lee, Martha F. *Earth First! Environmental Apocalypse.* Syracuse, NY: Syracuse University Press, 1995.

Roselle, Mike. *Tree Spiker: From Earth First! to Lowbagging: My Struggles in Radical Environmentalism.* New York: St. Martin's Press, 2009.

Zakin, Susan. *Coyotes and Town Dogs: Earth First! and the Environmental Movement.* New York: Viking, 1993.

Michael M. Gunter Jr.
Rollins College

Earth Liberation Front (ELF)/ Animal Liberation Front (ALF)

The Earth Liberation Front (ELF) and the Animal Liberation Front (ALF) are two activist groups that are found mostly in the United States and western Europe. Overall, they are distinct organizations that are concerned with human involvement with animals (ALF) and the environment (ELF), yet they have many similarities. On the whole, the ELF and the ALF and their members are critical of both humans and society and believe they have put their needs and desires above and beyond all other living and nonliving components of the Earth. To remedy this, both groups use direct action tactics that aim to bring awareness to many self-defined injustices that they view as detrimental to humans, to other living animals, and to the Earth. Both groups are controversial in the United States and abroad in that some view the groups as revolutionary, whereas others have labeled group members as terrorists.

The origins of the ALF are multifaceted in that there is no specific individual or event that is credited with its formation. Its beginning is linked to the early 1960s and the actions of a journalist who organized a group whose aim was to sabotage an annual deer hunting event in the United Kingdom. This group enjoyed initial success and popularity, but soon fell to infighting among key group members. Some of these members broke away and created a new group (first known as the Band of Mercy, which was later renamed the ALF) that supported the use of more controversial tactics, such as property damage, during its activities. Overall, members of the ALF are highly critical of the way in which humans have treated and used animals throughout society. In particular, they condemn the use of animals for food, in laboratories, and for any other profit-minded activity (such as zoos or the circus). The ALF views these activities as speciesism (the assumption that humans are superior to all other living things) and believes that all other living animals should not be considered the property of human beings.

The origin of the ELF is still disputed today, due to the secrecy built into the group membership, but generally the founding is attributed to the early 1990s in the United Kingdom when several former members of the Earth First! movement decided to form an alternative group. These members felt that direct action techniques were better suited to saving the environment than Earth First!'s commitment to use only legal means of protest. Overall, members of ELF (known as "elves") believe that humans have lived in ways that have caused much detriment to the Earth and the chance for survival of all other living things. To combat this, elves support biocentrism (the inherent worth of all living and nonliving things) and deep ecology (the harmonious symbiosis between all that exists on Earth) as ways for humans to interact with the world in more a sustainable way.

To achieve both their stated goals, the ALF and the ELF have several similarities when it comes to organizational structure, tactics employed, and communication techniques. Both groups are organized through the technique known as leaderless resistance. Originating from the radical right (e.g., the Ku Klux Klan and Aryan Nations movements), the basic premise behind this type of organization is that there is no centralized authority or "leader" who dictates orders to the rest of the organization. Instead, the ALF and the ELF are made up of numerous small autonomous groups ("cells") whose aim is to follow a set of general guidelines put forth by either the ALF or the ELF. The guidelines call for members to (1) inflict economic damage on those profiting from the exploitation of animals (ALF) or the environment (ELF); (2) reveal and educate society about these systems of exploitation; and (3) take all necessary precautions to prevent harm to any living organism (human and nonhuman). Both groups maintain that this approach toward organization offers several advantages. First, it allows each cell to be autonomous and free to fulfill the general guidelines creatively in any way it sees fit. Second, since many of the actions taken are illegal, the lack of a central command creates numerous difficulties for law enforcement to infiltrate and/or prosecute any members of either group.

Both the ALF and the ELF have a deep commitment to the philosophy of direct action. Direct action refers to activities undertaken by a group used to achieve a stated goal that is outside the social norms in a given society. There are many types of direct action, such as strikes and sit-ins, but those favored by the ALF and the ELF are vandalism and other forms of sabotage (such as arson and bombings). Each cell picks a target that it believes is exploiting animals (ALF) or the environment (ELF) and takes some form of direct action against the target. For example, the ALF has previously targeted labs that use animal testing and farms that supply furs to coat makers; the ELF has targeted logging operations, ski areas, and SUV auto dealerships. To achieve the third guideline, upon completing an action against a target, both groups make their involvement known through the use of communiqués. This occurs by members sending communications to predetermined media outlets stating the action taken and discussing why they did it.

Both the ALF and the ELF are controversial groups wherever they strike. Labels assigned to both groups have ranged from "revolutionaries" to "terrorists." Much of the controversy stems from the use of direct action tactics, especially arson and bombings. Supporters of both organizations argue that they are raising awareness of many injustices in the world while putting great effort into actions to assure that no one (human or nonhuman) is hurt during the action. Opponents of the ALF and the ELF see their actions as terroristic and argue that they have mindlessly destroyed property and cost property owners millions of dollars. To date, no actions claimed by the ALF and the ELF have been linked to the death of a human being.

See Also: Alternative Communities; Animals (Confinement/Cruelty); Earth First!; Individual Action, National and Local; Social Action, National and Local.

Further Readings

Animal Liberation Front. http://www.animalliberationfront.com (Accessed September 2010).

Earth Liberation Front. http://earth-liberation-front.org (Accessed September 2010).

Joose, Paul. "Leaderless Resistance and Ideological Inclusion: The Case of the Earth Liberation Front." *Terrorism and Political Violence*, 19/3 (2007).

Leader, Stephen H. and Peter Probst. "The Earth Liberation and Environmental Terrorism." *Terrorism and Political Violence*, 15/4 (2003).

Justin Schupp
The Ohio State University

Ecocriticism

Reverberating from various academic epicenters is the claim that Western culture is now in the throes of a "New Renaissance." The central intellectual force behind this renewal is a boom in formulating and merging literary theories, and their surge through the humanities, social sciences, and even in what some favor as the hard sciences—physics, chemistry, geology, meteorology, and so forth. As the initial wave of theories, suffused with philosophical deconstructionism, made its way across the well-entrenched boundaries of the hard sciences, an approach to reading and understanding the literary text emerged in 1991 at a special session of the Modern Language Association, where Harold Fromm announced the "greening of literary studies." Shortly after, the Association for the Study of Literature and Environment (ASLE) formed with the specific mission of "promoting the exchange of ideas and information pertaining to literature that considers the relationship between human beings and the natural world." The cadres would later call the approach "ecocriticism."

Literary Theory

Ecocriticism is an approach to rather than a theory of reading and analyzing and writing about certain kinds of literary experiences. An ecocritic thus may apply and integrate any number of theories in any one critique. There are perhaps as many ways to read artistic literature such as a poem, novel, or epic as there are literary critics. Traditional and contemporary literary critics who study texts approach them with a point of view, similar to readers who carry attitudes, values, and beliefs that affect their own interpretations, the difference being that critics base their critiques on well-thought-out theoretical constructs or contexts of ideas from which they determine the quality, meanings, and functions of texts.

No matter which theory or theories a literary critic chooses to apply to a given text, every theory presupposes one or more of four fundamental critical orientations: mimetic, pragmatic, expressionistic, and objectivist. Each orientation prioritizes one of the primary elements of criticism, which are text, author, audience, and world. All critical judgments about a text are based on the element prioritized in a specific theory.

Literary theories that are mimetic judge the value of a text by its trueness to the world. Texts are analyzed and interpreted in relation to the world or reality. In contrast is the pragmatic orientation, which focuses on the relationship between text and reader as opposed to text and world. Pragmatic theorists want to know what it is about a work, for example, its descriptions of a natural landscape, that enables it to invoke emotions in readers and convey information such as environmental values.

Another orientation is expressionistic. Here the value of the work lies in the individuality of the author, in his or her unique creative and imaginative abilities to synthesize images, thoughts, and feelings in a poem or short story. Opposing this critical orientation is the objectivist orientation, which views the written work as an autonomous object projecting its own more or less self-contained reality. Works of art are considered to be a unique system of interrelationships having structural properties unresponsive to the immediacies of culture and the natural world.

One fundamental difference between eco- and nonecocriticisms is the starting point for analyzing relationships between elements of criticism. For the traditional literary critic, the "world" represents society. For the ecocritic, the world is an explicit or implicit starting point, but the meaning expands beyond the borders of society to include environmental processes such as orogenesis, precipitation, and glaciation that can be viewed as cultural and/or not of culture. This choice of perspective, and there are more than two here, depends on the context of ideas the ecocritic adopts, be it biographical, feminist, Marxist, postcolonial, aesthetic, mythical, ritual and symbolic, semiotic, or a combination of these.

Geocritical Antecedents

At a time when adventurous academics were crossing disciplinary boundaries in their research, or perhaps more fitting, erasing these boundaries, ecocriticism ("eco" derives from the Greek *oikos* meaning house) was delineating an exclusive territory of ideas and assumptions. This becomes apparent when one acknowledges the fact that prior to the greening of literary studies, geographers (Greek *geo* meaning "Earth"; *graphien,* "description") had been researching and writing on these same human/nature relationships for over a century. In fact, studying the Earth as the home of humankind in which these relationships take place has been geography's main focus in Western tradition since its inception, and the goal of understanding these relationships is now used as a justification for resuscitating in academia this so-called interdisciplinary discipline.

Geographers first saw the literary text as a worthy object of analysis more than 30 years ago with the emergence of humanistic geography. The adoption of methods and theories of the humanities and the focus on the culture of artistic expression in the form of literature and painting was a reaction to a "quantitative revolution" and crude economism dominating the discipline. During the late 1970s and early 1980s, major scholarly journals of geography published a number of literary critiques. The influential book *Humanistic Geography and Literature*, which is a compilation of critical analyses carried out by professional geographers, began circulating in 1981. These critiques of fiction and poetry were primarily mimetic in orientation.

Critics dissected literary works to determine whether they reflected factual geographies predetermined either by the geographers' understanding of their collection of other texts deemed more objective or by their exceptional experiences of place. Similarly, the comparative method revealed whether a particular fiction writer of the past could be considered a good geographer. This was done by juxtaposing his or her work with factual geographies recovered through historical or geographical monographs. In one other approach, critics sought to extract a sense of place or of rootedness in place. The search was not for positive information or a palpable reality but rather for the glory of a unique mind's eye contemplating its surroundings. And who else could so eloquently communicate the internalized experience of place, thereby confirming an authentic sense of place, but the novelist or poet?

When "ecocriticism" arrived on the scene by the 1990s, geography's critics had already veered toward deconstructing the sacred texts of nonfiction, including those of its own discipline. The analysis of fiction became more sophisticated in that, following more the objectivist orientation, geographers examined the ways literary artists constitute spaces within their works and how these function as psychological, historical, and cultural arenas. But the lure of (de)constructionism, with its "no facts, just interpretations" credo, proved too potent for many even in the more standardized, traditional subfields of economic and political geography. The latter's "critical geopolitics" applied literary theory to political speeches, theoretical treatises, and even popular media sources such as magazines and film to reveal the hidden assumptions about people and places that galvanize international and domestic policies, contests for place, and even initiate warfare. In cultural geography, landscape became a text constituted through other texts: fiction, nonfiction, paintings, maps, postcards, etc. Extra-textual reality, a reality existing beyond or before words and images, slipped through their fingers or, more radically, was never even grasped. Nature, with a capital N, could only vanish from the scene.

Although ecocriticism's intellectual apples did not roll far from the tree, as an institutionalized mode of reasoning and writing about literature that like humanistic geography tumbled out of the revolts of the 1960s and 1970s, the cadre did not give credit where credit was due. In *The Ecocriticism Reader, Landmarks in Literary Ecology* (1996), arguably the most significant tome published on the subject, Cheryll Glotfelty and Harold Fromm attempt to establish a canon of fiction and nonfiction essays and books for the benefit of budding ecocritics. Of 333 entries, only one from the discipline of geography, *Landscapes: Selected Writings of J. B. Jackson* (1970), was considered suitable reading on the subject of man-and-nature. It is true that the term *ecocriticism* was coined in 1978 by William Rueckert in an essay titled *Literature and Ecology: An Experiment in Ecocriticism* and that well-known Western scientists, philosophers, and writers of the distant past were occasionally inspired to ponder the ecological nexus in literary text of well-known fiction writers. But, it seems, the instinctual passion and exclusivity accompanying paradigm building, the burgeoning of academic journals and niche publications, and the amplified complexity of academic lingo delayed recognition of the fine critical works rendered in the grand ol' field of geography, works such as Yi Fu Tuan's internationally renowned *Topophilia* (1974).

Ecocriticism Today

However, the canon has expanded. Rueckert's neologism, along with critiques from the distant past, has become retroactively ecocritical heritage. Over the past 10 years, ecocritics have produced a profusion of sophisticated and elegant expository critiques on such topics as wilderness, nature/culture dualism, anthropocentrism, symbolism of nature, and animal rights. The quarterly journal *Interdisciplinary Studies in Literature and Environment*, first issued in 1993 by the Association for the Study of Literature and the Environment (ASLE), is a major platform for many of these publications. The association—established in 1992 during a special session of the Western Literature Association in Reno, Nevada—and especially its website, is the nerve center of literature and the environment, conducting seminars, stimulating debates, and linking a plethora of relevant resources. ASLE has recently joined with affiliates in, among other countries, the United Kingdom, Japan, New Zealand, Taiwan, and India.

With expansion of the canon come discussions and debates concerning issues of representation, constructionism, real versus imagined, and with these, a search for the meaning of the term *ecocriticism*. Such deliberations have already taken place in geography and led to a full-blown identity crisis. A similar crisis inflicts postmillennial ecocriticism as an unparalleled sympathy for the lives of nonhuman beings of this world can no longer be the quirk that keeps together a mutual disposition. Geography, at least in the United States, has de facto resolved the crisis by depreciating the "interdisciplinarities" crossing into the humanities through an exchange rate that favors specific data sources, techniques, and algorithms serving surveillance society's immediate affection for homeland security, climate change, and profit maximization.

The crisis in ecocriticism signals an early onset of maturity for a young academic perspective that only recently set the boundaries for a circumscribed scope of knowledge within the humanities, particularly the literary arts, even though the guiding principle is "promoting the exchange of ideas and information pertaining to literature that considers the relationship between human beings and the natural world." The definition of "literature" continues to broaden, to say nothing of the meaning of "relationships between human beings and the natural world," which involves just about everything under the sun. Yet ecocritcism flourishes in a vibrant atmosphere that prolongs the survival of an intellectual diversity in which philosophy, science, criticism, history, poetry, fiction, and nonfiction can intermingle to create new forms of knowledge and understanding. Ecocriticism may be the last refuge for a truly interdisciplinary paradigm.

See Also: Human Geography; Novels and Nonfiction; Poetry; Space/Place/Geography and Materialism.

Further Readings

Brosseau, Mark. "Geography's Literature." *Progress in Human Geography*, 18/4 (1994).
Coupe, Lawrence, ed. *The Green Studies Reader: From Romanticism to Ecocriticism*. London: Routledge, 2000.
Garrard, Greg. *Ecocriticism*. New York: Routledge, 2004.
Glotfelty, Cheryll and Harold Fromm, eds. *The Ecocriticism Reader: Landmarks in Literary Ecology*. Athens: University of Georgia Press, 1996.
Ingram, Annie Merrill, et al., eds. *Coming Into Contact: Explorations in Ecocritical Theory and Practice*. Athens: University of Georgia Press, 2007.
Mazel, David, ed. *A Century of Early Ecocriticism*. Athens: University of Georgia Press, 2001.
Pocock, Douglass C. D., ed. *Humanistic Geography and Literature*. London: Croom Helm, 1981.

Ken Whalen
University of Brunei Darussalam

EcoPopulism

EcoPopulism is a term attributed to community-based environmental movements whose mobilization patterns extend beyond their initial concern or localized causes in order to address wider ecological or political concerns, be they regional, national, or global. Andrew Szasz has detailed the U.S. anti-toxics movement in his text *EcoPopulism* (1994, 1998), and argues that campaigns can move away from their NIMBYist (not in my backyard) inception into ecopopulist politics. According to Szasz, a distinction can be made between the localized impact of a toxic waste "event" and any subsequent representation of this event as a media issue at the national level. For Szasz, the media's role in getting an impact nationally from a local environmental issue is crucial to the success of that issue as a target of movement grievance. This impact can be measured by the airtime given to the Love Canal toxic dump in New York when the story broke in the late 1970s; according to Szasz, "The network's nightly news programs alone carried about 190 minutes of news about hazardous waste."

This large amount of prime-time news coverage led to an overall shaping of the issue of toxic waste into what came to be a societal understanding of what toxic waste problems are about in the United States. As such, local activism was picked up by the national media, which then played a major role in regard to building consensus about toxic waste issues. One result of this increased coverage was an increase in localized activism. After Love Canal, contamination protests that had started earlier but had floundered suddenly took off after 1978.

A sense of momentum grew around the grassroots anti-toxics movements as people found a sense of inspiration from the success of that movement. By 1979, the waste disposal industry was facing a "major barrier" from public resistance to toxic dumping, whereas previous to the Love Canal campaign, the issue had not captured much public attention. What resulted was a major siting problem for the toxic waste industry in the face of "inflamed public opinion." The strength of public opinion bolstered the nationwide antitoxics movement in the United States, including groups such as the Citizens Clearinghouse for Hazardous Wastes (CCHW), which itself evolved from the Love Canal protests. The CCHW endeavored to bring together regional groups that were embarking on campaigns and provided them with start-up information about conducting a protest. Furthermore, a National Toxics Campaign (NTC) based on localized source reduction was established. The NTC also targets toxic waste incinerators, and offers "organizing help, technical assistance and laboratory testing," according to Szasz. In this way, the NTC has parallels with the CCHW, as both groups have become a major resource for anti-toxics campaigns in their early stages.

The utilization of technology is maximized through the input of the academic sector, which provides information on the most recent scientific research into toxics. This information is broken down by ecopopulist groups into a set of prioritized findings and posted in newsletters such as the *Remote Access Chemical Hazards Electronic Library (RACHAEL) Hazardous Waste News*. The utilization of technology and expertise has facilitated anti-toxics movements across class and racial divides, in both U.S. and global settings. This has allowed anti-toxics movements to maintain an immediacy that formal political structures would hamper. Another benefit of the dissemination of ecopopulist anti-toxics information has been the altering of activist demographics in relation to class and racial profiles. Historically, toxic waste disposal facilities were sited in poor areas of high black, Latino, or Native American populations. These areas were deliberately targeted as, states Szasz, "some policy analysts began to advocate a strategy of siting in communities that are least capable of politically resisting or more amenable to accepting some form of financial compensation in exchange for accepting the facility."

The labeling of anti-toxic agitation as NIMBYist is seen by writers such as Szasz to be an attempt to reduce the impact of the professionalized links that have become a feature of these ecopopulist movements. The success of the U.S. anti-toxics movement in "inflaming" public opinion has invariably led to an increase in the exportation of toxic wastes to poorer states in the Southern Hemisphere.

See Also: Grassroots Organizations; Public Opinion; Social Action, National and Local.

Further Readings

Leonard, Liam. *Politics Inflamed: Galway for a Safe Environment and the Campaign Against Incineration in Ireland*, Ecopolitics Series Vol. 1 Galway, Ireland: Greenhouse Press, 2005.

Szasz, Andrew. *EcoPopulism: Toxic Waste and the Movement for Environmental Justice (Social Movements, Protest and Contention)*. Minneapolis: University Of Minnesota Press, 1998.

Liam Leonard
Independent Scholar

ECOTOURISM

As the ecotourism industry grows, a lack of English literacy by indigenous rainforest dwellers means that fewer locals become guides; as a result, tourists like this man in the Ancient Mayan ruins at Caracol, Belize, may have a guide that is a better English speaker but holds less local knowledge.

Source: iStockphoto

Ecotourism (also interchangeably used as "ecological tourism") has often been presented as a model of sustainable development. It is often defined as a form of responsible tour or travel to ecologically and naturally sensitive, pristine, and protected areas. It is an economic, social, and ecological enterprise or means that aims at protecting land and important biological diversity while at the same time providing long-term social and economic benefits, particularly to the local communities through sustained resource use.

Ecotourism combines natural conservation, community well-being, and sustainable travel. This means that those who implement and participate in ecotourism activities should follow some ecotourism principles such as minimizing impact on natural habitats, building ecological and social consciousness and respect, providing positive experiences for both visitors and hosts, providing direct financial benefits for conservation, providing financial benefits and empowerment for local people, and raising sensitivity to host countries' political, environmental, and social climate.

According to the World Tourism Organization, tourism is one of the fastest-growing industries in the world. As the largest business sector in the world economy, the travel and tourism industry is responsible for over 230 million jobs and over 10 percent of the gross domestic product worldwide. In 2006, travel and tourism (consumption, investment, government spending, and exports) is expected to grow 4.6 percent and total $6.5 trillion. Tourism is a principle "export" (foreign exchange earner) for 83 percent of developing countries, and the leading export for one-third of the poorest countries. For the world's 40 poorest countries, tourism is the second most important source of foreign exchange, after oil.

While the history of tourism dates far back, the concept of ecotourism was introduced and popularized in the late 1980s when the notion of sustainable development was officially introduced by, among others, the World Commission on Environment and Development in its report titled *Our Common Future* in 1987. "Sustainable development" was defined in the report as "development that meets the needs of the present without compromising the ability of future generations to meet their own needs." The notion of sustainable development was very appealing at that time as it presumed that economic growth and environmental protection could be reconciled. *Our Common Future* identified three common goals of sustainable development: economic growth, environmental protection, and social equity. As a new paradigm of development, sustainable development was widely embraced by different development institutions, organizations, and states as both marketing tool and fashion. In this context, ecotourism has been heralded as a model for sustainable development by international aid agencies, national governments, nongovernmental organizations, and indigenous groups. Currently, in many countries such as Belize, Nepal, Madagascar, Costa Rica, Ecuador, and Kenya, ecotourism is not a simple attempt to preserve nature and its biodiversity but a major industry for the national economy.

While the goals and principles of ecotourism in theory are quite appealing, critics claim that it is largely a marketing tool in practice to promote tourism related to nature. The processes and practices of ecotourism therefore need to be examined in the context of some fundamental questions: Who are the key beneficiaries of ecotourism? Whose economic growth is maximized? Which social groups make decisions about development? To what extent do locals and states benefit? How are benefits distributed? What is protected and at what costs? In short, can ecotourism, as a form of sustainable development, really provide for sustained economic growth, sustained social equity, and sustained environmental protection?

Research in various ecotourism sites show that the notion of ecotourism has largely been used as a greenwashing device that occludes many inherent practices that are ecologically and socially unfriendly and go against the very principles of ecotourism. Tour guide is one of the best jobs in the ecotourism industry. In most ecotourism sites, nation-states—often with financial loans and academic and private sector tourism consultants from the north—have developed a system of certification and regulation of nature tourism guides and designed a system to guarantee the quality and ability of the tour guides. States therefore established a series of mandatory courses leading to legal certification of nature tourism guides. Upon completing the required courses, tour guides are required to pay a license fee in order to be legally employed in the industry. Tour guides without appropriate certification and licensing are subject to fines and other forms of punishment. Research in Belizean ecotourism, for example, shows that those with limited English reading and writing skills are less likely to pursue the required formal coursework and to pass the courses if they try. The program therefore excludes many guides with extensive local environmental knowledge in favor of those with solid literacy but perhaps little local knowledge. Therefore, in practice, the value of the indigenous knowledge has been decreased, while the value of formal academic skills has been increased. As the program has been implemented, fewer rainforest dwellers have been employed as guides, while more formally trained guides explain rainforest ecology to nature consumers. This structure subverts the indigenous environmental knowledge as well as the opportunity of livelihoods for the indigenous people.

Many governments may have policy commitments to utilize tourism to protect the environment rather than to destroy it and so contribute to sustainable development; however, such government commitments may prove to be unsustainable in the face of growing transnational pressures to open their rich natural resource base to transnational corporations (TNCs) and to repay a mounting international debt. Most ecotourism resorts are attractive not only to the nature consumers but also to the transnational corporations. Most ecotourism requires a considerable amount of reshaping of nature in order to provide comforts for the nature consumers. Roads, sometimes amenities like highways and airports, hotels, resort centers, tourism offices, food courts, and swimming pools are built to attract tourists and to keep all necessary facilities ready for wealthy nature consumers. As a result, nature itself often loses its natural shape and biodiversity, while increasing infrastructural developments and human activities create more wastes and pollution. Therefore, protection of the environment under ecotourism increasingly decreases. Ecotourism then becomes a vehicle of environmental disruption in otherwise undisturbed ecosystems, making it more difficult to view ecotourism as environmentally protective. This new infrastructure then makes other potentially less-sustainable economic activities such as logging, mining, or agriculture viable in previously inaccessible locations.

Ecotourism will remain unsustainable if environmental protection and social equity are subordinated to only economic gain. Most ecotourism destinations are experiencing a new form of imperialism by multinational corporations, as less than 5 percent of revenues go to the local population. Local populations also lose control over their deeply connected natural resources. Community-based natural resource management aiming at small scale, slow growth, and local control is of paramount importance to realize the real goals and principles of ecotourism.

See Also: Biodiversity Loss/Species Extinction; Environmental "Goods"; Green Consumerism; Greenwashing.

Further Readings

Gould, K. A. and T. L. Lewis, eds. *Twenty Lessons in Environmental Sociology*. New York: Oxford University Press, 2009.

Honey, M. *Ecotourism and Sustainable Development: Who Owns Paradise?* Washington, DC: Island Press, 2008.

International Ecotourism Society. http://www.ecotourism.org/site/c.orLQKXPCLmF/b.4832143/k.BD87/Home.htm (Accessed June 2010).

McMichael, P. *Development and Social Change: A Global Perspective*, 4th ed. Thousands Oaks, CA: Pine Forge Press, 2007.

Md Saidul Islam
Nanyang Technological University

Education (Climate Literacy)

Learners at all levels commonly hold misperceptions about climate science (e.g., the workings of the carbon cycle, the degree of influence of humans on Earth systems, the meaning of scientific uncertainty). But a correct understanding of climate science is necessary for society to respond well to the challenges of a changing climate system. Since climate science is an interdisciplinary field that draws from all Earth systems, including human interactions and impacts, few educators, communicators, or citizens have engaged in a cohesive formal study of the discipline. Because of the need for people to navigate this complex topic, the climate education and science community sought to provide guidance by defining the "big ideas of climate science": those ideas and concepts that, if understood, would provide an adequate body of knowledge for a person to be considered "climate literate." The effort, inspired in part by the Essential Principles of Ocean Literacy, resulted in the program Climate Literacy: The Essential Principles of Climate Science (EPCS). The framework was first released in spring 2008 for the K–12 education community and then revised as a "guide for individuals and communities" in spring 2009.

A climate-literate person should study the following:

- Understand the essential principles of Earth's climate system
- Know how to assess scientifically credible information about climate
- Communicate about climate and climate change in a meaningful way
- Be able to make informed and responsible decisions with regard to actions that may affect climate

Why does it matter if people are climate science literate? "Climate science literacy is an understanding of the climate's influence on you and society and your influence on climate." As described in the EPCS, scientific observations and climate model results indicate that human activities are the primary cause of most of the ongoing increase in Earth's globally averaged surface temperature and other related impacts such as species extinction and sea level rise. Because of this, society needs citizens who understand the climate system and know how to apply that knowledge in their careers and in their engagement as active members of their communities.

The Essential Principles of Climate Science (EPCS)

To foster that knowledge, the climate community, in a workshop sponsored by the National Oceanographic and Atmospheric Administration (NOAA) and the American Association for the Advancement of Science (AAAS), began the development of a coherent framework of "the big ideas" in climate science, along with the fundamental concepts that underpin them. This framework, Climate Literacy: The Essential Principles of Climate Science, has been adopted by the U.S. Global Change Research Program (USGCRP) as an outreach and communication document and is endorsed by many science and education organizations and agencies. The EPCS is being used to guide new funding, new education projects, and communication with decision makers and the public. Each essential principle is undergirded by several fundamental concepts, which, if understood, support an understanding of the essential principle.

The Essential Principles of Climate Literacy includes a guiding principle and seven essential principles. The guiding principle for informed climate decisions is that humankind can take actions to reduce climate change and its impacts. The essential principles of climate science are the following:

- The sun is the primary source of energy for Earth's climate system.
- Climate is regulated by complex interactions among components of the Earth system.
- Life on Earth depends on, is shaped by, and affects climate.
- Climate varies over space and time through both natural and man-made processes.
- Our understanding of the climate system is improved through observations, theoretical studies, and modeling.
- Human activities are impacting the climate system.
- Climate change will have consequences for the Earth system and human lives.

A Community-Based Resource

The EPCS is intended to be a framework of foundational principles that may be used to guide the content of climate education and communication. However, a set of foundational principles can only be considered foundational if they have been thoroughly grounded in the perspectives of the professional community in that field. While the climate community was inspired by the previous Ocean Literacy framework, and the Climate Literacy effort was followed by other geosciences literacy frameworks (e.g., Atmospheric Sciences, Earth Sciences), a distinction of the Climate Literacy framework is that it is built upon the research-based AAAS *Atlas for Science Literacy* as a primary scaffold. Many hundreds of comments were received and considered in a formal review process in order to provide a thoroughly credible resource.

Climate Literacy for Educators

The first published draft of the Essential Principles of Climate Literacy was meant for a pre-college education audience. The people who attended the workshop and provided comments on drafts included climate scientists, formal and informal educators, representatives from agencies, nongovernmental organizations, and other institutions involved with climate research, education, and outreach. Teachers are often accountable for teaching content as defined by state education standards. Thus, the essential principles and fundamental concepts were aligned within a matrix of the benchmarks so that teachers could easily cross-reference climate literacy concepts with the fundamental concepts they are required to teach in their classrooms. The document was accepted as a helpful road map for climate education, but it was clear that more was needed.

Climate Literacy for Individuals and Communities

After receiving feedback from the community, a revised version, "Guide for Individuals and Communities" was issued, with the addition of a guiding principle for informed climate decisions: "Humans can take actions to reduce climate change and its impacts." The "Guide for Individuals and Communities" has been adopted by the U.S. Global Change Research Program as an official outreach and communications document.

Fostering Climate Literacy

"These frameworks are all really great and everything and we need them, but what we really need is the curriculum and professional development to go with them," commented a Florida educator of preservice teachers on the Climate Literacy framework in 2008. No document, however worthy, can lead to a transformative change unless it is used. The climate literacy community, like the Florida teacher quoted above, recognized that it would take a committed and inclusive organization to provide the resources, expertise, and leadership necessary to foster climate literacy for the nation's public, learners, and decision makers. To this end, the Climate Literacy Network (CLN) was founded in 2008 as a leadership and action organization. The CLN includes contributors from universities, agencies, and education organizations who meet regularly to coordinate efforts. The overarching goals of the CLN are the following:

- Encourage adoption of climate literacy goals in state education standards
- Promote educational initiatives at local, state, national, and global levels
- Develop and disseminate strong curriculum materials
- Provide professional development for teachers and other educators

- Identify appropriate informal education populations
- Tailor, develop, disseminate, and apply materials for informal education
- Evaluate and refine educational materials and practices
- Share best practice ideas and resources
- Enable collaborations among scientists, educators, community leaders, and policy makers to achieve these goals

Through the continued efforts of the CLN and the agencies that support climate science education, projects intended to foster climate literacy are becoming a reality. The National Science Foundation (NSF), NOAA, and the National Aeronautics and Space Administration (NASA) have all issued requests for proposals in which fostering climate literacy is one of the program goals.

New Resources for Climate Literacy

Resources are becoming available that are aligned with the EPCS. These include online courses, reviewed and aligned curricula, teacher professional development, museum exhibits, film products, and more. An online climate literacy course is available through the University of Colorado, the NASA Global Climate Change Education program funds multiple climate literacy projects, an online handbook of climate literacy is available through the Encyclopedia of Earth, and more resources are emerging as time goes on.

Next Steps

As the climate literacy community evolves, challenges emerge. How should climate education remain sensitive to the emotional needs of young learners grappling with difficult and charged problems? How should solutions and decision making be incorporated into science classes? What are the best strategies for responding to controversy and misinformation? Given that the need is so great, what are the most important things to do? It will take the entire climate-attentive community to answer these questions.

See Also: Demonstrations and Events; Environmental Communication, Public Participation; Global Warming and Culture.

Further Readings

American Association for the Advancement of Science (AAAS). *Atlas of Science Literacy.* Washington, DC: 2001.

McCaffrey, Mark, et al. 2010. "Climate Literacy Handbook." In *Encyclopedia of Earth.* http://www.eoearth.org/article/Climate_Literacy_Handbook (Accessed January 2010).

U.S. Global Change Research Program. "Climate Literacy: The Essential Principles of Climate Science: A Guide for Individuals and Communities," Version 2, 2009. http://www.globalchange.gov/resources/educators/climate-literacy (Accessed January 2010).

Susan M. Buhr
University of Colorado, Boulder

ENERGY

Nuclear power, created in plants like this one in France, has been seeking rebirth as a clean energy source, since it does not produce greenhouse gasses. However, safe disposal of potentially harmful radioactive waste remains an environmental problem.

Source: iStockphoto

The main energy drivers of human sociocultural evolution have been wood (which fueled growth from the time fire was used to burn wood for cooking, heating, and light over 50,000 years ago), the combustion of coal for early industrializing countries in the 18th and 19th centuries, and the myriad applications of oil in the 20th century. An additional contemporary energy source from combustion is generated in waste-to-energy plants. But the use of these energy resources also generates environmental degradation, such as the destruction of the Earth from surface mining for coal, catastrophic oil spills, and emission of greenhouse gases and toxic particulates from burning fossil fuels and waste. In the case of energy generated from nuclear fission, releases of radiation from accidents and the need for long-term storage of radioactive waste are serious concerns. We now appear to be on the brink of a new energy shift as the era of "peak oil" will inevitably cause significant changes in our relationship to energy and to the biosphere.

Wood: One of the First Fuels

Burning wood releases the energy stored in carbon, and also cellulose, although the latter generates much less energy. The uses of wood as a fuel source include cooking, heating, metallurgy, and manufacturing. While it was replaced as a primary energy source in more-developed countries by the early 20th century, wood remains a significant energy resource in many developing countries. And in areas with relatively high degrees of poverty, wood consumption can be essential to life. Today, China and India consume the most wood and crop residues in total volume, followed by Brazil and Indonesia, but with these resources supplying over 80 percent of the fuel for sub-Saharan Africa (primarily for cooking and heating), the region is the largest consumer relative to its total demand.

But wood, compared to fossil fuels, is quite inefficient as an energy source; wood has a typical energy density (energy output per weight) that is about 80 percent of steam coal and 45 percent of crude oil, depending on the amount of water in the wood, which can vary widely and reduces the energy density. This deficiency is heightened when wood use is not coupled with modern technology. With an energy density similar to coal and with less smoke than wood, charcoal, created from the combustion of wood to remove water and other volatile components, fares better.

While a renewable resource, wood is not immediately reproducible, creating shortages or elimination of wood resources in many locations, and its acquisition has resulted in the destruction of forests and their ecosystems, with resultant soil erosion. Silviculture (managing wood for use) is a potentially sustainable practice that has been utilized for centuries, but the damage to native ecosystems can still be problematic as monocultures typically form that reduce complexity and foster the spread of disease. The burning of wood also produces both carbon monoxide (CO) and carbon dioxide (CO_2); the former is toxic to humans and animals and an indirect greenhouse gas, and the latter directly contributes to anthropogenic climate change. Burning 1 gram of carbon requires 2.66 grams of oxygen, producing 33 kilojoules of energy, and 3.66 grams of CO_2. (It should be noted that the burning of biomass to produce fuel is generally considered to be carbon neutral by the Intergovernmental Panel on Climate Change [IPCC] if the source of wood is allowed to regenerate, acting as a carbon sink.) The use of wood, then, like the fossil fuels and waste-to-energy production discussed below, has both positive energy contributions to human societies but also negative impacts on the biosphere.

Coal: The Energy Source of Industrialization

Coal is a family of sedimentary rocks formed as organic matter—primarily plants—that is partially decomposed in water, then transformed anaerobically under the pressure of encasing sediment layers. Much of the coal we now have came from the Carboniferous period (around 380 to 286 million years ago). As the transformation takes place, peat is first formed; the substance then progressively hardens, first as lignite, or brown coal, then sub-bituminous, bituminous, steam coal, anthracite, and finally as graphite—the latter not typically used as an energy source. The primary elements in coal are carbon, then some mixture of hydrogen, nitrogen, oxygen, and sulfur, with the concentrations of carbon increasing and the other elements decreasing with the hardness. Coal is also used in the form of coke, created by baking off the volatile components of bituminous coal, resulting in much less smoke when burned. The hydrocarbons that are produced during the coking process (coal-gas and coal-tar) are now often used in a separate process to generate additional energy sources such as synthesis gas, one type of coal-to-liquid (CTL) production.

Coal is extracted by mining deposits found variously underground, at the surface, and in hills and mountains. The need for coal to fuel the steam engine and the demand for coke were causes for the dramatic expansion of coal mining that took place in the 18th century. Until the middle of the 20th century, when machine technology spread and began replacing some human labor, coal mining required a substantial amount of human and also animal (horses and donkeys) energy to extract. Using modern technology, the net energy return on invested energy (EROI)—the energy generated divided by the energy used in extraction and production—for coal ranges from 200 to 50, making it fairly efficient to produce, and this increase in available energy over wood and earlier sources fueled rapid population and agricultural expansion beginning in the 18th century. But coal mining remains a dangerous profession, as the potential for calamitous explosions and collapses within mines and dire health effects from the ingestion of coal dust and other toxins linger. Mines are also subject to flooding. The expansion of surface mining has reduced many of the negative effects for miners, but the cost for the biosphere remains high.

While it was not used extensively in any ancient or medieval societies, coal's early uses include iron production in the Han Dynasty (206 B.C.E. –220 C.E.). By the 12th century C.E., coal was being commercially produced in Europe, and—as peat—it fueled the Golden Age in the Netherlands during the 17th century. But England was the first country to shift from phytomass (primarily wood) as the primary energy source to coal, beginning in the late 16th century, and the country dominated global coal extraction until the end of the 19th century, fueling Britain's dominance of the global economy during the 18th and 19th centuries. Coal remained the primary global energy source until it was replaced by oil and natural gas in the 1960s in the more-developed countries. Today, the United States, Russia, and China currently contain over two-thirds of known reserves; Australia also has large reserves and is the largest exporter of coal in the world, while the South American continent as a whole contains only around 2 percent of the world's coal.

Yet coal remains an important energy source, with most of its consumption used for electricity generation. According to the International Energy Agency, coal is the single largest input for electricity generation—in coal-fired power plants—representing 48 percent of the total electricity inputs in 2007, and its share is expected to increase over the coming decades. Coal's share of the global primary fuel mix is also expected to increase during the same period by 2 percent to 29 percent. Demand for coal is expected to grow at close to 2 percent per year on average from 2007 to 2030, to nearly 7,000 metric tons carbon equivalent (MTCE). China is projected to account for 61 percent of the growth in global production as its demand nearly doubles; the remaining increase will come mostly from India, where demand is projected to more than double, with other developing countries accounting for the rest. The main types of coal produced today are steam coal (three-fourths of global production), coking coal (16 percent), and brown coal and peat (8 percent).

Coal is a nonrenewable resource when measured in human time frames, creating the potential for exhaustion (relative to available technology and market price-to-cost). According to the International Energy Association, reserves of coal are large and could sustain growth at current (2008) levels for approximately 146 years. But coal production and use has many negative environmental effects, including—as with all fossil fuels—contributing to climate change from emissions of CO_2, methane (CH_4), and nitrous oxide (N_2O). Using current technology, the burning of coal generates more CO_2 than oil, and both produce almost twice as much as natural gas. Coal combustion also produces sulfur dioxide (SO_2) and nitrogen oxides (NOx), which also produce acid rain and particulates harmful to health—the latter is also an indirect greenhouse gas. This has led to the promise of "clean coal" technology to reduce the negative environmental effects of coal transformation, including increasing the efficiency of coal-fired plants (so less harmful emissions are produced) and carbon sequestration (capturing the emissions and pumping the by-products into the sea or into storage chambers underground or into other locations). The cost and large-scale feasibility of the solutions leave their broad application uncertain.

Oil: The Source and Scourge of Contemporary Life

Oil is a hydrocarbon made from molecules containing carbon and hydrogen. (Natural gases such as methane and propane are also hydrocarbons.) As an energy resource, oil is typically broken into conventional sources (primarily crude oil and natural gas liquids) and nonconventional sources (primarily oil shales, tar sands, very heavy oil, gas-to-liquids, and coal-to-liquids). Crude oil exists in liquid form in reservoirs in the Earth. It is created by the transformation of marine biomass through pressure and heat; a similar process creates coal as described earlier; however, the carbon chains in oil are much smaller and less ridged, producing liquid instead of solid forms. Crude oil is then refined into products such as naphtha, kerosene, diesel oil, asphalt, tar, and most important for our auto-dominated times, gasoline.

One of the earliest uses of crude oil took place in the Middle Ages in what is now Azerbaijan, taking oil from shallow wells for use as kerosene in lighting. The first commercial oil distillery was set up by Russia in 1837, and the first exploratory oil well drilled in 1847. Commercial production of oil began to spread in the late 19th century, used mainly for illumination and as a lubricant. However, the modern oil era began in the United States with the use of a steam engine for percussion drilling in 1859; this leading position in the exploitation of oil and the development of an oil-based infrastructure helped propel the United States to economic dominance by the middle of the 20th century. Demand increased during the 20th century with the spread of the internal combustion engine and the growth of petrochemical industries. Oil is now heavily used in modern agricultural production, including the manufacture of fertilizers and pesticides. Technological changes in drilling (e.g., horizontal, directional, and deepwater drilling), transportation (e.g., ever-larger ships and longer pipelines), and advances in refinery processing have spurred the dramatic growth of oil, which is now the single largest source of global primary energy.

The supply of oil was relatively stable and the price relatively low for the first two-thirds of the 20th century; however, this changed with the oil crises in 1973 and 1978. The 1973 crisis occurred when countries in the Organization of Petroleum Exporting Countries (OPEC) raised prices on oil and placed an oil embargo on countries supporting Israel in its war with Egypt and Syria. In 1978, the Iranian revolution that re-installed the Moslem leader Ayatollah Ruhollah Khomeini to power resulted in shrinking oil exports from Iran and an increase in world oil prices. These crises led to a dramatic expansion of oil exploration and development, and coupled with reduced demand, created a relative oil glut that helped fuel rapid economic growth for oil-consuming countries in the 1980s, with prices remaining low through the end of the century. Currently, the Persian Gulf-Zagros Basin contains approximately 60 percent of the world's crude oil reserves. Much of the undiscovered oil is predicted to exist in offshore undersea fields (which now produce around one-third of all crude oil) and small onshore reservoirs.

According to the International Energy Agency, the largest consumers of oil are currently (as of 2008) the United States and China, which consumed 21.8 percent and 9.1 percent of the global total, respectively. Demand for oil is expected to grow by 1 percent per year on average from 2007 to 2030, to 105 million barrels a day; however, oil's share of the global primary fuel mix is expected to drop four percentage points to 30 percent as other fuel sources, including coal, natural gas, nuclear, and renewables make up more of the total. Developing countries will likely be the source of increasing demand, with the demand in developed countries declining. The transport sector will produce almost all of the growth in demand, mainly from the use of refined oil as gasoline in internal combustion engines.

The net energy return for oil ranges from 10,000:1 for Middle Eastern crude oil to less than 10:1 for refined fuel, making it fairly efficient to produce. Production of non-crude sources is less efficient and has higher environmental costs; this is particularly true for oil shale and tar sand production, which require massive amounts of energy inputs to produce an equivalent output as crude. The combustion of oil, like coal and the incineration of waste, also produces greenhouse gases, and oil's dominance as a fuel source, particularly one with which consumers frequently interact, makes its use a ripe target of environmental concern. Possibly equally as devastating for the environment may be the consequences of oil spills, such as the 2010 British Petroleum/Deepwater Horizon disaster in the Gulf of Mexico, which is the largest oil spill to date.

Peak Oil: Nearing the End of the Oil-Fueled Era

Peak oil is the point at which oil production reaches maximum output, which signals the decline that follows, a phenomenon first predicted by M. King Hubbert in 1956. Hubbert correctly predicted the peak of U.S. oil production in 1970, and his theoretical model estimated the global peak around the year 2000. Hubbert's insight was based on the production cycle of nonrenewable resources, which forms a logistic curve. At first, production rises to meet demand, taking the highest quality and easiest extractable sources, and the cost of production declines as efficiencies increase. As the resource becomes scarcer, production levels off, declining as the resource continues to be depleted and it becomes increasingly difficult and costly to produce.

According to the International Energy Agency, increasing global oil demand will be met by OPEC—whose members have more than half of the remaining proven reserves and ultimately recoverable resources—while oil production in non-OPEC countries was expected to peak around 2010. According to a British Petroleum report, the reserve to production ratio of oil is currently around 42 years, an increase from its low of 20 years after World War II. Others predict that world peak oil may have already been reached, or that it soon will be. The range of estimates stems mainly from the difficulty in forecasting new discoveries and in determining accurate oil reserves; the latter are subject to much debate since there is no uniform and unbiased assessor, and estimates can include reserves that have a 50 percent or less probability of actual production. While uncertainty surrounds the date of peak oil, it is undoubtedly going to occur, and the shift from an oil-based energy regime will not only be costly but finding resources that can provide the same level of energy output as oil and in its many current uses—while also necessarily protecting the environment—seems a formidable task.

Waste-to-Energy Plants: A Renewable Solution?

Waste-to-energy (WtE) plants generate electricity and/or heat from the combustion of waste sources and can also produce fuel sources such as methane and, through fermentation, ethanol. This relatively old technology can reduce the amount of municipal waste that would otherwise be dumped in landfills or incinerated without any energy recapture. Waste-to-energy is a growing industry, with Europe leading the way in the number of WtE plants and many others located in the United States, Canada, Japan, and China.

This technology is not without environmental and health concerns. Incinerators produce particulates, ash, heavy metals, and dioxin—which are highly toxic to humans—and acidic gas emissions, although modern incinerators typically use technological means such as lime scrubbers in smokestacks to reduce the by-products. Waste-to-energy production also generates greenhouse gases; however, the climate change potential from the incineration of waste may be less than would occur from uncaptured methane produced during the decomposition of the same waste in a landfill. According to a report produced for Friends of the Earth, recycling the nonbiogenic sources of CO_2 (such as plastics) to remove them from the waste stream and anaerobic digestion rather than incineration of waste are remedies to problems created during the search for sustainable energy from waste.

Nuclear Power: Energy From Atoms

The fission of uranium atoms was first demonstrated in 1939, and the first sustained chain reaction occurred in 1942. Energy from nuclear fission is harnessed in reactors that convert the heat produced during the reaction to steam that drives turbines. The first large-scale commercial reactor became operational in Great Britain in 1956. Between 1956 and 1975, nuclear power plant construction and use increased, particularly in Europe and the Soviet Union. Nuclear power began to lose favor by the 1980s for a number of reasons, including a reduction in the growth rate of electricity demand and the high cost of reactor construction. More damaging to the expansion of nuclear plants were the nuclear accidents in Three Mile Island in Pennsylvania in 1979 and the Chernobyl reactor in the Ukraine in 1986. In the former, the partial core meltdown released a relatively small amount of radioactive gases, while in the latter, the core exploded, releasing nuclear particles into the atmosphere. The accidents, along with concern of nuclear weapon proliferation, were a catalyst for the anti-nuclear movement and the resultant decline or cessation of nuclear power growth.

Nuclear power currently accounts for less than 10 percent of the global primary energy supply (about the same as all renewables) and around 25 percent of the electricity input. France is by far the largest proportional user, with over 75 percent of its electricity coming from nuclear generation; the Ukraine and Belgium follow at around 50 percent. The industry has been seeking a rebirth as a "clean" energy source since it does not produce greenhouse gases. Yet nuclear power suffers from the problem of safe disposal of radioactive waste that remains potentially harmful for thousands of years. An additional limitation is the depletion of the relatively scarce uranium isotope 235 U used in thermal fission reactors. "Fast breeder reactors" that can use the more prevalent 238U are expensive and have not been proved to be viable for large-scale commercial production. Moreover, despite significant research funding and its potential sustainability, nuclear fusion, the fusing of two nuclei to generate energy (the same process that fuels stars), remains unavailable as a commercial energy source due to technological and resource constraints.

See Also: Alternative/Sustainable Energy; Cooking; Global Warming and Culture; Hybrid Cars; Kyoto Protocol.

Further Readings

British Petroleum. "BP Statistical Review of World Energy, June 2009." http://www.bp.com/productlanding.do?categoryId=6929&contentId=7044622 (Accessed June 2010).

Deffeyes, Kenneth S. *Beyond Oil: The View From Hubbert's Peak*. New York: Hill and Wang, 2005.

Hogg, Dominic. "A Changing Climate for Energy From Waste? Final Report for Friends of the Earth." Eunomia Research & Consulting Ltd. (2006). http://www.foe.co.uk/resource/reports/changing_climate.pdf (Accessed June 2010).

Intergovernmental Panel on Climate Change. "IPCC Frequently Asked Questions." http://www.ipcc-nggip.iges.or.jp/faq/faq.html (Accessed June 2010).

International Energy Agency (IEA). *Coal Information*. Paris: Organisation for Economic Co-operation and Development/IEA, 2009.

International Energy Agency (IEA). *World Energy Outlook, 2009*. Paris: Organisation for Economic Co-operation and Development/IEA, 2009.

Podobnik, Bruce. *Global Energy Shifts: Fostering Sustainability in a Turbulent Age*. Philadelphia, PA: Temple University Press, 2006.

Smil, Vaclav. *Energy in Nature and Society: General Energetics of Complex Systems*. Cambridge, MA: MIT Press, 2008.

Kirk S. Lawrence
University of California, Riverside

Environmental "Bads"

The term *environmental "bads"* refers to a broad category of things or events that primarily endanger or damage environmental ecosystems. They correspond to negative consequences or derivatives of production, distribution, consumption, and disposal, and are typically opposed to economic "goods," as both are created through parallel technical systems and often unjustly distributed via similar socioeconomic or geocultural patterns. Central discussions occur essentially within economic grounds, where major concerns are ways of assessing financial and social costs, along with exploring solutions for penalties and later reparations. Nevertheless, environmental inequality and environmental justice movements are equally significant topics in this field, with debates ranging from uneven sharing of environmental burdens to local resistances and global transits of "bads."

These "bads" are always the result of human actions, whether accidental such as oil spills and food contaminations or expected such as by-products like e-wastes and nuclear toxic deposits. By definition, air and water pollution, trash, dumps, and so forth, are cited as key examples of environmental "bads." But this class may equally contain other events and things socially or biologically linked to ecological troubles such as environmental refugee migrations or environmentally induced maladies. There are also disputes within this category about including some organic movements that damage human, animal, or vegetal settings, like pests or floods, due to potential human accountability in their progress.

Between the liaisons connecting these "bads" to the "goods" of our modern worlds, there are symmetrical inversions that we need to be careful with, such as exploring one in places regularly inhabited by the other, that is, to regard "goods" as "bads" and vice versa, as if their qualities were frequently interchangeable. For instance, huge increases in the amount of some "goods" may trigger their observation as "bads" through expected quick boosts in future toxic dejects due to obsolescence trends. On the other hand, "bads" can also be the subject of role inversions. It happens when we add up the depletion of natural resources to positive parcels of "goods," while seldom subtracting values of lost biodiversity, or yet, when we pinpoint a few "bads" for enabling economic benefits. Regarding environmental diseases, for example, many events tend to be summed up as good financial inputs, amid medical treatments, waste eradications, court expenses, family relocations, and so forth.

Nevertheless, environmental "bads" are seen as negative aspects for the most part, even when they are not factored into the pricing systems of goods and become absent from their final values. In fact, the undesirability that comes with this negative essence is the nurturing point of leading discussions about their existences in economic domains and what primarily constitutes them as externalities. With severe costs, they regularly become disassociated from the economic routes that define particular goods within markets or daily life realms. But while this responsibility is repeatedly bypassed by producers using unsustainable practices, their costs do not simply disappear. In the end these costs, sometimes incommensurable, may be internalized through other goods and taxes, or become visible as troubling social or ecological costs in the short or long term.

Additionally, the negative aspects of environmental "bads" are obvious triggers in any search for ways to cease their effects, or at least mitigate and balance their impacts, from reactive market and government mechanisms to proactive solutions entailing micro or macro technical and lifestyle changes. Green tax shifts and analogous courses gained strength in the last decades by this token, since various government and nongovernment organizations began to support taxing "bads" instead of goods, with demands for fiscal penalties on damaging things such as vehicle emissions, abandoned terrains, etc., and cutbacks on beneficial ones, like savings, labor, etc. At the same time, national and international regulatory trends of "bads" were ordered over their quantities and qualities, along with their exchangeability, delocalization, and so forth. We can find them within municipal dispositions with price rankings of energy or water consumption levels, in the carbon trade regulations that became landmarks in supranational climate treaties such as the Kyoto Protocol, or yet, in efforts to discontinue environmental dumping acts such as the Basel Convention.

Ecological modernization frameworks have been at play within this domain as head runners in proactive paths to deal with "bads." Even with a treadmill of production critics observing that all goods will eventually generate negative outputs, defenders of this approach counteract with forms of production that help benefits to surpass costs, while healing ecological injuries and improving salubrity. Accordingly, reorganizing invention, production, distribution, consumption, or disposal activities toward green efficiency standards became a route to neutralizing environmental "bads." Adopted plans range from green consumerism and sustainable practices, by consumers, to alternative technologies or the assembly of environmental goods, by producers and technical agents. Such goods, like green vehicles, public parks, environmental education projects, etc., are even interpreted as a type of anti-bads, and many see them as a track for a decoupled economy where growth occurs without ecologically pressuring people and milieus previously endangered.

These solutions have, however, a long trail to cover regarding the environmental inequalities linked to unjust distributions of "bads." By following patterns also implicated in the provision of goods, "environmental bads" tend to be imposed in disproportionate ways throughout class, gender, and race groups. Moreover, due to territorial gatherings of unprivileged populations in already deprived areas, the weight of "bads" frequently becomes higher, whether by north–south globalized tendencies or urban-suburban local divisions. Many of these problems have been addressed by environmental justice movements concerned with ending discrimination in sharing the burden of "bads" and subsequent fair distributions of both "goods" and "bads." The NIMBY (not in my backyard) and NIABY (not in anyone's backyard) oppositions may be cited as examples of their milestones in the past decades. But nowadays, the capability of concerned communities to engage in collective action against environmental "bads" has been expanded toward global scenarios, with explored events from transnational misuses of fuel reserves to the "Great Pacific Garbage Patch."

See Also: Air Pollution; Environmental "Goods"; Environmental Justice Movements; Green Consumerism; Kyoto Protocol.

Further Readings

Beck, Ulrich. *World Risk Society.* Cambridge, UK: Polity, 1999.

Beder, Sharon. *Global Spin: The Corporate Assault on Environmentalism.* London: Green Books, 2002.

Fischer, Frank. *Citizens, Experts, and the Environment: The Politics of Local Knowledge.* Durham, NC: Duke University Press, 2000.

Latouche, Serge. *Farewell to Growth.* Cambridge, UK: Polity, 2009.

Alexandre Pólvora
University of Paris, 1 Pantheon-Sorbonne

Environmental Communication, Public Participation

People, places, and planning—public participation and communication practices related to physical environments have captivated scholars and practitioners for centuries. Recent work in this area has emphasized public involvement in sustainability planning and decision making. Complex environmental issues ranging from national energy policy to local waste and recycling programs have spurred broad public participation initiatives. Public participation requirements are now incorporated into many pieces of federal environmental legislation in the United States, with responsibilities for engaging the public often falling to local government agencies. Although there is considerable debate about the nature and merits of various public participation models, commonalities include an orientation toward particular communication practices such as conflict resolution, public deliberation, and collaboration. Still, difficult questions related to expertise, public knowledge, and the definition and assessment of meaningful participation persist. Central to these questions is a concern about how to meaningfully engage culturally diverse populations in environmental decision making.

Although conceptions of "the public" vary, the term is typically used to refer to individuals acting as citizens and/or as representatives acting on behalf of particular organizations or constituencies. The term *stakeholder* is also frequently used to refer to parties directly impacted by or interested in environmental decisions. "The public" is not a given entity but a concept that is constructed and interpreted in different ways. In some cases, "the public" may refer to people selected purposively based on a particular set of desired skills or characteristics. In other cases, "the public" may refer to special interest groups or stakeholders directly impacted by a particular issue. In many cases, use of the term *public* may serve to include or exclude particular populations depending on how it is employed and understood. There has been significant scholarly debate about the extent to which any one conception of "the public" is sufficient to capture the multiplicity of publics with environmentally related interests. Historically, particular groups such as women, minorities, reform organizations, and working-class publics have been neglected by traditional conceptions of "the public" used to identify potential participants.

Public participation related to environmental issues typically involves activities such as assessment, planning, decision making, conflict resolution, and evaluation. Public participation also includes social activism designed to influence government decisions—including the mobilization of public concern, lobbying, or interfering with the implementation of decisions. Parties use a wide variety of techniques for structuring participation, and communication processes shape and are shaped by ideologies, structural requirements (e.g., laws and regulations), personal and institutional preferences and resources, and situational factors (e.g., scope of issue). When information acquisition or exchange is the primary goal of public participation, the process is based on a simple sender-receiver model of communication, typically from government to publics for comment. When social or ideological learning and transformation are sought, communication processes become more complex.

In recent decades, public surveys have captured growing public distrust of government, and government-managed public participation processes have frequently generated cynicism. Public and business administration programs have often promoted the management benefits of strategically engaging stakeholders so that terms like *buy-in* and *political consensus* have become associated with carefully managed campaigns rather than open-ended public participation processes. Even well-intentioned community and global development organizations have frequently mandated participation models without sufficient attention to the complexity of everyday collaborative work confronting diverse communities of citizens and practitioners. Criticisms typically relate to questions of legitimacy, quality, and efficiency of public participation processes.

Public Participation Models

Since the 1980s, public participation has moved from concerns about relatively infrequent acts of "expressive" participation such as voting, political campaigning, and protest to conceptualizing participation as an ongoing, social aspect of democratic citizenship and governance. The shift from participation conceived as an individual "act" to participation as collective "acting" has been rapid, and diversifying conceptions of what it means to be active in public life have complicated the conceptual and practical field. Public participation in its broadest sense has become a catchall term among scholars, government officials, and publics for a wide variety of practices, processes, and perspectives on citizen engagement in policy discussions and decision-making processes.

Scholars and practitioners who engage in environmental public participation work are scattered across fields like communication, sociology, public affairs, political science, environmental science, and development studies. This makes it difficult to characterize the scholarly work in a succinct way. Even a narrower focus on environmental communication and participation involves both scholarly research and practitioner work across numerous fields. Such diversity offers benefits associated with the existence of many different models of participation, multiple perspectives on public-government interaction, and numerous facilitation techniques for all types of groups, issues, and processes. A drawback to this diversity, however, is that literatures often talk past each other, terms and concepts are saddled with multiple and sometimes conflicting meanings, and few models of public participation receive empirical attention. Existing public participation research has not always lent itself to comparisons, and much research on public participation and communication focuses on highly general aspects of public participation design or, alternately, on localized case studies.

Public Participation and Deliberation

Despite fragmentation, much of the public participation theory and research has emerged from within a liberal democratic framework. Attention to environmental communication and participation issues such as representation, voice, and deliberation frequently draw on John Rawls's notion of justice as fairness and Jürgen Habermas's conceptions of the public sphere and the ideal speech situation. Habermas put forward the notions that (1) discourse in the public sphere affects public policy by establishing the terms of legitimate public debate, (2) public deliberation should be guided by communicative rules (e.g., claims must be acceptable to all potentially affected), and (3) equitable deliberation requires certain conditions be met (e.g., every person must be allowed to question, introduce, and express opinions). Although debates about the nature and efficacy of public deliberation continue, it is generally conceptualized as a process unmediated by power differentials during which information is exchanged so that participants may consider as many sides of an issue as possible.

Two dimensions of public deliberation help to frame the concept—deliberation's stated goals and its purported effects on participants. Three goals are commonly associated with public deliberation: (1) education, where the purpose is for participants to learn about the various dimensions and perspectives on public issues (e.g., climate change literacy forums); (2) consultation, in which participants deliberate and offer an informed opinion or proposal to policy makers who use it as one more data point when considering their final decision (e.g., land use planning advisory groups); and (3) decision making, a process in which participants have ultimate decision-making authority. Because decision making in the public deliberation mode is extremely rare, the categories of educative and consultative deliberation have received the most attention from researchers.

Chief among claims about public deliberation's benefits are (1) deliberation increases participants' understanding, complex thinking, and flexibility about issues; and (2) involvement in public deliberation increases engagement in other methods of public participation. Several reviews of deliberative democracy research have concluded that public deliberation can affect individuals' understanding of an issue and increase participation, but in the absence of longitudinal studies it is difficult to determine whether these effects persist over time.

A popular deliberative forum is the citizen jury, in which a randomly selected group of stakeholders is convened for the purpose of learning about and discussing policy options with the goal of presenting a consensus statement to the convening party (e.g., government agency). Citizen juries have been used around the world. One citizen jury in India brought together indigenous farmers for four days of deliberation concerning genetically modified organisms (GMOs) and the future of farming in their communities. The participants heard from and questioned expert witnesses in government, business, and agriculture who presented a wide range of positions on GMOs, allowing "citizen jurors" to navigate the complex realities of biotechnology in light of social, political, and environmental concerns. A report from the Environmental Mainstreaming Initiative concluded that the diversity and lack of disciplinary boundaries among the participants in this process demonstrated that deliberation practices could enrich democracy and ensure greater accountability relative to food and agricultural decisions.

Alternative Communication Approaches

Unlike deliberative models, which seek to reduce conflict and pursue consensus, critical approaches to environmental communication and public participation emphasize contestation and difference based on the assumption that conflict is both unavoidable and potentially productive. Those who favor this approach critique traditional consensus-based deliberation processes as neglecting or minimizing the influence of power differentials on social interactions. Critical communication theorists suggest that liberal democracy's reliance on public deliberation mechanisms is insufficient to ensure meaningful participation from diverse publics. Public deliberation processes typically assume that all members of the public have equal access to participate meaningfully in deliberative forums. Scholars and environmental justice activists argue that existing participation systems favor particular cultural norms and frequently exclude populations without the economic means, technical knowledge, or social access to engage in deliberation about environmental problems and inequities.

Many scholars and practitioners also adopt a pragmatic approach to addressing public participation in environmental problem solving based on early work by John Dewey. This approach is future oriented and attentive to the social construction of reality. The practical consequences of environmental theories and measures are important not in relation to some complete or perfectly verifiable truth but in terms of their capacity to guide future actions related to environmental policy formation and social behavior. In recent decades, a number of pragmatic scholars have worked with practitioners to develop new communication practices to help resolve tenacious environmental conflicts. In some cases, formal models have been crafted in response to environmental legislation, and in other cases, conflict resolution efforts have relied on more informal communication related to dialogue and/or collaboration.

Public Participation and Dialogue

Public dialogue is grounded in the work of physicist David Bohm, existential philosopher Martin Buber, and psychotherapist Carl Rogers, among others. Public dialogue takes as its starting point the assumption that understanding self and other is central to constructing better public policies and ultimately better publics. Whereas public deliberation is focused on developing public policy that is responsive to the greatest number of people, public dialogue's primary purpose is to heal social rifts between parties so that progress can be made on policy disputes.

Unlike deliberation, dialogue has not been widely embraced by governments and the public as a form of policy discussion for two reasons. The first reason is revealed clearly in the title of Daniel Yankelovich's influential 1999 text, *The Magic of Dialogue*. Dialogue, as the term is used by participation scholars in this tradition, cannot be manufactured or mandated; one cannot be ordered to dialogue and, even if one enters into a possible dialogic space, dialogue itself may not manifest. In this sense dialogue is "magic," something that happens spontaneously in the moment, cannot be re-created, and is not necessarily bound to last. All dialogic theorists agree on this point: only the conditions for dialogue can be created, not the dialogic experience itself. The conditions for the possibility of dialogue vary from author to author, but those who have given thought to a systematic approach generally agree that dialogue is marked by (1) an acknowledged equality among participants, (2) the bringing of assumptions to the fore of discussion, (3) listening with empathy, and (4) a disavowal of communication structures (i.e., rules). Many advocates of the dialogic approach suggest that these characteristics make it particularly effective for intercultural participation initiatives.

The second reason dialogue has not made significant headway in contributing to environmental participation practices is its lack of emphasis on decision making. Decision making, which invariably follows procedural rules, closes off routes to dialogue by promoting movement in a linear fashion orientated toward a final goal. As pre-interactional aspects of communication, decision-making rules can be structured to give one or more parties advantages over others, violating the equality principle of dialogue. More importantly, the goal of decision making places participants in an antagonistic relationship, the result of which is likely to be debate, not dialogue, which violates dialogue's empathy principle. Since dialogue is conceptualized as a way of gaining an understanding of others and questioning one's own assumptions, encountering others and interrogating our own assumptions cannot be a linear process with objectifiably measurable outcomes. Dialogue is an open-ended phenomenon directed at continually reopening questions and relations. This makes it unlikely that dialogue will contribute directly to environmental policy making unless it is coupled with other communication and participation models that facilitate time-sensitive decision making.

Public Participation and Collaboration

Environmental collaboration efforts have become increasingly common in recent decades. Most work in this area has originated from projects and studies related to conflict resolution or environmental resource management. Although various collaborative models exist, environmental collaboration efforts typically share several common assumptions and characteristics. In most cases, these efforts involve participants who do not normally interact on a regular basis, and, in some cases, these participants are already engaged in adversarial relationships. Collaborative models typically rely on voluntary and uncompensated participation. Rather than attempting to minimize differences, most collaborative processes seek to engage differences in order to generate new perspectives and solutions to environmental problems. Theorists typically advocate collaboration models that engage public participants early in a process, assuming that it is important to allow participants to constructively explore differences prior to establishing agendas or identifying solutions. In this sense, environmental collaboration is not a vehicle for achieving consensus around existing solutions but a process that allows for more complex understandings of environmental problems and for more creative solutions.

Advocates for environmental collaboration models often point out that successful collaboration has the potential to generate outcomes that individuals or homogenous groups would be unable to achieve on their own. However, a reliance on voluntary participation coupled with attention to difference also makes collaborative processes tenuous and easily disrupted. If participants are unable to identify shared goals or develop and sustain trust, a collaborative process may be untenable. For this reason, many collaborative efforts rely on skilled facilitators and mediators to help design and implement effective communication practices. According to Barbara Gray, effective collaborative processes should involve all participants in defining problems, establishing a direction, and implementing and monitoring agreements.

In the mid- to late 1980s, a number of governmental agencies began adopting collaborative models for engaging the public in long-term resource planning. Initially, these efforts were promoted most heavily at local and regional levels where policy makers were faced with difficult decisions related to land use planning and watershed management. Over time, a number of federal agencies adopted various environmental collaboration models. For example, the U.S. Forest Service has used outside facilitators to help direct and mediate collaborative roundtables tasked with revising forest management plans. Environmental collaboration efforts are most often implemented at the local level in relationship to specific projects or issues, but new communication technologies are contributing to expanded conceptions of collaboration.

Public Participation Challenges

Although the focus so far has been on points of agreement among the models of public participation, there are also enduring conflicts. Two common points of contestation related to public environmental participation and communication involve questions of knowledge and legitimacy. There is significant disagreement over the degree to which nonexperts should be involved in activities and decisions typically assigned to experts. Many scholars and community activists have argued that traditional approaches to environmental planning and problem solving neglect the potential contributions of local culture and knowledge. They argue that particular ways of knowing and communicating are given preference in ways that exclude important stakeholders. Critics point out that certain forms of public participation in environmental decision making have the potential to supplant scientific facts with opinion or conjecture. Some critics suggest that comprehensive participation processes cost too much and take too much time when responses to environmental crises are time sensitive. There is also debate about when the public should get involved in addressing environmental issues. Despite these challenges, new approaches to public participation and communication may offer alternative ways of engaging diverse populations in a wide range of environmental decision making and planning processes.

See Also: Communication, National and Local; Public Opinion; Social Action, National and Local.

Further Readings

Calhoun, Craig, ed. *Habermas and the Public Sphere*. Cambridge, MA: MIT Press, 1991.
Cox, Robert J. *Environmental Communication and the Public Sphere*. Thousand Oaks, CA: Sage, 2009.
Dietz, Thomas and Paul C. Stern, eds. *Public Participation in Environmental Assessment and Decision Making*. Washington, DC: National Academies Press, 2008.
Gray, Barbara. *Collaborating: Finding Common Ground for Multiparty Problems*. San Francisco, CA: Jossey-Bass, 1989.
Rawls, John. *A Theory of Justice*. Cambridge, MA: Belknap, 1971.
Ryfe, David M. "The Practice of Deliberative Democracy: A Study of 16 Deliberative Organizations." *Political Communication*, 19 (2002).
U.S. Environmental Protection Agency. "Public Involvement." http://www.epa.gov/publicinvolvement/index.htm (Accessed January 2010).
Webler, Thomas and Seth Tuler. "Fairness and Competence in Citizen Participation." *Administration and Society*, 32/5 (2000).
Yankelovich, Daniel. *The Magic of Dialogue: Transforming Conflict Into Cooperation*. New York: Simon & Schuster, 1999.

Sara McClellan
Stephen Konieczka
University of Colorado

Environmental "Goods"

How we understand environmental "goods" and what they mean to our decision making as people on Earth is critical to our collective and individual health as well as that of the planet. We have always had a material, and thus an economic, relationship with our environment. It is one that both informs and, conversely, is informed by our particular relationship vis-à-vis nature. How we understand and determine economic values for the physical natural world that surrounds us is based on our worldview. Do we save a mountain from mining or highway construction because it has a spiritual connection to our gods or because it has great recreational value? If we perceive it to have tangible public value—how do we decide or know this? Do we develop sustainably harvested forests for the health of the forest and wildlife or because they create longer-term market strategy for us as consumers? The struggle with advantaging environmental goods with a concurrent positive economic impact is the current economic systems' need for competition. Negative impacts are counted as valuable. Our economy operates on a destructive model. Disasters are good. Work and resources are used. Nondisaster, nonresource, and labor-intensive activities are not "productive."

Economic Definition

A "public good" is an "economic good" that is available for everyone and, when consumed by one person, it does not become less available for others. A "public good" is not privately owned. The responsibility for and the management of many "environmental goods and services" is held by the government. There are rivalrous and nonrivalrous goods as well as excludable and nonexcludable goods. Rivalrous goods are goods whose consumption by one consumer prevents simultaneous consumption by other consumers. One consumer may consume nonrival goods without preventing simultaneous consumption by others. Excludable goods are goods or services that, when possible, prevent people who have not paid for them to have access. It is not possible to prevent access to nonexcludable goods or services, as they are available for everyone.

Environmental Goods

Environmental goods is the term used to describe a subsection of the economic concept of public goods. This concept is used to define how we see and understand the natural world around us. It is used to describe the goods and services we call clean air, clean water, flora, fauna, mountains, forests, beaches, rivers, scenic views, landscapes, urban parks, town squares, and green transport infrastructure (footpaths, cycleways, greenways). These goods are tangible and yet not to be confused with "common goods," which may be depleted with use (fishing stocks, timber, coal, and so forth). Environmental goods are not perceived to be depleted by use.

Public Domain

Our cultural awareness and economic needs shape these environmental goods. How we protect them in the case of air and water, flora and fauna, or develop them in the case of green infrastructure and urban parks, shapes and reflects who we are and what we hold dear.

History is full of privately built gardens that are now in the public realm (examples range from Hyde Park in London to the Imperial Palace in Beijing), and many of our conservation lands and national forests were once held as private hunting grounds by some prince or lord; Central Park in New York, on the other hand, was built as a public space with the public welfare in mind. Today, when private lands transfer to public ownership, for example in the conservation of farmland or bear habitat, there is often a real economic loss in terms of tax income to a municipality, but also less-tangible benefits to the community that are unaccounted for in terms of health, safety, and welfare.

Other environmental goods such as water have a long and complicated history of ownership, access, negotiation, and control. The Salton Sea, now a natural area and bird migration flyover in southern California, was created by a water breach in the cross-desert channeling of the Colorado River. The ongoing pollution of our clean waters by urban and agricultural runoff, oil spills large and small, and ownership of waterways, dams, and irrigation systems is depleting our collective "environmental good" and brings into question how we build, regulate, or maintain environmental goods within the public domain.

Beaches, especially as public amenities, carry large environmental impacts. We expect our municipal places to accommodate all abilities and ages of people and therefore ask for shoreline protection, sandy footing, breakwaters for safety, boardwalks, toilets, and so forth, often at the expense of natural systems, weather patterns, and the natural ebb and flow of sandbars.

Current Practices

Enhancing, building, and maintaining the realm of environmental goods is an ongoing concern. There is considerable design effort going into increasing the range and quality of environmental goods within our cities as a means of making them more sustainable. Taking back highways such as Portland, Oregon, did a decade ago to make a public park along the river, or main streets for marketplaces, and riverfronts as festival places such as in San Antonio, Texas, and Providence, Rhode Island, can increase the environmental good while also having other positive environmental and economic impacts. On the other hand, the viewsheds and airspace once highly regarded and strictly kept are being lost to increasing building density and aeronautical needs. For example, the density of high-rise buildings in Hong Kong is so great that the natural air movement between buildings is blocked, so they have begun to cut donut holes out of existing buildings to improve airflow and ventilation into apartments.

Other practices influencing environmental goods are the growth of community gardens within a city, rain gardens and constructed wetlands as storm water mitigation efforts, street tree planting programs, and new trails such as the High Line in New York City. The development of green transport infrastructure, such as footpaths, bicycling paths, and greenways, is occurring at many scales and in many places, but one model that stands out is the tree-lined pedestrian road built by Enrique Peñalosa when he was mayor of Bogotá, Colombia, from 1998 to 2001.

Conclusion

Cultural awareness, place in history, technology, and industry evolving over the centuries and around the globe impact our practices toward our environment and our assumptions about environmental goods.

See Also: Environmental "Bads"; Environmental Justice Movements; Nature Experiences; Vacation; Weekends.

Further Readings

Chander, Parkash, Jacques Drèze, C. Knox Lovell, and Jack Mintz, eds. *Public Goods, Environmental Externalities and Fiscal Competition: Selected Papers on Competition, Efficiency, and Cooperation in Public Economics, Essays by Henry Tulkens*. New York: Springer, 2010.

Phyper, John-David and Paul MacLean. *Good to Green: Managing Business Risks and Opportunities in the Age of Environmental Awareness*. Hoboken, NJ: Wiley, 2009.

Smithsonian. "Colombia Dispatch 11: Former Bogotá Mayor Enrique Peñalosa." http://www.smithsonianmag.com/travel/Colombia-Dispatch-11-Former-Bogota-mayor-Enrique-Penalosa.html (Accessed December 9, 2010).

Dianne Elliott Gayer
Independent Scholar

Willa Antczak
Vermont Design Institute

Environmental Justice Movements

For decades, sites for toxic waste disposal targeted minority and low-income rural areas; environmental justice movements strove to change this through litigation and by using tactics from the Equal Rights Movement.

Source: iStockphoto

The U.S. Environmental Protection Agency Office of Environmental Justice defines environmental justice (EJ) thus: "fair treatment and meaningful involvement of all people regardless of race, color, national origin, or income with respect to the development, implementation, and enforcement of environmental laws, regulations, and policies."

Environmental injustice exists when disadvantaged populations, usually racial minorities and economically underprivileged groups, disproportionately bear the burdens of environmental degradation, and are denied both participation in and access to justice in environment-related matters. Advocates of environmental justice movements (EJMs) agree that the root causes of environmental injustice are "institutionalized racism; the commodification of land, water, energy and air; unresponsive, unaccountable government policies and regulation; and lack of resources and power in affected communities."

Birth of a Movement

Long before the term *environmental justice* was coined, farm workers, organized by César Chávez in the 1960s, demanded more rights in the workplace and protection from pesticides and other harmful chemicals used by the agriculture industry. Shortly thereafter, Rachel Carson's *Silent Spring* (1962) became a turning point in bringing issues of environmental deterioration and the resulting impact on humans to the fore. Both these events were instrumental in bringing environmental concerns into public discourse, which created a need for fighting environmental injustice.

Environmental injustice, also known as environmental racism, is defined as "enactment or enforcement of any policy, practice, or regulation that negatively affects the environment of low-income and/or racially homogeneous communities at a disparate rate than affluent communities."

Similar to other social movements that aim for equitable living standards, EJMs use the same approach as the movement from which they take inspiration—the civil rights movement—and address the inequitable distributions of environmental burdens, namely, pollution, industrial facilities, environmental crime, etc. In the following years, using their newfound courage from the passing of the Civil Rights Act in 1964, the African American community fought against several incidences of environmental racism, all of which did not result in victory. Nevertheless, it put EJMs on the map with other social movements and highlighted the need for institutions that kept environmental racism in check.

In the United States, EJMs led to creation of institutions like the Environmental Protection Agency (EPA) and the Council for Environmental Quality (CEQ). Today, the EPA manages a variety of issues that address environmental justice, for example, "federal research, standard setting, and enforcement activities to ensure environmental protection."

Despite the establishment of the EPA and the CEQ, the first protest to be recognized nationally and to crystallize the presence of the environmental justice movement, also driven by people of color, came over a decade later in 1982, when the primarily African American community in Warren County, North Carolina, rose against dumping of toxic PCB-laced soil. A year later, the concerns of EJ advocates were testified to in a report by the General Accounting Office (GAO). The GAO study, urged by Walter Fauntroy, District of Columbia congressional delegate and then-chair of the Congressional Black Caucus, revealed that "three-quarters of the hazardous waste disposal sites in eight southeastern states are in poor and African American communities."

These findings had already given momentum to the EJMs, when in 1984, toxic fumes due to a leak at the Union Carbide plant, a pesticide manufacturing plant in Bhopal, India, killed at least 6,000 people in the first few weeks. The death toll and the number of those affected continue to rise more than two decades later. The event made international headlines and made the concerns of EJMs international concerns. The same year, the *Los Angeles Times* broke the story that the California Waste Management Board advised governments and companies looking to dump hazardous waste to target "small, low-income and rural communities with a high percentage of people who are old or have little education."

Several studies published in the 1980s and early 1990s lent credibility to the charges of environmental racism. Notable evidence came through the efforts of the United Church of Christ (UCC) Commission for Racial Justice (CRJ). In 1987, Reverend Benjamin Chavis, then-director of CRJ, published *Toxic Wastes and Race in the United States*. The report provided empirical data that race was the most important factor in determining the sites for toxic waste in the United States and became a significant tool in garnering support for the cause of environmental justice. The report, by the UCC's director of research Charles Lee, established a strong statistical correlation between race and the location of hazardous wastes sites, implying that the siting of these facilities in communities of color was not an accident but rather the "intentional result of local, state and federal land-use policies."

In 1990, sociologist Robert Bullard, known as the "Father of Environmental Justice," continued with the theme of environmental racism in his book *Dumping in Dixie: Race,*

Class, and Environmental Quality, which brings to light that race was a significant factor in the siting of unwanted toxics-producing facilities.

Environmental Justice Around the World

Though EJMs in the United States deal with the issues of race and inequality in distribution of resources, EJ campaigns around the world have shifted their focus to address issues in the context of their respective countries. For example, in the United Kingdom, the Environmental Justice Foundation, a nongovernmental organization (NGO), has sought to link the need for environmental security and the defense of basic human rights.

In South Africa in the 1990s, due to a history of apartheid and the intention of removing discrimination from all policies, the Environmental Justice Networking Forum (EJNF) was created as a nationwide umbrella NGO designed to coordinate the activities of environmental activists and organizations interested in social and environmental justice. The EJNF, which included as many as 600 organizations, linked "poverty and environmental degradation." In 1994, the newly elected African National Congress (ANC) connected "poverty and environmental degradation" and ensured that issues of environmental injustice were addressed as a part of post-apartheid reconstruction and development. South Africa's 1996 constitution grants South Africans "the right to an environment that is not harmful to their health or well being, and to have environment protected, through reasonable legislative and other measures that: prevent pollution and ecological degradation; promote conservation; and secure ecologically sustainable development and use of natural resources while promoting justifiable economic and social development."

The 2000 movie *Erin Brockovich* was based on the true story of a woman investigating the adverse health effects of exposure to chromium 6, a toxic substance that caused numerous serious ill-health effects for residents of Hinkley, California.

See Also: Chávez, César (and United Farm Workers); Environmental Protection Agency; National People of Color Environmental Leadership Summit; United Church of Christ.

Further Readings

Bullard, R. D. *The Quest for Environmental Justice: Human Rights and the Politics of Pollution.* San Francisco, CA: Sierra Club Books, 2005.

Fortun, K. *Advocacy After Bhopal: Environmentalism, Disaster, New Global Orders.* Chicago, IL: University of Chicago Press, 2001.

Rhodes, Edwardo L. *Environmental Justice in America.* Bloomington: Indiana University Press, 2005.

Schlosberg, D. *Defining Environmental Justice: Theories, Movements, and Nature.* Oxford, UK: Oxford University Press, 2009.

Smith, M. J. and P. Piya Pangsapa. *Environment and Citizenship: Integrating Justice, Responsibility and Civic Engagement.* New York: Zed Books, 2008.

Szasz, Andrew. *EcoPopulism.* Minneapolis: Minnesota University Press, 1994.

Charu Uppal
Karlstad University

Environmental Law, U.S.

Over the past century, a body of law centered upon the regulation of activities with environmental impacts has developed in response to the growing recognition of the environmental damages associated with an increasingly industrialized American society. Predominantly developed in the post-1970 era, these environmental laws have provided a foundation designed to regulate economic activities in an effort to protect air, water, and other vital resources, and moreover, have fundamentally transformed American society's relationship to its natural environment and have begun a shift away from unbridled economic growth toward an ideal of economic and environmental balance.

The Early Years: Environmental Law Before Earth Day

Common Law

Although U.S. environmental law came into its own during the 1970s' Earth Day–era ferment (the so-called Environmental Decade), this body of law did not lack precedents from earlier periods. As environmental issues began bubbling up, courts applied common law doctrines with varying impact as well as whatever local regulations or ordinances that existed in a given area. The doctrine of nuisance law, which allowed a landowner whose reasonable use of his or her land was unreasonably impacted by the use of another to bring suit against the offending individual, was the most prevalent common law theory applied to environmental ills. This approach was able to address highly localized environmental issues (i.e., a neighboring hog farm spilling manure into your yard), but lacked any mechanism for addressing larger and more national environmental problems. This paradigm also relied on concerned individuals to actually bring a lawsuit to remedy the particular local environmental concern—which led to inconsistent and uneven results depending on resources and willingness to proceed to litigation.

The Great Depression and Rise of the Administrative State

However, the Great Depression and rise of the administrative state began the profound shift in the nature of the relationship of the federal government to the environment. Again, prior to the New Deal, environmental regulations, if any, emanated from local or state governments, as environmental issues were seen as a largely local concern. However, by the 1930s, the pace of industrial development and the associated risks had accelerated to the degree that local regulation of these issues was no longer practicable or possible. To this end, New Deal–era planners developed large-scale planning and coordination activities aimed at altering land use or encouraging voluntary natural resource planning activities. Although New Deal–era legislation did begin to rein in unbridled agricultural and industrial activity, such efforts were also of limited scope or impact and generally did not go as far as to impose the affirmative regulatory mandates associated with environmental law.

The Environmental Decade: The 1970s

As the 1970s dawned, awareness of the environmental impacts of contemporary society reached the point where action was necessary in order to place some affirmative limitations on industrial activities and growth. Again, common law, resource planning/management, and voluntary solutions had largely not checked these concerns, and the quickening pace of environmental issues facing society had largely outpaced the ability of available legal options to address such issues. A few very prominent environmental disasters—such as the Cuyahoga River fire in 1969—made this inability patently clear.

In response to the failure of common law and the difficulty in applying many common law doctrines to the expanding problem of environmental issues, affirmative legislation became the mechanism through which society attempted to address these problems. Beginning in the late 1960s through approximately 1980, the U.S. Congress passed a number of landmark legislative enactments intended to address many of the most severe environmental issues. This wave of legislation certainly was not perfect, and efforts have routinely attempted to increase or minimize their application depending upon political vagaries—but regardless, these enactments have proven to be the bedrock of U.S. environmental law. The following are the principal legislative enactments and typify the field of this body of law.

National Environmental Policy Act

In 1969, the National Environmental Policy Act (NEPA) was enacted. NEPA required governmental agencies to consider the environmental impact of their actions on a number of scenarios prior to taking action that could have environmental or societal impact. Although NEPA did not affirmatively require an agency or funded entity to alter its activities based upon the findings of its investigation, NEPA did, for the first time, force governmental actors to consider the impacts of their proposed actions and to evaluate whether less harmful alternatives could be conducted to minimize the environmental impacts.

Clean Air Act

NEPA was followed with the enactment of the Clean Air Act (CAA). The CAA was passed in response to increasingly harmful levels of air pollution and focused on both stationary sources of pollution (e.g., a building or structure emitting air pollution) and mobile sources of pollution (e.g., automobiles). The CAA has been critical in regulating air pollution from industrial sources and in regulating the amount of output of regulated pollutants from a given source or in a given jurisdiction. The CAA also set national ambient air quality standards in an attempt to establish baseline air quality levels in given localities.

Clean Water Act

In 1972, the Clean Water Act (CWA) established a system for regulating the discharge of pollutants into the waters of the United States. Overall, the CWA provided a basic program for developing pollution control programs—including regulation of wastewater discharges from industrial sources and establishment of water quality standards for contaminants in given waterways. To accomplish its regulatory goals, the CWA established a comprehensive permitting system tailored to limit such discharges.

Toxic and Hazardous Substances

During this same period, Congress also passed regulations governing toxic or hazardous chemicals. In 1976, the Resource Conservation and Recovery Act (RCRA) was enacted to create a life-cycle regulatory process governing all aspects relating to the management of hazardous waste. This enactment recognized the growing issue related to the harmful impacts associated with hazardous waste in landfills and provided a separate operating platform designed to mitigate and control these chemicals.

In 1980, the Comprehensive Environmental Response, Compensation, and Liability Act (CERCLA) was enacted to provide for federally funded remediation of sites impacted by hazardous waste. This act created a "Superfund" to fund these remediation efforts and also tasked EPA with tracking down responsible parties who contributed to the environmental impacts associated with their hazardous waste. Overall, CERCLA provided a meaningful method to locate and to restore areas impacted by the improper disposal of hazardous waste.

Conclusion

As demonstrated by the sampling of U.S. environmental laws discussed above, since 1970, a body of regulatory law has been developed to address environmental quality issues on a topical area basis. This body of law admittedly is not perfect, but it has fundamentally transformed American society's relationship to the environment. But beyond this, the enactment of these provisions reflects a shift in values and a growing recognition that environmental regulation is indeed necessary and beneficial as society grows ever more industrialized.

See Also: California AB 32; Clinton, William J. (Executive Order 12898); Environmental Protection Agency; International Law and Treaties; Kyoto Protocol; Law and Culture.

Further Readings

Brooks, Karl B. *Before Earth Day: The Origins of American Environmental Law, 1945–1970*. Lawrence: University of Kansas Press, 2009.

Henderson, H. and W. Woolner, eds. *FDR and the Environment*. New York: Palgrave Macmillan, 2005.

Kraft, Michael E. *Environmental Policy and Politics*. New York: Pearson Longman, 2007.

Lazarus, Richard J. *The Making of Environmental Law*. Chicago, IL: University of Chicago Press, 2006.

Jess Phelps
Independent Scholar

Environmental Media Association

The entertainment industry has the power to change people's behaviors, introduce new cultures, make people aware of new trends and problems, and prompt the masses both via its products and its heroes. The Environmental Media Association (EMA) in Los Angeles, California, aims to channel this power toward establishing an environmentally friendly and sustainable lifestyle and making people aware of these problems.

The EMA is a premier nonprofit organization founded in 1989 by Cindy and Alan Horn and Lyn and Norman Lear. The EMA collaborates with the entertainment industry to encourage green production and raise the public's environmental awareness. The EMA targets environmental awareness through celebrity and the power of the media and seeks to mobilize the entertainment industry in a global effort to educate people about environmental issues and inspire them into action. The EMA was created with a simple yet powerful concept—through music, television, and film, the entire entertainment community could influence the environmental awareness of millions.

Studies carried out by EMA have various dimensions. The EMA arranges green campaigns with the participation of Hollywood celebrities on the one hand, and on the other, tries to make the show business industry greener. Its primary effort to this end is to get producers to take environmental consciousness into account during production of movies, music, and TV shows. The association remarks on issues such as recycling and energy usage during entertainment production and makes recommendations on how environmental damages could be eliminated. To this end, it cooperates with the Green Seal organization, which evaluates companies with respect to environmental responsibility. As a result of this cooperation, a new organization, EMA Green Seal, was established.

EMA Green Seal in 2004 established an award to recognize and express approval of film and television productions that incorporate green initiatives into production practices. Green Seal, Inc., and EMA have officially partnered in this groundbreaking work in greening the entertainment industry. The Green Seal Award recognizes environmentally responsible production efforts behind the scenes, such as set construction, energy usage, resource conservation, recycling, and purchasing policies. The first movie to display the EMA Green Seal in its credits was *The Incredible Hulk*, which made specific efforts during its 2007 filming to reduce carbon emissions and waste during production. The EMA also hosts the annual Environmental Media Awards, an awards ceremony that celebrates the entertainment industry's environmental efforts. The annual EMA Awards recognize writers, producers, directors, actors, and others who include environmental messages in their work.

In addition to these awards, a second dimension of EMA's mission is to collaborate with the environmental community and famous singers, actors, and actresses. The EMA has partnered with some environmental communities, including the Endangered Species Coalition, the Alaska Rainforest Campaign, the Sierra Club, and the World Wildlife Fund. It has also sponsored some environmental projects, for example, "An Evening in Brazil." This project benefits the work of the Rainforest Foundation. Bruce Springsteen, Sting, Billy Crystal, and many other artists have given support, and raised over $1 million in 1990. By weaving environmental messages into entertainment programming, utilizing "celebrity" for positive role modeling, and positioning itself at Hollywood parties and celebrity events such as the EMA Awards on the E! Channel and the annual Golden Globes Green Party, EMA serves as a valuable link between the entertainment industry and the environmental community.

Another working area of EMA is to encourage people to adopt a green lifestyle. It carries out studies in this area in a different dimension. The first is to cooperate with various companies to encourage them to join the social responsibility campaign, using celebrities to give a face to this campaign. Ed Begley, Jr., Danny DeVito and Rhea Pearlman, Daryl Hannah, Salma Hayek, Edward Norton, and Brad Pitt celebrated the BP Solar Neighbors Program with EMA and the Enterprise Foundation. EMA recruits celebrities to purchase solar power for their homes. For every home, BP Solar donated a system to a low-income family in Los Angeles. The second method is based on the programs developed within the association. The target group of these studies is young people, and the supporters of the studies are popular Hollywood stars who are role models. The EMA calls this program "gen e." The EMA launched the "gen e" program as an opportunity for young Hollywood to become involved with environmental concerns and to model sustainable lifestyles. The first "gen e" public service campaign is with Amy Smart. The EMA agenda includes assembling a group of young celebrity ambassadors to support green behavior among their fans and launching an organic garden program in public schools.

The EMA also creates and sponsors TV shows to promote the green lifestyle and increase environmental awareness. *Green Power Family Hour* was one of these programs that celebrated the environment and teaches adults and children tips on ways to keep a greener lifestyle. The EMA also sponsored a special screening of HBO's environmental documentary *Earth and the American Dream* in Washington, D.C., with a delegation of actors lending their support, and arranged a briefing with Bill Clinton, Al Gore, congressional members, environmental leaders, and top environmental officials to discuss environmental issues facing the administration.

See Also: Begley, Jr., Ed; Film, Product Placement; Institutional/Organizational Action; Nongovernmental Organizations; Television, Cable Networks.

Further Readings

Environmental Media Association. http://www.ema-online.org (Accessed May 2010).

Environmental Media Association "gen e." http://www.ema-online.org/gen_e.php (Accessed November 2010).

Environmental Media Association Green Seal. http://www.ema-online.org/green_seal.php (Accessed November 2010).

Şule Yüksel Öztürk
Anadolu University

Environmental Peacemaking

The emerging field of environmental peacemaking seeks to address in a positive manner the impacts of living in a shared environment with finite resources. The implications of issues like climate change, energy shortages, drought, and environmental pollutants have steadily moved into mainstream political and social thought. It is generally presumed the nations and peoples of the world will, for the foreseeable future, be wracked by competition and conflict over scarce and dwindling resources, exacerbated by the environmental challenges that we presently face. The resultant "resource wars," as Michael Klare surmises, will likely be fought over the possession and control of vital economic goods, especially those resources most needed for the functioning of modern industrial societies. By most accounts, in the years ahead, resource wars will be a dominant feature of the global security environment.

Resource conflict recently has been understood as both a leading cause and a consequence of warfare, and the environmental impacts of militarism have been included among the

Peace parks, generally along contentious borders like the demilitarized zone (DMZ) between North and South Korea, are regions that have endured past conflict and are transformed into protected areas as a way to promote peaceful relations between nations and nature. The DMZ has become a massive wildlife and bird sanctuary since the end of the Korean War.

Source: Wikimedia Commons

casualties of war by scholars in the field since at least the Vietnam War. In her book *The Ecology of War*, Susan Lanier-Graham observes that the environment has always been a victim in warfare, dating back at least to Old Testament descriptions of Samson burning the Philistines' crops and Abimelech sowing the ground with salt after a military victory in order to render it infertile—with the former tactic repeated in the Peloponnesian War in 429 B.C.E. and the latter by the Romans in Carthage circa 150 B.C.E. Despite often being perceived as "good wars," Lanier-Graham points out that the American Revolution and World War II both possessed elements of "environmental warfare" (the destruction of the enemy's resources as a conscious tactic of war) and "scorched-Earth tactics" similar to those described in the biblical narratives. Indeed, many scholars likewise have noted the use of practices of deliberate degradation of the environment for hostile military purposes in the modern era as well.

Inquiries of this sort generally have looked at issues such as the enormous cost of warfare and its resource-draining tendencies; the resultant despoliation of resources, including infrastructure, water systems, forests, transportation, and agricultural sites; impacts upon animal populations and habitats; and the long-term toxifying and disease-causing effects of warfare on environments and peoples. In 2008, the United Nations Environment Programme (UNEP) launched a post-conflict and disaster management branch designed to focus on these concerns in an attempt to forestall future conflicts. UNEP is framed by a view suggesting that the exploitation of natural resources and similar environmental stresses are implicated in all phases of the conflict cycle, contributing to the outbreak and perpetuation of violence and further eroding the prospects for peace. In seeking to redress these issues, UNEP argues that the recognition that environmental issues can contribute to violent conflict conversely indicates their potential significance as pathways for cooperation, transformation, and the advent of peace in societies around the world. In response to these challenges and possibilities, an emerging perspective suggests that people and cultures will often find ways of managing ecological concerns that not only work to avoid conflicts but that can also serve to promote peaceful relations among human communities and with the environment itself. This line of reasoning increasingly falls under the heading of environmental peacemaking, indicating a dualistic sense of the phrase in which the environment can be a source of peace rather than strife, and likewise that humans can exist peacefully with the balance of nature.

Forestalling Conflict, Promoting Peace

The seminal work in this field is perhaps Ken Conca and Geoffrey Dabelko's book by the same name, *Environmental Peacemaking*, in which the authors systematically demonstrate precisely how environmental cooperation can help forestall conflict and promote peace. In brief, there are two necessary elements for an environmental peacemaking effort to be successful: (1) it must create minimum levels of trust, transparency, and cooperative gain, and (2) it must strive to transform the nation-state itself, which is often marked by dysfunctional institutions and practices that become further obstacles to peaceful coexistence and cooperation. The basic premise is that an environmental crisis or conflict can be transformed into an opportunity for peace when it can be demonstrated to participants that there is more to be gained by cooperating than by competing, and further when the essence of peaceful cooperation transcends the interests and aims of nation-states that are generally focused on security as a function of control.

These trends toward ecologically cooperative conflict transformation have been on the rise in recent years at various "hotspot" locations around the world, including disputed territories between Peru and Ecuador and along the China–Vietnam border. In such conflict-ridden regions, scholars and activists see opportunities to bring nations and peoples closer together. In fact, as Alexander Carius notes in a report on environmental cooperation, the uniquely mutual dependence on environmental resources can help to promote the creation of communities of diverse users and stakeholders and thus yield advantages for all participants through the cooperative management of natural resources. This potential for environmental conflicts to create opportunities for peace is enhanced by the fact that environmental problems ignore political borders and often require a long-term perspective with the active participation of local communities and nongovernmental organizations in order to manage them. In this light, the shared management of resources both among and within nations can be seen as a crucial aspect of peace building that is often omitted from the analysis of how to confront our present crises.

In his work on "peace parks," Saleem Ali further emphasizes the positive sense of how environmental issues can play a role in cooperation, regardless of whether they are in fact part of the original conflict. Peace parks are regions, generally along contentious national borders, that have endured conflict in the past and are being transformed into protected areas. Examining international conservation efforts around the world—including those between India and Pakistan and those between Iraq and Iran—Ali observes that positive exchanges and trust-building gestures are a consequence of managing common environmental threats and that a focus on common environmental harms can be very successful in leading to cooperative outcomes.

A specific example of this thesis is the demilitarized zone (DMZ) between North and South Korea, which has become a massive wildlife and bird sanctuary in the decades since the end of the Korean War. Having remained relatively untouched by human hands during this postwar period, the DMZ is a rich habitat made up of marshes and grasslands, inhabited by many rare and endangered species, including Asiatic black bears, leopards, lynx, and a significant portion of the world's population of red-crowned cranes. The implications of this case are that demilitarization can lead to species diversification and a thriving ecosystem, and that formerly warring nations have an equal interest in cooperating to maintain the integrity of unique regions such as this, further illustrating the mutual benefits and responsibilities of environmental peacemaking.

In a landmark 1967 article on ecology and peace, Gutorm Gjessing looked at the intersections of historical resource innovations such as agriculture and the domestication of horses with conflict and warfare. Positing that modern war is centered more upon the command of raw materials and markets than upon ideologies, Gjessing concluded that in order to alleviate conflict we would need to develop values emphasizing cooperation rather than competition. These insights helped to begin a process of viewing peacemaking as a concept that links human concerns with environmental issues. Richard Matthew and Ted Gaulin subsequently elaborated on these themes, pointing out that early environmentalists often opposed war because of its potential for grave ecological destruction, and that some of the key figures in early American peace movements were likewise motivated by environmental concerns in their work. They concluded that it is by now a widely accepted notion that conservation measures and programs geared toward sustainable development can reduce the likelihood of violence and help preserve conditions of peace, and likewise that protecting the environment can also be an essential part of the groundwork for rebuilding peace in environmentally stressed areas.

Vandana Shiva has similarly observed that conflict and war have become pervasive, with severe impacts on human communities and the environment alike, but that there are cooperative methods in which we can turn the ways of war into a chance for peace. She notes that the path to peace comes through nourishing ecological and economic democracy and nurturing diversity in both ourselves and nature and concludes that the creation of peace requires us to resolve water wars, as well as wars over food, biodiversity, and the atmosphere. In the end, Shiva reminds us that, as Gandhi once said, "The Earth has enough for the needs of all, but not the greed of a few."

Wangari Maathai, who won the 2004 Nobel Peace Prize for her groundbreaking Green Belt Project in Kenya linking women's rights, economic self-sufficiency, resource conflict resolution, and environmental restoration, has echoed these themes in describing the nature of her efforts aimed at environmental peacemaking. Maathai highlights the linkages between the ways in which we manage our resources and the ways in which we govern ourselves. She emphasizes that there is an inherent link between environmental sustainability and the management of our shared planetary resources. As Alan Weisman subsequently concurred, the environment unites us all, people of every nation and creed. If we fail to save it, we may all perish in the process; but if we rise to the task, we will survive together in a world defined by the promise of peace.

Environmental Peacemaking and Inherent Interconnectedness

The practice of environmental peacemaking thus implicitly emphasizes an inherent interconnectedness, as Martin Luther King, Jr., alluded to in his call for a holistic concept of justice. While he was referring to social and political concerns at the time, the larger principle of interconnectedness reflected in his teachings continues to influence the views and work of environmental peacemakers. Indeed, a central tenet of these efforts is that peacemaking must occur at all levels in order to be effective, yet it often remains the case that environmental peacemaking as a practical matter is taken to imply that the world's problems primarily can be resolved through more pacific intercourse among nation-states and international entities. Such bodies might well play an important role in potentially moving the world toward peace and sustainability, yet environmental peacemakers remind us that individuals and local communities also possess the power to promote these changes. As Thich Nhat Hanh counsels in his recent book on Buddhism, ecology, and peace, what most of us presently lack are concrete ways of making a commitment to sustainable living a practical reality in our daily lives. While we tend to blame governments and corporations for pollution, violence, and wars, we oftentimes fail to account for the power and possibility that comes through taking action in our own lives as well.

Environmental peacemaking is an important concept in the emerging effort to create a more just and sustainable world, seeking to turn theory into practice, crisis into opportunity, and war into peace. By linking human concerns with ecological well-being, this perspective is becoming a critical tool in global conflict resolution practices, as well as in environmental preservation and restoration efforts. The impetus to connect culture and nature in a pragmatic manner stands at the cutting edge of interdisciplinary engagement. Indeed, environmental peacemaking is a potentially transformative way of managing the myriad challenges confronting humankind. It will likely continue to grow in its influence as we strive to navigate a course away from a world of degradation and depletion and toward one of conservation and cooperation.

See Also: Antiwar Actions/Movement; Environmental Justice Movements; Human Geography; International Law and Treaties; Maathai, Wangari.

Further Readings

Ali, Saleem H., ed. *Peace Parks: Conservation and Conflict Resolution.* Cambridge, MA: MIT Press, 2007.

Carius, Alexander. "Environmental Cooperation as an Instrument of Crisis Prevention and Peacebuilding: Conditions for Success and Constraints." Report commissioned by the German Federal Ministry for Economic Cooperation and Development, January 2006.

Clayton, Mark. "Environmental Peacemaking." *The Christian Science Monitor* (March 4, 2004).

Conca, Ken and Geoffrey D. Dabelko, eds. *Environmental Peacemaking.* Baltimore, MD: Johns Hopkins University Press, 2002.

Democracy Now! "Unbowed: Nobel Peace Laureate Wangari Maathai on Climate Change, Wars for Resources, the Greenbelt Movement and More." (October 1, 2007). http://www.democracynow.org/2007/10/1/unbowed_nobel_peace_laureate_wangari_maathai (Accessed November 2010).

Gjessing, Gutorm. "Ecology and Peace Research." *Journal of Peace Research,* 4/2 (1967).

Hanh, Thich Nhat. *The World We Have: A Buddhist Approach to Peace and Ecology.* Berkeley, CA: Parallax Press, 2008.

Klare, Michael T. *Resource Wars: The New Landscape of Global Conflict.* New York: Owl Books, 2002.

Lanier-Graham, Susan D. *The Ecology of War: Environmental Impacts of Weaponry and Warfare.* New York: Walker & Co., 2003.

Matthew, Richard A. and Ted Gaulin. "The Ecology of Peace." *Peace Review,* 14/1 (2003).

Shiva, Vandana. *Water Wars: Privatization, Pollution, and Profit.* Cambridge, MA: South End Press, 2002.

Westing, Arthur H., et al. "Environmental Degradation as Both Consequence and Cause of Armed Conflict." Working Paper prepared for Nobel Peace Laureate Forum (PREPCOM Subcommittee on Environmental Degradation), June 2001.

Randall Amster
Prescott College

Environmental Protection Agency

To consolidate government oversight of myriad environmental laws, President Richard Nixon created the U.S. Environmental Protection Agency (EPA) in December 1970. The EPA's mission is to regulate air, water, and solid waste hazards in order to protect human health and safeguard the environment. Each presidential administration has put its stamp on the agency.

The establishment of the EPA followed a decade of rising concern with environmental problems. Rachel Carson's *Silent Spring* (1962) heralded this concern with criticism of DDT and other synthetic organics. The Santa Barbara oil spill and Cuyahoga River fire sealed a decade of concern. While there had been previous oil spills and burning rivers, public outrage marked a sociopolitical change. On the tide of this concern, Nixon signed the National Environmental Quality Act into law in January 1970, requiring environmental assessments and environmental impact statements for federal agency decisions. The popularity of the first Earth Day kept environmental issues in the forefront, and Congress passed additional laws such as a new Clean Air Act, which required new standards for six critical pollutants, including sulfur dioxide and lead.

As the first EPA administrator, William Ruckelshaus organized and staffed a new agency fragmented in offices throughout Washington, D.C. In his first week, Ruckelshaus gave Cleveland, Detroit, and Atlanta mayors six months to comply with water pollution standards. After reviewing health studies, the EPA mandated a phase-out of the gasoline additive tetraethyl lead. Delayed by Ethyl Corporation's lawsuit, the EPA prevailed, phase-out began in 1976, and by 1991, Americans' blood lead levels dropped significantly. On other issues such as the DDT ban, legal challenges by the Environmental Defense Fund forced agency action.

During President Jimmy Carter's term, Congress authorized a Science Advisory Board (SAB) to support the EPA with outside scientific advice. This helped defuse criticism that agency decisions were politicized and reflected a new understanding of science as a transparent process instead of a "black box." In legal challenges to the EPA's new ozone standard, courts upheld the agency decision based on scientific evidence—from the SAB and other outside experts—of harmful effects.

In 1980, the EPA's role and scope increased dramatically with passage of the Comprehensive Environmental Response, Compensation, and Liability Act (Superfund)—a "polluter pays" response to hazardous sites such as Love Canal, a residential subdivision built on a former chemical dumpsite. If they existed, responsible parties were to pay cleanup costs. For truly abandoned sites, a "superfund" created by a tax on oil and chemical companies paid costs. The tax was not renewed after expiring in 1995, and the Superfund is now seriously underfinanced.

Originally, Superfund was to implement remedies by identifying hazardous materials and then permanently removing or isolating them from people and the environment. As the sciences of toxicology, epidemiology, and risk analysis developed, it became clear that thresholds for cleanup and models for exposure pathways could not always be defined a priority. In recent years, adaptive management has been used as an alternative to predetermined remedy. Along with increasing scientific complexity and uncertainty, new demands for public participation and social justice have increased difficulties for site managers, though evidence shows that public involvement produces better technical and political remedies.

During President Ronald Reagan's first term, scandal rocked the EPA. Appointee Anne Gorsuch (later Anne Burford) followed Reagan's mandate to downsize government and roll back environmental regulations. Gorsuch replaced SAB scientists with political loyalists, cut budget and staff, and was noncompliant with court orders to standardize chemical testing. A subsequent congressional investigation found Gorsuch in contempt for withholding agency documents, and she resigned following investigation of agency mismanagement. Courts convicted Gorsuch's assistant administrator Rita Lavelle on felonies of perjury and obstructing federal inquiry. Reagan reappointed Ruckelshaus, who improved agency operations. Still, Ruckelshaus opposed stringent standards such as new arsenic regulations that might cause economic hardships for copper smelters and other polluters.

President Bill Clinton's appointee, Carol Browner, served both terms and took a strong enforcement stance—doubling the rate of cases referred to the Justice Department for enforcement. The Energy Star ecolabeling program promoted consumer choice as a conservation tool. Based on SAB recommendations, the EPA declared a new arsenic standard for drinking water during the administration's final days.

Like Reagan, President George W. Bush sought to reduce the economic burden of environmental regulations on industry. After Bush suspended Clinton's arsenic rule, appointee Christine Whitman reinstated it, and also presided over a global warming report with which the administration disagreed. Because of these and other conflicts, Whitman resigned in mid-2003. The Bush-led EPA was criticized for cutting funding for the Office of Environmental Justice, refusing to regulate carbon dioxide as a greenhouse gas, closing agency libraries, and suppressing studies showing the need to control mercury emissions.

Challenges for the Barack Obama administration and his EPA appointee, Lisa Jackson, include greenhouse gas regulation, oil spill prevention and response, and making headway on cleanup at Superfund sites.

See Also: Air Pollution; Environmental Justice Movements; Environmental Law, U.S.; Global Warming and Culture; Love Canal and Lois Gibbs; Organics; Toxic/Hazardous Waste; Water Pollution.

Further Readings

Hays, Samuel P. *Environmental Politics Since 1945*. Pittsburgh, PA: University of Pittsburgh Press, 2000.

Hird, John A. *Superfund: The Political Economy of Environmental Risk*. Baltimore, MD: Johns Hopkins University Press, 1994.

Landy, Marc A., Marc J. Roberts, and Stephen R. Thomas. *Environmental Protection Agency: Asking the Wrong Questions, From Nixon to Clinton*. New York: Oxford University Press, 1994.

Pat Munday
Montana Tech of the University of Montana

Fair Trade

Fair Trade is a social movement and an everyday market-based approach to trade that attempts to address the structural inequalities found in free trade markets. Predominantly affecting trade relations between the geopolitical north and south, Fair Trade is usually promoted to consumers in the north as a more just approach to international trade. Proponents of Fair Trade juxtapose fair-trade practices against the current global trading system in an attempt to reveal the historical reproduction of inequitable terms of trade for southern producers.

Responding to various issues inherent to free-trade practice, fair-trade principles typically include stable fair prices, safe and fair labor conditions, direct trade with producers, democratic and transparent organizations, community development, and environmental sustainability. It focuses in particular on exports from developing countries to the developed world. Most notably, goods such as coffee, cocoa, sugar, tea, bananas, honey, cotton, wine, fresh fruit, chocolate, and flowers are traded. Fair Trade's strategic intent is to work with marginalized producers in an attempt to assist them toward economic self-sufficiency and stability. It also aims to provide a path for producers to become greater stakeholders in their own organizations, as well as gain broader representation in international trade. U.S. participation in Fair Trade markets continues to grow, and the consumption of fairly traded goods is typically framed as a moral alternative to conventional market practice. Although Fair Trade participation is commonly associated with altruism, many criticisms of Fair Trade exist.

History of Fair Trade

The origins of Fair Trade can be traced back to the early 1940s when North American and European organizations provided economic relief to post-war poverty-stricken communities and refugees throughout Europe. These Fair Trade markets manifested from a sense of justice and began to sell producer handicrafts to North American markets outside the conventional trading structures. In establishing fair and direct trade, these Alternative Trade Organizations (ATOs) gave higher returns to producers throughout various parts of the world. Networks of church-based groups and charity organizations working in the global south helped found partner cooperatives and associations in southern nations that organized disadvantaged groups to export their products. While the total volume of this trade was minor, sales grew rapidly, and the profile of alternative trade increased to the point of movement status.

Fair trade became prevalent in the United States during the 1980s. The emergence of the Fair Trade movement in the United States is said to have arisen in response to the negative effects conventional globalized capitalism had on much of the world's small-scale producers. Neoliberal economic policies in the form of deregulation created instability for many traders and distributors participating in global markets. Coffee farmers in particular found themselves vulnerable in these harsh economic conditions, in part due to a lack of self-sufficient coffee unions. Historically, coffee farmers have formed cooperatives, but a lack of sufficient production facilities and sales experience has lead to their historical inability to successfully unionize. This helps to explain why in 1988, when world coffee prices began a sharp decline, Fair Trade activists founded Equal Exchange in West Bridgewater, Massachusetts. Responding to the negative effects of a self-regulated market, Equal Exchange provided a unique change in momentum for human rights activists in the Fair Trade movement. As a subsequent result of Equal Exchange's formation, a variety of nongovernmental organizations (NGOs), universities, and consumer activist organizations began to unite in an effort to improve small-scale and factory producer conditions. The year 1988 also brought forth the first Fair Trade certification initiative. It was called the Max Havelaar initiative, named after a fictional Dutch character that fought against the exploitation of coffee pickers in Dutch colonies. In 1997, the Max Havelaar label expanded its certifying process under the Fair Trade Labeling Organization (FLO) in order to reach farmers and roasters around the globe. Today, TransFair USA is a certified member of the FLO and is currently represented in over 19 countries across Europe, North America, and Japan. Building on the success of Max Havelaar and the ATOs, TransFair USA began certifying Fair Trade coffee in 1999. Since then, TransFair USA has also made Fair Trade tea and cocoa available to U.S. markets. Similarly, the formation of the International Federation for Alternative Trade (IFAT) in the Netherlands, the Fair Trade Federation (FTF) in Washington, D.C., and the FLO International stationed in Germany were all established during the late 1980s and 1990s to secure Fair Trade as the first "ethically distinctive" commodities market.

Fair Trade Today

In order for producers to officially enter the Fair Trade market they must first become Fair Trade certified. The mechanics for determining license and certification involves multiple organizational levels and members. Fair Trade labeling initiatives license the Fair Trade certification mark on products and advocate Fair Trade standards in their territory. Currently, there are 19 labeling initiatives covering 23 countries. Similarly, Fair Trade marketing organizations are national organizations that market and promote fair trade in their country. The FLO directly licenses companies in these countries to use the FAIRTRADE certification mark. Producer networks are also a key element to the process and help to bring producer feedback to the larger labeling body. They are regional associations that Fair Trade–certified producer organizations may join if they wish. They represent small-scale producers, workers, and other producer stakeholders. Currently, there are producer networks on three continents: Africa, Asia, and Latin America and the Caribbean. It is the expressed intension of the FLO to run the certification and licensing process in a way that enables its members and other stakeholders to contribute to strategy and standard setting. That said, the Fair Trade labeling initiatives and the producer networks are full members of the FLO and have the right to vote at the annual General Assembly. There are also two associate members, Fair Trade Label South Africa and Comercio Justo México. All members and certified producer organizations participate in the FLO's decision making through the General Assembly and their respective assemblies. In addition to the board being composed of representatives from the Fair Trade labeling initiatives and certified producer organizations, representatives also include certified traders and up to three external independent experts. Together, this organizational body aims to ensure a transparent, equal process for determining Fair Trade certification standards.

On the consumer end of Fair Trade, participation has grown and shifted, extending the participating population beyond activists to include more and more everyday consumers. As more North American consumers become involved, consumer demands for corporate responsibility continue to rise. As a result, corporations like Starbucks are engaging in Fair Trade practices in an attempt to meet public demands for more ethical business practices. The phenomenon is growing, and U.S. consumers can now purchase fairly traded food and coffee goods at a variety of retails stores and supermarkets across America. Furthermore, there is no indication that Fair Trade consumption is on the decline. On the contrary, according to the Fair Trade Federation, U.S. and Canadian sales for Fair Trade grew over 100 percent between 2004 and 2007, with global sales topping $4.12 billion in 2008.

Culture of Fair Trade

Historically, the Fair Trade consumer has been typified as an activist-oriented individual, more concerned with the environmental, social, and political impact associated with their purchase than with making a fashion statement. Perhaps that is why Fair Trade apparel has largely been limited to low-fashion items such as hemp bags, belts, and other accessories sold outside mainstream retailers. Today's Fair Trade consumer has a new, hip face. Fair Trade consumerism has become trendy among young progressives and their mainstream counterparts who cater to this new, cool, green consumer. Perhaps a good example would be American Apparel, a trendy youth-focused company based in Los Angeles. Recognized as a Fair Trade business, American Apparel is best known for garments like organic cotton t-shirts, leggings, and scarves. Some of its most popular accessories are terry cloth headbands and wristbands aimed at young urban buyers. But American Apparel is not the only one tapping into this new, hip generation of green consumers. Manufactures like Nike, Tommy Hilfiger, and others now post codes of conduct that bar child labor and mandate legal minimum wages along with other labor laws. Celebrities like Bono, of U2, are also in the Fair Trade business. In 2005, Bono introduced Edun, a Fair Trade fashion brand sold through high-end department stores such as Saks and Nordstrom. Fair Trade as a fashion statement can also be found exclusively online through web catalogs like Fair Indigo. Fair Indigo markets its Fair Trade apparel to mass-market consumers who want guilt-free, upscale, trendy clothes at prices comparable to other department stores. According to market researchers, Fair Trade is just one of the ways young urban buyers are voting with their dollars. The overall sentiment seems to reflect that as long as "being green" is viewed as cool, Fair Trade markets will continue to experience rising sales and exposure.

Criticisms of Fair Trade

Some scholars argue that there is a fundamental dilemma at the root of Fair Trade's current growth and success. In order to gain mainstream visibility and make a dent in the market,

Fair Trade activists need to work with large corporate traders. These corporate entities often bring to the table a very different set of interests and often fall short of fully committing to Fair Trade ideals. One such criticism has been made regarding Starbucks' commitment, or lack thereof, to Fair Trade standards. In 2007, only 3 percent of Starbucks coffee was fairly traded, yet its corporate image became that of socially responsible soon after announcing its participation in the Fair Trade market. Because Starbucks is the largest international coffee chain, it is responsible for purchasing more Fair Trade coffee than any other retailer in the world. However, data suggests that only a small percentage of its total coffee is Fair Trade. Some criticize the FLO, responsible for providing Fair Trade certification, for not requiring higher standards from large multinationals before giving them certification. In short, critics believe the growing co-optation of the Fair Trade brand without fully executed Fair Trade principles weakens the overall goals of Fair Trade ideals and distorts the practice of corporate responsibility.

Another criticism of Fair Trade argues that the certification process producers must go through reproduces the same kind of insider-outsider market dynamics inherent to the free trade model. Fair Trade certification requires producers to go through a rigorous application process that mandates broad requirements for social, socioeconomic, and environmental development, labor and safety condition standards, and the ability to monitor and implement new and/or changing amendments. Although these requirements are generally viewed by both producers and consumers as long-term investments toward sustainability, the resources needed to gain certification are not free. Critics argue that small producers face a disadvantage compared with larger producers when applying for certification due to disproportionate access to resources. Consequently, some scholars have criticized the Fair Trade market for reproducing the same kinds of problematic insider-outsider markets that occur in the conventional free trade market.

A third criticism of Fair Trade argues that the alternative market is paternalist and relies on colonial notions of first world and third world modernity. Tangent to this criticism is the argument that the core of Fair Trade's approach to market is derived from a northern consumerist view of justice that southern producers do not participate in setting.

The last major criticism of Fair Trade consumerism highlights the issue that consumption, regardless of market source, still uses energy, produces waste, and in some cases perpetuates the same kinds of capitalist consumption cycles that led to acute needs for conservation in the first place. The literature generalizes some basic tenets of green consumerism to involve an effort to recycle, reduce, and reuse. Also tangent is an effort to minimize energy usage, avoid the production of waste, eat more sensibly and sustainably, and consider human rights issues when making a purchase. However, consumerism and the desire to establish one's identity with the things he or she chooses to buy is viewed by some to be counterintuitive to the expressed goals of the green movement. The products people buy have to be manufactured from materials and resources we take from the Earth. Manufacturing usually requires the use of energy. Shipping items to various retail stores uses energy. Some longtime environmental activists warn that one risk associated with buying into green marketing could lead to falling prey to what is commonly called "greenwashing." Greenwashing occurs when individuals excuse corporations for their endless pursuit of waste and consumption because of a few environmentally themed advertisements. This rationale, combined with the sentiment that "if it's green, it's guilt-free," has some Fair Trade supporters concerned. The problem might be summed up with the critique that if people believe they are saving the planet by purchasing organic, Fair Trade cotton instead of polyester, they might be lulled into a false sense of security and righteousness when cutting overall consumption should be the real goal.

The Future of Fair Trade

As the market for sustainable, socially responsible production and consumption grows, so does the demand for fairly traded goods. It seems that northern consumers become informed about Fair Trade primarily through package labels, sales brochures, magazine articles, farmers markets and community supported agricultural (CSA) projects, campaigns, protests, community organizations, and word-of-mouth sources. Recent evidence shows how ethical consumers, those who purchase goods with concern for certain ethical issues (human rights, environmental sustainability, labor conditions, and so forth), see a direct correlation between their purchasing behavior and the associated ethical concern. According to market research, the purchase of fairly traded goods demonstrates one of the most typical examples of conscious consumer behavior. It seems as if what started out as a niche market may be growing into a burgeoning force in today's market.

See Also: Corporate Social Responsibility; Environmental Justice Movements; Fashion; Green Consumerism; Greenwashing; Popular Green Culture; Shopping.

Further Readings

Fairtrade Labelling Organization International. "Resources: Fair Trade at a Glance." http://www.fairtrade.net/fileadmin/user_upload/content/2009/resources/Fairtrade_at_a_Glance_Jan10.pdf (Accessed February 2010).

Jaffee, Daniel. *Brewing Justice: Fair Trade Coffee, Sustainability, and Survival*. Berkeley: University of California Press, 2007.

Jaffee, Daniel, Jack Kloppenburg, and Mario Monroy. "Bringing the 'Moral Charge' Home: Fair Trade Within the North and Within the South." *Rural Sociology*, 69 (2004).

Levi, Margaret and April Linton. "Fair Trade: A Cup at a Time?" *Politics and Society*, 31 (2003).

Oxfam America. "Advocacy Groups and Shareholders Persuade Procter & Gamble to Offer Fair Trade Coffee." Press release. Boston, MA: 2003.

Prasad, Monica, Kimeldorf Howard, Rachel Meyer, and Ian Robinson. "Consumers of the World Unite: A Market-Based Response to Sweatshops." *Labor Studies Journal*, 29 (2004).

Simpson, Charles and Anitta Raone. "Community Development From the Ground Up: Social-Justice Coffee." *Human Ecology Review*, 7 (2000).

Kristine Harris
California State University, Sacramento

Fashion

Critics do associate many ideas with the fashion industry: fast fashion, McFashion, high-speed production and high-volume consumption, impermanence and ephemerality, and individuality and self-absorption. Few of these would be immediately associated with the ideas of sustainability. Nonetheless, since the early 1990s, a significant segment of the fashion industry has embraced elements of sustainability.

Sustainability in fashion refers not only to the fiber, fabric, and manufacturing of clothing but also to its environmental footprint during use (laundering) as well as its disposal; for instance, the shoes this Air Force officer is helping a girl try on were donated instead of thrown away.

Source: U.S. Air Force

Green fashion refers to clothing whose design and fabrication incorporates sustainable production techniques, that is, the use of materials and processes that exhibit environmental and social responsibility and that reflect that responsibility through an appreciation of and response to an assessment of the entire life cycle of the products. The basic ideas have been reflecting in a variety of terms, including *ecofashion*, *sustainable fashion*, *eco-friendly clothing*, *environmentally conscious design*, *ecodesign*, *environmentally friendly clothing*, and *ecochic*. Moreover, dimensions of green fashion share their philosophies, approaches, and objectives with a number of related movements such as those of Fair Trade, social justice, sustainable design, organic agriculture, and corporate social responsibility.

Sustainability concerns ideally should extend through the entire supply chain and throughout the life cycle of garments. In each stage of production (fiber, fabric, and product manufacturing), green issues extend to a variety of considerations. Firms have adopted innovations in choice of fibers and other materials, production processes (such as the use of unbleached fabrics and biodegradable enzyme washes), designs to enhance product durability and recyclability, energy use (including the reduction of greenhouse gases), and waste disposal. Natural or recycled fibers have been substituted for conventional fibers. For example, firms have substituted organically grown for conventionally grown cotton, which has been criticized as a chemically intense consumer of pesticides and fertilizers; polymers have been developed from renewable sources (e.g., corn); and polyester fleece garments are manufactured from recycled beverage bottles.

Sustainability considerations extend to garments after their purchase; for example, the greatest proportion of a garment's carbon footprint occurs during its use, mainly due to laundering. Manufacturing choices affecting the durability, resistance to shrinkage, and colorfastness, as well as the stain, dirt, and odor resistance of textiles can extend the use of garments. Some fibers can be recycled or reclaimed and processed into new yarn; however, challenges exist in retaining fiber quality. Recycled, reclaimed, surplus, and vintage fabrics and garments arguably represent the most sustainable choice—there is a global trade in secondhand clothing.

The fashion supply chain is particularly vulnerable to criticisms due to its complex, fragmented, and global character. Delocalized global manufacturing has led to concerns over the provision of fair working conditions, fair trade, and energy use in shipping. Except in the case of the largest companies involved, individual firms often face great challenges in greening their extended supply chain. Elements of green fashion incorporate a range of ethical issues, including those involving animal welfare, fair working conditions, and responsible human resource management. Notable brand names in fashion have suffered as result of the exposure of workplace conditions in garment manufacturing. In response to the issue, progressive codes of conduct such as that of the Clean Clothes Campaign have been developed.

Driven by regulatory compliance, the search for competitive advantage, and corporate responsibility, the first adopters in the green fashion movement were often small, specialized firms, but since then, its principles to a greater or lesser degree have been embraced by large textile and clothing manufacturers, prominent fashion labels, fashion event planners, and the editors of fashion media. Embracing and promoting a green image may be a cost-effective means of marketing for new entrants into the industry and a mechanism to refresh an existing brand image. Prominent clothing lines have adopted a green image in their marketing, perhaps first notably among performance clothing labels such as Patagonia and fashion labels and designers such as Marc Jacobs and Stella McCartney. Major companies such as Diesel's "Global Warming Ready Collection" also reflect the orientation. Clothing label Edun has partnered with the Wildlife Conservation Society to create the Conservation Cotton Initiative, which encourages African farmers to employ organic production methods and supports fair trade. Global brands such as H&M and Levi's have incorporated organic cotton and other environmentally friendly materials into their merchandise mix. Major Fashion Week events such as those in London and Amsterdam have added eco-designer showcases and competitions.

Critics of green fashion have labeled it as oxymoron: the inherent focus on consumerism is seen as antithetical to the core values of environmentalism and sustainability. It has been characterized as a marketing ploy—"greenwashing"—without any alteration of the fundamental character of the industry or any substantial internalization by any major segment of the industry of the principles of sustainability. Overall, green impulses have not been fully embraced by the industry.

The industry is also faced with a continuing challenge: will consumers in sufficient numbers respond to green innovations in the industry and are they ready to pay a premium to purchase green fashion? To date, the evidence is mixed concerning whether issues related to sustainability have any significant effect on consumers' fashion purchases.

See Also: Corporate Social Responsibility; Fair Trade; Greenwashing; Organic Clothing/Fabrics; Popular Green Culture; Shopping; *Vanity Fair* Green Issue.

Further Readings

Fletcher, K. *Sustainable Fashion and Textiles: Design Journeys*. Sterling, VA: Earthscan, 2008.

Hethorn, J. and C. Ulasewicz, eds. *Sustainable Fashion: Why Now?* New York: Fairchild, 2008.

Palmer, A. and H. Clark, eds. *Old Clothes, New Looks: Second Hand Fashion*. Oxford, UK: Berg, 2005.

Michal Bardecki
Anika Kozlowski
Ryerson University

Film, Documentaries

Documentaries are a form of nonfiction film that attempts to capture and document reality. A documentary offers its audience a window into a part of the world or life that is not readily available for them to participate in firsthand. As a proxy for direct experience, documentaries are made for aesthetic, entertainment, emotional, social, and political purposes. Producers of environmental documentaries use film to bring voice to environmental injustices to humans, wildlife, and the planet. As such, the power of images and stories narrated and revealed in documentaries have lead to personal and political change in response to details they exposed. The documentary *An Inconvenient Truth* provides a strong example of the power of the genre to bring about environmental change as it successfully inspired both political and personal action to combat the issues of global warming and climate change.

Environmental documentaries, sometimes referred to as political documentaries, convey a point of view or document a reality, often with the intention of influencing public opinion or motivating action. Environmental films are used to document and preserve nature as it exists, whether pristine or destroyed. Documentaries commonly strive to expose the results of negative environmental impacts and identify their source or cause. In doing so, the films can create awareness and public outcry that gains media attention and influences public policy. This type of public agenda setting through documentary has yielded positive results for environmental issues, including whaling and the use of drift nets in the fishing industry, for example.

An Inconvenient Truth (AIT) is the 2006 Academy Award–winning documentary narrated by Al Gore and directed by Davis Guggenheim. The film documents Gore's town hall meeting slide show, which he had been presenting worldwide since 2000 in order to educate the public about global warming. The documentary presents scientific data and predictions on climate change, combined with an autobiographical thread of Gore's life and personal challenges connected to the subject.

How It Began

Producer Laurie David saw a short version of Gore's slide show in 2004 as part of a panel presentation on global warming timed with the release of Hollywood's science fiction blockbuster *The Day After Tomorrow*. Inspired by the clarity with which the slide show depicted global warming and the powerful impact of the data, David set things in motion for a film adaptation of the presentation to bring the message to a much larger audience.

The film promotes global warming as an imminent anthropogenic disaster: the world is getting hotter due to the burning of fossil fuels. At the time of the film's release, the Associated Press contacted over 100 climate scientists to check on the validity of the science quoted in AIT. While the scientists represented a range of viewpoints on global warming, including skeptics, those who had seen the documentary all reported that it captured the important scientific aspects necessary for understanding the subject, that the information was fundamentally correct, and that minor factual errors did not negate the message of the film. For example, Gore states in the film that Antarctic ice cores show the results of the Clean Air Act. While scientists point out that this is an error of fact in the film, and there is no evidence of the Clean Air Act present anywhere in the Antarctic ice, they concede that the evidence is, in fact, recorded clearly in Greenland ice.

The film came under fire for many of its scientific details, for example, its choice to use sensitive Hurricane Katrina imagery to illustrate a scientific connection between global warming and hurricanes. Some critics felt that the science was too controversial to be included when other, equally disturbing effects of climate change that are more agreed upon in the scientific community could have been used instead.

An Inconvenient Truth is widely debated for both scientific and political reasons. Websites and media coverage swing the pendulum back and forth between agreement and disbelief, depending on the source. Critics of the film and global warming skeptics tend to point to errors of fact in the film to discount the message in its entirety.

The film triggered a number of oppositional films featuring a variety of minority opinions challenging Gore's point of view, including *An Inconvenient Truth . . . Or Convenient Fiction?*, *Not Evil Just Wrong*, and *The Great Global Warming Swindle*. None of these appear to have gained traction in the documentary genre, but have sometimes been used to present counterarguments to global warming through educational companion screenings with *An Inconvenient Truth*.

The film, like many environmental and scientific documentaries, provides a venue for translating technical information to the public—an audience it rarely otherwise reaches. Because of that, *An Inconvenient Truth* has had a greater impact on public awareness of global warming and climate change issues than reports on scientific issues alone had been able to accomplish.

What It Accomplished

In its opening weekend, *An Inconvenient Truth* broke records for previous documentaries in gross income. It was shown at the Sundance Film Festival to standing ovations, and it still ranks fifth in the top-grossing documentaries in the United States. It was, for the most part, positively reviewed by the critics, some pointing out that its importance in message outweighed its "dryness," while others reported being riveted by the drama of its implications.

The personal narrative introduced into the film was included as a device to connect the movie-going audience with Gore without his being present in the room. During his public presentations of the slide show, audiences could connect with him because they were physically in the same space. That connection was replaced on camera by Gore telling his own stories, becoming the main character of the film.

Reviewers critical of the film saw it as a political move toward election, claimed it exploited public fears, and stated that the personal narrative exposed Gore as an intellectual fraud. But exit polling showed the film to be a success with the large majority of its viewers. Paramount Classics surveyed audiences across the country, including conservative suburban populations, to gauge the public's response. Results showed 92 percent of respondents rated *An Inconvenient Truth* highly (60 percent is typical of a successful movie) and 87 percent said they would recommend it to a friend (the average response for a successful film is 47 percent).

Among its many awards, *An Inconvenient Truth* was the first documentary to receive two Academy Awards—one for Best Documentary Feature and the other for Best Original Song. The film was instrumental in drawing attention to global warming and led, in part, to Gore and the Intergovernmental Panel on Climate Change (IPCC) receiving the 2007 Nobel Peace Prize.

An Inconvenient Truth not only informed the public, it inspired people to change their behavior. After the film, Gore founded the Climate Project and trained motivated activists around the world to give the slide show presentation in their communities. An international Internet survey conducted by the Nielsen Company in 2007 reported that 66 percent of viewers said the film changed their opinion about global warming, and 74 percent of viewers reported that they changed some of their habits after seeing it (although the type of action was not reported).

There was a broad range of political response to the film. While some world leaders refused to meet with Gore or to take environmental action, the film inspired other political world leaders to take positive action on behalf of climate issues. Positive responses included convening entire governmental offices for screenings of the film and creating national reforestation campaigns.

The film continues to educate as an adopted element in school science programs and through community screenings. Like other documentaries, *An Inconvenient Truth* has been incorporated into classroom curricula. It is often shown to communities and groups to encourage awareness of climate change and motivate individual and community-wide environmental action.

See Also: Censorship of Climatologists and Environmental Scientists; Education (Climate Literacy); Film, Drama/Fiction; Global Warming and Culture; Gore, Jr., Al

Further Readings

Borenstein, Seth. "Scientists OK Gore's Movie for Accuracy." *Washington Post* (June 27, 2006).

Garofoli, Joe. "Gore Movie Reaching the Red States, Too." *San Francisco Chronicle* (July 8, 2006).

Gore, Jr., Al. *An Inconvenient Truth: The Planetary Emergency of Global Warming and What We Can Do About It*. New York: Rodale Books, 2006.

Nolan, Jessica M. "*An Inconvenient Truth* Increases Knowledge, Concern, and Willingness to Reduce Greenhouse Gases." *Environment and Behavior*, 42/5 (2010).

Dyanna Innes Smith
Antioch University New England

FILM, DRAMA/FICTION

Hollywood has a propensity for fictionalizing environmental concerns into apocalyptic scenarios, stretching both scientific theory and common sense for the purpose of high drama. While this provides a cultural opportunity to bring important issues to the forefront of public dialogue, storylines built on exaggerated fact confuse salient topics, making it difficult for audiences to separate reality and fiction. Roland Emmerich's 2004 Hollywood blockbuster *The Day After Tomorrow* is an example of just such a film. The plotline shows a series of cataclysmic events blending scientific fact and fiction in a global warming scenario that ushers in a modern ice age in a rapidly condensed timeline. The film's release generated broad-scale political and social debate around the issue of climate change, causing a surge in media and public attention for the important topic. Key elements of the debate included the film's impact on public perception of climate change, the scientific accuracy of the film's content, and its political implications.

Media is a large part of our lives and influential in our decision making as a society and as individuals. Media selects and powerfully promotes events and stories that influence attitude and behavior on a massive scale. This gives film and other popular, widespread sources of information the opportunity to create highly impactful environmental messages. In *The Day After Tomorrow*, environmental themes presented fictionally simultaneously drive the issue of climate change forward in the common dialogue while popularizing scientific misinformation.

The plot follows main character Jack Hall (played by actor Dennis Quaid), a National Oceanic and Atmospheric Administration (NOAA) paleoclimatologist, whose theory of abrupt climate shift holds the predictive key to increasingly violent weather events around the globe. Hall presents his theory at a world climate conference, where political leaders voice their skepticism of global warming. At the same time, signs of shifting climate rapidly escalate: Antarctic ice has fractured, buoys are registering massive drops in ocean temperatures, giant hail kills people in the streets of Tokyo, and Los Angeles is destroyed by multiple tornadoes. Hall and other researchers use computer modeling to forecast the rapid destabilization of Earth's climate, which culminates in three massive storm systems across the Northern Hemisphere.

Much of the movie's personal story line depicts Hall's relationship with his son (played by Jake Gyllenhaal). One of the three super storms is centered over the northeastern United States, causing a storm surge that floods Manhattan, where Hall's son is located. At the climax of the film, Hall travels in the storm from Washington, D.C., to New York City, rationalizing that he has traveled farther distances on foot in the Antarctic. He and a colleague snowshoe in to rescue his son and the others who manage to survive the storm's freeze-on-contact temperatures, ushering in the new ice age.

The Day After Tomorrow was praised for its special effects. Among them were the iconic images of a tidal wave consuming the Statue of Liberty and an ocean tanker floating down Fifth Avenue. Effects were integral to the elevation of the super storm into a villainous main character of the film.

Public and Media Reaction

The Day After Tomorrow was released worldwide in May 2004, grossing over $500 million. Although clearly a box-office hit, it was received with mixed reviews.

It is estimated that the film was seen by approximately 10 percent of the U.S. adult population. While this percentage is not high enough to affect public opinion on a national scale, the film did have a significant effect on that 10 percent. Researchers studying the effect of the film on its audience found that it raised concern of global warming as an issue, but also increased confusion over understanding of the possible effects of climate change.

The most significant accomplishment of the film from an environmental perspective is the way it stimulated mainstream dialogue about climate change. Before the film's release, there was a lack of engagement by the public on the issue of climate change. Out of 13 environmental issues of importance to Americans, a 2000 Gallup poll showed climate change ranked next to last. In 2001, a report released by the Intergovernmental Panel on Climate Change (IPCC) providing a summary of scientific consensus on global climate change received little attention in media coverage.

But for three months surrounding the film's opening in 2004, *The Day After Tomorrow* generated a large spike in both print and multimedia coverage of the film and global warming controversy around the film. The total number of substantive new stories provided more than 10 times the coverage of the IPCC report, and of those stories, fewer than a quarter of them were entertainment stories. This was most likely due to the combination of massive marketing efforts for the film and the controversy over scientific and political issues in the film's content.

Also timed with the film's release was a marked increase in online activity seeking information about global warming issues that continued for almost three weeks afterward. Environmental organizations seized the opportunity to present more global warming information on their websites, expanding the online resources available to the public.

Scientific Reaction

Although the film's genre is labeled science fiction, its representation of the science around climate change caused broad debate. For example, the film opens with the Larsen Ice Shelf calving, or fracturing off, the Antarctic Peninsula. As point of fact, two of the three shelves that make up the Larsen Ice Shelf have already disintegrated—Larsen A in 1995 and Larsen B in 2002—due to localized warming. The film takes this factual climate event and shows it as the first in a chain of increasingly exaggerated catastrophes that happen within hours of each other, leading to an ice age that covers half the planet in a matter of days.

Many environmental groups and leaders saw the controversial science as an opportunity to educate the public, holding rallies and town hall meetings and distributing informational flyers to moviegoers. These activities added momentum to the discussion on climate change.

Critics, however, felt that the extreme misrepresentation of climate science combined with the rapid speed of the film's timeline would either lead to an audience convinced that the impossible scenarios depicted could eventually come true, or cause them to dismiss the entire climate change issue as fictional. But surveys showed the majority of viewers left the film feeling increased concern over climate change, not alarmist or extreme opinions or denial of the existence of global warming. In fact, viewers reported that their increased concern translated into a willingness to change certain personal behaviors, like switch to a more energy-efficient vehicle, and to discuss their opinions on climate change with family, friends, and political representatives.

In addition to the debated scientific facts, the film created informal knowledge of climate change, such as the connection between the issue and certain governmental agencies like NOAA and the National Aeronautics and Space Administration (NASA). And although exaggerated, the depiction of the ability of large climate events occurring in one part of the world to dramatically influence weather patterns at a location thousands of miles away may have helped to connect the concept of global climate on a local scale.

Political Reaction

The Day After Tomorrow makes a point of attaching at least partial blame for climate change on the effects of human action. This anthropogenic causality is a commonly debated central issue around the public acceptance of climate change. The film drives this home through a narrative arc converting U.S. Vice President Raymond Becker (played by actor Kenneth Welsh) from skeptic, rejecting Hall's plea for urgent governmental action, to regretful national leader. The film ends with Becker as the new president making a public apology for his inaction and for humans' negative impact on the planet in general. Emmerich admitted to intentionally casting Welsh in the role because of his physical resemblance to then–Vice President Dick Cheney. This drew heavy criticism, pointing to an overt political agenda, fueling a common dismissal of the film as propagandist.

Another political response to the film was a report that the White House had issued a memo ordering NASA officials not to discuss the film. Once the report was publicized in the media, it was apparently rescinded.

See Also: Censorship of Climatologists and Environmental Scientists; Ecocriticism; Environmental Media Association; Film, Documentaries; Global Warming and Culture.

Further Readings

Ingram, David. *Green Screen: Environmentalism and Hollywood Cinema*. Exeter, UK: University of Exeter Press, 2004.

Ivakhiv, Adrian J. "Green Film Criticism and Its Futures." *Interdisciplinary Studies in Literature*, 15/2 (2008).

Leiserowitz, Anthony. "*Day After Tomorrow*: Study of Climate Change Risk Perception." *Environment: Science and Policy for Sustainable Development*, 46/9 (2004).

Lowe, Thomas, et al. "Does Tomorrow Ever Come? Disaster Narrative and Public Perceptions of Climate Change." *Public Understanding of Science*, 15/4 (2006).

Dyanna Innes Smith
Antioch University New England

FILM, PRODUCT PLACEMENT

Through the increasingly pervasive tactic of product placement, companies attempt to showcase goods or services by giving them visibility within movies, television shows, video games, or other media. This advertising approach, also known as embedded media, is a subtle and often effective alternative to traditional commercials. Product placement is often intertwined within the storyline of a show and not disclosed as an ad. Critics worry that viewers can be exposed unknowingly to dozens or even hundreds of messages every day. While product placement is often associated with selling familiar products such as cars, soft drinks, and high-tech devices, it can also be used to highlight ecofriendly products and behaviors.

Product placement has long been shown to shape the decisions of viewers. When Reese's Pieces were highlighted in *E.T.: The Extra-Terrestrial* in 1982, sales of the candy jumped by 65 percent. Similarly, after actor Gene Hackman suggested to Tom Cruise in the 1993 movie *The Firm* that he "grab a Red Stripe," demand for the Jamaican-brewed beer sharply increased, enabling the company's owners to sell a majority stake in the brewery to Guinness Brewing Worldwide. Even before such calculated placing of products within scenes, movies wielded enormous influence on society, sometimes in troubling ways. Scenes of hooded Klansmen in the 1915 movie *The Birth of a Nation* contributed to a resurgence of the Ku Klux Klan, while shots of Hollywood stars enjoying cigarettes spurred increased smoking among viewers.

Although calculated product placement can be dated back at least as far as the 1951 movie *The African Queen*, which highlighted boxes of Gordon's Gin, the technique took off in the first decade of the 21st century, spurred partly by the increasing use of TiVo and other digital video recorders that enable viewers to avoid having to sit through commercials. By 2006, about two-thirds of advertisers were using product placement, with 80 percent of the activity taking place on commercial television. As technology evolves, production companies are turning to virtual product placement that uses computer graphics to insert a product into a scene after the show has been shot, sometimes altering the sales message depending on a particular audience.

For companies that produce green materials, product placement offers an opportunity to highlight ecofriendly goods, ranging from hybrid cars to compact fluorescent light bulbs. NBC Universal (NBCU) has been particularly prominent in promoting environmental messages. In some cases, this takes the form of "behavior placement," in which characters engage in environmentally oriented activities such as recycling and bicycling. The tactic can bring in advertisers who want to associate their message with a socially aware show, while swaying viewers to begin modeling actions that take place in the script. If, for example, Jennifer Aniston as Rachel in *Friends* could influence women across the country to get their hair cut in the shaggy, highlighted, layered look known as "the Rachel," could Rainn Wilson as Dwight Schrute in *The Office* set an example of recycling for millions of viewers?

NBC executives in 2007 began asking producers of almost every prime-time and daytime show to give a green plot to a script at least once a year. Such storylines could focus on getting rid of plastic water bottles, eating organic food, or getting more exercise. The messages can be delivered with comedy or drama so as not to seem forced, such as the portrayal of a tycoon in the Bravo reality series *Millionaire Matchmaker* who "flies off the handle" when his blind date orders red meat, or of *The Office*'s Schrute as a recycling-obsessed superhero, replete with cape.

Such ecofriendly messages draw in money from manufacturers and retailers. Seventh Generation, Inc., the Vermont-based manufacturer that specializes in cleaning products, paid NBC for a vignette about organic gardening featuring the stars of Oxygen's reality series *Tori and Dean: Home Sweet Hollywood*. Walmart ran an ad highlighting locally grown produce in conjunction with an episode of *Trauma* in which a man in a hybrid vehicle alerts medics to an emergency situation involving a window washer on the verge of falling off a building. Environmentally oriented television campaigns such as a Green Week of programming have garnered tens of millions of dollars in traditional ad revenue and drawn in new advertisers who focus on the ecofriendly market.

The demand for product placement, whether with a green theme or not, can create tensions between movie and TV executives and their writers and directors. In some cases, the messages can be integrated into the show without significantly altering the production, drawing on themes already in the script or simply adding a product to the set (or even inserting it later through digital effects). In other cases, the placement can distort the original intent of the show. The Writers Guild of America has warned that its members sometimes have to write thinly disguised ad copy. Product placement has also drawn criticism from media analysts who say it plants messages insidiously and amounts to deceptive advertising.

See Also: Advertising; Green Consumerism; Television, Advertising.

Further Readings

Chozik, Amy. "What Your TV Is Telling You to Do." *Wall Street Journal*. http://online.wsj.com/article/SB10001424052702304364904575166581279549318.html?mod=WSJEUROPE_hpp_MIDDLEFourthNews (Accessed September 2010).
Cohen, Nancy. "Virtual Product Placement Infiltrates TV, Film, Games." http://www.technewsworld.com/story/business/48956.html?wlc=1284492534&wlc=1285522057 (Accessed September 2010).

David Hosansky
Independent Scholar

Fish

Overfishing has become a major ecological concern; for example, bluefin tuna used to be sold for pennies on the pound. But after it became a popular sushi ingredient, a fisherman could feed his family for a year by the sale of a single large bluefin like this one, and the stock dropped by 80 percent.

Source: National Oceanic and Atmospheric Administration

When environmental awareness and the concept of "green culture" emerged out of the countercultural movement of the 1960s, many fish populations had already become severely depleted. A history of consumer demand to eat specific fish, combined with increasing advancements in their mechanical harvest and habitat destruction by human activity, led to overfished waters. However, the 1960s green movement was no failure. In fact, it is more "hip" than ever these days to catch, farm, market, and consume ethically grown fish. Indeed, many trend markers flag positive social shifts over the past 40 years, and they show no sign of slowing down, despite continued practices of overfishing that still plague the industry.

The worldwide consumption of fish has greatly increased in the past three decades as a result of many consumer-end factors. There is an increasing realization that fish consumption may contribute to a healthy diet. The world's overall population has increased. Higher standards of living in both the developing and the developed world allow for augmented fish consumption—in many societies, such as western Europe and the United States, fish is a more expensive dinnertime choice than other possible high-protein items. Fish has also enjoyed a positive overall image among consumers, especially green consumers in western Europe and the United States.

Sometimes, the appealing factors of fish conflict with the realities of habitat destruction caused by human pollution. Over the past 30 years, fish has become associated with healthy living because omega-3—an unsaturated fatty acid found in fish—has been shown to reduce cancer and promote longevity; this has increased fish consumption in the United States, especially among upper-middle-class, health-conscious consumers. While fish consumption can contribute to a healthy diet, ironically, eating too much fish may be harmful. The emergence of the Industrial Revolution projected a path of escalating fossil fuel use and dumping heavy metals, such as lead and mercury, into the oceans. Thus, certain types of fish are beginning to capture unsafe levels of lead and mercury in their flesh. Paradoxically, many of the varieties of fish that are high in omega-3—such as tuna and sardines—are the same fish that often have the highest amounts of toxins in their system. When people overconsume certain types of fish, they may become ill. A story of mercury poisoning through fish consumption garnered much fanfare when actor Jeremy Piven decided to leave the set of a Broadway comedy 10 weeks before the end of production, citing illness from mercury poisoning. According to Piven, he became sick with mercury poisoning from consuming two daily helpings of fish. Dr. Carlon Colker stated, "Jeremy has been an avid sushi eater for many years, regularly eating sushi twice in one day," which apparently gave him mercury poisoning.

Northwest salmon stocks have also been affected by pollution. Wild-caught Northwest coast salmon—such as chinook, coho, chum, pink, and sockeye— have become some of the preferred fish species found in upscale U.S. restaurants and upper-class-consumer family dinner tables. Again, ironically, their populations are also being diminished rapidly by habitat destruction from agricultural runoff of suburbanization. Suburbs and strip-mall construction of coastal forest areas in California, Oregon, and Washington rechannel creeks and streams, destroying the home of the salmon creatures these same consumers love so much.

Fishing High-Tech Style

While consumer demand and habitat destruction are on the rise, so too are advances in fishing technology and machinery over the last century. Giant nets called purse seines introduced in the 1930s can be thrown around entire schools of fish and then gathered up with drawstrings, functioning as huge laundry bags. After World War II, factory-freezer trawlers grew so large they resembled seafaring towns. Trawlers dragged along shallow ocean floors or through waters collected whole schools of fish and anything else in the way, wreaking damage to target and nontarget fish and the bottom-dwelling ecosystems. In the second half of the 20th century, many fleets adopted echo-sounding sonar that detects schools of fish from far distances long before they surface. Currently, fishing fleets employ specially designed buoys called "fishing aggregating devices" (FADS), that attract species such as bluefin tuna, yellowfin tuna, and marlin. Equipped with a global positioning system (GPS) and sonar, FADS boat operators can detect when they are actually surrounded by fish.

All the practices highlighted above have led to the depletion of fish stocks around the globe. Commercial fishing of abalone began at the turn of the 19th century and continued to the first half of the 20th; abalone is a small to very large edible sea snail that once grew abundantly along the California coast. Abalone is considered a delicacy by Japanese and Chinese consumers. At the time, much of the commercially harvested abalone went to these Asian consumer markets. By the 1960s, abalone populations suffered greatly from overharvesting, which resulted in restrictions on abalone diving and gathering. Today, one must buy an expensive permit to fish these creatures, and only three abalone are allowed per diver per season. Now, most of the abalone sold on the international wholesale market are farmed in Mexico and China. As recently as the 1930s, Atlantic bluefin tuna sold in the

United States for pennies per pound. The bluefin was mostly ground into cat food. By the mid-1960s, the Japanese developed a taste for sushi made with bluefin tuna. Quickly, the fishing of bluefin for the Japanese market became extremely lucrative (a fisherman could feed his family for a year with the sale of a single large bluefin). Overfishing ensued, and bluefin stocks fell 80 percent in the last 40 years (1970–2010). In this same time period, overfishing has also depleted the stocks of sea bass, sole, turbot, and swordfish. Atlantic halibut became a U.S. consumer favorite in the 1800s, but by mid-century, inshore halibut stocks off the Western Atlantic had collapsed and have not recovered since. The most extreme case of the depletion of a fish species is the near extinction of the Atlantic cod. Over the past four centuries, but again especially in the past 40 years, cod has been overfished in the North Sea, off the coasts of Newfoundland and the northeast United States, to such a large degree that this nonsustainable fishing practice has led to the devastating collapse of the cod stocks in this mega-fishing region. As a result, the Canadian government banned the fishing of cod in 1992, resulting in the loss of 40,000 jobs and the elimination of a livelihood and culture that had existed along Canada's east coast for 400 years.

Sustainable Fish?

These depletions have not gone unnoticed among green consumers and advocates of environmental and especially agricultural sustainability. In tandem with the slow food movement, associated with locavore groups and organic food production and consumption, awareness of the negative material effects of harmful fishing and consumption practices has led to the mobilization of the green culture movement vis-à-vis the fish industry. Day-boat scallops, line-caught halibut, and environmentally sustainable farmed salmon, clams, and oysters comprise the menus of often expensive, slow food–designated restaurants and display cases of fishmongers in expensive farmers markets and Whole Foods grocery stores. Green culture advocates have promoted a rise in the awareness of overfishing, unsustainable farm-fishing practices, and increased demand for sustainable fish production and consumption. They have encouraged the mobilization of projects to "Save the Oceans" and guidelines for green consumers on how to pick sustainable seafood species harvested and also farmed in an "environmentally correct" manner for consumption. The Monterey Bay Aquarium funds a save the oceans program promoting sustainable fish consumption. The aquarium provides a website (www.montereybayaquaruim.org) that publishes information on sustainable seafood recipes; a Seafood Watch Pocket Guide that informs consumers which fish varieties are ethically fished; sustainable seafood sources; a list of partner restaurants and retailers; and information on sustainable seafood aquaculture resources and educator resources.

Along with the Monterrey Bay Aquarium's initiative to promote the consumption of sustainable wild-caught fish, other organizations such as Clean Fish (www.cleanfish.com) are educating green consumers on how and where to purchase environmentally friendly and even sustainably farmed fish. According to Clean Fish, as soon as a decade from now, farm-fish will replace wild-caught fish as the majority of the product consumers eat. Currently, global aquaculture—the farming of aquatic organisms such as fish—includes over 100 species, with salmon and carp being the most important species, earning global revenues of $10.7 billion in 2007. Global aquaculture employs farming systems ranging from traditional earthen ponds to high-tech tank systems. Each farming method has its own distinct environmental footprint. By choosing seafood from better farms and production systems, consumers can play a positive role in reducing aquaculture's potential negative impacts.

Farm-fish techniques consist of open net pens and cages, ponds, raceways, recirculation systems, and shellfish culture. The most environmentally contaminating farm-fishing practices are pens and cages found in open seas or lakes and contained ponds. Cages and pens are considered a high-impact aquaculture method on the environment. This occurs because waste from the fish passes freely into the surrounding environment, polluting wild habitat. In the wild, diseases and parasites are normally found at low levels, and kept in check by natural predation on weakened individuals. This is not the case for fish trapped in large numbers in cages and pens. In crowded net pens, fish can become ill from parasites and diseases, which subsequently transfer from farmed to wild fish populations. Farmed fish can also escape and compete with wild fish for natural resources or interbreed with wild fish of the same species, compromising the wild population. While pond farm-fishing can prove less environmentally damaging, the discharge of untreated wastewater from the ponds can pollute surrounding environments and contaminate groundwater. Irresponsible and unregulated construction of ponds in mangrove forests has destroyed more than 3.7 million acres of coastal habitat important to fish, birds, and humans.

Farm-fish recirculation systems are the least harmful to the consumer, environment, and wild fish species. This method raises fish in tanks in which water is treated and recycled through a recirculation system that removes unwanted parasites, bacteria, and any harmful food components given to fish. Almost any finfish species, such as striped bass, salmon, and sturgeon, can be raised in recirculation systems. Recirculation systems address many environmental concerns associated with fish farming; fish cannot escape, and wastewater is treated, but they are costly to operate and rely on electricity or other power sources. As more green consumers choose to eat fish due to its health benefits, more green farm-fish circulation systems will need to be built to meet this demand while not contaminating the oceans and waterways or jeopardizing the health of wild fish populations. Tim O'Shea, chief executive officer and cofounder of Clean Fish, believes this is already occurring and sees a bright future in the world of farmed fish: "It all connects: taste connects with stewardship, which connects with ecosystem sustainability. People get it!"

The emergence of green practices of eating sustainably caught and farmed fish are wrought by class, consumption patterns, and geography. Restaurants and retailers that partner with the Monterey Bay Aquarium's Seafood Watch program are more often than not situated in more affluent neighborhoods in primarily urban centers. The Seafood Watch Pocket Guide is divided regionally, but is limited to U.S. consumers, and the listed collaborators are found in more prosperous communities. That green culture–sustainable fish consumers wield large pocketbooks has bled into other areas of high-end consumption—such as the wine industry—invigorating cross-scalar alliances to protect fish environments to ensure a sustainable product for green culture fish consumers.

One project currently under way motivated by green culture consumers is the restoration of salmon fisheries along the Russian and Napa Rivers in northern California—a project called Fish Friendly Farming—representing how deep-pocketed green consumers concerned with rebuilding salmon habitats can influence grape growers and winemakers in the famous Napa and Sonoma County wine regions. In these locales, grape growers have been persuaded by vocal agents in the green-culture-cum-fish movement to environmentally improve grape-growing techniques and to eventually restore salmon spawning sites. Organic grape growers—a sector of the wine industry particularly keen to attract and endear itself to green consumers—have especially embraced the restoration of river systems. Tom Piper of Fetzer Vineyard—a winery that participates in organic grape production—states that "the Fish Friendly Farming program rounded out our operations by bringing erosion control practices on roads and environmental enhancement and restoration projects to our streams. We've learned to go beyond accepting what we have to enhancing it for fish and wildlife."

Conclusion

Four centuries of rapid depletion of fish stocks—with accentuated exhaustion just in the past 40 years—occurred due to increased consumption coupled with the modernization and mechanization of harvesting fish and habitat destruction. The emerging green culture movements have taken notice of these practices. In so doing, the green culture movement is now contributing to creating spaces for responsible and sustainable fish consumption, harvesting, and farming. As green culture consumers often wield great economic influence over sustainable, organic, and environmental consumer practices and marketing habits, the increase in interest in the consumption and harvesting of fish united with attempts to preserve and restore fish habitats informs the materialization of cross-scalar environmental alliances exemplified by the Fish Friendly Farming program, as well as the Green Fish initiative and efforts made by the Monterey Bay Aquarium to promote green fish farming, harvesting, and consumption. To ensure continued sustainable fish consumption, farming, and harvesting, and to encourage and maintain the conservation and restoration of fishing habitats requires that green culture interests move beyond the upper-class demographics where these practices currently primarily occur. While the green cultural movement came out of the countercultural movement of the 1960s that was concerned with civil rights, women's rights, and gay rights influenced environmental movements, the green culture in relation to fish consumption must also embrace bridging the class gap in relation to fish consumption, harvesting, and the protection of fish habitats.

See Also: Biodiversity Loss/Species Extinction; Diet/Nutrition; Green Consumerism; Locavores.

Further Readings

Bay Nature Institute. "Looking Ahead on the Napa River." http://baynature.org/articles/web-only-articles/looking-ahead-on-the-napa-river (Accessed December 2010).
Greenberg, Paul. *Four Fish: The Future of the Last Wild Food.* New York: Penguin Press, 2010.
Helvarg, David. *Saved by the Sea: A Love Story With Fish.* New York: St. Martin's Press, 2010.
Iles, Alastair. "Making the Seafood Industry More Sustainable: Creating Production Chain Transparency and Accountability." *Journal of Celaner Production*, 15 (2007).
Jaquet, Jenifer L. and D. Pauly. "The Rise of Seafood Awareness Campaigns in an Era of Collapsing Fisheries." *Marine Policy*, 31 (2007).
Kurlansky, Mark. *The Last Fish Tale: The Fate of the Atlantic and Survival in Gloucester, America's Oldest Fishing Port and Most Original Town.* New York: Riverhead Books, 2008.
"The Sustainability of Seafood Production and Consumption." Special issue. *Journal of Cleaner Production*, 17/3 (February 2009).
Verbeke, Wim, et al. "Perceived Importance of Sustainable Ethics Related to Fish: A Consumer Behavior Perspective." *Ambio*, 36/7580-585 (2007).

David M. Walker
Ohio Wesleyan University

Forest Service

The emergence of the U.S. Forest Service in the fragmented milieu of conservation within a growth-oriented capitalist sociocultural system is a seminal moment for green culture. As a dynamic cultural artifact, the service—founded as the Division of Forestry in 1891 in the U.S. Department of Agriculture and established as the Forest Service in 1905—represents evolving contradictions not only in the public–private sphere of national life but also in broader meanings of conservation and green thought.

From the outset, the Division of Forestry (from 1898) and later the Forest Service under Gifford Pinchot expressed values of scientific forest management under government leadership. Forests across the country, ravaged by "cut and run" lumbering and fires during the last half of the 19th century, became a rallying point for conservationists interested not only in preserving wood for future generations but also in protecting water supplies and preventing floods.

Pinchot—the nation's first professionally trained forester and a conservation movement and progressive Republican Party leader—leveraged concerns about natural resource depletion to build the case for forest preservation. The service articulated his vision, encouraging efficient and wise use of forest resources based on commonwealth, utilitarianism, and pragmatism. In contrast to John Muir's wilderness-for-wilderness'-sake ideal (expressed in national parks) and the rapacious cutting by timber firms, Pinchot took a middle path. Carefully managed forests would assure wood supplies for current and future human needs. Forests could help stimulate economic growth while benefiting the greatest number of people under the eyes of watchful government experts.

For about six decades, the Forest Service operated under the 1897 Organic Act (Sundry Civil Appropriations Act) to administer national forests to protect water and provide a continuous timber supply for the use and necessity of U.S. citizens. But national forests found themselves under increasing pressure as the nation's larger culture changed, especially after World War II. Rising population and booming construction boosted demand for wood products. More automobiles and increasing disposable income and time fueled demand for recreation. Areas in and around national forests became vacation spots, natural havens from urban and suburban life. A growing environmental consciousness also entered the complicated sociocultural landscape.

During these years, the Forest Service wrestled with an identity crisis. Increasing demand for timber fostered closer relationships with the private citizens whose land was interspersed with federal forests. Conditions also demanded enhanced communication with state and local governments for fire control and other ways to protect timber, soil, and water.

Senator Hubert Humphrey's unsuccessful introduction of a wilderness bill in 1956 was a harbinger of significant sociocultural change. The Forest Service was facing multiple and conflicting demands, pitting profit-centered timber interests against others who had a different, often aesthetic vision for national forests. "Multiple use" became increasingly integral to Forest Service activities in theory and practice, an indication of increasing economic, social, and cultural demands on forestlands.

In 1957, for example, the service launched Operation Outdoors, a five-year plan to improve recreational facilities. In 1958, however, the service published *Timber Resources for America's Future,* noting a shrinking timber base because of urban and agriculture expansion related to population pressures, technology, consumption, mobility, and increased leisure. Meanwhile, the Outdoor Recreation Resources Review Commission inventoried national forest resources; between 1958 and 1962, it developed policies that underpinned recreation plans in the coming decades.

The 1960 Multiple-Use-Sustained-Yield Act codified existing sociocultural trends. While seeking to protect timber and watersheds, the measure also put recreation, rangeland, and wildlife into the Forest Service mission. It called for the "most judicious use of land," suggesting that use does not necessarily imply commodity harvesting. Added pressure on the service stemmed from the controversial 1964 National Wilderness Preservation System Act. This law was in the spirit of John Muir; it prevented most commercial development of certain federal forestlands. The Rare and Endangered Species Act and the National Historic Preservation Act of 1966 further widened the service's responsibilities.

Policy debates and legislation of the 1950s and 1960s sought to make national forests more accessible to citizens with varying needs. They helped bolster differing interpretations of what these forests were, could be, and should be, part of a larger, evolving, and hotly debated conflict about the environment's ecological future.

The contradictory world of the Forest Service expresses larger sociocultural patterns of human relationships with the land. The service's current motto, "Caring for the land and serving people," recognizes important ties between humans and nature, but in practice, the service faces ecological, economic, political, and sociocultural fragmentation.

See Also: "Agri-Culture"; Environmental "Goods"; Leisure; National Park Service; Nature Experiences.

Further Readings

Collins, Timothy. "Changing Forests, Changing Forest Service: Chronology and Quotes." ASPI Critical Issues Series No. 4. Livingston, KY: Appalachia—Science in the Public Interest, 1992.

The Forest Society. "U.S. Forest Service History." http://www.foresthistory.org/ASPNET/Policy/Agency_Organization/index.aspx (Accessed August 2010).

Frome, Michael. *Whose Woods These Are: The Story of the National Forests.* Garden City, NY: Doubleday, 1962.

Hays, Samuel P. *Conservation and the Gospel of Efficiency: The Progressive Conservation Movement.* New York: Atheneum, 1980.

Sparhawk, W. N. "The History of Forestry in America." *Trees: The Yearbook of Agriculture.* Washington, DC: U.S. Department of Agriculture, 1949.

U.S. Forest Service. "100 Years of Caring for the Land and Serving People." http://www.fs.fed.us/ (Accessed August 2010).

U.S. Forest Service. *Timber Resources for America's Future.* Forest Resource Report No. 14, 1958.

Timothy Collins
Western Illinois University

G

GARBAGE

Today, there are about 6,000 landfills in the United States. Waste is compacted to reduce its volume, and covered with layers of soil, which are renewed frequently.

Source: Michigan.gov

Garbage is waste, especially household, consumer, and office waste—what the waste management industry calls municipal solid waste or urban solid waste, constituting the trash collected from households and commercial buildings but not including industrial hazardous wastes, medical waste from hospitals, and so forth. Municipal solid waste includes recyclable materials like paper, glass, cardboard, plastics, cans, and bottles; household toxic or hazardous waste like batteries, shoe polish, paints, light bulbs, aerosol cans, medication, and electronic waste; composite wastes like broken toys and discarded clothing; and biodegradable waste in the form of paper, food, and kitchen waste. Waste is collected by a waste management service and brought to a facility where it may be separated and processed, set aside for transfer to another site, or added to a landfill.

Landfills are sites where waste materials are buried. Burying waste is the oldest and most common form of waste treatment, and the sites used for such burial were originally known as "middens." Today, there are about 6,000 landfills in the United States. Waste in landfills is compacted to reduce its volume and covered with layers of soil, which are renewed frequently (usually daily). A given landfill is staffed with inspectors who assure that the load being trucked in is free of wastes the landfill does not accept, such as hazardous materials that are supposed to be processed separately. Bulldozers spread the waste along the working face of the landfill, compacting it in the process. In some cases, chipped wood, spray-on foams, or blankets may be used as daily landfill cover materials instead of dirt.

The environmental impacts of landfills are significant. The soil, groundwater, and aquifers can become polluted through substances leaching out of the landfill. As they decay, organic wastes generate methane, a greenhouse gas that is both dangerous to human health in large concentrations and more potent than carbon dioxide in its atmospheric effects. By attracting rats and other vermin, landfills enable potentially serious disease vectors. This is a special concern in the developing world, where scavenger communities develop around landfills, exploring the waste material in search of usable or sellable goods. In some parts of the developing world, waste from the industrialized world is brought in for landfills—sparing the industrialized country from needing to find room for it and, as a by-product, providing more valuable scavenge. Much electronic waste is handled this way, and though it is dangerous to scavenge because of the toxic and radioactive materials present, it is also especially valuable because of the copper, gold, platinum, and reusable components that it contains. Many of these communities consist of adult overseers who monitor children who do the bulk of the scavenging.

As organic materials in landfills break down under the attention of microbes and chemical reactions, methane builds up—landfills are in fact the largest source of anthropogenic methane emissions in the United States, contributing at least 500 billion cubic feet annually. Because landfills are constantly compacted as new waste is added, the gas production is unpredictable, but the gas is released into the atmosphere as pressure is applied to the mass. Landfill gas is about half methane, with the remainder primarily carbon dioxide, with amounts of water vapor, oxygen, nitrogen, sulfur, inorganic compounds like mercury, radioactive contaminants like tritium, and organic compounds like benzene, carbon tetrachloride, and chloroform, depending on the contents of the landfill. From time to time, instead of being released into the atmosphere, the gas is "squeezed" through the landfill and migrates to nearby locations, leading to several attested accidents and deaths. When landfill gas explodes below ground, it can come into contact with groundwater, contaminating it with the above toxins. Sometimes landfill gas will be collected, and the methane can be converted into methyl alcohol for use as a fuel—but the unpredictability makes this difficult. For safety reasons, landfill gas is sometimes flared off (combusted deliberately), but this results in releasing toxins into the air.

There are about 500 landfill gas projects in the United States, dealing with the gas in various ways. The city government of Sioux Falls, South Dakota, for instance, operates a landfill gas collection system that compresses the gas into an 11-mile pipeline that powers an ethanol plant. Such projects reduce local energy costs and reduce greenhouse gas emissions but are opposed by many environmental groups because they profit from a phenomenon—the overabundance of garbage and landfills—that should be curtailed and because technically, garbage cannot be considered a renewable resource. The Sierra Club and the Natural Resources Defense Council have both taken strong positions opposing government incentives for landfill gas projects.

Going a step further than the landfill scavengers of the developing world is the business of landfill mining and reclamation, wherein landfills are excavated and processed. One of the benefits is to free up space for further landfilling, or for other uses—the process aerates the soil, making it easier to reclaim and convert to other purposes. The landfill that is removed is sieved and sorted, with the use of excavators, conveyor belts, and both coarse and fine rotating trommels. The fine rotating trommel lets soil pass through, leaving the non-biodegradable material behind to be sorted. Electromagnets then remove any ferrous material, and an air classifier separates light organics from heavy organics. The sorted materials are either processed on-site or sold to third parties that do the processing, depending on the project. The easiest landfill mining is performed on construction landfills, where there are only a few sorts of materials to deal with, little or none of it hazardous. Municipal landfill mining is considerably more challenging but has the additional benefit of reclaiming hazardous materials that may otherwise begin someday to leach into the soil.

See Also: Plastic Bags; Recycling; Shopping; Theoretical Perspectives; Toxic/Hazardous Waste.

Further Readings

Redclift, Michael. *Sustainable Development*. New York: Routledge, 1987.

Rogers, Elizabeth and Thomas Kostigen. *The Green Book*. New York: Three Rivers Press, 2007.

Torgerson, Douglas. *The Promise of Green Politics: Environmentalism and the Public Sphere*. Durham, NC: Duke University Press, 1999.

Bill Kte'pi
Independent Scholar

Gardening and Lawns

Today, there is a return to sustainability driven in part by the loss of environmental quality and diminishing resources resulting from unsustainable practices. The result is the shift toward sustainable practices and a natural (organic) approach to gardening and lawn care in the United States. This transformation is taking place, in part, through a government and public that is increasingly focused on carbon emissions and global climate change concerns. People are concerned about the quality of the environment, the food consumed, and water quality issues. As a consequence, sustainable practices in the American lawn-dominated landscape are increasingly becoming more mainstream. This article discusses a few ways in which green culture is adapting and changing traditional approaches to gardening and lawn care.

Landscape professionals and gardeners alike are working more closely with the environment through alternatives to harmful pesticides and fertilizers, or going pesticide free, composting, growing vegetables, and adopting practices that protect the environment such as water conservation and using native plants. This sector focuses on natural and organic land care. Here, the emphasis is on building healthy soil that teems with living things. The philosophy is that in the long term, healthy soils reduce the need for watering and produce more robust plants; concomitantly, costs are reduced with little input. Without the use of synthetic fertilizers and pesticides, these practitioners advocate the use of natural materials that are less likely to pollute water because they release nutrients slowly. The emphasis is to increase landscape diversity through the use of native species. Together, all these practices improve the health of the land and the people who live on it.

The lawn is a good place to start in moving toward a more environmentally friendly stance. Garden writer Joan Lee Faust writes that U.S. homeowners use 10 times more chemical fertilizers per acre than farmers or golf courses. Best-selling author Michael Pollan asserts that lawns are a symptom of, and a metaphor for, our skewed relationship to the land. They teach us that, with the help of petrochemicals and technology, we can bend nature to our will. If you are concerned about the human impact on our planet's climate, reducing the amount of lawn mowed each week is one of the best ways to reduce your family's carbon emissions. If we add in the environmental costs of mowing lawns—50 million acres in the United States—the benefits of rethinking the lawn mentality become evident. According to Professor Douglas Tallamy, on average, mowing your lawn for one hour produces as much pollution as driving 650 miles, and we now burn 800 million gallons of gas each year in lawnmower engines to keep our lawns at bay. Converting lawn to trees or garden would not only save money, it would additionally create much-needed food and habitat for wildlife; it would also have the dual benefit of producing less and absorbing more carbon dioxide.

To combat this lawn dilemma, there are some simple steps to adopting a more green approach to lawn care. A simple solution is to reduce the lawn area. This automatically cuts the time needed for mowing, raking, fertilizing, and watering. Cut the grass as high as possible—this reduces maintenance. Leave the clippings on the lawn to return nutrition to the soil. Use ground covers instead of turf on steep slopes and shady sites under large trees. Ground covers such as native grasses or conventional ground covers require less maintenance than grass. Treat the borders of the property as buffer areas and screens where the best care is the least care. Here, grassy areas can be mowed infrequently. A natural buffer between your property and the neighbors' yards provides a good place to rake or blow leaves. This translates into less time and money spent bagging leaves, provides natural mulch to reduce chemical maintenance, and is a means to conserve water.

If your sense of aesthetics dictates that you must have an expanse of lawn, get with the current movement utilizing grasses that require less care. In the Northeast, these are the fescues, and they include the fine-leaf fescues—Creeping Red Fescue, Chewing's Fescue, and Hard Fescue. Other durable species are the improved turf-type tall fescues adapted to athletic fields and high-traffic areas. Once established, they require less water and fertilizer. An additional perspective is offered by the advocates of the "freedom lawn." This philosophy forgoes pesticides and fertilizers and encourages the existing vegetation of grasses and weeds to flourish with infrequent mowing. Another growing movement of lawn aficionados is moving toward native species of grass, sedges, or moss to form a mantle of green, replacing conventional lawns.

A good way to cut down on the use of pesticides, fertilizers, and maintenance is to use native plants. Today, there has been a renewed interest in incorporating native plants with an ever-increasing gardening public that is interested in an ecological approach to gardening. Native species are members of a community that includes other plants, animals, and microorganisms. In this manner, native plants provide a natural balance that keeps each species in check, allowing them to thrive in suitable conditions but preventing them from running amok. Native species rarely become invasive unless a major disturbance disrupts the natural balance of the community. Native plants provide food and shelter for birds, butterflies, and other desirable wildlife.

Many of these native plants are xerophytic plants, requiring little water on dry landscape as well as other landscape sites. These are low-maintenance species of lawn grasses, trees, shrubs, and perennials that will help to create a more ecologically sound landscape. Water has become a limiting factor in many communities, especially during hot, dry spells. Landscaping to minimize water use includes careful planning, using drought-resistant plant varieties, and improving soils using mulches to help retain moisture. These types of plants will eliminate the need for watering and wasting this precious resource once they are established; they also require little care and input once they are planted. Planting a wide spectrum of species from local wildflowers to native plants to noninvasive plants will invite local pollinators, beneficial insects, and hummingbirds to visit gardens again and again.

Transforming yard waste into yard wealth is nothing more than recycling all organic matter, including leaves, grass clippings, and yard trimmings that can be reused, recycled, or even reduced. Recycling as much as possible in your yard eliminates the need for outside inputs of fertilizers, whose excessive use in the suburban landscape can be a source of runoff or nonpoint source pollution. Recycling kitchen vegetable and fruit waste in a compost pile mixed with leaves and grass clippings produces rich humus for garden and lawn use. There is a renewed interest in growing vegetables, and what better spot to grow them than in your own backyard utilizing your compost? According to Bruce Butterfield of the National Gardening Association, since 2008 there has been a renewed interest in household participation in food gardening. Thirty-one percent of all U.S households (an estimated 36 million households) participated in food gardening, including growing vegetables, fruit, berries, and herbs.

See Also: Green Consumerism; Home; Soil Pollution; Sprawl/Suburbs/Exurbs.

Further Readings

Butterfield, Bruce. *Personal Communication*. South Burlington, VT: National Gardening Association, 2008.
Faust, Joan Lee. "Greensward Care Without Pesticides." *New York Times* (May 15, 1988).
Pollan, Michael. *Second Nature: A Gardener's Education*. New York: Dell, 1991.
Tallamy, Douglas W. *Bringing Nature Home: How You Can Sustain Wildlife With Native Plants*. Portland, OR: Timber Press, 2006.

Carl A. Salsedo
University of Connecticut

Global Warming and Culture

When Rachel Carson published her paradigm-shifting research and analysis on the effects pesticides were having on the planet and those who call the planet home, she could not have imagined that she was ushering in the dawn of an environmental movement that would span the globe and fundamentally challenge the way that we had come to think about our relationship with the planet. Now, some 50 years later, the information, pictures, slogans, and products that relate to "saving the planet" are everywhere and there is, perhaps, no issue that better epitomizes the relatively sudden omnipresence of "the environment" and environmental discourse than global warming and climate change.

The story of climate change really does read like something that was proposed as a plot for a Hollywood film. What if the very phenomenon that keeps Earth warm enough to be habitable—the greenhouse effect—were changed by the essence of industrial society (i.e., the burning of fossil fuels) such that the more fossil fuels we burned, the hotter the planet would become? As the plot for a film, it seems almost too far-fetched. Yet this is exactly what is at the heart of climate change: the very essence of industrial society is changing the very climate of the planet.

The enormity of what the issue of climate change proposes, surfacing at a time when the environment had become a widespread social movement, meant that climate change quickly became a global environmental issue of such immense scope that it is, in many ways, unlike any other environmental issue. And perhaps the best way to understand the uniqueness of climate change is to explore how it has infiltrated our very culture—everywhere in the world.

Climate change is an issue that is on the political agenda of every country on the planet. Global organizations such as the Intergovernmental Panel on Climate Change (IPCC) and the United Nations Framework Convention on Climate Change (UNFCCC) have members from countries all over the world and speak to citizens everywhere in the world. The Kyoto Protocol, perhaps the best-known document associated with how the international community has come together around climate change, was originally developed as part of a UNFCCC meeting in Kyoto, Japan, in 1997 (and came into effect in 2005). According to the UNFCCC's *Kyoto Protocol: Status of Ratification* (2009), there are now 187 countries that have signed on to the protocol. The IPCC has similarly had a great global impact as indicated, perhaps most succinctly, by the organization being awarded the Nobel Peace Prize in 2007.

Political awareness of an issue does not, however, necessarily indicate the way that an issue has had broad cultural resonance. Such resonance can be measured in various ways, one being a country's media attention to the issue. Numerous academic studies have illustrated both the prevalence of communication about climate change (in various kinds of news outlets) as well as national differences in the way in which countries cover the issue. For example, the Pew Center on Global Climate Change lists a range of publications that explore the way in which climate change has been communicated around the world.

Of course, while politics and news can indicate the ways in which an issue such as global warming and climate change has infiltrated a country's culture, in some ways, popular culture is an even better measure of the cultural resonance of an issue. Al Gore's climate change documentary *An Inconvenient Truth* perhaps epitomizes how the issue of climate change has become a part of U.S. and global popular culture. Gore's work on climate change won him the Nobel Peace Prize (with the IPCC) in 2007, and *An Inconvenient Truth* won an Oscar for Best Documentary in 2007. As a result of such attention, Gore made appearances on such popular culture iconic programs as *The Simpsons, Futurama,* and *South Park*.

Other climate change documentaries have drawn on star power to call attention to the issue. Perhaps best known is Leonardo DiCaprio's *The 11th Hour*, a documentary about climate change that he directed, produced, and narrated.

Climate change–related movies have also been made into fictional feature-length movies. In 1995, Kevin Reynolds directed *Waterworld*, which explored a post–ice cap melted world. The 2004 movie by director Roland Emmerich, *The Day After Tomorrow* (2004), tells the tale of planet Earth after it has been hit by a barrage of freak, climate change–like weather events. A more recent post-apocalypse movie by the same director, *2012*, tells a similar tale of extreme events changing the world as we know it.

And while it is not a climate change movie per se, the 2009 blockbuster *Avatar*, which director James Cameron insisted be released on DVD on Earth Day 2010, made it abundantly clear that stories based on what might happen when a dying Earth sends its unsustainable tentacles to other planets resonate with audiences.

Global warming and climate change has infiltrated popular culture and the marketplace in other ways. Advertisers have come to rely on "green" pitches for all kinds of products, and some of the most successful campaigns have been for products, such as cars, that might seem irrevocably linked to climate change. Toyota's hybrid car (a hybrid car being a car that can make use of more than one power source, usually gas and some version of an electric motor) Prius has come to be synonymous with environmental, and climate change, salvation, not degradation. The Prius has been so successful at infiltrating popular culture that celebrities such as Leonardo DiCaprio have made a point of affiliating themselves with the car and sharing why it is important to them to be driving a Prius.

With all the exposure that the issue of global warming and climate change has had in popular culture, one might think that awareness of and concern about it would similarly be very "popular," but one would be wrong to conclude that. For example, in an October 2007 *Newsweek* poll, 39 percent of Americans said "there is a lot of disagreement among climate scientists," 42 percent said "there is a lot of disagreement that human activities are a major cause of global warming," and less than half of Americans, 46 percent, said that climate change is being experienced today. So it is not surprising that less than half of Americans indicated in a GlobeScan, Inc., survey that personal action would be necessary to address climate change.

Part, if not all, of the explanation for such skepticism around the causes of, even the existence of, global warming and climate change is that it is not only organizations that are concerned about climate change, such as the IPCC and UNFCCC, that have made it into the news and infiltrated our popular culture. Scientists, companies, and organizations that question the existence of and science around global warming and climate change—sometimes referred to as climate change deniers—have also had a significant presence in the news and in popular culture more generally in the past few years. Such hesitancy also plays out in national political culture.

For example, according to Forecast Earth, in March 2006, then-President George W. Bush stated that the "fundamental debate is whether climate change is 'man-made or natural.'" Then, in 2007, President Bush indicated that the Kyoto Protocol "would have wrecked our economy." And while it is easy to propose that the issue of climate change tends to fall along political lines such that more conservative politicians lean toward denying, or at least resisting, climate change science and related solutions while liberal politicians embrace the science and solutions, it is also true that the (arguably) most energy-intensive country in the world, the United States, is now governed by a liberal, President Barack Obama, and the United States continues to be one of the only countries in the world to have not signed the Kyoto Protocol.

As well, while global warming and climate change and concerns about the environment have become prevalent in such things as advertising, some have raised questions about the motives behind such messages. The term *greenwashing* has been developed to address messages that seem "environmentally friendly," but are really about supporting "business as usual."

In some ways, this leaves us where we began this exploration of climate change and culture. The reality is that what global warming and climate change proposes—industrial society's burning of fossil fuels is changing the planet's climate—hits at the heart of the way in which we, as humanity, have created our modern lives. The solutions to climate change—the real solutions—therefore will be equally fundamentally profound. Yes, popular culture has given us glimpses of what the future might hold if we do not address climate change, and glimpses of what some of the change might look like, but it would seem that we are still very far from embracing the essence of what climate change calls for. Climate change seems to demand not that we consume differently but that we consume much, much, less.

See Also: DiCaprio, Leonardo; Education (Climate Literacy); Film, Documentaries; Film, Drama/Fiction; Gore, Jr., Al; Greenwashing; Hybrid Cars; Kyoto Protocol; Popular Green Culture; Public Opinion; Theoretical Perspectives.

Further Readings

Begley, Sharon. "The Truth About Denial." *Newsweek* (August 13, 2007).

Mauch, Christof. *Shades of Green: Environmental Activism Around the Globe*. Lanham, MD: Rowman & Littlefield, 2006.

McKibben, Bill, Jonathan Isham, and Sissel Waage. *Ignition: What You Can Do to Fight Global Warming and Spark a Movement*. Washington, DC: Island Press, 2007.

Pew Center on Global Climate Change. http://www.pewclimate.org/communicating (Accessed September 2010).

Jennifer Good
Brock University

Gore, Jr., Al

A recipient of the 2007 Nobel Peace Prize, Al Gore has focused on creating a well-informed citizenry through movies, books, and speeches, which include a mixture of science communication, personal anecdotes, and explanations of solutions.

Source: Embassy of the U.S. in Norway/U.S. Department of State

Former U.S. Vice President Al Gore and the Intergovernmental Panel for Climate Change (IPCC) were jointly awarded the 2007 Nobel Peace Prize "for their efforts to build up and disseminate greater knowledge about man-made climate change, and to lay the foundations for the measures that are needed to counteract such change." In 2006, Al Gore featured in the Academy Award–winning documentary *An Inconvenient Truth*, directed by Davis Guggenheim, a documentary on Gore's campaign to make the issue of climate change internationally and politically more prominent. According to the Nobel Prize committee, the film presented the accumulated science of climate change and the results of the IPCC's work to a worldwide audience.

Al Gore has become a household name of the environmental movement. In media discourse, 2006 has been described as the year when public discourse on climate change shifted from whether global climate change was real to what humanity needed to do about it. This shift is commonly ascribed to the cumulative effect of Hurricane Katrina, Nicholas Stern's report on the economic impact of climate change, and Al Gore's efforts to create awareness of climate change (particularly with the film *An Inconvenient Truth*, which is said to have profoundly influenced the world's attitude toward climate change). This optimism in public discourse, however, was not matched in the field of politics and public policy. The United Nations climate conference in Copenhagen 2009, which was to agree on a treaty to replace the Kyoto Protocol, failed to achieve a global, binding contract. Also, driven by the vested interests of the fossil fuel lobby, climate change denial has increasingly gained media and political traction. This political paralysis and politicization of climate change is the main topic of Gore's current environmental advocacy.

Al Gore has been active in the communication of environmental issues and development of environmental policies throughout his career. The environment and a well-informed citizenry are key issues of his activities. Gore worked as an investigative journalist during the Vietnam War and then in Tennessee, covering public affairs, before winning a seat in the U.S. House of Representatives in 1976. In 1984, he was elected to the U.S. Senate. In 1992, he was elected vice president of the United States. As part of the Clinton-Gore administration, Gore was involved both in environmental policies and those policies that enabled the development of the Internet and of information technologies. After leaving the White House in 2001, Gore entered the business world, starting new businesses as well as becoming adviser to established high-tech businesses. He is a member of the board of directors of Apple and a senior adviser to Google.

Gore also is the author of many best-selling books on environmental and political issues. His first, *Earth in the Balance* (1992), described the dimensions of a global environmental crisis and offered a blueprint for political action; it put him firmly on the map as an environmental politician and advocate.

Gore's writings and public appearances follow a characteristic formula: a mixture of science communication, personal anecdotes, and explanations of solutions (technological and political). These elements are combined to educate and to inspire, and make a plea for moral responsibility and global political change.

Gore says that to solve the climate crisis we need to solve the crisis of democracy. The focus of his work is on how we gather information about the state of the environment, how we communicate information, and how we educate people about the environment and democracy to enable active citizenship and political action. Gore's emphasis has shifted from communicating the crisis to offering solutions. He focuses increasingly on promoting renewable energy technologies and sustainable capitalism. He is cofounder and chairman of Generation Investment Management, for example. Gore's environmental campaigning uses an integrated approach, spanning business and alternative energy technologies (e.g., www.ourchoicethebook.com, Virgin Earth Challenge), new media technologies (e.g., Digital Earth, www.climatecrisis.net), entertainment and media (e.g., Current TV, Live Earth, appearances with other celebrity advocates), politics (e.g., the Alliance for Climate Protection), and education initiatives (e.g., the Climate Project).

Controversy

Criticism leveled at Gore revolves around questions regarding his personal investment in, and the way he benefits from, the climate change crisis. Climate change deniers have decried Gore as a climate change doomsayer. He has been accused of hypocrisy because of his personal energy use and his activities as investor in alternative energies. Similarly, Gore has been criticized as a pop icon. His many awards, including major U.S. entertainment awards (e.g., an Emmy, a Webby, a Grammy, and the prestigious Quill Award) have been used cynically to raise questions about the environmental value of greening entertainment culture. This reflects the wider question of the accuracy of scientific facts when taking messages across to the public, particularly via entertainment media. Gore has been accused of misrepresenting scientific facts. *An Inconvenient Truth* caused some controversy and resulted in a British High Court case. These criticisms result from the question of whether Al Gore is capitalizing on an ever-increasing and self-perpetuating green discourse—"climate porn," as some scientists call it—or whether he is helping to create a discourse that informs concrete action and policy development. As part of the latter, he has been criticized for his promotion of carbon market schemes by, most famously, James Hansen (director, NASA Goddard Institute for Space Studies and climatology researcher). Hansen claims that carbon market schemes failed in the context of Kyoto and Copenhagen and espouses a carbon tax as an alternative. He criticizes Gore (and President Barack Obama) for failing to meet the moral challenge of our age because of the compromises of politics.

See Also: Clinton, William J. (Executive Order 12898); Environmental Communication, Public Participation; Film, Documentaries; Global Warming and Culture; Kyoto Protocol.

Further Readings

Gore, Al. *Earth in the Balance: Forging a New Common Purpose.* London: Earthscan, 1992.
Gore, Al. *Our Choice: A Plan to Solve the Climate Crisis.* Emmaus, PA: Rodale, 2009.
Lewis, Judith. "Everything's Gone Green: Filmmakers Unleash a New Kind of Scary Movie." *LA Weekly* (March 14, 2007). http://www.laweekly.com/film+tv/film/everythings-gone-green/15909/ (Accessed September 2010).

Angi Buettner
Victoria University of Wellington

Grassroots Organizations

Grassroots organizations (GROs) are collective attempts by ordinary people to address problems that impact local communities. Driven by a politics of civic participation that places the engagement of rank-and-file citizens at the center, these voluntary organizations are radically participatory and empower lay members to take an active role in the political process. Often emergent and spontaneous, grassroots organizations tend to engage global social and environmental issues at the local level. A major focus of these organizations is to serve as watchdog and put pressure on the regulatory process and to address concerns of corporate or political injustice rather than advocate new policy.

The goal for most grassroots environmental organizations centers on creating a sustainable and just world. Rather than work through the formal political bureaucracy for political reform, however, these organizations work directly and informally with citizen groups at a local level to address pressing social and environmental justice concerns. Though each grassroots organization has its own mission, ideology, and approach, most aim to educate and empower the public toward the redress of environmental issues while changing the culture of civic engagement in the process. In doing so, they engage politics outside official channels of communication and outside official positions of authority to strengthen civil society and democratize the political process by including greater citizen participation. Formed in opposition or as an alternative to bureaucratic state activity, these organizations place citizens at the center of the political process and coalesce around struggles for community, citizenship, civil rights, and government accountability. A conservative estimate places the number of grassroots organizations in the United States at around 10,000, with significantly more emerging around the globe.

Organizational Structure and Aims

Grassroots organizations have emerged organically from the bottom of the political system to address contradictions or failures in a liberal democratic state. Their "bottom-up" approach grows from a lack of faith in traditional political and commercial institutions to adequately and quickly address pressing environmental concerns. Often, their emergence is spurred by the failure of traditional state mechanisms of social, economic, or environmental control, which lead to pollution or other forms of environmental risk. These groups see the neoliberal state as either blind to, complicit in, or unresponsive to existing forms of social and environmental inequality. Critics of the U.S. environmental policy pioneered in the 1970s charge that these laws, while ambitious and guided by sound environmental research, were often weak or inadequately enforced and often fell victim to corporate pressure. It is through the construction of alternative strategies and institutions that these organizations seek the redress of these environmental regulatory concerns.

Grassroots organizations are developed in contrast to traditional organizations, which are defined by a hierarchal organizational structure, permanent roles, offices, and responsibilities, concentrated power and authority, and formal mechanisms of organizational governance. Instead, grassroots organizations feature a relatively flat or inverted organizational pyramid, emergent leadership structure, flexible roles and responsibilities, and informal and participatory mechanisms of governance. Whereas traditional organizations rely on individuals with specialized training or expertise to fulfill key roles, grassroots organizations rely largely on lay volunteers and informal "nonexperts" who often develop their own forms of expertise and build unique skill sets through their service to the organization. To be sure, many expert professionals contribute their training in government, legal, or academic pursuits to further the efforts of these groups—however, their role in the organization is often not limited to their specific vocation or area of expertise as would likely be the case in a traditional bureaucratic organization.

Membership and Networking

Grassroots organizations, by their nature, seek to maintain a small-scale, decentralized authority structure and fluid organizational formation. Often driven by radically democratic political values, these organizations are characterized by a diffuse membership with a flexible organizational structure that can accommodate the skills and expertise that each individual brings to the organization. Volunteer members may come and go as interest in or urgency of the target problem waxes and wanes, but there usually persists a core group of dedicated group members. Though there are seldom official "leaders" to these groups, core members and those with the greatest experience and seniority in the organization often take on a quasi-leadership role as a "point person" or "committee chair" who organizes the activities of a specialized working group. However, these positions often rotate through the ranks of the membership, giving leadership opportunities to lay members and providing them opportunities to further develop their skill set within the organization. When the size of these organizations begin to grow, internal pressures, differing ideologies, and economies of scale often necessitate that they break into smaller working groups or splinter into parallel organizations to retain their vitality and connections to the grassroots. Because these organizations are directed at a specific and often local issue or problem, their goals and missions change as the nature of the problem changes. In fact, when a key issue is satisfactorily resolved or reaches a point of stasis, the organization may disband, diffuse, or reemerge as a new organization altogether.

Through strategic networking and coalition building, grassroots organizations create alliances with other groups working for environmental and social justice. These partnerships often include the mobilization of groups that have been marginalized politically, economically, and socially, including indigenous peoples, trade unions, women's groups, farmers' movements, and human rights organizations that have a stake in particular policy solutions. These groups also build coalitions across issue areas to pool resources, people power, and energy to fight common enemies as witnessed during the World Trade Organization protests in Seattle in the late 1990s (discussed below). Existing at the bottom of the political pyramid, their relative political marginality is often a strength, as they can "fly under the radar" of government agencies and operate more agilely in addressing changing political landscapes than more established organizations.

Grassroots organizational activities may include lobbying local and state representatives, writing letters to local officials, submitting editorials to local newspapers, writing grant applications, canvassing, organizing community forums and press conferences, and participating in the public process through public hearings. Some organizations also enlist the assistance of public interest lawyers to initiate legal action to challenge legislation or encourage stronger enforcement of existing environmental laws. More radical grassroots organizations participate in direct-action protests, which may include rallies, protest marches, picketing, civil disobedience, street theater, monkey-wrenching, or eco-sabotage to raise public awareness about environmental justice issues.

Mobilizing Against Environmental Pollution: Love Canal, NIMBYs, and LULUs

One of the first and most notable grassroots organizations came about in Niagara Falls, New York, in the late 1970s in response to the contamination of local working-class neighborhoods. Industrial chemical dumping in the 1940s and early 1950s by Hooker Chemical Corporation contaminated the Love Canal landfill and surrounding land, which was later capped and closed. To deal with an expanding population, Hooker Chemical (which neglected to clean or detoxify the contaminated site) sold the site to the local school board, which was seeking to build a new school building. In the years that followed, the surrounding area became home to a working-class housing development and a public school in 1955. At the time, many residents were either unaware of the contamination or believed local officials who downplayed the potential risk of living near the former landfill site.

After years of documented health problems (including miscarriages, birth defects, seizures, abdominal pain, and skin disorders), odors, vapors, and fumes and visible substances leeching from the ground, lay citizens mobilized to seek redress from politicians, governmental agencies, and corporate representatives who the citizens felt repeatedly ignored their concerns. Pressure from this grassroots citizens group led to the eventual cleanup of the site and reimbursement and relocation of more than 800 families. Because of the work of this organization, the U.S. Congress passed legislation that holds polluters accountable for their waste and damages incurred by environmental contamination.

There are countless examples of grassroots community groups who have mobilized to protest strip mining, logging, dam building, chemical processing, or environmental risks, including oil spills, chemical leaks, or various other industrial disasters. Many of these citizen groups have organized against locally unwanted/undesirable land uses (LULUs), including the siting of prisons, power plants, landfills, and big-box chain stores like Walmart. Often referred to as NIMBY (not in my backyard) movements, these campaigns seek to block the siting of these LULUs in their local neighborhood. However, these groups often fail to recognize that their efforts to block this unwanted development displace these projects onto communities that are politically marginal or poorly equipped to mobilize against them. Also, NIMBY movements often operate at a point of relative privilege. These communities may be in a position to pass up the potential benefits of these projects (jobs, tax revenue, public infrastructure), whereas less advantaged communities may not. In these cases, one community's sustainability becomes another's nightmare. Moving beyond NIMBY, environmental organizations are increasingly embracing notions of environmental justice, which encourage organizations to consider the impact of their efforts on other communities and landscapes, therefore taking a more holistic, ecological approach to their advocacy work.

Networking for Change: Antiglobalization, WTO Protests, and Friends of the Earth

As mentioned above, grassroots organizations have formed in a political context of widespread resistance to neoliberalism, especially in global commercial activities and international trade. A flashpoint in this grassroots movement occurred during the World Trade Organization (WTO) Ministerial Conference held in Seattle, Washington, on November 30, 1999. The goal of the WTO is to liberalize international trade and broker the regulation of trade between participating countries; its ministerial conferences bring together representatives from member nations to negotiate trade partnerships. The meetings themselves were quickly eclipsed by the protests that emerged among hundreds of grassroots organizations, student groups, labor unions, religious groups, and international nongovernmental organizations (NGOs) that had a stake in resisting the outcomes of the WTO.

The protests brought together groups representing labor, women's rights, indigenous peoples, consumer rights, and various other environmental causes whose membership felt their interests were in conflict with the objectives of the conference.

Brought together by a word-of-mouth and Internet networking campaign, various grassroots groups mobilized their membership to attend the peaceful protest. Joined by thousands of college students, they took to the streets outside the hotels and the Washington State Convention and Trade Center to exercise their right to gather and to express their concerns about the direction of the WTO. While most of these organizations engaged in peaceful protest, some anarchist and direct-action groups chose to engage in more deliberate forms of civil disobedience to disrupt the meetings and prevent delegates from entering the meetings. The most extreme of these actions resulted in intentional vandalism of corporate properties. The protests turned ugly, with Seattle police resorting to the mobilization of a SWAT team, which made use of tear gas, pepper spray, stun grenades, and rubber bullets to disperse crowds. In the end, over 600 protesters were arrested. Given its global media attention, the "Battle in Seattle" became a public symbol pitting average citizens against corporate and political interests and became an organizing tool for many organizations in the years that followed.

The work of grassroots organizations has also facilitated the creation of alternative institutions, the most important of which have been independent media. Due to their independence from corporate sponsorship or mainstream media control, grassroots journalists, filmmakers, bloggers, and radio programmers provide these organizations with a vehicle for expressing their perspectives about pressing environmental issues. The Independent Media Center, for example, brings together a "decentralized and autonomous network" of independent media professionals and audiences from around the globe to provide "accurate" and "truthful" and on-the-ground portrayal of current events, especially those stories that are poorly covered or altogether ignored by mainstream outlets. Independent media played a central role providing grassroots coverage of the WTO protests in Seattle in 1999. Emerging simultaneously with the protests themselves, the Independent Media Center provided a clearinghouse for unaffiliated and amateur photographers to pool audio, video, and images through its website (www.indymedia.org); distributed its own newspaper during the event; and broadcast radio programming through Real World Radio (www.radiomundialreal.fm), featuring up-to-the-minute and on-the-ground coverage. Today, both websites contain a searchable database of stories, videos, news segments, articles, commentary, and research, with links to thousands of grassroots organizations. This grassroots collaborative also supports local media center affiliates whose work focuses on local issues and promotes various social and environmental justice campaigns. The work of these grassroots media professionals can often be heard and seen on independent radio stations and cable news programs across the country, including Pacifica, Democracy Now!, Real World Radio, Free Speech TV, and Link TV.

In recent years, the rise of social networking facilitated through the Internet and mobile technologies as well as growth of social media like Twitter, Facebook, and MySpace have revolutionized the ability of grassroots organizations to mobilize their membership and recruit new participants. Whereas previously grassroots organizations relied on a "boots to the ground" mentality, which required face-to-face or written communication with the communities they represented, a decentralized and dispersed network of members sharing information and organizing events in real time now supplements these efforts. For example, Friends of the Earth is an international grassroots environmental organization that supports local chapters in 77 countries to address local concerns through its broad social network. The organization provides a wealth of resources and taps into a global storehouse of knowledge through its quarterly magazine, e-newsletters, and various Facebook pages.

Forget FEMA: Grassroots Response to Hurricane Katrina

After landfall of Hurricane Katrina along the Gulf Coast states of Louisiana and Mississippi in August 2005, media outlets for several months continuously circulated media images of human suffering. The most notable images showed thousands of largely poor and minority residents stranded on rooftops and lined up outside the New Orleans Convention Center and Superdome football stadium in dire need of food, water, and shelter. Several days after the storm made landfall, citizens began criticizing the federal response as poorly coordinated, woefully inadequate, and ill timed. For many, the help was too little and too late. From this ferment of governmental mistrust grew a number of grassroots organizations that sought to address the needs of local residents that were going unmet. In the tremendous void left by the failed governmental response led by the Federal Emergency Management Agency (FEMA), these organizations stepped in to serve hot meals; provide shelter, water, and clothing; supply legal advice and advocacy; and assist in the clearing and removal of possessions from flooded properties.

One particular community-initiated volunteer organization, Common Ground Relief, emerged out of the storm's aftermath to address the loss of housing and personal possessions of the Lower Ninth Ward. This majority African American neighborhood was ground zero for the failed Industrial Canal Levy, which gave way, sending a torrent of water throughout this low-lying community. While the growing uncertainty about whether homes should or could be rebuilt in this community stalled official planning and rebuilding efforts, Common Ground sought to provide a vehicle for residents to rebuild their community from the ground up. Cofounder and former Black Panther Malik Rahim, along with fellow community organizers, mobilized resident volunteers to advocate the right of the residents to return and rebuild.

Their early efforts included repurposing one of the few standing houses in the neighborhood as a community center to distribute canned food; prepare hot meals; provide access to the Internet; supply tools, boots, gloves, and work clothing through their lending library; and disseminate information about the evolving legal challenges facing these flooded communities. Volunteers from around the country assisted with these efforts and worked with local residents to "muck and gut" over 3,000 homes, which would enable them to rebuild and move on with their lives more quickly. As the relief effort moved into the recovery and rebuilding phases, Common Ground retooled to provide a free legal clinic (staffed by area law students and a licensed Louisiana attorney) to remediate victims of wrongful demolition and contractor fraud and to assist with property and title claims and mortgage applications. Additional projects have included the establishment of an independent health clinic, a women's shelter, a job training program, and a wetlands restoration initiative. In further efforts to promote self-sufficiency for impacted residents, their Lower Ninth Ward Urban Farming Coalition provided an urban gardening program that offered various forms of assistance, including soil testing, raised bed construction, and a local garden club for residents seeking to grow their own food. To date, they have provided assistance to over 25,000 residents impacted by the storm, and they continue to serve.

Global Grassroots Activity: Sustainable Development

At the global level, grassroots organizations have emerged to address the growing needs of a growing world. In various regions of the developing world, grassroots organizations work closely with local populations who are seeking greater access to a clean and safe environment. Realizing the environmental limits of growing unchecked, these organizations are working in local communities to ensure that development occurs in a way that is both environmentally sustainable and socially just.

For example, international grassroots network organizations like GROOTS (Grassroots Organizations Operating Together in Sisterhood) are driven by an ecofeminist ethic that seeks to empower poor and impoverished women to address problems in their community. Their view is that development issues are simultaneously environmental and human justice issues—class exploitation, male dominance, and treatment of girl children are inextricably linked to access to clean water, clean air, farmable land, shelter, and quality healthcare (including treatment and prevention of human immunodeficiency virus and acquired immune deficiency syndrome [HIV/AIDS]). In particular, GROOTS provides an international network to link leaders and grassroots groups in poor rural and urban areas around the globe. Through information sharing, coalition building, and community development, their goal is to empower women to solve problems within their own communities and to develop their capacity to reach out to other communities of women worldwide.

When their husbands and male children leave to find work elsewhere, women are often left to tend to village and community affairs. However, they often face cultural and social barriers to effectively voice their concerns and initiate actions to ameliorate environmental challenges. Organizations like GROOTS work to enable women to develop a viable, credible, and legitimate voice in the political process and empower them to challenge the very traditions that have left them outside the political process in the past. This is not to say that these women are defenseless or powerless, but that their roles have been historically undervalued and fragmented. The role of these organizations is to partner with and work within these communities to overcome gender discrimination to stimulate political self-sufficiency and empower women to enact proactive change within their own communities.

See Also: Antiglobalization Movement; Environmental Communication, Public Participation; Green Anarchism; Hurricane Katrina; Nongovernmental Organizations; Social Action, National and Local.

Further Readings

Azad, Nandini. "Grassroots Women: A Response to Global Poverty: Excerpts, Statement by Grassroots Organizations Operating Together in Sisterhood (GROOTS), 9 September 1995." *Women's Studies Quarterly*, 24/1–2, Beijing and Beyond: Toward the Twenty-First Century of Women (Spring–Summer, 1996).

Cable, Sherry and Michael Benson. "Acting Locally: Environmental Injustice and the Emergence of Grass-Roots Environmental Organizations." *Social Problems*, 40/4 (November 1993).

Castells, Manuel. *The City and the Grassroots: A Cross-Cultural Theory of Urban Social Movements*. Berkeley: University of California Press, 1983.

Castells, Manuel. *The Power of Identity, The Information Age: Economy, Society and Culture, Vol. II*. Oxford, UK: Blackwell, 1997.

Jacobs, Jamie Elizabeth. "Community Participation, the Environment, and Democracy: Brazil in Comparative Perspective." *Latin American Politics and Society*, 44/4 (Winter 2002).

Middlemiss, Lucie. "Influencing Individual Sustainability: A Review of the Evidence on the Role of Community-Based Organizations." *International Journal of Environment and Sustainable Development*, 7/1 (2008).

Norris, Lachelle and Sherry Cable. "The Seeds of Protest: From Elite Initiation to Grassroots Mobilization." *Sociological Perspectives*, 37/2 (Summer 1994).

Robert Owen Gardner
Linfield College

Green Anarchism

Green anarchy (or green anarchism) is a distinct branch of anarchism that is concerned with matters that arise from human interactions with the environment. In broad terms, it is viewed as the synthesis between the political philosophy of anarchism and the social practice of environmentalism. Green anarchy is highly critical of the current cultural and social structures devised by humans. It views these structures as placing human needs and desires above and to the detriment of all other living creatures. Green anarchists believe that placing humans above all others consistently devalues and devastates all other living beings and the well-being of Earth. In contrast, green anarchy promotes the harmonious symbiosis between all that exist on Earth, where humans strive to live as simply as possible with as little impact on the rest of the Earth, both living and nonliving. Many different groups practice the main tenets of green anarchy, which has led to much confusion on what exactly a green anarchist is to outsiders.

Green anarchism is considered a part of the broader anarchist umbrella because, though focused on the environment, it follows the main principles of the anarchist philosophy. Several important writers have contributed to the development of green anarchy. Peter Kropotkin, Murray Bookchin, Emma Goodman, and Alexander Berkman have all written influential works. Briefly, anarchy, as a political philosophy, promotes the absence of an elite centralized body of rule makers responsible for governing all others of a given society. For anarchists, this belief is much different than the common usage of anarchy as being synonymous with lawlessness, chaos, and disorder. In place of centralized authority, anarchists believe that every individual should be allowed to participate in the construction of the morals, ethics, and social practices that are responsible for regulating the collective body. This type of social regulation is generally known as consensus governing and/or the "rule" of collective decision making. In particular, anarchists are critical of the current structure of the economy and politics, believing these are formulated in a way that a small body of individuals (an "elite") are rewarded at the expense of everyone else. To combat this, anarchists promote, and highly value, individual freedom and the ability of economic and political self-rule.

Transcendentalism and Green Anarchy

The transcendental work of Henry David Thoreau is a significant influence to the environmental portion of green anarchy. Transcendentalism is broadly viewed as a challenge to the way in which human beings have organized themselves since the Industrial Revolution. Thoreau and other transcendentalists were highly critical of the way in which industrial and post-industrial societies were constructed in relation to the rest of the natural world. Particularly, this line of thought was highly critical of the way in which humans utilized nature with little regard to any other living being. As a way to "transcend" this perceived problem, this philosophy argued that humans should live as simply as possible and strive for self-sufficiency. Green anarchy highly values transcendentalism's commitment to simple living and self-sufficiency.

Combining the traditions of anarchy and transcendentalism, green anarchists are critical and question what they view as society's presumed superiority of humans over all other life forms on Earth. Green anarchists view that this assumed dominance is a fallacy and that this assumption has led to numerous environmental and social problems. In particular, they focus on the negative by-products of continued industrialization and urbanization, such as air and water degradation, and feel they need to be halted. Because of this, green anarchists are viewed as being highly concerned with the effect of human-created structures on the environment, especially those being produced in rich countries of the Western world. For example, green anarchists note that modern nation-states require hundreds of inputs from other parts of the world to keep functioning, and many of these inputs are taken without much thought about the effect on the rest of the world. Green anarchists feel this an important topic to address because it causes a significant imbalance in the natural world's ecological functioning—the current relationship benefits members of some nation-states at the cost of environmental degradation across many other eco-regions of the world.

As a way to begin to solve this problem, green anarchists believe that Earth's land, air, and water should be viewed as the birthright of every being (human and nonhuman) and not as commodities to be bought and sold in the marketplace for profit. Green anarchists argue that one of the most important ways to do this is through the commitment toward the absence of hierarchical structures, which would be the absence of relationships that place any one individual actor in direct power over the actions of others. To value the rest of the world, green anarchists commit themselves to creating communities based on self-regulation through consensus-based decision making that considers all other things on Earth, both living and nonliving. They argue that this can begin to address the environmental degradation they feel exists all around them.

Green Anarchy's Ideal Community

Thus, green anarchists argue for a radical reorganization of human beings in relation to the rest of the planet. One such proposition put forth by green anarchists is the return to smaller communities, both in population size and land area occupied, that are more closely aligned to the living principles of communities that existed prior to the Industrial Revolution. The ideal community for green anarchists would be organized by ecological region, which honors the diversity, decentralization, and autonomy of each entity. While there is disagreement within green anarchy on the ideal size of these communities, the most complex civil-society entity would be an independent, self-regulating, autonomous city (no more than a couple of thousand people) based on total self-sufficiency.

To do this, green anarchy believes that these communities should focus primarily on the production of socially necessary goods (i.e., food and shelter) using environmentally sensitive techniques. For example, green anarchists liken their desires to the coral reefs of the world, where the reef thrives via symbiotic relationships between numerous species that foster its continued growth and success—without any external requirements or internal outputs that need to be outsourced to other parts of the environment. Thus, green anarchism supports following nature as a guide when suggesting how communities should organize themselves—especially highlighting that the community can survive by itself and when left undisturbed. Green anarchists feel that creating communities in this way can reconstitute a more symbiotic relationship where humans can continue to benefit as a species, but in a way that does not come at the expense of the rest of Earth. Thus, an important principle of green anarchy is to create cultures and societies that allow for a balance between humans and all other living things, where all have the ability to prevail.

There is considerable confusion and debate on what exactly it means to subscribe to green anarchy in contemporary society. One of the main reasons for this is that there is no specific dogma that one can subscribe to wholly, like one can subscribe to Christianity or vegetarianism. With the lack of a specific set of absolute principles, there is a deficiency in what green anarchy is exactly, and there are many groups that claim to be the primary advocate of green anarchism. Some of these groups are known more commonly as anarcho-primitivists, deep ecologists, social ecologists, and ecofeminists.

Overall, these groups all believe and practice the same general structure of green anarchism; fractions only begin to occur between the groups over more specific points of contention. In particular, a hotly contested issue is what the implementation of green anarchism should look like. Though minor for participants of green anarchy, this contributes significantly to the difficultly for outsiders to understand what green anarchy is and is not. Many have argued that with the lack of absolute principles, all of these groups subscribe to green anarchism, but practice it differently by focusing on different parts of the philosophy. Anarcho-primitivists focus on promotion of ways in which humans could live more in harmony with Earth, deep ecologists emphasize the equality of all living things on Earth, social ecologists focus their efforts toward building a community that incorporates all living things, and ecofeminists endorse the recognition of instability of patriarchal structures as a way to repair environmental degradation.

See Also: Alternative Communities; Earth First!; Earth Liberation Front (ELF)/Animal Liberation Front (ALF); Environmental "Bads"; Social Action, Global and Regional; Social Action, National and Local.

Further Readings

Berkman, Alexander. *ABC of Anarchism*. New York: Vanguard Press, 1929.

The Green Anarchist Infoshop. "The Vanguard of the Apocalypse." http://www.GreenAnarchy.info (Accessed June 2010).

Purchase, Graham. *Anarchism and Ecology*. Montreal: Black Rose Books, 1997.

Purchase, Graham. *Anarchism and Environmental Survival*. Tucson, AZ: Sharp Press, 1994.

Ward, Colin. *Anarchism: A Very Short Introduction*. New York: Oxford University Press, 2004.

Justin Schupp
The Ohio State University

Green Consumerism

Green (or conscious) consumerism is both a philosophy and a movement that reconciles consumers' reflexive choices toward environmentally friendly and socially responsible commodities and profit motives of businesses. It is also part of the ecological modernization paradigm that advocates a reflexive modernity in which consumer behavior will ultimately move or drive capitalism toward a "green" one. The idea is that if consumers are environmentally and socially conscious, purchase only products or commodities that do not have troubled environmental and social legacies in production and processing, and boycott harmful commodities, then producers will change environmentally and socially harmful processes. Consumers are therefore a proactive and powerful agent of change. A green consumer is someone who is environmentally and socially conscious, which leads him or her to purchase products that are ecofriendly. Products with little or no packaging, products made from natural ingredients, and products that are made without causing pollution are all examples of ecofriendly products. A green consumer is likely to drive a hybrid vehicle and buy products made with hemp or those made from recycled materials.

Broadly speaking, green consumerism is not just what consumers buy, but how they live. It is therefore an overarching worldview or philosophy of consumers toward greening their lives and their choices and thereby shaping businesses. According to green consumerism, a consumer does not just buy a commodity, but also everything that went into its production, processing, impacts, and everything that will happen in the future as a result of that product. The concept of green consumerism also focuses on businesses and their survivability as they respond quickly to demands of consumers for products and services that are also environmentally friendly, socially responsible, and culturally viable.

Development and Evolution of Green Consumerism

Green consumerism emerged partly due to the development of a conscious consumer sector with high levels of disposable income and good access to media, and partly through campaigning by environmental and social justice nongovernmental organizations (NGOs) and pressure groups from both the global south and north. Because of these factors, consumer concern has centered not only around long-established issues like safety and environmental degradation but has also included questions of labor exploitation and gender disparity in countries that produce a commodity. In the food sector, for example, European countries have led the way in this transition, largely in response to various food scares (e.g., BSE or mad cow disease) and wariness of genetically engineered (GE) foods. The United Kingdom's 1990 Food Safety Act, which outsourced government oversight of food safety to retailers, provides a symbolic marker for the shift. Consumers are therefore motivated to demand increasingly more specific products, and they have information technology that supports their ability to seek, identify, and procure these products. Businesses have therefore looked into the green process, which is not just green as a superficial label but a whole array of processes that include generating corporate environmental profiles, monitoring and evaluating green performance, and improving corporate image, as a result. Green products have increased competition among businesses to generate more environmentally friendly products and technologies. It has also generated a unique governing structure of business—partly shaped and driven by the neoliberal private governance.

Traditionally, government agencies had the responsibility for monitoring safety standards and other quality attributes. However, the recent emergence of privately regulated supply chains organized more around principles of "green quality" has precipitated a shift in governance that is sometimes known as the private regulation of the public. In the case of food, for instance, previously, the notion of quality was understood as freshness and taste; however, recent movements have extended this notion to include other social and environmental attributes. Local and global environmental and civil rights movements have launched campaigns to address social justice issues by making sure that commodities are environmentally friendly, socially responsible, and have meaningful community participation. These are sometimes known as "credence" or nonmaterial characteristics of "quality," characteristics that consumers cannot detect after purchase in the same way that they detect freshness and taste. These credence qualities include the environmental and ethical conditions of production. A green consumer is therefore concerned about, for example, whether a food commodity is produced organically, whether beef cattle are fed hormones, whether crops contain genetically modified varieties, whether coffee producers were paid a fair price, whether tuna are caught by methods that minimize dangers to dolphins, what conditions animals were raised in, and so forth. Because these qualities are process based and not readily apparent in the physical product that reaches the consumer, a green consumer can respond only if the product has trustworthy labels. The effect is to make the regulation of food production a way of producing new quality-based values that can be marketed to consumers. Green consumerism is therefore seen as a driving force for green business.

This shift toward a broader definition of "green" (e.g., appearance, size, consistency, taste, freshness, safety, fair trade, and environment) is driven, partly, by the neoliberal turn in commodity regulation, with states ceding responsibility for regulation just as corporate actors step in to provide a proliferating array of quality assurance schemes and voluntary standards, a trend that is often characterized as "privatizing regulations" or "certification regimes." Among the important mechanisms used in certification schemes are identity preservation (IP), segregation, and traceability systems allowing for "field-to-plate" monitoring of supply chains. One of the most significant trends in businesses in recent years has been not only the attempt to deregulate and liberalize international trade in commodities, but also the proliferation of new forms of reregulation. *Re*-regulation has emerged largely in response to the concerns of consumers in the wealthy nations of the global north. Some of this has happened at the governmental and intergovernmental level, for example, the deliberations of the Codex Alimentarius, but there has also been a marked increase in the number of private regulatory schemes. In aquaculture, for example, recent years have witnessed an increasing interest in certification largely based on private regulations for aquaculture products, and an increasing number of schemes covering ecolabeling, organic certification, and, recently, fair trade. The growing number of certification programs, and possible competition among certification schemes, has the potential to result in confusion among buyers and consumers. As "green quality" becomes the basis under which certification agencies are competing with one another, issues of labor and gender are currently incorporated in certification codes. As large seafood buyers such as Walmart, Darden, and Lyons have already committed to buy only certified seafood, for example, it is now realistic to imagine that significant portions of global industrial aquaculture production could come to be certified within the next few years. Therefore, green consumerism is going to be a powerful business and consumer trend.

Impact of Green Consumerism

Because of, among other things, green consumerism, many corporations have embraced voluntary aspects of green development. Corporations have created a number of voluntary agreements, such as the CERES principles. CERES principles stand for Coalition for Environmentally Responsible Economics. Corporate signers pledge to participate in voluntary environmental reporting and ongoing environmental developments and improvements. Businesses such as American Airlines, Bethlehem Steel, General Motors, and Sunoco have signed the CERES principles. Associating with such voluntary principles provides positive public relations benefits without requiring dramatic shifts in actual corporate behavior and signals that there is little need for government intervention.

While green consumerism has a positive impact in terms of creating awareness in environmental sustainability and social equity, the shift to "green quality" involves new dilemmas for agricultural and other commodity sectors, regions, and individual producers, as privately regulated supply chains define different sets of winners and losers. For producers, the exacting contract specifications required for participation in quality chains can increase costs and tend to marginalize smaller, less-sophisticated producers. Aside from creating industry consolidation, the trend will potentially generate a consumer democracy undermining small producers and sidelining the nation-states.

While green consumerism can be one of the ways to address negative environmental and social legacies of different commodities, critics of this approach think that there are other structural factors and class relations that need to be addressed. To them, green consumerism is sometimes presented as the only way to address environmental and social catastrophes of modern capitalism by occluding the structural factors that are the sole cause of these problems.

See Also: Antiglobalization Movement: Ecotourism; Popular Green Culture; Recycling; Shopping; Social Action, Global and Regional; Social Action, National and Local.

Further Readings

Gould, K. A. and T. L. Lewis, eds. *Twenty Lessons in Environmental Sociology*. New York: Oxford University Press, 2009.

Hawken, P. *The Ecology of Commerce: A Declaration of Sustainability*. New York: HarperBusiness, 1993.

Islam, M. S. "From Pond to Plate: Towards a Twin-Driven Commodity Chain in Bangladesh Shrimp Aquaculture." *Food Policy*, 33/3 (2008).

McMichael, P. *Development and Social Change: A Global Perspective*. Thousand Oaks, CA: Pine Forge Press, 2008.

Vandergeest, P. "Certification and Communities: Alternatives for Regulating the Environmental and Social Impacts of Shrimp Farming." *World Development*, 35/7 (2007).

Md Saidul Islam
Nanyang Technological University

Green Jobs

Contemporary world society faces profound environmental challenges, above all, global climate change and the threats to ecosystems, human health, and social justice that accompany global climate change. At the same time, economic globalization has meant a dramatic increase in the gap between rich and poor worldwide and a loss of high-paying, career-track jobs for unskilled and semi-skilled workers in the United States and other developed nations. The concept of green jobs, and the economic development paradigm that follows from this concept, proposes to simultaneously solve both the environmental and economic crises human society currently confronts. The green jobs model suggests that national economies in the developed world must undertake radically redesigned and rebuilt physical infrastructure in order to neutralize the emission of carbon and other greenhouse gases into the atmosphere. This push to renew our physical infrastructure will also require a large semi-skilled workforce, the development of which will renew economic vitality within the working and middle classes.

Biofuels, such as fuel created from algae, is the largest renewable energy sector and could employ as many as 12 million people over the next decade.

Source: Randy Montoya/Sandia Corporation

What Are Green Jobs?

Green jobs are well-paid, career-track positions that contribute to environmental sustainability or enhance the quality of the natural environment. Renewable energy, sustainable transportation, electrical grid enhancement, water provision, waste management/sanitation, and green construction are the key industries for green job creation. The growth of green industries also creates indirect employment in sectors that provide inputs for green production such as the plastics industry and steel industry, which provide inputs for the fabrication of wind turbines and solar panels.

The bulk of green jobs are in the provision and transmission of renewable energies such as wind, solar, hydro, and geothermal energy. Within the renewable energy industries, biofuels constitute the largest sector at present. Following biofuels, the production of solar energy is the second most significant subsector of renewable energy production.

Our vehicles and systems of transport are responsible for a massive amount of fossil fuel consumption and thus greenhouse gas emissions into the atmosphere. The transition to sustainable transportation, a key dimension of the green economy model, consists of updating the public transportation infrastructure in the United States to consume energy more efficiently, and increasing the energy efficiency of personal vehicles on the road.

The smart grid, an energy-efficient overlay of the current electricity transmission and distribution infrastructure, holds the potential to contribute significantly to green jobs. The smart grid digitizes the control, coordination, and metering of the electric grid, and it allows energy to travel in two directions so that surplus energy can be sold back to the utility. Also, alternative energy production is intermittent and has different infrastructural requirements. The smart grid enables more effective management of energy from alternative sources.

Green building refers to new building design and the rehabilitation of existing structures with an emphasis on environmental sustainability, energy efficiency, and human health. Green building seeks the efficient use of water and energy resources and the reduction of waste and other pollutants. Green buildings achieve energy efficiency through active and passive solar, efficient windows and appliances, and rooftop gardens (which insulate while sequestering carbon). Rain gardens, permeable parking lots, and rainwater capture contribute to water efficiency. These buildings reduce waste by recycling grey water for irrigation and industrial uses and converting black water into fertilizer. Almost all commercial green buildings are constructed to the specifications of the third-party certification system Leadership in Energy and Environmental Design (LEED).

The green economy model relies on an initial investment push of public and private funding to launch workforce training programs to develop the vast workforce necessary to transform our energy infrastructure. The focus is, therefore, not just on job creation, but on the training of unskilled workers to launch them into career trajectories that mean increased wages and benefits for themselves and their families. Green jobs can thus have a substantial environmental justice component. A key dimension of the green jobs model is the opportunity for advancement in terms of skills, salary, and responsibility. The green job model is also noteworthy for its incorporation of a diverse workforce, including highly skilled and educated designers, planners, engineers, and scientists along with many semi-skilled workers such as carpenters, electricians, machinists, mechanics, welders, and many others. Typically, low-wage, unskilled labor with little opportunity for further professional development is not considered a green job even if it is in a green industry.

The Scope and Economic Benefits of Green Jobs

Although estimates vary widely, most research agrees that the green jobs model offers a significant source of job creation. Some estimates suggest that there are currently around 8.5 million green jobs in the United States, the majority of which are in industries related to the production and distribution of renewable energy. Research suggests that over the next couple of decades that figure could increase to approximately 40 million. A report by the United Nations Environment Programme (UNEP) estimates that 2.3 million people have begun careers in the renewable energy sector alone over the past few years. Wind and solar energy may grow to employ roughly 8 million people over the next two decades. Farming for biofuel production could employ as many as 12 million people over the next decade. Additional millions of jobs could be created in building new energy-efficient buildings and rehabilitating existing buildings to be more energy efficient. Recycling and waste management will also grow, particularly in the developing world, which will also produce millions of new positions. Countries around the world are generating green jobs at a rapid rate. China has 600,000 people producing solar energy; Nigeria may create 200,000 jobs in biofuel production; and India may create a million new jobs in biofuels over the next several years.

The production of biofuels is unique and important within the renewable energy industries because it is the only source of renewable energy that incorporates farmers on a large scale. Successfully integrating farmers into the green economy will be a boon to some rural areas. Further, biofuel production plants are often located in proximity to areas where biofuel crops are cultivated, which further enhances the development implications of the biofuel industry for certain rural areas.

Proponents of the green economy model suggest that beyond job creation, this model further contributes to economic development by reducing energy and water consumption and thus saving money, which is then reinvested in the economy. Governments also save money, which can then be invested in public works or social programs for the larger public good. The Apollo Alliance estimates that the city of Los Angeles, California, for example, could save up to $10 million a year simply by retrofitting existing buildings to make them more energy efficient.

The model offers noneconomic benefits as well. More environmentally sustainable buildings and industries improve worker health and quality of life as well through increased access to natural light, higher-quality air, fewer toxins, and so forth. Finally, some argue that the green economy model could contribute to the national security of the United States and help mitigate violence and instability in oil-rich regions of the world by reducing American dependence on fossil fuels and thus promoting American energy independence.

Another advantage that green jobs have over the blue-collar jobs of bygone eras is that these jobs, because most are tied to specific geographies, are less exportable to nations with lower costs of production than jobs in industries such as textile or automobile manufacturing. Certainly, some parts of these industries are mobile. Wind turbines, for example, are manufactured around the world. Europe, China, Korea, and the United States are the most active turbine-producing economies. And crop production for biofuels is carried out around the world as well. But hydroelectric dams, electrical grid construction and maintenance, and the installation and maintenance of solar panels and wind turbines are service jobs that cannot be exported.

The History of the Green Jobs Model

Green jobs are often also referred to as "green-collar jobs." Although both the concept and the term have been used intermittently since the 1970s, the green economy model first gained popularity in 2007 when Representatives Hilda Solis (D-CA) and John Tierney (D-MA) introduced the Green Jobs Act of 2007 in the U.S. Congress. This act proposed to amend the Workforce Investment Act of 1998 to make approximately $125 million available to establish an energy efficiency and renewable energy-training program for workers. In this incarnation, the bill never became law, but key elements of the bill were subsumed into the Energy Independence and Security Act of 2007, which became law in December 2007. Also in 2007, the UNEP and the International Labor Organization launched a joint Green Jobs Initiative, which has further contributed to the growing public profile of this model. The green jobs model was further popularized during the 2008 presidential campaign when both Hillary Clinton and Barack Obama made the provision of green jobs a centerpiece of their campaign platforms. These high-profile endorsements have led to a dramatic upsurge in its popularity since 2008. Today, many cities across the United States are rolling out green jobs programs, numerous nonprofit organizations have begun to promote the model, and many politicians advocate a green economy.

The Appeal of the Green Jobs Model

The green jobs model has become widely supported over the past several years for many reasons. First, it is a model that appeals to both conservative and progressive political perspectives and therefore can achieve bipartisan support in Congress. Crudely speaking, the model appeals to conservatives because of its emphasis on market mechanisms to produce environmental goods and because of its promise as a means to redouble national security through energy independence. It appeals to political progressives because it offers a path out of poverty to many underemployed Americans while simultaneously fighting global climate change and strengthening the environment. The "two birds with one stone" rhetoric of the green economy model also deepens the appeal to politicians and civic leaders.

Furthermore, greening is just good business. Over the past two decades, environmental values have increasingly entered the mainstream. This has created an opportunity for the makers of consumer products, and it has pushed firms to reconceptualize their business models to take environmental planning and impact assessment more seriously. According to some estimates, the global market for green goods and services will double from $1.37 trillion per year to $2.74 trillion over the next decade.

Because green products constitute a novel and exciting market, and because their market share is projected to grow significantly as environmental values increasingly permeate the mainstream and the climate crisis grows increasingly urgent, venture capitalists and governments are eager to invest in green technologies. In the United States, green technology is presently the third most significant destination for venture capital, following the biotech and information technology sectors. In China, venture capital investment in green industries has more than doubled in recent years.

The Dangers of Green Jobs

As a model of economic growth first, and environmental sustainability second, the green jobs model runs the risk of reproducing many of the aspects of contemporary capitalism that ironically are partially responsible for the dire environmental and economic situation that the green jobs model purports to address. In other words, devising and constructing means of performing the same behaviors more energy efficiently does not challenge the ideological foundations of modern consumer culture. Many environmentalists argue, for example, that our time and energy should be spent expanding the accessibility and desirability of public transportation systems rather than building more energy-efficient cars, which rather than ultimately bringing about a net reduction in carbon dioxide emissions may just encourage drivers to drive longer distances more frequently for the same amount of gasoline. Similarly, the aesthetic and marketing appeal of constructing new LEED-certified green buildings may encourage a family or a company to erect a new building entirely when a retrofit of an existing structure may have used less energy overall. No matter how energy efficient, the construction of new, unnecessary structures does not constitute a net gain for the natural environment. And again, no matter how energy efficient the structure, there is a great deal of energy embedded in its construction.

Building the massive infrastructure necessary for the new green economy will require a great deal of additional energy, and the economic loss to society of the current functioning-but-less-energy-efficient infrastructure is rarely taken into account when the green jobs model is advocated. Furthermore, the construction of the green infrastructure is, in simple terms, a temporary economic boom spurred to a large extent by public sector investment. Once the infrastructure is built, the additional jobs created by the boom will begin to fall off.

There are additional shortcomings with specific sectors within the green jobs model. Hydroelectric dams, for example, often displace human settlements and flood large areas, which leads to many concerns about human rights violations in the construction of large hydroelectric dams. The production of biofuels is also controversial. Many scholars suggest that the petroleum-based inputs (such as fertilizer) required to produce biofuels largely negate the environmental value of their production. Additionally, there is concern that diverting valuable agricultural land away from food production may exacerbate hunger and malnourishment among the chronically poor, particularly as the human population continues to grow. Finally, scholars point out that even if all of the arable land in North America were devoted to the production of biofuels, it still would be insufficient to satisfy our energy needs. For these reasons, many environmentalists are critical of the biofuel industry, and increasingly, policy makers are looking to wind and solar power to form the backbone of our alternative energy infrastructure.

An additional problem with the green economy model is the ostensible urban bias. Cities like Los Angeles, California; Chicago, Illinois and Milwaukee, Wisconsin; that have the capital to invest significantly in public works using green-building techniques and vast workforce development programs are at a competitive advantage over rural communities that lack the capital to invest. Unfortunately, rural communities are disproportionately underdeveloped compared to urban communities and thus arguably stand to benefit to a greater extent from the new green economy model. The federal government could help disrupt the urban bias in the green jobs sector by incentivizing the investment in rural communities, but rural communities must also be proactive in recruiting green industries and training their workforces to attract new investors. Furthermore, because certain regions of the country lend themselves to certain green industries, the distribution of the economic benefits may not be even. For example, the Midwest has a comparative advantage in biofuel production, while the Southwest lends itself to solar energy production, and coastal states such as Texas and California have an advantage in wind energy.

In sum, the green jobs model, as it has developed over the past several years, purports to present a single solution to our current economic and environmental crises through massive public and private investment in a radically transformed energy infrastructure worldwide. The model has promise, particularly as a strategy for community economic development, yet its rhetorical appeal may ultimately be greater than its transformative potential.

See Also: Alternative/Sustainable Energy; Corporate Green Culture; Environmental Justice Movements; Recycling; Sprawl/Suburbs/Exurbs.

Further Readings

Apollo Alliance. "Green-Collar Jobs in America's Cities: Building Pathways out of Poverty and Careers in the Clean Energy Economy" (2008). http://www.apolloalliance.org/downloads/greencollarjobs.pdf (Accessed July 2010).

Green, Gary Paul and Andrew Dane. "Green-Collar Jobs." In *Berkshire Encyclopedia of Sustainability, Vol. 2: The Business of Sustainability,* Karen Christensen, Daniel Fogel, Gernot Wagner, and Peter Whitehouse, eds. Great Barrington, MA: Berkshire Publishers, 2010.

Jones, Van. *The Green-Collar Economy: How One Solution Can Fix Our Two Biggest Problems.* New York: Harper One, 2008.

United Nations Environment Programme. "Green Jobs: Towards Decent Work in a Sustainable, Low-Carbon World" (2008). http://www.unep.org/civil_society/Publications/index.asp (Accessed July 2010).

Michael L. Dougherty
University of Wisconsin–Madison

Greenwashing

The term *greenwashing* is normally used as a pejorative, referring to the practice of construing an activity as more environmentally friendly than it really is. In that, it likens its precursor term *whitewashing* used to signify money laundering (i.e., the creation of value based on using resources that were illegally gained). Greenwashing references the concept of "whitewashing" (superficially painting over unsightly blemishes so as the object appears more valuable than it actually is), extended to the nonenvironmental activities of an organization.

Actors in Charges of Greenwashing

A charge that greenwashing is taking place involves individuals and organizations who question whether information regarding an activity's environmental friendliness is truthful. Typically, the charge is published with the intention of drawing public awareness to the responsible organization. To illustrate, consider autonomous activists who claim that energy providers using lignite as a resource greenwash the inherent destructiveness of their fossil-fuel activities. The charge in this case would be that energy providers are channeling public attention to a new technology (carbon capture and storage [CCS]), supposedly rendering the activity of burning lignite sustainable by storing greenhouse gas emissions away from the atmosphere. Rather than contacting the energy provider, activists make public their reasons for questioning whether CCS is really environmentally friendly in itself, as well as whether CCS really renders lignite-based energy production ecologically sustainable. Thus, critics who make a public charge against greenwashing share an (implicit) analysis, suggesting that the organization that presumably is actually environmentally unfriendly would only consider a change in its practices under public pressure. Often, the charge against an organization's activity is transformed into a charge against the organization itself. This indicates that not only the truth value of the claim that an activity is green is at stake, but also the claim or assumption that an organization is credible.

The organization—often a business—responsible for the activity that is being questioned may react to such charges in diverse ways. When the organization holds on to its construal of the activity's being environmentally friendly, the charge is transformed into a contested issue. Sometimes, the organization provides further information to demonstrate why its portrayal depicts the truth. However, the organization may also choose to ignore the charge or intensify its exercise in rendering the contested activity green. Alternatively, an organization sometimes (implicitly) accepts the charge.

The third collective actor needed to make the charge significant is "the public." If publics are not interested in the charge, the polluting organization may safely ignore the charge. Therefore, those charging usually actively disseminate their information to specific social systems, such as media or the judicial system.

Finally, organizations have evolved that support either side of these charges—normally, but not necessarily, staying on their sides. On the one hand, entities like the World Business Council for Sustainable Development focus their voice to support the construal of organizations (and their activities) as green. On the other hand, groups like Corporate Watch specialize in charging businesses. Such entities continually invest their resources in research and public relations to affirm their respective interpretations.

Varieties of Greenwashing

As indicated above, the term *greenwashing* is used to refer to a variety of situations. There are at least the following four types of greenwashing under attack.

Products

An organization may depict a product as being environmentally friendly itself or as being more ecologically sound than competitive products. For example, organic food products are often questioned in this respect.

Processes

This type of greenwash includes statements of organizations claiming that some of their processes have positive effects and, thus, reduce their environmental impacts. This type refers to both end-of-pipe processes as well as integrated processes. "End-of-pipe" greening indicates that environmental protection measures are put in place as add-ons to existing production processes. Thus, the original destructive processes are not altered in themselves. To illustrate, a mining company may advertise its processes of reforestation after a landscape has been stripped of its resources. In contrast, an organization may transform its processes to render them more environmentally friendly (integrated environmental protection). To exemplify, an organization may claim to have carried out such integrated environmental protection when it reduced the energy consumption of a process, for example, by using more efficient machines.

Symbols

An organization may claim to act in an ecologically sustainable manner based on symbolic action that—from the point of view of critics—does not affect an actual material problem. An example for this type of greenwash charge is the case of the European Union's trying to induce a mass change of light bulbs: In September 2009, the European Union (EU) started a ban on "traditional light bulbs." This policy is seen by critics as a symbolic greenwash, shifting attention from more destructive environmental dynamics and politics within the EU to a far less significant issue.

Structures

Finally, greenwashing may take place at a structural or systemic level. This points to any kind of grand structures that, as respective claims suggest, are transformed into being environmentally friendly. An instance of this type is the emerging carbon markets (for the trading of carbon emissions/carbon emission reduction credits). Some proponents claim that these markets are effectively key to greening capitalism, while critics deconstruct these markets as intensifying capitalist dynamics by appropriating still another "playground" for capitalist relations. In this case, the grand structure "capitalism" is understood as inherently blind to environmental interests. Then, green markets can be conceptualized as greenwashing.

Understanding Greenwashing: What Is Under Charge?

As indicated above, classically, a charge is made against the construal of an activity or a thing as being environmentally friendly. The issue at hand, thus, may be, for example, whether or not a car (based on gasoline consumption) is green. However, the picture can easily be more complex. We are able to differentiate at least three levels of charges and, respectively, three types of objects under charge.

First, an organization claims that something it does (i.e., a process) or produces (i.e., a product) is green. The object under charge is the relation between the claim and the significance of the claim. The first-level charge challenges this relation, suggesting that the claim does not hold true. To illustrate, reconsider the car produced. Claiming that the car ought to be categorized as green(washed) refers not to the car itself, but rather to its relation to some standard according to which the categorization is right or wrong. The same pattern exists for cases in which an activity is under scrutiny, rather than a material thing: A company may claim that it runs an environmental management system that will transform the organization into being environmentally friendly.

Second, an organization may imply with the first level that not only the product or activity is green, but the total organization is green. The corresponding charge, then, targets the relation between the implied claim and the total organization. Thus, the second-level charge questions whether indeed it is true that the total organization is green because of the first level's product/activity. As an example, we may recall advertisements showing a specific car in some romantic natural environment and pointing out its allegedly low carbon emissions. The message is, of course, that not only the car shown but its general form is green: All cars of the same model are depicted as green and, if the company is mostly active in the automobile industry, the company itself is portrayed as environmentally friendly.

Another version of a second-level charge of greenwashing, challenging the understanding that the organization in itself is working in an ecologically sustainable way, can arise when the organization simply encourages connotations of being green by showing consumers/publics that they feel green. For that, no (first-level) claim that actual activities or products are green is really necessary. An illustration is the critique of Corporate Watch against E.ON's relation to the 10:10 campaign. The activists, in this case, suggested that E.ON portrays itself as a green business by endorsing a campaign suggesting that actors ought to act in environmentally friendly ways.

Finally, organizations publicly engage with partners who are commonly perceived as defending ecological sustainability. A third-level charge, then, takes as its object of criticism that another entity, for example, a nongovernmental organization (NGO), lends credibility to the for-profit organization. The charge, again, points to the relation between the other entity and the for-profit organization, suggesting that credibility cannot be earned by simple encounters between both organizations. World Wildlife Fund (WWF), for example, lends its credibility to many companies. Correspondingly, then, the NGO receives much criticism by more radical environmental pressure groups.

Epistemological and Ontological Issues of Charges of Greenwashing

"This car is environmentally friendly"; such a statement implies a realist ontology, that is, an understanding of a reality existing independently of our understanding of it. The organization making such a claim has to assume it knows the truth about this reality. If an organization was not convinced of the correspondence of its statement with reality, then its statement would be dishonest. The charge of greenwashing questions whether such a correspondence actually exists. Thus, as described above, critics may question whether the claim that a product, process, or structure (including the organization in itself) is environmentally friendly is true.

Thus, normally, both actors assume a singular truth to exist. This, often, fits well into public discourses and hegemonic knowledge systems. Both mass media and judicial systems may try to establish a decision as to who is right and who is wrong. However, as indicated by social constructivist "takes" on knowledge claims—for example, originating from science and technology studies—what counts as fact is often in itself contested. To some degree, then, we can conceptualize conflicting interpretations of reality, that is, regarding whether the environmental friendliness claim is justified or not, as grounded in diverging theories/ideologies about the world. The conflicting parties may subscribe to opposing sets of assumptions and, thus, derive different conclusions about the truth value of the original statement. To illustrate: a widely shared assumption holds that greening capitalism is possible by economically internalizing all goods and bads (e.g., resources and pollution). Then, this approach suggests, markets would allocate their traded entities optimally. The case of carbon markets is a case in point. Actors sharing the assumption often support the establishment of carbon markets. Those actors, however, who do not share the assumption "debunk" such emerging markets as greenwash. Both groups may then easily construe the respective other's understanding as ideology.

Critics of an environmental friendliness statement are also able to problematize its truth claim without making an alternative definite truth claim themselves. This would be the case when they show that an environmental friendliness claim simplifies reality to a degree that, according to their understanding, is unjustified. An example of this is the question of whether nuclear energy should be considered carbon neutral. While nuclear energy proponents often endorse this source of energy as carbon neutral, critics may point out that the picture changes depending on which sources of emissions are included in the picture. Organizations may try to avoid discussions that put truth values at stake. Alternative moves may include statements that do not address specific products or processes, but rather create the impression that the organization is green (discussed above as different objects under charge). In these latter cases, the organization suggests or implies that it is feeling, and thus acting, environmentally friendly without having to engage with its specific products or processes. Hence, critics suggest that greenwashing can exist by organizations' being deceptive.

Presuppositions of Greenwashing

Finally, then, it is necessary to explicate the rationale of an organization's constructing itself and its products as green. The advantage of selling cars as green cars assumes consumers prefer greener products over less green ones. This preference order is widely spreading globally—in specific social milieus as well as across societies, consumers prefer buying green products over non-green products, especially when both products share the same price. Marketing research aims to establish how much more consumers are willing to pay for green products.

For an organization, then, it is rational to advertise its products as environmentally friendly. For a variety of classes of products, however, consumers show some loyalty to brands. For that reason, producers may aim to spread the green connotation from some products to their brand/organization itself.

The communication of a for-profit organization to consumers can be understood as advertisement. If an advertisement suggests that a product or an organization is green, then it is safe to assume that the organization intentionally designed this suggestion.

A wider social problem is based on this communication structure. Companies produce advertisement to support their economic interests. Scholars, however, have conflicting analyses of what this practically affects. It can be argued that companies have an interest in providing true information about themselves in order to earn the trust of consumers. On the other hand, other scholars conceptualize companies as communicating strategically with consumers. Such strategies may sacrifice the trust of some consumers, while successfully portraying the organization as green to the majority of consumers. Then the issue of how true an advertisement has to be is decided according to prospective sales rather than truth values. Nevertheless, in both accounts of the conduct of companies, consumers are dependent on the producers. For many, then, the discussion about greenwashing provides an argument in support of media independent of for-profit organizations.

To summarize, greenwashing emerged as a side effect of public sensibilities toward environmental issues. For specific instances of greenwashing, it is relevant to consider both the description of products, processes, and organizations involved and the identification of the normative standards drawn on by the diverging parties in the dispute. Many actors agree that greenwashing exists and should be avoided. The critical questions revolve around whether greenwashing can be avoided under the pressures and incentives of competitive markets.

See Also: Advertising; Corporate Green Culture; Corporate Social Responsibility; Environmental Communication, Public Participation; Media Greenwashing.

Further Readings

Jermier, John M. and Laura Forbes. "Greening Organizations: Critical Issues." In *Studying Management Critically*. London: Sage, 2003.

Laufer, William S. "Social Accountability and Corporate Greenwashing." *Journal of Business Ethics*, 43/3 (2003).

Rikhardsson, Pall and Richard Welford. "Clouding the Crisis: The Construction of Corporate Environmental Management." In *Hijacking Environmentalism: Corporate Responses to Sustainable Development*. London: Earthscan, 1997.

Ingmar Lippert
University of Augsburg

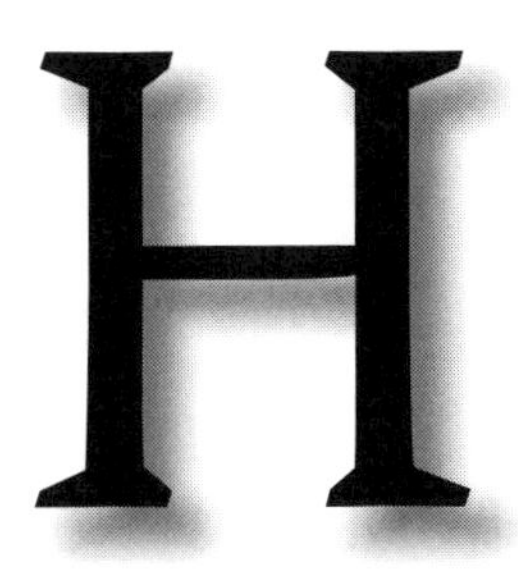

Home

The places we construct are physical definitions of who we are, the work we do, and the beliefs we hold. These places, from buildings and infrastructure to art and technology, reflect our knowledge of the world and our relationship to geography, climate, heritage, and history. Yet today our cities, towns, and villages are becoming generic, 24-hour, climate-controlled stations more aligned with the global marketplace than the place-identity we traditionally have called "home."

What makes a home? A home is something we create through a series of personal and cultural decisions. We care for our homes as living space in which we invest meaning and identity, time and energy, and work and money. There are many factors that go into defining and creating a home, from physical size, location, materials, energy usage, and landscaping to living practices and type of ownership. Every home is different, whether it's an urban apartment, a small house, a nomadic tent, or a country estate. A home reflects who you are and how you live, whether intentionally or not.

Physical Size

The size of the average American home has doubled since 1970, while family size has decreased. The organization of spatial uses is critical to the way a home functions and has a direct relationship to the quantities of material and energy used and the landmass needed; in other words, the larger the home, the greater the amount of land, energy, and materials consumed. In designing the functions of a house or apartment, one can arrive at a tighter, more energy-efficient, and ecologically beautiful model by thinking three-dimensionally in addition to reviewing two-dimensional plans. Rather than having a room for each specific activity, such as media or exercise, integrating activity areas will make the home feel more hospitable and cohesive. Computer modeling or cardboard models are a great help for understanding spatial usage and functional relationships.

Location

The location of a home is critically linked to access and mobility. While we are not all free to choose where we live, most of us do self-determine the location based on work, friends, family, affordability, or identity. How we get to and from our home has an impact on our daily lives and the surrounding urbanization. For downtown, urban neighborhoods and suburban locations we can define walkability to schools and shops, proximity to neighbors, type and distance of commute to work, and access to trains and buses as a means of comparison. For rural sites, we can measure the same things, but added to these patterns are the ability to grow food, conserve open space, buffer shorelines, preserve wildlife habitat, and maintain a dark sky.

Site/Climate/Orientation

When building a new house, understanding the site is critical. In residential neighborhoods, it is important to assess the density and style of the neighboring houses. Through this assessment, design criteria can be developed that will help guide decisions such as orienting the house for optimum sun exposure, inclusion of a front porch, style and technical specifications for windows, siding, and roofing materials, and exterior landscaping. Ultimately, a house designed for a specific site is more responsive to climate and context than a generic house built for Anywhere, U.S.A. For example, a passive solar house with south-facing windows and a concrete floor for thermal storage would not make sense in a rainforest climate on the equator.

Natural ventilation uses natural breezes to remove excess heat from buildings, reducing energy costs. Strategically placed windows in passively cooled homes create cross-ventilation. Good airflow occurs when the inlets and outlets are approximately the same size. A larger area of outlet than of inlet will allow for faster airflow and therefore better ventilation. This process works very well in many climates, especially if the incoming air comes from a shaded area of the yard, which can be created through landscaping, and the planting of trees and shrubs to direct the breeze toward the house. Observing the seasonal airflow across the contours of a specific site ensures that natural ventilation is used effectively.

Materials

With so many materials available for construction, the simplest way to begin is to think about what the intended use is and compare that to the best use of a particular material. It is important to research which materials you would like to use and find where to source them.

There are many types of wood, but where and how the wood is harvested is very relevant to highest and best use of a particular natural resource. Recycled materials (wood, glass, brick), recycled-content materials (fly-ash in concrete, sheetrock), and natural materials (straw bales, stone, sustainably harvested wood) all have different embodied-energy and transport implications. For example, a well-known college shipped its locally and sustainably harvested wood across the country to have it milled and fabricated and then returned the distance to be installed in the new environmental center. Clarity of decision making on material usage is also linked to available technical skills and construction practices.

Energy

The energy usage of a home is related to external as well as internal factors. External energy factors that influence one's home are orientation and protection with regard to sun, wind, rain, shadow zones, and so forth. A home can use the sun and wind to heat and cool the interior living space. Understanding the climate where one lives allows for appropriate decision making with regard to energy needs and construction techniques. A house in a cold climate is defined by a longer heating than cooling season and the need for compact, equally distributed distances from the heating source; a house in a temperate zone can make good use of east-west orientation and southern exposure; a house in a hot-humid climate needs air movement and makes good use of deep overhangs with many openings for cross-ventilation; and a house in a hot, dry climate uses deep overhangs for protection from the sun, masonry materials to create a time lag in solar gain, and water-ponding on the roof or in courtyards for cooling effects.

Internal factors include the number of residents and the ages of those living in the home; type of foundation, wall, and roof construction; location and types of doors and windows; heat generated or energy consumed by the household; and ownership.

Actions that affect our human comfort and overall energy consumption vary in cost and labor, but are effective over time. These include proper insulation and weatherization in cold climates; shading devices in hot climates; buying energy-efficient refrigerators and washers; upgrading heating/cooling equipment; installing water-saving devices; buying electrical service from green energy options; adding a grid-tied solar array or wind generator when appropriate; turning off lights and plug loads; and using a line to dry clothes.

Water

Freshwater is becoming a critical resource, and many cities as well as rural areas are finding themselves in difficult negotiations over access, seasonal availability, and quality. Reduction and conservation are known ways to mitigate the compounding problems of agriculture, industrial uses, urban sprawl, and residential consumption. Means to reduce water usage in the home include low-flow toilets, showerheads, and faucets and water-efficient washing machines and dishwashers. Reuse of household greywater is gaining acceptance for nonpotable uses, and rain gardens are a proven method of allowing water runoff from roofs, driveways, and compacted lawn areas to be absorbed into the subsoil. Rain gardens cut down on the pollutants that reach streams and rivers and use wetland plants in their design.

Landscape

Residential landscaping is often about creating outdoor rooms for particular needs or functions. In typical temperate zone conditions, deciduous trees are good for shelter from sun because they provide shade in summer but also allow sun to penetrate through their branches in winter. For winter protection, a buffer of evergreens works best in the U.S. Northwest. While each site will have its microclimate, summer breezes tend to be from the U.S. Southwest and winter winds from the U.S. Northwest.

Planting a home garden is a means to grow food as well as to provide an aesthetically pleasing and healthy environment. Composting is an effective way to reduce food waste and build soil fertility. Including food and flowers in dense urban places through the use of rooftop gardens, small balcony spaces, and indoor potted plants and herbs is useful in helping local air quality as well as food security. Accessing empty lots for community gardens, garden plots at the perimeter of the city, and consumer-supported agriculture are other means of creating garden spaces.

Inside the House

The products and furnishings we use inside the home affect the atmosphere we breathe and have implications for our health and welfare. New carpets and cabinetry are often culprits of outgassing, requiring substantial ventilation before being safe, but opening windows and doors to allow proper ventilation may not be feasible depending on the climate or location. Additionally, until recently, household cleaning products have been quite toxic and our cleaning habits have not caught up with old-fashioned, low-cost replacements such as the use of borax or baking soda instead of bleach. Environmentally friendly products are available, or we could make our own. There are many recipes for everything from window cleaner to hand soap that can be made right in the kitchen with easily available ingredients.

Living the Practice

There are many ways to create a home, from remodeling an apartment to building a house. The most important part is creating it so as to reflect oneself and family while "tuning in" to one's life work on the planet. How we choose to live sustainably depends on where we live and what kind of work we do, but home, however simple or elaborate, is where we go to be nourished. Our home is shelter, hearth, and heart. We ascribe meaning to home beyond the physical, and we carry that intangible bond with us as we go.

See Also: Commuting; Gardening and Lawns; Indoor Air Pollution; Organics; Social Action, Global and Regional; Social Action, National and Local.

Further Readings

Adams, Anthony. *Your Energy-Efficient House: Building and Remodeling Ideas.* Charlotte, VT: Garden Way Publishing, 1975.

Chiras, Daniel. *The Solar House.* White River Junction, VT: Chelsea Green Publishing, 2002.

Connell, John. *Homing Instinct: Using Your Lifestyle to Design and Build Your Home.* New York: Warner Books, 1993.

Day, Christopher. *Places of the Soul.* New York: HarperCollins, 1990.

Gayer, D. and J. Petrillo. *Groundswell.* Burlington, VT: Vermont Design Institute, 1998.

Olgyay, Victor. *Design With Climate: Bioclimatic Approach to Architectural Regionalism.* Princeton, NJ: Princeton University Press, 1963.

Dianne Elliott Gayer
Independent Scholar

Willa Antczak
Vermont Design Institute

Huerta, Dolores

Dolores Huerta is the cofounder (with César Chávez) of the United Farm Workers of America, AFL-CIO (UFW). She is also the president of the Dolores Huerta Foundation, a nonprofit organization founded to help build active communities in the pursuit of fair and equal access to healthcare, housing, education, and jobs. She is the mother of 11 children and continues to play a major role in the U.S. civil rights movement.

Dolores Huerta, primarily an advocate for farmworker rights, was also one of the first activists to speak out against toxic pesticides. Here, she answers questions after a keynote addressing human rights.

Source: Eric Gui/Wikimedia Commons

Dolores Huerta was born on April 10, 1930, in Dawson, a mining town in northern New Mexico, to Alicia Chávez and Juan Fernández, a miner, field worker, union activist, and state assemblyman. When she was 5, her parents divorced, and her mother, a businesswoman, bought a restaurant and a 70-room hotel in the San Joaquin Valley town of Stockton. A kind and generous woman, Alicia Chávez often housed farmworker families for free. After high school, Huerta attended Delta Community College and received a teaching certificate, becoming the first of her family to receive a higher education. Although she taught grammar school for a while, she was disheartened to see so many farmworkers' children in need of shoes and coming to class hungry. Convinced she could make more of an impact on their lives by organizing their farmworker parents, she decided to leave teaching.

In 1955, she helped found the Stockton chapter of the Community Service Organization (CSO). The CSO fought against segregation and police brutality, led voter registration drives, pushed for improved public services, and fought to get new legislation passed. Acutely aware of the needs of farmworkers, Huerta founded the Agricultural Workers Association in 1960. In 1961, she succeeded in getting citizenship requirements removed from pension and public assistance programs. She also played a key role in getting legislation passed to allow citizens the right to vote in Spanish and to take the driver's license examination in their native language. In 1962, she lobbied Congress for an end to the Bracero program, a series of laws and agreements permitting the entrance of temporary contract laborers from Mexico into the United States.

Through her work with the CSO, Huerta met César Chávez. Realizing the need to organize agricultural workers, Chávez and Huerta both resigned from the CSO (which had already turned down Chávez's request to organize the farmworkers) in 1962 and formed the National Farm Workers Association (NFWA), the predecessor to the UFW.

Huerta continued her lobbying efforts, and in 1963, she was instrumental in securing Aid to Families with Dependent Children (AFDC) for the unemployed and underemployed, as well as disability insurance for farmworkers in the state of California.

By 1965, Huerta and Chávez had recruited many farmworkers throughout the San Joaquin Valley. On September 8 of that year, Filipino members of the Agricultural Workers Organizing Committee (AWOC) demanded higher wages and went on strike against Delano grape growers. Although Huerta and Chávez had intended to organize farmworkers for several more years before confronting the grape industry, they could not ignore the strikers' request to get involved. Thus, on September 16, the NFWA joined in the now-famous Delano Grape Strike, in which over 5,000 grape workers walked off their jobs. The strike would continue for five years.

In 1966, AWOC and NFWA merged to form the United Farm Workers Organizing Committee (UFWOC). Taking the plight of the farmworkers to the consumers, Huerta directed the UFW's national grape boycott. In the end, the entire California table grape industry signed a three-year collective bargaining agreement with the United Farm Workers.

In 1973, the grape contracts expired, and grape growers chose to sign contracts with the Teamsters Union. In response, Huerta organized picket lines and continued lobbying. In 1974, she was instrumental in securing unemployment benefits for farmworkers. In the meantime, the UFW continued to organize not only grape workers but also workers in the vegetable industry—until violence erupted and farmworkers were killed. To galvanize public opinion, the UFW decided to organize a consumer boycott. Huerta directed the East Coast boycott of grapes, lettuce, and Gallo wines, which resulted in enactment of the Agricultural Labor Relations Act, the first law of its kind in the United States. This act granted farmworkers the right to collectively organize and bargain for better wages and working conditions. In 1975, Huerta lobbied against federal guest worker programs and spearheaded legislation granting amnesty to farmworkers who had lived, worked, and paid taxes in the United States for many years but were unable to enjoy the privileges of citizenship. Such activism resulted in the Immigration Act of 1985.

Known primarily as an advocate for farmworker rights, Dolores Huerta was also one of the first activists to speak out early and often against toxic pesticides that threaten farmworkers, consumers, and the environment. Aware that farmworkers and their families suffered from many illnesses and physical problems due to their contact with pesticides, she and the UFW documented the existence of cancer clusters in many communities in the San Joaquin Valley. Additionally, under Huerta's guidance, members of the UFWOC managed to get agreements signed requiring growers to stop using such dangerous pesticides as DDT and Parathion.

In her long career as an activist, Huerta has received many awards. The California State Senate, for example, gave her the Outstanding Labor Leader Award in 1984. In 1993, Huerta was inducted into the National Women's Hall of Fame, received the American Civil Liberties Union (ACLU) Roger Baldwin Medal of Liberty Award, the Eugene V. Debs Foundation Outstanding American Award, and the Ellis Island Medal of Freedom Award. In 1998, Huerta received the Eleanor D. Roosevelt Human Rights Award from President Clinton. She is also the recipient of the Consumers' Union Trumpeter's Award.

See Also: "Agri-Culture"; Chávez, César (and United Farm Workers); Environmental Justice Movements; Organizations and Unions; Social Action, Global and Regional; Social Action, National and Local; United Farm Workers (UFW) and Antipesticide Activities.

Further Readings

Chávez, Alicia. "Dolores Huerta and the United Farm Workers." In *Latina Legacies: Identity, Biography, and Community*. Vicki L. Ruiz and Virginia Sánchez Korrol, eds. New York: Oxford University Press, 2005.

Doak, Robin Santos. *Dolores Huerta: Labor Leader and Civil Rights Activist*. Minneapolis, MN: Compass Point Books, 2008.

The Fight in the Fields: César Chávez and the Farmworkers' Struggle. Video recording produced and directed by Ray Telles and Rick Tejada-Flores. New York: Cinema Guild, 2003.
García, Mario T., ed. *A Dolores Huerta Reader.* Albuquerque: University of New Mexico Press, 2008.

Danielle Roth-Johnson
University of Nevada, Las Vegas

Human Geography

There are numerous disastrous environmental consequences of war and conflict on both human geography and the environment. The modern tactics and weapons of warfare have resulted in profound environmental and subsequent human consequences that have attracted the attention of environmental activists around the globe, who often ally themselves with other groups such as peace activists. International development organizations and environmental activists have noted that the environmental damage of conflicts and the neglect of environmental issues during the course of war often result in long-term damage that threatens a nation or region's environmental development and human suffering outlook well into the future.

Early modern warfare often featured "scorched earth" policies, in which retreating forces laid waste to the landscape as well as transportation and military facilities to prevent advancing enemy forces from utilizing the resources. One of the best-known examples is Union General William Tecumseh Sherman's March to the Sea through Georgia during the American Civil War of 1861–65. These tactics destroyed regional ecologies as well as agricultural bases. Modern warfare tactics also include the use of waterways to control enemy troop movements, such as the deliberate destruction of dams and flooding of areas to prevent enemy advances. Such tactics result in profound changes to the area's ecosystems and human geography.

In 20th- and 21st-century warfare, landmines, chemicals, nuclear devices, and many other explosives that lay waste to agricultural land and cities have had devastating environmental impacts, with the atomic bomb having perhaps the most damaging results. More than 100 million landmines as well as countless more undetonated bombs and live cartridges from previous wars remain in Europe, Africa, Asia, and numerous other locations around the globe. As environmental and peace activists such as the International Campaign to Ban Landmines (ICBL) note, landmines alone kill or maim approximately 26,000 people annually, with most victims being civilians in countries at peace.

Biological, chemical, or nuclear agents can seep into the air, soil, and underground aquifers and other drinking water sources. They can also contaminate plants and animals consumed by humans, causing illness and death or forcing relocations or diet changes. The clearance of forests for road construction and troop movements as well as the elimination of potential enemy cover destroys natural habitats, harms the climate, and often involves the use of hazardous chemical defoliants with human and environmental risks. The results have included deforestation, agricultural disruptions, massive refugee movements or forced relocations, and ecosystem and habitat destruction. The resulting air, water, and soil contamination often lasts well beyond the original conflicts.

Although nuclear weapons have rarely been detonated on civilian populations, as in World War II, there have been a number of nuclear tests, and the development of the weapons themselves leaves hazardous environmental by-products such as nuclear waste. Nuclear waste can travel throughout the world, carried by air and ocean currents. Underground nuclear tests in the Pacific, for example, have destroyed atolls and coral reefs, killed ocean plants and animals, and inadvertently resulted in earthquakes and tsunamis. Countries and regions affected by nuclear testing include the South Pacific, the United States, Russia, China, India, Pakistan, and Kazakhstan. Uranium mining has resulted in environmental contamination in Canada, Namibia, Australia, and Germany. Cleaning up such contamination has proved costly, with one 1991 U.S. government estimate for nuclear decontamination in that country alone placed at approximately $300 billion through the year 2070. Green activists have targeted nuclear technologies and their toxic by-products as a key environmental movement target. The cleanup and monitoring of nuclear sites will be a costly effort.

Russia has experienced some of the worst nuclear environmental contamination of both the soil and the water. Large segments of its population have had to relocate from contaminated areas such as that surrounding the Chernobyl nuclear power plant, site of one of the world's worst nuclear accidents in 1986. Many environmental scientists believe Russia's Lake Karachay, located at Chelyabinsk, to be the world's most radioactive body of water. Since the end of the Cold War and breakup of the Soviet Union in the late 20th century, much of Russia's nuclear and other weapons stockpiles have been unaccounted for or stored in inadequate safety conditions, while contaminated military bases and testing ranges remained a threat. For example, armed nuclear submarines sit neglected in Murmansk.

Refugees and the Environment

A final key challenge to human geography and the environment due to war that has attracted the attention of the green culture movement is the problem of forced population dislocations and refugee camps. Refugees are often forced into environmentally sensitive areas such as national parks or preserves or into areas with already scarce natural resources such as food, fuel, arable land, and potable water. Influxes of large refugee populations, often over-crowded into unsanitary camps, strain already scarce resources, hurt native plant and animal populations, and introduce contaminants into the environment.

Another key aspect of human geography and the environment involves the conflict-related displacement of large refugee populations, who often relocate to submarginal or environmentally sensitive lands. Refugees also do not initially consider the environmental consequences of their resettlement, as they are most concerned with basic survival issues such as shelter, food, water, health, and safety. For example, they may cut down trees for fuel and shelter without regard for long-term deforestation or soil erosion. Environmental degradation caused by the conflict within their homelands is also an issue, as it lengthens population displacement, as refugees are often unable to return to environmentally compromised areas even after the conflict has passed. Refugee populations can also come into conflict with local populations for scarce natural resources or create demand-induced scarcity.

The United Nations (UN) has recognized the global problems posed by refugee populations since the post–World War II period, as evidenced in the passage of the 1951 Convention on the Status of Refugees and its update the 1967 Protocol Relating to the Status of Refugees. The overall rise in environmental awareness has resulted in the awareness that sustainable development and resource management must be considered when dealing with refugees. The United Nations Refugee Agency (UNHCR) has recognized the environmental threat posed by large refugee populations or those forced to stay out of their homelands for extended periods of time. As a result, the UNHCR has taken steps to reduce the negative environmental impact of refugee camps and populations.

One notable example of the impact of refugee populations on environmental degradation and sustainability is the large Palestinian refugee population of the Gaza Strip, an area of marginal land with limited potable water resources. Most of Gaza became part of Israel after the 1948 Arab–Israeli War, resulting in an influx of Palestinian refugees to the Egyptian-controlled portion, which became known as the Gaza Strip. The refugees almost tripled the area's population, with approximately one-third residing in UN refugee camps. Population pressures are also the result of high fertility rates, which offset the area's high mortality rates. Israel later gained control of the Gaza Strip and its land and water resources after the 1967 Six Day War. Inequitable water allocation policies have led to threats of desertification and water shortages, harmed agriculture, and led to salt water intrusion from the Mediterranean Sea. Other environmental issues have included the increased use of chemical fertilizers and pesticides and improper waste disposal. In addition to environmental degradation and health hazards, these environmental problems have also served to aggravate the ongoing Arab–Israeli conflict.

Individual Conflicts and the Environment

Individual conflicts have also created their own environmental hazards, drawing the responses of green movements. World Wars I and II introduced new weapons and techniques of modern warfare that left significant environmental damage. Trench warfare left huge trenches dug into the soil and disrupted plants and animals, while logging and road building left deforestation and soil erosion. The Germans and the British used large amounts of gases in World War I, and the retreating Japanese army buried millions of chemical weapons across northeastern China in World War II. The use of chemical weapons such as chlorine and mustard gases released these toxic chemicals into the air, soil, and waterways.

The U.S. explosion of atomic bombs over the cities of Hiroshima and Nagasaki in the summer of 1945 as World War II drew to a close left dust particles, radioactive debris, fires, drinking water contamination, destroyed crops, and animal and human radiation sicknesses and deaths for years to come. These devastating consequences became a touchstone of the environmental movement against the development and use of nuclear weapons. In the aftermath of World War II, peace activism changed as groups such as the World Congress of Intellectuals for Peace, European Nuclear Disarmament, the British Campaign for Nuclear Disarmament and its breakaway Committee of 100, and larger environmental groups such as Greenpeace lobbied against nuclear weapons based on their actual and potential threats to the environment as well as the humans that depend on it.

The revolutions that brought communist governments to power in the Soviet Union in 1917 and in China in 1949 both featured movements toward industrialization and the collectivization of agriculture that profoundly affected both human populations and the environment. The Chinese Cultural Revolution of the 1950s through the 1970s under Mao Zedong destroyed forests and reduced arable land due to industrialization and fuel needs, constructed unsafe dams as part of hydroelectric projects, and encouraged overhunting and overfishing as well as population growth that strained the country's natural resources and food and water supplies. Outcomes included floods, periods of famine and starvation, desertification, and the industrial contamination of air, water, and soil. Similar events occurred within the Soviet Union. Movements against agricultural collectivization and other environmental changes were often brutally repressed, and environmental movements are still slow to take a foothold among the populations of China and the former Soviet Union.

The Vietnam War of the 1960s and 1970s devastated that country's jungles and waterways as well as the people dependent upon them, with both short- and long-term impacts. U.S. military forces used large amounts of the chemical defoliant code-named Agent Orange to destroy the jungle cover used by North Vietnamese Communist fighters and their South Vietnamese allies, the Viet Cong. Agent Orange's main ingredient was the toxic herbicide dioxin, a potent carcinogen that can kill humans and animals as well as plants. Neighboring Cambodia also felt the war's effects as well as those of a civil war around the same time period after the Maoist extremist Khmer Rouge brought communism to that country. The Khmer Rouge enacted damaging environmental policies, including extensive timber logging, to help finance the war effort and extensive forced agriculture that exhausted the soil.

Environmental impacts have included massive deforestation of both inland jungle and coastal mangrove forests, the shrinking of rice seed sizes and damaged rice crops, the contamination of soil and water due to chemical runoffs, and severe floods related to deforestation. Estimates have shown that over 2 million acres of Vietnamese land may have been affected and approximately 35 percent of Cambodian forests destroyed. Human health impacts have included headaches, vomiting, chest pain, and birth defects. Other human impacts include food shortages from the dangers of eating fish, a staple of the Vietnamese diet, as dioxin runoff reaches the rivers and agricultural disruptions due to contaminated soil, and the dangers of live landmines and bombs that remain hidden beneath the ground. U.S. soldiers were also affected by the use of Agent Orange, with many Vietnam veterans reporting adverse health effects.

The late 20th century saw similar environmental damage resulting from numerous conflicts that arose in eastern Europe and the former Soviet Union after the collapse of communism and the rise of independence movements. One of the worst examples of environmental damage occurred as a direct result of the province of Chechnya's war of independence against Russia, a series of conflicts and ongoing tensions that began in 1994. Bombings destroyed oil plants and refineries, contaminating the soil and water with spilled oil. Radioactive waste and airborne radiation have also polluted the surrounding environment. The consequences have included the destruction of agricultural land; the pollution of rivers; human health risks; possible food and water shortages; the deaths of plants, animals, and fish; and long-term effects due to the destruction of the region's natural ecosystems.

The Balkan region was a key site for numerous independence conflicts as national boundaries were reshaped in the late 20th century. The Kosovo War of 1996–99 included conflicts between Serbia and Kosovo and between Kosovo and troops of the North Atlantic Treaty Organization (NATO). The latter included a months-long NATO bombing campaign of Serbia. A subsequent UN Environment Programme (UNEP) investigation revealed the environmental damage left behind. This included oil leaks from refineries and storage depots, blast damage to industrial sites and petrochemical plants that released toxic chemicals such as sulfur dioxide and carbon monoxide into the environment, the destruction of waste treatment facilities, damaged habitat in national parks, and the contamination of the Danube River with mercury, petroleum, liquid ammonia, and other toxic chemicals that resulted in fish kills. The thick black smoke led to black rain. Human consequences included health problems from chemical poisonings and large numbers of refugees living in overcrowded and unsanitary conditions in refugee camps.

During the Gulf War in the early 1990s, Iraq dumped approximately 1 million tons of oil into the Persian Gulf and the forests in one of history's largest oil spills. Retreating Iraqi forces also deliberately set hundreds of Kuwaiti oilfields on fire during the war's last stages. Toxic black smoke filled the sky, visible for miles around, while oil droplets scattered many miles distant. The oil fires devastated the once-thriving Kuwait desert ecosystem, killing numerous plants, birds, and animals. They also resulted in a thick polluting smog and acid rain that blanketed the region in soot. Oil that traveled into the nearby ocean killed marine birds and animals and damaged coral reefs. Gulf bird and shrimp populations noticeably declined, and scientists believe that some species of otters, rats, bats, and fish have likely become extinct. Similar oil fires occurred on a lesser scale during the Iraq War in the early 2000s.

Other environmental damage related to both the first and second Gulf wars includes the destruction of dams and sewage treatment plants and the bombing of chemical production plants, allowing raw sewage and toxic waste to spill into the Tigris and Euphrates Rivers. The resulting contamination of drinking water has led to high illness rates among both humans and animals. During the Iraq War, the mishandling of toxic radioactive materials led to potential soil, water, and food contamination. Reported human health consequences include kidney damage, lung infections, cancers, and birth defects. The movement of tanks and other heavy military equipment across the desert terrain led to soil erosion and changes to the landscape.

The Afghanistan War of the early 21st century also took an environmental toll on the region, as U.S. forces invaded to topple the Taliban government and suppress ongoing Taliban resistance and terrorist threats. Destruction of the water infrastructure led to water contamination by bacteria, shortages of potable water, and incidents of water theft. Taliban logging and the illegal trade of timber as well as U.S. bombings led to heavy deforestation. Bombs and other weapons have left chemical contamination of the soil, air, and water, resulting in thyroid gland damage, cancer, and other illnesses. Landmines pose continuous threats and disrupt populations and agriculture. The use of mountains as military refuges disrupted the habitats of predatory animals such as leopards, which had also sought sanctuary there when their original habitats were destroyed. The number of migratory birds passing through the country has decreased approximately 85 percent.

There have been numerous other Middle East conflicts in the 20th and 21st centuries that have left their mark on that region's environment. These include the Iran–Iraq War and ongoing tensions between the two neighboring countries. The largest environmental toll occurred along the confluence of the Tigris and Euphrates Rivers, home to numerous ecosystems such as marshes, lakes, lagoons, and mudflats. The opposing military forces both utilized dikes to aid troop movements and floods to hinder enemy troop movements, resulting in changes to the area's natural ecosystems and the desertification of some areas.

The Israeli–Lebanon War of 2006 resulted in the bombing of oil storage tanks that spilled 20,000 tons of oil into the Mediterranean Sea, killing fish, threatening the habitat of the endangered green sea turtle, washing oil ashore on Lebanon beaches, and sparking oil fires that polluted the atmosphere. Hezbollah bombings in northern Israel sparked fires that burned thousands of acres of forests, causing deforestation and endangering bird sanctuaries. The ongoing Israeli–Arab disputes have also created a refugee crisis with environmental consequences, as crowded refugee camps often face unsanitary conditions that can harm the environment as well as their inhabitants. These and other Middle East conflicts have altered the region's environment and threatened its natural ecosystems, resulting in loss of habitats, desertification, potable water shortages, human and animal illnesses, and human population displacements.

Numerous African conflicts in the late 20th and early 21st centuries resulted in environmental degradation, human health problems, and the disruption of ecosystems, industries, and food sources. Examples have included the Somali Civil War of 1991, the Rwandan genocide of 1994, and the civil wars in the Congo and Sudan as well as the border dispute between Ethiopia and Eritrea in the late 1990s. The Somali Civil War resulted in overfishing to meet dietary needs, with the Somalis eventually ignoring international fishing regulations and battles over property rights. During the Rwandan genocide campaign, thousands of refugees displaced from their homes fled to nearby forests and the Virunga National Park, where they damaged natural ecosystems and impinged on the habitat of the mountain gorillas. During the Congo civil war, national parks became victims of mineral exploitation.

Results include infrastructure damage or destruction and agricultural disruption. Large sections of forests were destroyed for fuel or shelter materials, logging, or the clearance of additional farmland as older farmlands were damaged or mined. The use of landmines, often unmarked or left behind after conflicts, has disrupted agriculture because of the threat of potential contact and explosion and leaves behind numerous victims who are often unable to work or attend school as a result of their injuries. National parks and other wild habitats suffered irreparable environmental damage due to human encroachment, refugee encampments, habitat loss, and mineral exploitation and conflicts such as those over diamond mining. Elephants have been heavily hunted for meat and the ivory in their tusks. Refugee displacement also resulted in agricultural disruption, desertification, drought, and periodic famines, such as that in Darfur, which claimed approximately 95,000 lives.

Much of the environmental or green culture movement to repair the damages left by war and the environmental consequences of other human geography issues has come from international environmental groups and organizations such as the Sierra Club, the Worldwatch Institute, Reaching Critical Will, the Peace Pledge Union, International Physicians for the Prevention of Nuclear War, and the Institute for Energy and Environmental Research. Such groups have gathered and disseminated research and case studies to show the dire environmental consequences of human actions such as war, the need to repair the damage left behind, and the need to prevent future conflict in order to guarantee a sustainable future.

See Also: "Agri-Culture"; Antiwar Actions/Movements; Environmental "Bads"; International Law and Treaties; Toxic/Hazardous Waste.

Further Readings

Adley, Jessica and Andrea Grant. "The Environmental Consequences of War." Sierra Club of Canada. http://www.sierraclub.ca/national/postings/war-and-environment.html (Accessed October 2010).

Ayittey, George B. N. *Africa in Chaos*. New York: St. Martin's Press, 1998.

Bennis, Phyllis and Michel Moushabeck. *Beyond the Storm: A Gulf Crisis Reader*. Brooklyn, NY: Olive Branch Press, 1991.

Black, Richard. *Refugees, Environment and Development. Longman Development Studies* series. New York: Prentice-Hall, 1998.

Bright, Chris. *Life Out of Bounds: Bioinvasion in a Borderless World*. New York: Norton, 1998.
Cameron, Maxwell A. and Robert J. Lawson. *To Walk Without Fear: The Global Movement to Ban Landmines*. New York: Oxford University Press, 1998.
Carlson, Richard C. and Bruce Goldman. *Fast Forward: Where Technology, Demographics, and History Will Take America and the World in the Next Thirty Years*. New York: HarperBusiness, 1994.
Closmann, Charles E., ed. *War and the Environment: Military Destruction in the Modern Age*. College Station: Texas A&M University Press, 2009.
De Blij, Harm J. *Why Geography Matters: Three Challenges Facing America: Climate Change, the Rise of China, and Global Terrorism*. New York: Oxford University Press, 2005.
Deweerdt, Sarah. "War and the Environment." Worldwatch Institute, December 6, 2007. http://www.worldwatch.org/node/5520 (Accessed October 2010).
Diamond, Jared M. *Collapse: How Societies Choose to Fail or Succeed*. New York: Viking, 2005.
Diamond, Jared M. *Guns, Germs, and Steel: The Fates of Human Societies*. New York: W. W. Norton, 1998.
Ehrlich, Paul R. *The Cold and the Dark: The World After Nuclear War: The Conference on the Long-Term Worldwide Biological Consequences of Nuclear War*. New York: Norton, 1984.
Ehrlich, Paul R. and Anne H. Ehrlich. *The Dominant Animal: Human Evolution and the Environment*. Washington, DC: Island Press, 2008.
"Environmental Effects of French Nuclear Testing." Report prepared by the International Physicians for the Prevention of Nuclear War and the Institute for Energy and Environmental Research. http://www.cyberplace.org.nz/peace/nukenviro.html (Accessed October 2010).
Enzler, S. M. "Environmental Effects of Warfare." Lenntech. http://www.lenntech.com/environmental-effects-war.htm#Africa (Accessed October 2010).
Gallagher, Winifred. *The Power of Place: How Our Surroundings Shape Our Thoughts, Emotions, and Actions*. New York: Poseidon Press, 1993.
Goodson, Larry P. *Afghanistan's Endless War: State Failure, Regional Politics, and the Rise of the Taliban*. Seattle: University of Washington Press, 2001.
Harwell, Mark A. *Nuclear Winter: The Human and Environmental Consequences of Nuclear War*. New York: Springer-Verlag, 1984.
Hawley, T. M. *Against the Fires of Hell: The Environmental Disaster of the Gulf War*. New York: Harcourt Brace Jovanovich, 1992.
McNeill, John Robert. *Something New Under the Sun: An Environmental History of the Twentieth-Century World*. New York: W. W. Norton, 2000.
Monin, Lydia and Andrew Gallimore. *The Devil's Gardens: A History of Landmines*. London: Pimlico, 2002.
Peace Pledge Union. "War and the Environment." http://www.ppu.org.uk/learn/infodocs/st_environment.html (Accessed October 2010).
Polk, William Roe. *Understanding Iraq: The Whole Sweep of Iraqi History, From Genghis Khan's Mongols to the Ottoman Turks to the British Mandate to the American Occupation*. New York: HarperCollins, 2005.
Reaching Critical Will. "The Environment and the Nuclear Age." http://www.reachingcriticalwill.org/technical/factsheets/environmental.html (Accessed October 2010).

Risen, Clay. "The Environmental Consequences of War: Why Militaries Almost Never Clean Up the Messes They Leave Behind. *Washington Monthly* (January–February 2010). http://www.washingtonmonthly.com/features/2010/1001.risen.html (Accessed October 2010).
Shapiro, Judith. *Mao's War Against Nature: Politics and the Environment in Revolutionary China*. Cambridge, UK: Cambridge University Press, 2001.
United Nations Refugee Agency (UNHCR). http://www.unhcr.org/3b039f3c4.html (Accessed November 2010).
Vanasselt, Wendy. "Armed Conflict, Refugees, and the Environment" (July 2003). EarthTrends. http://earthtrends.wri.org/features/view_feature.php?theme=10&fid=43 (Accessed October 2010).
Williams, Jody, Mary Wareham, and Stephen D. Goose, eds. *Banning Landmines: Disarmament, Citizen Diplomacy, and Human Security*. Lanham, MD: Rowman & Littlefield, 2008.
Williams, Michael. *Planet Management*. New York: Oxford University Press, 1993.

Marcella Bush Trevino
Barry University

Hurricane Katrina

Before Hurricane Katrina swamped New Orleans in 2005, the city that moderation forgot was known for almost everything but its environmental image. But in the devastating wake of the disaster, jazz music, voodoo, gumbo, and Mardi Gras gave way to solar panels, organic farming, and scores of hybrid city buses as the green movement took root in the nation's party place. Five years after the destruction, the citizens of New Orleans are still rebuilding, using the once-in-a-lifetime opportunity Hurricane Katrina presented to turn the city into a model for green rebuilding.

Given the enormity of the environmental damage to the Gulf Coast region, rebuilding has been painfully slow. Entire neighborhoods in the New Orleans area alone were deluged with petroleum and other chemicals, washed ashore from the deep-sea rigs that dot the Louisiana coastline. Gasoline stations abandoned during the storm emptied their contents into the streets of the city and surrounding suburbs, causing serious long-term harm to the soil. Although much of the estimated 7 million gallons of oil that spilled into the region was naturally dispersed or cleaned up, experts cannot accurately gauge how much damage the catastrophe will cause in the long term.

In addition to the petroleum problem, countless other toxic hazards—including pesticides, human waste, and heavy metals—permeated the soil and contaminated an enormous area of the region's groundwater. In this treacherous environment, the city of New Orleans and the entire region have struggled to rebuild and thrive. Considering the high probability of another massive storm in the Gulf Coast region, cities, governmental bodies, and private businesses alike have seen the sensibility in "going green."

But what exactly is green building? According to a Sierra Club report issued in 2009, it is planning, designing, and constructing low-impact buildings, which are a critical component of sustainable redevelopment. But green building is not just about using environmentally suitable materials for constructing dwellings and other property; it is also about focusing on maintaining social fabrics and providing good job opportunities. In other words, building green is about uniting neighborhoods together in a communal effort to sustain and cohere as well as to protect and provide. With this in mind, New Orleans—which lost more than 200,000 homes to Hurricane Katrina—has set out to hold itself up as a national model for green development.

Today, those same hybrid city buses buzzing around New Orleans boast the slogan "Cleaner, Smarter" on their sides. This is a far cry from the state of affairs following the Category 3 (having weakened from a Category 5) behemoth that overwhelmed the city in August 2005. Back then, 80 percent of New Orleans stewed underneath fetid floodwater that had gushed from broken levees. As the city began to rebuild, it used a combination of government tax credits for green building and the specter of high energy prices to urge a sustainable approach. What once was the butt of jokes from Louisiana politicians became the cause de jour as green rebuilding spawned many innovative, creative, and profitable ideas. For instance, homes destroyed in the infamous Lower Ninth Ward of New Orleans are now being rebuilt with high-tech fittings such as solar panels, which have cut the residents' electric bills substantially.

New Orleans has pushed solar energy aggressively, and in 2007, it became one of a dozen U.S. cities named a "Solar American City" and earned a grant of almost half a million dollars to install solar panels in homes and business. In the Lower Ninth Ward, arguably the area hardest hit by the storm, efforts to shore up former wetlands that had protected the area from flooding are under way. There, city planners are using a combination of trees and other vegetation to create a bayou that will restore habitats to their preindustrialization–era days.

Solar technology is booming now because of new state and federal tax credits. As a result, jobs are being created and, for those interested in training for green rebuilding, opportunities await. In New Orleans, there is a 2,000-square-foot facility that offers free green building workshops, replete with its own replica of an energy-efficient home that has toilets and showerheads that use less water.

The New Orleans city government is committed to the green building effort. In 2009, it approved a program to install compact fluorescent light bulbs, weather stripping, and insulation in more than 2,500 properties. On average, homeowners can quadruple their solar energy investments thanks to tax credits issued by the state and federal governments.

Environmental Justice, Cap and Trade, and the Green Movement

Community leaders in minority neighborhoods have shown great enthusiasm for green rebuilding efforts because it helps ameliorate years of neglect and inequity between wealthy areas of the city and the poorer districts. Unlike racial profiling or blatant institutional discrimination, environmental injustice is more insidious, that is, lead-based paint and rat poison used in public housing or waste management facilities clustered in socioeconomically impoverished areas. While climate change obviously affects the delicate ecosystems of Earth unequally, killing susceptible animal populations at various flashpoints more than others (e.g., polar bears), most of the unseen recipients of environmental destruction are the poorest among us. As Hurricane Katrina illustrated, those people who lacked the resources to flee the storm suffered the most.

Today, more than ever, these formerly forgotten victims of environmental injustice have a voice in the rebuilding of their neighborhoods and an advocate in the highest halls of power. In fact, the leader of the nation's most powerful environmental agency is a Gulf Coast resident. President Barack Obama's administration has put much effort into the green rebuilding of New Orleans, which got a big boost when he named Linda Jackson, a former Lower Ninth Ward resident, to lead the Environmental Protection Agency (EPA). Since then, the city has increased the health opportunities for its citizens by building more bike paths, which in turn has decreased traffic congestion from automobiles. Public transportation is also playing a key role in the renewal. Three dozen city buses, purchased with a $15 million federal grant, get citizens of New Orleans to their destinations around the city using a mixture of electric power, biodiesel, and gasoline. For the first time in decades, the plight of the city's most vulnerable citizens is being considered first and foremost as planners and developers move forward to renew the city of New Orleans. But it is not only the city that has benefited. The entire state of Louisiana stands to prosper under new environmental legislation under way in Congress.

Although the state of Louisiana has long profited from the oil companies that dot its Gulf Coast, more and more efforts today are incorporating some of Louisiana's other natural resources—namely, sun, wind, and water. This trinity of resources is the perfect combination for legislation aimed at reducing carbon emissions, and a renewed effort to pass meaningful cap-and-trade legislation during President Obama's second term, should he win one, would have a lasting impact on the state of Louisiana and the entire nation. Cutting emissions profits companies under cap-and-trade legislation, which aims at reducing pollution. Those companies and utilities that reduce their emissions sufficiently can then sell carbon offsets to companies that are exceeding the emissions cap.

Global Warming and Hurricanes

Although New Orleans and the Gulf Coast region in general have embraced the idea of green building and ecological and urban sustainability, the fact remains that it won't prevent another natural disaster. And these types of disasters are sure to continue if the current trend of global warming continues. From 1995 to 1999, 33 hurricanes—an all-time record—ravaged the Atlantic basin. And as global warming models predict increasing average world temperatures, scientists warn that the trend could continue to worsen. This means that another Hurricane Katrina could happen next week, 100 years from now, or never. If it does, those who learned from the mistakes of the past and went green won't be doomed to repeat them.

See Also: Environmental "Bads"; Environmental Justice Movements; Institutional/Organizational Action; Soil Pollution; Toxic/Hazardous Waste; Water Pollution.

Further Readings

Freudenburg, William R., Robert B. Gramling, Shirley Laska, and Kai Erikson. *Catastrophe in the Making: The Engineering of Katrina and the Disasters of Tomorrow*. Washington, DC: Shearwater Press, 2009.

McQuaid, John and Mark Schleifstein. *Path of Destruction: The Devastation of New Orleans and the Coming Age of Superstorms*. Boston: Little Brown & Company, 2006.

Sierra Club Report. http://www.southernstudies.org/2009/08/post-68.html (Accessed September 2010).

Christopher Mapp
University of Louisiana at Monroe

Hybrid Cars

A Toyota Prius, the most popular hybrid car, gets nearly twice the gas mileage of a comparable conventional sedan. Here, a plug-in Prius that will allow short trips using electricity alone is tested at Argonne National Laboratory.

Source: Argonne National Laboratory, U.S. Department of Energy

In conventional automobiles, more than 80 percent of the energy that goes into the engine is lost. First, most of the energy value of the fuel is given off as heat; then, the engine continues to run while the car is slowing down or stopped. At those times, the fuel driving the engine is wasted. When a car accelerates, it gains kinetic (motion) energy, but during braking, all of that energy is dissipated as useless heat. The goal of hybrid automobile technology is to use electrical power as a buffer to reduce these wastes to a minimum.

Hybrid cars combine a conventional internal combustion engine with an electric generator and batteries to use fuel only when it is needed and to recapture some of the kinetic energy dissipated when a vehicle is slowing. There are two basic designs: parallel hybrids and serial hybrids.

In a parallel hybrid, such as the Toyota Prius, a gasoline engine drives the wheels directly through a mechanical transmission and also can charge storage batteries through a generator. The generator can also be driven by the car's wheels during braking, a process known as regenerative braking. Energy thus stored in the batteries is released when not very much power is need to drive the vehicle, such as low-speed driving or constant-speed cruising on level highways or slight downslopes. Under these conditions, the generator serves as an electric motor to drive the vehicle. When peak power is required, the engine and the electric motor work together to propel the vehicle. The engine is turned off when not required and automatically restarted to recharge the batteries or when power needs are too great for the electric drive alone. Restarting is done through the large electric motor powered by the propulsion battery pack, so it is silent and transparent to the driver. This can be disconcerting to new hybrid drivers—when the engine quits while the vehicle is moving. Eventually, the hybrid driver, when switching to a conventional vehicle, wonders why the engine is running when it is not needed. Some of the hybrid advantage is not realized for short trips because the engine runs continuously to warm it up upon starting.

A second variety of parallel hybrid, such as the Honda system, returns energy to batteries by regenerative braking. During acceleration, the gasoline engine is aided by the generator, which then functions as a motor. The engine does not directly charge the batteries, however, and it runs constantly during normal operation.

Serial hybrids have no direct connection from the gasoline engine to the wheels. Instead, the engine runs a generator, which then powers electric motors at the wheels. This design requires a larger generator and separate electric motors because they must provide enough power to run the vehicle during peak loads (acceleration and uphill operation). Unlike the parallel design, the generator cannot double as a motor when electrical power is required, because both generator and motor are in constant operation. The Chevrolet Volt, referred to as a series hybrid vehicle, was introduced in December 2010 and was named the "2011 Car of the Year" by *Motor Trend*. Another approach to the serial design is in diesel-electric locomotives, where the engine drives a generator that drives the wheels. Braking is partly regenerative to reduce wear on friction brakes, but in current locomotives the regenerated power is wasted by running it through air-cooled resistors on the roof of the engine.

Because of the extra weight of the larger generator and electric motor, the serial design is generally heavier than the parallel design. Both designs share the use of batteries that can be charged for assistance with peak loads, so that the gasoline engine can be smaller than would be required in a conventional automobile. A substantial portion of the weight of the engine in conventional automobiles is dedicated to peak-power capacity that is hardly ever required. Only when the accelerator is "floored" can a conventional automobile use all of the power available in the engine, and then the maximum power is available only at an optimal engine rotation speed, higher than that used in normal operation. Running such an engine at less than its maximal throttle reduces its efficiency and gasoline mileage. Hybrids, in contrast, can use maximum or optimum engine settings at any speed, so a much smaller and less powerful engine is required.

The most popular of the hybrids, the Toyota Prius, has sold over a million vehicles, and the technology has been licensed to several other manufacturers. The vehicle has several other energy-saving features in addition to the basic hybrid technology. It offers less wind resistance than any other automobile and is equipped with stiffer tires that give less rolling resistance than conventional tires. Software replaces hardware for many of the vehicle's functions, such as the instrument panel and cruise control. There is no mechanical reverse gear—the direction of rotation of the electric motor is simply reversed by the motor control software. Reverse is always a low-speed operation, so the vehicle is driven only by electric power when going backward. The delivery of electric energy to the motor's rotors is also managed by software, without mechanical brushes. The gasoline tank is smaller and therefore lighter than in other vehicles of similar size and weight because of the higher mileage rating. Engine cooling is passive when the engine is temporarily turned off, saving the energy that would otherwise be needed to run a water pump. The transmission consists of a planetary gear with only a handful of moving parts, in contrast to the hundreds of parts in a conventional automatic transmission. With all of these strategies, the Prius gets nearly twice the gasoline mileage of a conventional sedan of comparable weight.

The Road Ahead

The next step in hybrid automobile technology will be the plug-in hybrid, a vehicle with enough battery capacity to drive the car for short trips on electricity alone. The Volt claims a battery-only range of about 40 miles (65 kilometers); since most trips are shorter than this, the gasoline engine will seldom engage. The practicality of this design depends on high-efficiency batteries. All hybrids use batteries that offer more power for a given amount of weight than the traditional lead-acid battery of conventional vehicles. The only function of that battery is to start the engine and to run accessories such as lights and radio. Hybrids also have a small lead-acid battery that acts as a buffer to run the vehicle's 12-volt accessories. That battery is charged by the main batteries but has no role in propelling the vehicle. The propulsion batteries, in contrast, are linked in series to provide several hundred volts to the electric motor.

Nickel-metal-hydride (NiMH) batteries constitute the main battery pack of the Prius. They have substantially higher power density (energy stored per kilogram of battery weight) than lead-acid batteries and can handle thousands of discharge cycles if managed properly. Software in the Prius optimizes battery life to roughly correspond with the expected service life of the vehicle. Future hybrids will probably use lithium batteries that offer an even greater power density, enough to make electric-only operation practical for significant distances. Issues of safety and service life have been largely overcome by intelligent computer-controlled battery management schemes.

The hybrid is a transition technology, developed to conserve dwindling oil supplies. The technology required has been available for decades, but was introduced only when gasoline became so expensive that the hybrid scheme was justified economically. Inexpensive oil, meaning oil pumped from onshore wells not in the Arctic, peaked in 2005, and worldwide production as a whole will probably peak between 2010 and 2015. It is physically impossible to pump oil for much longer at the present rate of about a thousand barrels per second. Most oil currently produced comes from a handful of giant oil fields that are becoming depleted; the smaller fields that replace them cannot provide the volume previously available. Since the 1970s, more oil has been pumped each year than has been discovered in that year. Alternative sources such as tar sands cannot approach the current rate of consumption, and biofuels are an illusion—it takes nearly as much fossil fuel energy to produce them as is recovered from burning them. Hybrid technology makes it practical to continue current driving habits with gasoline at twice its traditional price, but even that situation will not last for very many years as oil supplies dwindle while the population continues to increase.

Another function of hybrids is as a test platform for electric drive technology. Alternative energy sources such as wind, solar, and geothermal will have to replace gasoline for powering vehicles; these sources provide electricity, not liquid fuel, so all-electric vehicles will become necessary. Wind and solar electricity are intermittent, so that batteries can be charged only while the sun is shining or the wind is blowing, and the power will be more expensive than present electricity, but for limited uses, it will be possible to propel vehicles far into the future.

See Also: Begley, Jr., Ed; Commuting; Green Consumerism; Weekends.

Further Readings

Anderson, Curtis D. and Judy Andersen. *Electric and Hybrid Cars: A History.* McFarland, 2010.

Hewitt, B. "Plug-in Hybrid Electric Cars: How They'll Solve the Fuel Crunch." *Popular Mechanics* (May 2007).

HowStuffWorks, Inc. "Hybrid Cars." http://auto.howstuffworks.com/hybrid-car11.htm (Accessed September 2010).

Rauch, Jonathan. "Electro-Shock Therapy." *Atlantic Magazine* (July/August 2008).

Bruce Bridgeman
University of California, Santa Cruz

I

INDIA

India's Green Revolution consisted of agricultural research and experimentation, causing high yields for Indian rice. Critics, though, cite the decrease from over 300 rice varieties grown before the revolution to only 10 grown after it.

Source: Wikimedia Commons

India benefited from the global green revolution that boosted agricultural output in developing countries in the post–World War II period, emerging as one of the world's leading rice producers. Government initiatives supporting green culture have emphasized some of India's chief environmental problems, including waste disposal and pollution. The public has supported green culture initiatives such as recycling, but sometimes for economic rather than environmental motivations. Ecofriendly businesses, schools, and communities are increasing in popularity. The Indian government's large dam construction projects have proved more controversial, stirring intense public debate.

The Green Revolution in India

India's main environmental movement, known as the green revolution, began in the late 1960s. Key goals of the green revolution focused on increased agricultural outputs. India's green revolution was part of the larger green revolution that occurred in developing countries in the post–World War II period, beginning in Mexico in the 1940s. This movement involved agricultural research and experimentation, such as the development of domesticated high-yield crops engineered to respond to fertilizers as well as the introduction of mechanized agriculture. Results have been mixed, but India is held up as one of the most successful countries with regard to the green revolution. For example, the development of the high-yield rice variety IR8 proved so successful that India emerged from the green revolution as one of the world's leading rice producers.

The green revolution has not been without its critics. They note the changes in agriculture that have resulted, including an increased need for fertilizers and pesticides and irrigation methods such as the controversial building of large dams that displace large populations. They also note the dramatic decrease in varieties of individual crops grown, such as India's decline from over 300 varieties of rice grown before the green revolution to approximately 10 varieties of rice grown afterward. This lack of variety has fueled the increased use of fertilizers and pesticides. They also note that increased food production and decreased threat of famine has led to overpopulation, which brings new environmental threats.

The movement toward green culture and the green revolution began after the British left India in 1947. India had suffered through its worst famine in recorded food history in 1943, known as the Bengal Famine. Four million people died of starvation in east India and parts of what is now Bangladesh. As a result, food security was at the top of the Indian government's "wish list" after the British departed. India, however, had a difficult time attaining food security between 1947 and 1967, as the population was growing much faster than food production.

The so-called green revolution came to India from 1967 to 1978. The green revolution concentrated on three key areas: the continued expansion of agricultural land, double-cropping existing farmland, and using seeds improved through biogenetics. Continued expansion of farmland had occurred prior to independence, so double-cropping existing farmland was given greater importance. Double-cropping enabled the planting of two harvest seasons per year instead of the more traditional single harvest. The one crop season was based on the fact that there was only one monsoon season per year. The second crop was watered and nurtured through the use of large irrigation facilities and dams. The dams captured water from the natural monsoon and farmers used it as an "artificial" monsoon when needed for the second crop. Any monsoon rainwater that was not used for irrigation was used instead for hydroelectric power.

The Indian Council for Agricultural Research spearheaded the movement for improved genetics among seeds, developing new strains of wheat, rice, millet, and corn seeds. Dr. M. P. Singh, credited with the development of these high-yield seeds, is widely regarded as the hero of India's green revolution. By the late 1970s, India had a record grain output of 131 million tons. The country quickly became one of the largest agricultural producers in the world as well as a leading exporter of grain products.

Economically, India's green revolution was a boost to the manufacturing sector. The crops not only needed more water and fertilizer, but also more pesticides and other chemicals that maintained the well-being of the crops. This was another boost to the industrial and manufacturing sectors. Increased economic output resulted in higher employment levels and allowed India to repay its loan obligations for the start-up money to fund green projects from the World Bank, increasing its international credit rating.

There were some limitations to India's adoption of green culture through the green revolution. Despite its successes, India was unable to become self-sufficient in terms of food production. In the 1980s, India faced severe droughts due to lack of monsoon rain. The country still had to import food well into the 1990s. Also, the impact of the green revolution was not universal, leaving some parts of the country behind. For example, the best high-yield seeds and crops remained in certain parts of the country, such as in the Punjab and Haryana regions.

Environmental Issues and Green Culture in India

India's green culture movement has also brought environmental issues surrounding waste disposal to light, which is one of the main crises facing India's environment in the 21st century. About 80 percent of India's urban waste ends up in the country's rivers, and the problem is worsening due to population increases and poor government oversight and regulation. Some bodies of water are floating garbage dumps. As an example, only 55 percent of New Delhi's 15 million residents are connected to the sewage system. Those who are not instead flush their wastewater and other trash down pipes and into drains, which funnel the waste directly into the rivers. Many of India's landfills are reaching capacity as well.

Recycling has become increasingly popular, although often for financial rather than environmentally conscious motivations. Millions of those living in poverty recycle, making it a large part of the informal economy. People sift through trash to collect recyclable plastic bottles, paper, glass, and metal, which can then be sold to a trash trader for money. Plastic bottles, for example, are brought to a processing plant where they are shredded, washed, and dried in the sun. The plastic is then brought to a factory and melted down to make clothes and toys, among other things.

India is also one of the largest international depositories of electronic waste, or e-waste, composed of computers and appliances that have been stripped of their parts. This can be hazardous to the recyclers because they can be exposed to radioactive tubes. Hundreds of thousands of tons of e-waste are informally recycled in India every year. Corporations produce the majority of e-waste, most without regard to ecofriendly recycling methods. Therefore, a large amount of toxic chemicals are released into the air, adding to air pollution.

One of India's leading environmental problems is pollution. The country has a heavy reliance on coal for power generation, with coal supplying more than half of the country's energy needs and nearly three-quarters of its electricity generation, resulting in large increases in carbon emission levels. Indoor pollution from the burning of coal also accounts for many deaths annually. India's population is among the world's highest at over 1 billion people, but only a fraction of their electricity is generated from nuclear power. The United States is providing assistance to India's civilian nuclear energy program through a treaty agreement, which will also expand U.S.–India friendship and cooperation in energy and satellite technology.

Vehicle emissions are also critical, accounting for approximately 70 percent of the country's air pollution. Exhaust from vehicles, as well as industrial air pollution, has greatly risen in the past 20 years. Pollution control has not kept pace with the growth of the economy over the same period. Urban areas are the hardest hit, and as many as 51 Indian cities have dangerously high levels of air pollution. India is attempting to build environmentally friendly communities. One goal is the use of public transport, walking, or bicycling as means of transportation. The country recognizes that in order to become more "green" it needs to reduce usage of cars, better utilize public transportation, and create aggressive fuel economy and clean emission standards. India also has plans to launch a satellite into space to monitor the country's greenhouse gas emissions. This satellite technology will also study the impact of climate change and the consequences of greenhouse gas emissions to the environment.

The Center for Science and Environment (CSE), based in New Delhi, added mining to the list of polluting industries, stating that mining was causing pollution as well as displacement, forest degradation, and social unrest. India's CSE also works to harvest rainwater in both urban and rural areas. CSE believes that rainwater harvesting is a technological solution that can be universally adopted to help India manage its water in a more effective way and perhaps provide solutions to water crises.

India has also adopted a Green School Program. Schools in this program seek to effectively manage their natural resources and be environmentally friendly. This environmental activism is not occurring so much in the elite schools but, rather, in the semi-urban and middle-class schools. The CSE environmentally audits the schools participating in the program and then hands out "green school awards" to those with the most effective programs. Some of the schools given awards have impressive environmental practices, including encouraging students to commute on foot or bicycle rather than in cars or buses. One school reuses its water to irrigate its playing fields and green areas. Another school acquires 15 percent of its energy from solar power, using it for water heaters, streetlights, and cooking.

In New Delhi, ecofriendly businesses are increasing in popularity. These include apparel lines for customers that are environmentally conscious. Ecofriendly clothes are made out of natural dyes, cotton, and linen fabrics. There is a variety of clothing that is made using organic cotton, recycled fibers, and bamboo. Although this industry is still in its beginning stages, the concept of organic clothing has been gaining popularity in some parts of India since 2007.

Indian Government Responses to Green Culture

The Indian government has responded to environmental problems and green culture calls for changes since the 1970s. In 1972, the National Committee on Environmental Planning and Coordination was established to investigate and design solutions to India's environmental problems. The Department of Environment was created in 1980. The National Environmental Engineering Research Institute has field offices throughout India to monitor environmental issues. The government implemented a green buildings program with a focus on components related to water and wastewater at the city level, headed by the Green Building Water Management (GBWM). The GBWM researches and promotes green policies, increasing public awareness of the benefits.

India's national government Ministry of Environment and Forests plans, promotes, coordinates, and oversees the implementation of the country's environmental and forestry policy programs. The ministry implements policies that conserve India's natural resources, which includes its lakes and rivers, as well as biodiversity, forests and wildlife, and animals. It also takes a stance against pollution. These objectives are all supported by legislative and regulatory measures.

The Indian government has implemented a national action plan on climate change. Over 200 scientists from over 100 research institutions are working on determining the impact that climate change has had on agriculture, water, health, and forests. They will then issue a report recommending actions necessary to implement sustainable goals. The country's budget has also taken into consideration the state of its environment. A National Clean Energy Fund backs research and innovative products in clean energy technology. "Mission Clean Ganges 2020" has also been set up under the realm of the National Ganges River Basin Authority. Its objective is to stop the dumping of untreated municipal or industrial sewage into the river.

One of the most controversial national government projects in terms of environmental issues has been the building of large dams since the mid-20th century. India entered the 21st century as one of the world's leading nations in dam construction. For example, over 3,000 dams have been constructed or are planned in the Narmada River Valley. The main reason for constructing dams is for irrigation, with other benefits including flood control, water supply issues, and generating hydroelectric power. The government argues that the dams will provide irrigation systems to drought-prone areas and give a greater number of people access to electricity.

Proponents and critics of the dam-building projects have addressed environmental as well as social and economic benefits and costs. Proponents use aggregate statistics to show the dams' overall benefits in terms of agricultural productivity, hydroelectric power generation, and flood control, and note that people displaced from their lands receive government compensation and resettlement aid. Proponents also argue that although there are other potential irrigation methods, none can effectively serve as large a population, and that the dams are cost-effective despite their high price tags. Critics counter that the dams are not cost-effective and that higher agricultural outputs are the result of other techniques such as the implementation of high-yield crops and the increased use of fertilizers and mechanization.

Critics note that the benefits tend to accrue disproportionately to those living in the command areas, while the costs tend to accrue disproportionately to those living in the catchment areas. They also look at issues involved in government resettlement and compensation for the hundreds of thousands displaced when their traditional agricultural homelands are flooded, most of whom are traditionally disadvantaged groups such as peasants and tribal peoples known as *adivasi*. Compensation is only provided to landholders, does not account for communal land, and is not always given as promised. Other negative impacts have included health issues, such as waterborne illnesses, the poor quality of resettlement sites, the diversion of funding from other projects, and the vulnerability of the displaced to loan sharks and other predators. Grassroots protestors include the Narmada Bachao Andolan (Save the Narmada Movement).

See Also: "Agri-Culture"; Chipko Movement; Global Warming and Culture; Grassroots Organizations; Green Consumerism; Population/Overpopulation; Recycling; Toxic/Hazardous Waste; Water Pollution.

Further Readings

Centre for Science and Environment. http://cseindia.org (Accessed September 2010).

"Environmental Pollution in India." http://www.gits4u.com/envo/envo4.htm (Accessed September 2010).

Faris, Stephan. *Forecast: The Consequences of Climate Change, From the Amazon to the Arctic, From Darfur to Napa Valley*. New York: Henry Holt, 2009.

"Green Revolution." IndiaOneStop.com. http://www.indiaonestop.com/Greenrevolution.htm (Accessed September 2010).

Grossman, Elizabeth. *High Tech Trash: Digital Devices, Hidden Toxics, and Human Health.* Washington, DC: Island Press/Shearwater, 2006.

Kamdar, Mira. *Planet India: How the Fastest-Growing Democracy Is Transforming America and the World*. New York: Scribner, 2007.

Kennedy, Paul M. *Preparing for the Twenty-First Century*. New York: Random House, 1993.

Kurzman, Dan. *A Killing Wind: Inside Union Carbide and the Bhopal Catastrophe.* New York: McGraw-Hill, 1987.

Lappé, Frances Moore, and Anna Lappé. *Hope's Edge: The Next Diet for a Small Planet.* New York: Jeremy P. Tarcher/Putnam, 2002.

Leslie, Jacques. *Deep Water: The Epic Struggle Over Dams, Displaced People, and the Environment*. New York: Farrar, Straus & Giroux, 2005.
Ministry of Environment and Forests, Government of India. http://moef.nic.in/modules/about-the-ministry/introduction (Accessed September 2010).
Postel, Sandra. *Last Oasis: Facing Water Scarcity*. The Worldwatch Environmental Alert series. New York: W. W. Norton, 1992.
Robb, Peter. *A History of India*. New York: Palgrave, 2002.
Romm, Joseph J. *Hell and High Water: Global Warming—The Solution and the Politics—and What We Should Do*. New York: William Morrow, 2007.
Sachs, Jeffrey. *The End of Poverty: Economic Possibilities for Our Time*. New York: Penguin, 2005.
Weir, David. *The Bhopal Syndrome: Pesticides, Environment, and Health*. San Francisco, CA: Sierra Club Books, 1987.

Marcella Bush Trevino
Barry University

Individual Action, Global and Regional

The work, energy, and passion of individuals has led to a flourishing of the green movement around the world. Though they occupy different social ranks, have varying degrees of education, come from humble beginnings or currently rule as royalty, the individuals discussed below have the power to sway the opinions of the public. Some cause change through leading by example while others, aware of their power to set trends and shape the behavior of their followers, use their celebrity status to make the world a more environmentally conscious and greener place.

In 2007, *Time* magazine created a list titled "Heroes of the Environment," which honored the following people as either a leader, a visionary, or a key activist in the green movement: Mikhail Gorbachev, David Attenborough, Lee Myung Bak, Al Gore, Jr., Janine Benyus, Tommy Remengesau, Jr., José Goldemberg, Prince Charles, James Lovelock, Robert Redford, David Suzuki, Barnabas Suebu, Angela Merkel, Frederic Hauge, Wang Canfa, Olga Tsepilova, Von Hernandez, Wangari Maathai, Christine Loh, Benjamin Kahn, Karl Ammann, and Hammer Simwinga. Their passion for and dedication to the environment is unwavering and has helped to spearhead many important actions on behalf of the green movement.

David Attenborough, famous for BBC nature programs like *Life on Earth* and *The Living Planet*, is one of the United Kingdom's most important environmentalists. He is primarily known for his staunch support of causes that protect wildlife (he is president of the Leicestershire and Rutland Wildlife Trust), but also argues that the destruction of the planet is a result of overpopulation. Attenborough believes that climate change is a direct result of overpopulation, and through various media (documentaries, books, speeches), he challenges his audiences to think carefully about the ramifications of overpopulation, both for the people in the world but, mainly, for the Earth.

Former U.S. Vice President Al Gore is respected around the world for his commitment to the environment and his unwavering commitment to educating the public about environmental issues. Gore founded the Alliance for Climate Protection, and his critically acclaimed documentary about the dire effects and future of global warming, *An Inconvenient Truth* (2006), helped secure him the Nobel Peace Prize in 2007. Skeptics question his claims and evidence, but millions of others have been educated due, in part, to Gore's groundbreaking work. Gore has also been criticized for some of his personal choices relating to green matters; some accuse him of acting irresponsibly by flying on private jets (a situation, obviously, where fuel is not optimized). Though Gore may not always make choices that align with his green platform, his influence on making the environment a priority in the United States cannot be denied.

Prince Charles of England has long been supportive of environmental causes, and his decades of loyalty to the green movement easily earned him a place on *Time*'s list of environmental heroes. Prince Charles's The Prince's Foundation for the Built Environment champions sustainable living and green building. According to the foundation's website, the mission of the organization is to "improve communities that are long lasting and healthy for people and planet." Members of The Prince's Foundation educate builders, architects, and the general public about the most effective and soundest green practices. The foundation also hosts conferences, facilitates classes, and oversees both apprenticeships and graduate programs that encourage environmental awareness.

Wangari Maathai, a Kenyan and a Nobel Peace Prize winner, remains a powerful force in the international environmentalist movement. Known for the Green Belt Movement (a grassroots organization she founded), which focuses on not just one aspect of the environment but examines how elements like the land and its people must coexist harmoniously, Maathai is considered one of the most prominent members of the ecofeminist movement. She has brought positive attention to the movements to which she dedicates herself whole-heartedly and has helped to educate the world about preserving the Earth and the links between the ways in which women and the Earth are debased by similar methods (a key tenet in ecofeminism).

People like Maathai undoubtedly influence the decisions of many, but celebrities, arguably, have even more influence on shaping the behavior of the general masses. Appealing celebrities like Brad Pitt, Cameron Diaz, and Leonardo DiCaprio, among others, are often idolized or mimicked because of their social standing and influence. In recent years, green celebrities have received a wealth of attention and have supported environmental awareness projects ranging from film documentaries to magazine covers to writing their own guides about living green. Brad Pitt, founder of Make It Right: Helping to Rebuild New Orleans' Ninth Ward, has garnered worldwide support. Pitt founded the organization after Hurricane Katrina destroyed thousands of homes in the Ninth Ward. Known as Hollywood's "golden boy," Pitt's actions are closely followed and considered fashionable. In turn, he has helped educate fans about building sustainable green homes while recovering from destruction and tragedy. Another celebrity who remains a powerful force in the green movement is Leonardo DiCaprio, who is praised for his green lifestyle (his home is very environmentally conscious) and dedication to the environment. DiCaprio was one of only a few celebrities asked to appear on the cover of *Vanity Fair*'s Green Issue; DiCaprio graced the May 2007 issue and spoke frankly about environmental concerns like global warming. DiCaprio also masterminded and produced the documentary film *The 11th Hour*, a film that examines the impending peril Earth faces due to contemporary environmental concerns and practices.

Other celebrities promoting (to a lesser degree) environmental respect and justice include people like George Clooney, Cameron Diaz, and Ed Begley, Jr., among others. George

Clooney is famous for his electric cars and as an activist involved in causes around the world (he has devoted a lot of time to the Save Darfur campaign and has received the Bob Hope Humanitarian Award for his commitment to helping people in need around the globe). People admire Clooney and his choices; seeing him driving electric vehicles entices others to do so also. According to the website The Daily Green and its featured article "14 Celebrities Who Walk the Walk," celebrities like Cameron Diaz, Ted Danson, Edward Norton, Woody Harrelson, and Ed Begley, Jr., not only ask others to behave in an environmentally conscious manner, but do so themselves. The Daily Green article praises Diaz for her involvement in the Live Earth concert series and commends *The Green Book: The Everyday Guide to Saving the Planet One Simple Step at a Time*, which she coauthored. Ted Danson, according to the article, founded the American Oceans Campaign to protect oceans from pollution, while Edward Norton testified before Congress in favor of green housing. Woody Harrelson founded the website Voice Yourself, which focuses on being environmentally conscious, and Ed Begley, Jr., designed an environmentally friendly household cleaner (in addition to championing various other green practices). The power of celebrity, especially on a global scale, is undeniable. Celebrities who make the environment an important aspect of their platform influence people around the world, no matter their country of origin.

Individuals like David Attenborough, Wangari Maathai, Leonardo DiCaprio, and George Clooney employ vastly different approaches to create awareness about environmental issues around the world, ranging from pleading with audiences to save the Earth to simply driving electric cars. Without the dedication—and influence—of individuals, however, the green movement would stall and the planet would suffer.

See Also: Begley, Jr., Ed; Communication, Global and Regional; Demonstrations and Events; DiCaprio, Leonardo; Energy; Film, Documentaries; Gore, Jr., Al; Green Consumerism; Home; Hurricane Katrina; Individual Action, National and Local; Institutional/Organizational Action; International Law and Treaties; Maathai, Wangari; Nelson, Willie; Public Opinion; Redford, Robert; Social Action, Global and Regional; Social Action, National and Local; Tree Planting Movement; ***Vanity Fair*** Green Issue.

Further Readings

Adam, David. "Earth Shakers: The Top 100 Green Campaigners of All Time." *The Guardian* (November 28, 2006).

Breton, Mary Joy. *Women Pioneers for the Environment*. Lebanon, NH: Northeastern University Press, 2000.

Dryzek, John, Daid Downs, Hans-Kristian Hernes, and David Schlosberg. *Green States and Social Movements: Environmentalism in the United States, United Kingdom, and Norway*. Oxford, UK: Oxford University Press, 2003.

"14 Celebrities Who Walk the Walk." The Daily Green. http://www.whatsmycarbonfootprint.com/index.htm (Accessed September 2010).

Gore, Al, Jr. *An Inconvenient Truth: The Planetary Emergency of Global Warming and What We Can Do About It*. New York: Rodale Books, 2006.

"Heroes of the Environment." *Time*. http://www.time.com/time/specials/packages/completelist/0,29569,1663317,00.html (Accessed September 2010).

Kerry, John and Teresa Heinz Kerry. *This Moment on Earth: Today's New Environmentalists and Their Vision for the Future*. Cambridge, MA: Perseus Books, 2007.
"Make It Right: Helping to Rebuild New Orleans' Ninth Ward." http://www.makeitrightnola.org/index.php (Accessed September 2010).
"The Prince's Foundation for the Built Environment." http://www.princes-foundation.org (Accessed September 2010).

Karley Adney
Independent Scholar

Individual Action, National and Local

Many Americas, if asked, may likely explain that environmental awareness did not become a prominent topic in the American consciousness until the 1970s. In contrast, the United States has a rich tradition of individuals passionate about environmental issues, including both famous and ordinary citizens. While a tradition concerning environmental justice has long existed, only the past few decades have seen an exponential growth in citizens' awareness of and commitment to reducing the impact each person has on the planet. Americans have recently taken significant and important measures toward becoming greener in their homes, and there has also been a positive and growing movement to reduce one's carbon footprint in respect to personal transportation. American employers have also made an effort to make the workplace greener.

Americans have made and continue to make notable contributions to environmental awareness. Many prominent environmentalists like Rachel Carson and Al Gore influenced not only environmental practices in the United States but around the world as well. Americans who shaped the green movement into its present form include George Perkins Marsh, John Muir, Theodore Roosevelt, Bill Clinton, and Bill McKibben. George Perkins Marsh is often considered the first environmentalist of the United States. Marsh's landmark work *Man and Nature* (1864) analyzed the effects of deforestation and encouraged readers to be more environmentally conscious. Also during the 19th century, John Muir encouraged Americans to develop a greater respect for the environment; Muir founded the Sierra Club, which still functions as a prominent force in the green movement today. President Theodore Roosevelt contributed significantly to the environmental movement by conserving a wealth of land and by bringing environmental issues to the forefront of the American consciousness. Another president, Bill Clinton, also made significant contributions to the green movement. Clinton fought tirelessly to protect and secure additional funding for environmental programs during his presidency, reminding Americans that no matter the financial situation of the country, the environment should remain a priority. Bill McKibben, a prolific writer, is best known for raising environmental consciousness through his best-selling and critically acclaimed books like *The End of Nature* (1990), *Deep Economy: The Wealth of Communities and the Durable Future* (2007), and, most recently *Eaarth: Making of Life on a Tough New Planet* (2010).

Some of the most influential American environmentalists are Gaylord Nelson and Denis Hayes, the founders of Earth Day. The inaugural Earth Day celebration occurred in 1970 (the same year President Richard Nixon created the Environmental Protection Agency), and within a few decades, Earth Day was celebrated around the world. Also in the year 1970, Deane C. Davis, former governor of Vermont, founded Green Up Day. This day, which occurs in May, appeals to citizens to clean up roadside refuse. The day has remained an important occasion in Vermont, but is now also celebrated in other states, including Massachusetts and New Hampshire.

Ordinary citizens have also made major contributions to the green movement in America, both at the local and national levels. Take, for instance, the story of a man who changed paper-recycling practices in his town. John Sheehan, of Maryland, founded a company called ShredCycle; Sheehan (an autistic man) was praised for his invention and passion by the Maryland Association of Community Services for Persons with Developmental Disabilities in 2004. ShredCycle works with small businesses, collects their used paper, shreds it, and then delivers it to a recycling center. Sheehan's business encourages recycling in the community and has also raised awareness about reducing waste.

Other citizens, like Pat Cosby, also made differences in their own towns and then became known as an inspiration to green activists across the country. Brandy Norleen reports Cosby's story, which began with his implementing a successful recycling program at Columbus Water Works in Columbus, Georgia. Cosby's dedication for recycling and conservation soon extended into surrounding communities wherein he implemented other recycling programs, one of which focused on recycling wrapping paper during the holidays. Each year, Cosby collects more poundage of used wrapping paper—according to Norleen, he collected over 700 pounds in 2008. Cosby's story, published on the website Earth 911, elicited many responses from people across the United States and around the world praising him for his work.

Americans in general have used various strategies in their homes to become more environmentally conscious. Some common practices include conserving energy and water and reducing waste. Due in part to the state of the economy and also in response to protecting the environment, citizens have become more conscious about the amount of energy they use; whether it means buying energy-efficient appliances or simply switching to green light bulbs, people can take steps, big or small, to reduce the amount of energy they use each day. Americans have also become more conscious of the amount of water they use. A common suggestion for water conservation is to turn off the faucet while brushing one's teeth. Some people conserve water by installing low-flow showerheads or toilets, or by merely running the dishwasher or washing machine only when the appliance is loaded to capacity. Others take more significant measures and even incorporate green plumbing into their homes. Americans have also become more mindful of recycling in an effort to reduce waste. According to a study published on the green blog Treehugger, more than three-quarters of Americans actively participate in recycling programs. Some people participate in recycling programs because in some cases monetary incentives exist for returning items such as glass or plastic bottles to stores. While no national laws require that Americans recycle, obviously, the vast majority of citizens do so in order to reduce waste. Another popular strategy for reducing the amount of waste produced by each person in the United States is to make a habit of using reusable containers. More people bring food to work in reusable bags and pack their meals in containers that can be used indefinitely.

Another important strategy Americans use to reduce their carbon footprint is to make environmentally sound decisions in regard to their mode of transportation. While some make it their goal to drive less to conserve gas, money, and emissions, others invest in hybrid vehicles. Still others use strategies like carpooling or ride-shares to aid in conservation; similarly, people rely more on mass transportation like buses, subways, and trains. Some citizens forgo using a vehicle altogether and either bike or walk to their planned destination. Websites like What's My Carbon Footprint? also encourage people to make sure they complete routine maintenance on their vehicles to reduce pollution and to get the best gas mileage.

Both employers and employees have also taken significant steps in becoming more environmentally conscious. Business owners pay more attention to energy usage at work; some install green lighting, and most provide recycling receptacles. Like homeowners, business owners also maintain their heating and cooling systems for optimum efficiency; similarly, people are more mindful of turning down the heat or air conditioning during times when their businesses are unoccupied by employees and customers. U.S. employees have also taken measures to green their work environments. The sayings "Think before you print" and "Consider the environment before printing this e-mail" are now common in business institutions and encourage all business team members to print less, thus conserving paper. Similarly, many businesses make the default setting (or the only setting) for printing both-sided instead of single-sided. Employees are also more mindful of conserving energy in the workplace. Small actions can have significant ramifications, and when a company's 100 employees remember to turn off their computer monitors and shut down their computers before leaving for the weekend, a significant amount of energy is conserved.

From the founders of the environmental protection tradition who made significant progress in creating awareness to ordinary contemporary citizens who take even small measures, American individuals continue to take action in honor of the environment. People in general, more environmentally conscious than in the past, have chosen to make their homes and places of work more environmentally friendly and thus improve America's status as an eco-aware nation.

See Also: Alternative/Sustainable Energy; Clinton, William J. (Executive Order 12898); Communication, National and Local; Commuting; Corporate Green Culture; Corporate Social Responsibility; Demonstrations and Events; Energy; Environmental Law, U.S.; Garbage; Gore, Jr., Al; Grassroots Organizations; Green Consumerism; Home; Hybrid Cars; Individual Action, Global and Regional; Institutional/Organizational Action; Public Opinion; Social Action, Global and Regional; Social Action, National and Local; Water (Bottled/Tap); Water Pollution; Work.

Further Readings

Cevasco, George A., Richard P. Harmon, and Everett I. Mendelsohn, eds. *Modern American Environmentalists: A Biographical Encyclopedia*. Baltimore, MD: Johns Hopkins University Press, 2009.

Chua, Jasmin. "23 Percent of Americans Don't Recycle." Treehugger.com. http://www.treehugger.com/files/2007/08/americans_recycle.php (Accessed September 2010).

Norleen, Brandy. "How One Man Started a Recycling Program." Earth911. http://earth911.com/news/2009/01/26/how-one-man-started-a-recycling-program (Accessed September 2010).

Walker, Dionne. "Agency to Honor Disabled Founder of Recycling Business." *The Capital* (January 2004).
"What's My Carbon Footprint?" FirmGreen. http://www.whatsmycarbonfootprint.com/index.htm (Accessed September 2010).

Karley Adney
Independent Scholar

Indoor Air Pollution

While ventilation can lower indoor pollutants by drawing cleaner air from outside in to dilute indoor toxins, HVAC units like this rooftop model can often fail to control the level of existing air pollutants.

Source: Building Systems Program, Pacific Northwest National Laboratory (PNNL) for the U.S. Department of Energy

Indoor air pollution is the degradation of indoor air quality by contaminants that pose various degrees of risk to human health. There is ample scientific evidence that human exposure to air pollutants within buildings is much greater than that outdoors. One reason for this is that enclosed spaces enable pollutants to accumulate in larger amounts, and consequently, the indoor concentrations of many pollutants are several, and sometimes a hundred, times higher than those outdoors. Exposure to a pollutant or a synergy of pollutants can cause a variety of harmful effects on human organisms from allergies, nose or throat irritation, and asthma to hearing loss, coma, and lung cancer. Given that most people spend up to 90 percent of their time indoors, of which 65 percent is spent at home, indoor air pollution represents an obvious health hazard. The past four decades have faced an increasing awareness of indoor air quality problems, and various studies have looked into major air pollutants, their sources and health impacts, factors affecting air quality, and measures to prevent and combat indoor air pollution.

Indoor air pollutants are unwanted airborne elements that may deteriorate the air quality and adversely affect the health of building occupants (people or animals). Air pollutants can be of biological or chemical origin. Biological pollutants are living organisms or their derivatives, for example, bacteria, fungi, mold, viruses, and dust mites. Chemical pollutants are gases and particles that are generated by tobacco smoking, combustion, household products such as cleaning and disinfecting, personal care products, different building materials, and so forth. Table 1 lists the main indoor air pollutants alongside their basic sources, pathways, and associated health effects.

Table 1 Main Pollutants, Sources, Pathways, and Health Effects

Pollutant (definition)	*Basic source(s)*	*Pathway(s)*	*Health effects*
Radon *(A colorless, tasteless, radioactive, naturally occurring gas formed in soil/rocks by the decay of radium—a decay product of uranium)*	• Local geology • Soil • Building materials • Water wells	• Inhaled into the lungs where it continues decaying and releasing radiation • Swallowed with water	• Lung cancer
Environmental tobacco smoke *(A smoke mixture with more than 4000 chemicals, of which around 40 are carcinogenic)*	• Cigarette • Pipes • Cigars • Smoke exhaled by the smoker	• Inhaled by active smoker • Inhaled by passive smokers	• Lung and other cancers • Cardiovascular diseases • Eye irritation • Nasal congestion • Throats/coughs • Headaches • Pneumonia • Increased risk of asthma, ear, and respiratory infections in children
Asbestos *(A mineral microscopic fiber that can pollute air or water that is banned from wide use by many countries)*	• Disturbed, damaged, or deteriorating with age building materials (e.g., floor/ceiling tiles, fireproofing, insulation, mastics, textured paints and pipe wrap)	• Inhaled	• Lung cancer • Lung diseases • Chest cancer • Abdominal cancer
Biological contaminants *(see above)*	• Microorganisms (e.g., molds, dust mites, bacteria, and insects) growing in/on: ○ Heating/cooling systems ○ Humidifiers	• Swallowed • Inhaled • Physical contact	• Various allergies • Asthma • Infectious airborne diseases (e.g., chicken pox, influenza and measles) • Sneezing • Watery eyes

(Continued)

Table I (Continued)

Pollutant (definition)	*Basic source(s)*	*Pathway(s)*	*Health effects*
	○ Water-damaged or soiled building materials • Plants' pollens • Viruses and bacteria carried by people/animals • Soil/plant debris • Saliva of household pets and animal dander		• Coughing • Shortness of breath • Dizziness • Lethargy • Fever (humidifier) • Digestive problems
Lead *(A gray-white, soft, resistant to corrosion, ductile metal; an extremely harmful, especially for children, toxic air contaminant; leaded paints are banned in many countries)*	• Lead-based paint • Industrial sources • Weathering of contaminated soil polluted dust • Contaminated drinking water	• Airborne lead particles or dust are inhaled or swallowed	High levels: • Convulsions • Coma • Death Lower levels: • Brain damage • Hearing loss • Mental/learning disabilities • Kidney dysfunction • Personality changes
Mercury vapor *(Gas evaporating from liquid mercury or its compounds)*	• Interior latex paints • Broken mercurial thermometers • Sprinklings for religious purposes • Contaminated food	• Inhaled • Swallowed • Absorbed through the skin	• Neurological dysfunction • Muscle cramps • Headache • Intermittent fever • Tachycardia • Personality change
Volatile organic compounds *(Carbon-containing gaseous chemicals that evaporate at room temperature and pressure; for example, formaldehyde, pesticides, phenol)*	• Household products (e.g., disinfectants, cosmetics, moth repellents, and air fresheners) • Building materials	• Inhaled • Swallowed • Absorbed	• Cancer • Problems to central nervous system • Skin allergies • Nausea • Headache • Dizziness • Kidney damage • Eye irritation • Nose/throat irritation • Loss of coordination

Factors of Indoor Air Pollution

Four factors are considered to fundamentally affect indoor air quality, namely, the following:

- Sources of indoor air contaminants that exist within or outside of the building, for example, combustion appliances, building materials and furnishings, pesticides, tobacco smoke, household products, outside vehicle exhaust, pollen, and exterior standing water
- Heating, ventilation, and air conditioning (HVAC) systems, which often fail to control the level of existing air pollutants and ensure thermal comfort for building occupants
- Pollutant pathways that connect the source of pollutants to the occupants and represent avenues for distribution of pollutants in the building through movement of indoor air and occupants (e.g., radon gas penetrates buildings from beneath through cracks in concrete walls/floors, water wells, and so on)
- Building occupants who are present in the building and affect the air quality by shedding pollutants, such as bacteria, viruses, and allergens

Strategies for Avoiding, Lowering, and Controlling Indoor Air Pollution

Source Management

Source management encompasses the following strategies:

- Eliminating pollutant sources (e.g., banning smoking in public buildings, banning the use of asbestos, and removing moldy materials)
- Isolating or encapsulating sources (e.g., isolating smokers outdoors, arranging separate areas for photocopying machines, using products with lower emission rates of formaldehyde, and encapsulating lead containing paint)
- Replacing pollutant sources (e.g., selecting less-toxic paints and varnishes)
- Source avoidance (e.g., selecting building materials that do not emit toxic substances)
- Local point exhaust that implies using fume hoods and exhaust fans to draw sources of contaminants outdoors prior to their dispersion (e.g., exhaust systems installed in kitchens, laboratories)

Ventilation

Ventilation is a prime method for lowering indoor pollutant concentrations by drawing in cleaner air from outside and thereby diluting contaminated air indoors. Regular ventilation is used to reduce the concentrations of such indoor pollutants as formaldehyde, radon, pesticides, and aerosols.

Air Cleaning

Air cleaning is aimed at reducing (rarely removing) various airborne particulates and/or gases using such methods as particulate filtration, electrostatic precipitation, and gas sorption. As air cleaners, for example, ion or ozone generators, mechanical filters, or electronic air cleaners, are often designed to be effective against only particular pollutants, they alone cannot resolve indoor air quality issues; for a better result, they can be used in conjunction with proper ventilation and source management measures.

Conclusion

Unless a heavy pollution or one that requires specific expert/technical knowledge is concerned (e.g., in case of asbestos or radon), standard wet-mopping and drying, ventilating, vacuuming, and cleaning, maintaining a relatively low humidity, and buying "greener" household chemicals can help reduce air pollutants indoors.

See Also: Air Pollution; Environmental Protection Agency; Home; Soil Pollution; Water Pollution; Work.

Further Readings

American Lung Association, American Medical Association, U.S. Consumer Product Safety Commission, and U.S. Environmental Protection Agency. "Indoor Air Pollution: An Introduction for Health Professionals." http://www.cpsc.gov/cpscpub/pubs/455.html (Accessed August 2010).

Hansen, S. J. and H. E. Burroughs. *Managing Indoor Air Quality*, 4th ed. Lilburn, GA: Fairmont Press, 2008.

Leslie, G. B. and F. W. Lunau, eds. *Indoor Air Pollution: Problems and Priorities*. Cambridge: Cambridge University Press, 1992.

U.S. Environmental Protection Agency, U.S. Consumer Product Safety Commission. "The Inside Story: A Guide to Indoor Air Quality" (1993). http://www.epa.gov/iaq/pubs/insidest.html (Accessed August 2010).

Maia Gachechiladze-Bozhesku
Central European University, Hungary

Institutional/Organizational Action

The green movement has grown in popularity and impact over the past few decades. Institutional and organizational action that supports sustainability and environmentally friendly action has played a significant role in this growth. Individual action certainly plays a role both in encouraging policymakers to enact legislation and regulations that support green actions and making individual decisions that promote sustainability and protect the environment. Institutional and organizational action also influences both policy and practice in a manner that supports green sensibilities. Unlike individual action, however, institutions and organizations have a greater ability to structure their environmental policies and practices so that they can contribute to environmental sustainability on local, national, and international levels. Institutional and organizational action also can publicize and promote green achievements more readily than most individuals can. Colleges and universities, trade groups, professional associations, nongovernmental organizations (NGOs), and other institutions and organizations have all taken actions that influence sustainability and environmentally friendly policies. These actions have ranged from those promoting sustainable development to those making environmentally friendly practices and behaviors easier to achieve. Institutional and organization action thus has both shaped public policy and molded individual behavior.

Push for More Sustainable Practice

Due to increased consciousness of sustainability and environmental degradation, more industries engage in environmentally friendly practices than in the past. Certainly, environmental laws and regulations have encouraged many organizations and individuals to act in a more environmentally responsible manner. Additionally, organizations, groups, and individuals that have seen economic advantages stemming from green initiatives and other sustainable practices have shared these with others. These actions have created a growing demand for initiatives that help large corporations, organizations, institutions, individuals, and smaller enterprises to meet sustainability challenges and goals. This has helped to create increased sharing of practices that are effective, a demand for more sustainable behavior on the part of organizations and institutions, and a climate that encourages environmentally friendly actions. Organizational and institutional actions have helped to create a greener planet, and have transformed the way many colleges and universities, NGOs, corporations, and other business enterprises conduct their affairs. Such actions have also greatly strengthened the impact of laws and regulations that encourage green behavior and sustainable practices.

As a result of a push from the public sector, for example, laws in many states now require all new public buildings to incorporate recycled or renewable materials and energy-saving green technologies. While this has of course proven beneficial, in and of itself the effect on building practices would have been minimal. Many organizations and institutions, however, have voluntarily adopted similar green building practices, which has allowed a certain mass of sustainable development to emerge that makes it economically viable for others to proceed in an environmentally friendly manner. Many new businesses have emerged to respond to the rapidly growing demand for green building components. Institutional and organizational action has encouraged a host of green business initiatives that assist our society in efforts to become sustainable and leave little or no carbon footprint. Biodiesel fuel, recycling programs, ecofriendly cleaning products, and biodegradable packaging are just a few products that have found a market due to institutional and organizational action to encourage their development and sales. Institutional and organization action will continue to promote green initiatives that might otherwise flounder.

Colleges and Universities

Colleges and universities have played major roles in providing the impetus to move toward green action. Many individual campuses have promoted policies and procedures that encourage sustainable development and environmentally friendly practices, including Columbia University and the University of Minnesota. While these initiatives have been significant, coalitions and associations of like-minded institutions have often allowed information regarding sustainable practices to be widely disseminated, giving them greater influence. The Higher Education Associations Sustainability Consortium (HEASC), for example, was developed in December 2005 to help higher education exert strong leadership in making sustainable education, research, and practices a reality. Leaders of several higher education associations formed the consortium to support and enhance the capacity of higher education to fulfill a role in producing an educated and engaged citizenry and to produce the knowledge needed for a thriving and civil society. These organizations recognize that fulfilling their mission in the 21st century requires a broader, systemic, collaborative approach to their work and that of the constituents they serve. The societal challenges to create vibrant, secure communities and strong economies while preserving the life support system on which we all depend upon are daunting and will only increase as the world's population and our need to increase economic output grows. HEASC's purpose is to learn from each other, work together on joint projects, get access to the best expertise and information on sustainability, and to keep a collective, ongoing focus on advancing education for a sustainable future over time.

International cooperation between colleges and universities has also encouraged action that promotes sustainable development and environmentally friendly policies. Universities and colleges in many countries have begun to increasingly examine their own roles and responsibilities. For example, a 1990 conference of university presidents from every continent in Talloires, France, under the auspices of Tufts University of the United States, produced the Talloires Declaration. This declaration of environmental commitment attracted the support of more than 100 universities from dozens of countries. In December 1991 in Halifax, Canada, the specific challenge of environmentally sustainable development was addressed by the presidents of universities from Brazil, Canada, Indonesia, Zimbabwe, and elsewhere, as well as by senior representatives of the International Association of Universities, the United Nations University, and the Association of Universities and Colleges of Canada. These conferences provided valuable venues for continued action, both at the institutional level and throughout the organizational structures they encouraged.

The Halifax meeting allowed institutions whose leaders were deeply concerned about the continuing widespread degradation of the Earth's environment, about the pervasive influence of poverty on the process, and about widespread unsustainable environmental practices to take action to encourage sustainable development and environmentally friendly policies. The meeting was guided by the belief that solutions to these problems can only be effective to the extent that the mutual vulnerability of all societies in all regions were recognized, and the energies and skills of people everywhere be employed in a positive, cooperative fashion. Because the educational, research and public service roles of universities enable them to be competent, effective contributors to the major attitudinal and policy changes necessary for a sustainable future, the Halifax meeting invited the dedication of all universities to take actions that would entail the following:

- Ensure that colleges and universities be clear and uncompromising in their ongoing commitment to the principle and practice of sustainable development within their operations, as well as at the local, national, and global levels.
- Utilize the intellectual resources of colleges and universities to encourage a better understanding on the part of society of the interrelated physical, biological, and social dangers facing the planet.
- Emphasize the present generation's ethical obligation to overcome current practices of resource utilization and widespread circumstances of intolerable human disparity that lie at the root of environmental degradation.
- Enhance the capacity of colleges and universities so that they teach and practice sustainable development principles, build environmental literacy, and develop the understanding of environmental ethics among faculty, students, and the public at large.
- Cooperate with one another and all segments of society to pursue practical capacity-building and policy measures that will help achieve the effective revision and reversal of current practices that contribute to environmental degradation, regional disparities, and intergenerational inequity.
- Employ all channels open to colleges and universities that permit communication of these undertakings to the United Nations Conference on Environment and Development (UNCED), governments, and the general public.

Nongovernmental Organizations

NGOs have become a growing source of action on sustainability issues. Some NGOs are trade groups representing companies that strive to reduce their environmental footprints while furthering business goals, while others are membership organizations for like-minded environmentalists, such as the Sierra Club or Greenpeace USA. NGOs have invested heavily in building expertise in a range of areas, while the animosity that once existed between NGOs and the business community has steadily thawed. Many now view NGO–business partnerships as the rule, rather than the exception. An example of this is Conservation International (CI). Since it was founded in 1987, CI has interacted with dozens of companies through its efforts to preserve and protect biodiversity. Initially adversarial in its relationships with business enterprises, CI has evolved to partner with corporations to build action for green causes. CI offers two main types of corporate partnerships: they invite companies to donate funds to projects not directly related to their business, such as forest protection, and they help companies create best practices to reduce their environmental impact. Examples of this kind of action include helping Alcoa add biodiversity conservation to its environmental policies; counseling Starbucks Coffee on a range of sustainable issues related to coffee; and advising Walmart on issues related to tracking manufacturing processes related to jewelry. With new tools on the horizon, CI is also interested in playing a larger role in influencing entire industries toward better green practices, rather than single efforts with individual companies.

Another example of an NGO is The Nature Conservancy. Formed in 1951, The Nature Conservancy is one of the world's oldest and largest environmental NGOs. It prides itself on its scientific capacity and skill in implementing conservation programs. This set of tools gives The Nature Conservancy the ability to advise companies on large-scale corporate practices and strategies that can be connected to projects on the ground. The Nature Conservancy often assists companies by helping them understand their global water impact, followed with providing them assistance in the field. Its work in the areas of fresh water, biodiversity, and forestry and land management have led to a variety of partnerships crossing sector lines. It has worked with many organizations, such as International Paper, on large-scale conservation deals. Another recent accomplishment involved an agreement between timber products companies and NGOs that protected a large area of Canadian Boreal forest. Separately, The Nature Conservancy has collaborated with companies such as IBM, Caterpillar, and the Monsanto Company to analyze data and protect the world's great rivers.

The GSM Association (GSMA), a trade group representing more than 750 global system for mobile communication (GSM) operators across 218 countries, has initiated action to help mobile operators in developing markets go green. The organization announced their Green Power for Mobile initiative in 2010, which will help the industry use renewable energy sources, such as solar, wind, and biofuel to power 118,000 new and existing mobile base stations in developing countries by 2012. The initiative is backed by 25 mobile operators and provides expertise and guidelines for operators deploying low-energy base stations or base stations that use renewable energy. Although this initiative certainly has financial advantages for GSM operators, it also provides sustainable development for developing regions.

Corporate Initiatives

Corporations and other business enterprises have also embraced the sustainability movement, in part to make more environmentally friendly decisions, but also to embrace consumer interest in green issues. For example, the Entertainment and Sports Programming Network (ESPN), a subsidiary of the Walt Disney Company, promoted a *Green Game* in 2009 and 2010. All advertising revenue generated during these days were used to sponsor renewable energy research. ESPN also has undertaken numerous other green initiatives, which it promotes as evidence of its desire to promote and resurrect a greener world. For example, its annual ESPY ceremony is touted as being virtually waste-free and carbon neutral. ESPN has also made operational actions to ensure that its corporate offices are ecofriendly. These initiatives include the following:

- Using compostable to-go containers, cutlery, and napkins in cafeterias
- Seeking LEED compliance for all new construction and renovations
- Recycling light bulbs, batteries, construction debris, and used cooking oil
- Donating or recycling old office furniture
- Reducing the environmental impact of and energy consumption at events

Such actions have had positive repercussions, as the Walt Disney Company was ranked 34th by *Newsweek* in their "green rankings" of the Standard & Poor's (S&P) 500 corporations. The Walt Disney Company offers its own actions to promote green behaviors. The company has launched a variety of resource conservation initiatives in addition to programs that educate theme park guests on the importance of a healthy environment. Most recently, Disney introduced new goals in the areas of waste, carbon emissions, energy, water, and ecosystems to substantially reduce its impact on the environment and further enact environmentally responsible behavior among employees, guests, consumers, and business partners.

Other corporations have embraced green action. A large number of consumers have indicated that they are more likely to purchase environmentally friendly products, and as a result, companies have sought to be considered ecofriendly corporations. For example, competitors Coca-Cola and PepsiCo have announced initiatives designed to combat climate change. Both companies are trying to top each other in the rush to embrace green principles. The methods that these major corporations have chosen to take action, however, are very different.

PepsiCo embraced sustainable practices earlier than Coca-Cola by purchasing its first renewable energy certificates (RECs) in April 2007. This propelled PepsiCo to the top of the list of green power buyers in the United States. PepsiCo will buy more than 1.1 billion kilowatt-hours of renewable energy over the next three years, which is enough to offset the electricity use of all of its United States manufacturing, distribution, and administrative offices. As a result of these initiatives, PepsiCo now tops the U.S. Environmental Protection Agency's quarterly list of America's 25 greenest energy users, a position that grants the company marketing leverage over its competitors. Another PepsiCo environmental initiative is its commitment to green building. PepsiCo's 950,000-square-foot Gatorade facility in Wytheville, Virginia, has received a gold level Leadership in Energy and Environmental Design (LEED) certificate from the U.S. Green Building Council (USGBC). LEED promotes a whole-building approach to sustainability by recognizing performance in five key areas of human and environmental health: sustainable site development, water savings, energy efficiency, materials selection, and indoor environmental quality.

The Coca-Cola Company has embarked on a very different path to sustainability, announcing in 2007 a commitment to accomplish the following actions:

- Reduce the amount of water used to make its beverages
- Recycle the water used in production processes such as cleaning, cooling, heating, and rinsing
- Replenish water through initiatives that protect, conserve, and improve access to water in communities around the world

The Coca-Cola Company seeks to use water more efficiently, to the point that every gallon of water used will result in a gallon of product. Coca-Cola has embraced water conservation as a primary action because of concern for the over 1 billion people around the world without access to safe drinking water, and projections that by 2025, an estimated one-third of the world's population will face severe and chronic water shortages.

The company has also taken steps to make their packaging more sustainable. Because Coca-Cola beverages are consumed more than 1.3 billion times a day, using environmentally responsible packaging makes a tremendous difference in the amount of waste generated. To facilitate this, Coca-Cola has sought to encourage a recycling-based society and to design consumer-preferred, resource-effective packaging that will encourage recovery and reuse. Coca-Cola's packaging initiative has included encouraging and expanding the use of sustainable bulk packaging systems, using refillable steel tanks and bag-in-box (BIB) systems. BIBs and steel tanks now make up 12 percent of Coca-Cola's global volume distribution. The standard five-gallon BIB container is composed of recyclable cardboard and a lightweight, five-gallon plastic syrup bag that produces 30 gallons of product when mixed with local water.

Coca-Cola has made other efforts toward becoming environmentally responsible and self-sustaining in the future. In 2005, they were one of the first companies to join the Global Greenhouse Gas Register of the World Economic Forum. The company also adopted the Greenhouse Gas Protocol, a joint initiative of the World Business Council for Sustainable Development, and the World Resources Institute. The GHG protocol aims to harmonize GHG accounting and reporting standards. In 2005, the Carbon Disclosure Project, a world registry of corporate GHG emissions made available to institutional investors concerned about corporate climate policies, recognized Coca-Cola as one of the most-improved company responses.

Conclusion

As interest in sustainability and environmentally friendly development and business practices has grown, institutional and organizational action have permitted gains to be made that address these goals. Institutions and organizations, including colleges and universities, NGOs, and corporations, have used their ingenuity and influence to take actions that promote a greener planet. These institutional and organizational actions have resulted in tangible and intangible benefits. Changes in policies and procedures have led to reduced use of water and packaging, and research has increased knowledge regarding best practices that encourage sustainable development. Institutions and organizations that have embraced environmentally responsible actions have also enjoyed positive assessments from the public that have increased their stature.

See Also: Alternative Communities; Corporate Green Culture; Environmental Communication, Public Participation; Grassroots Organizations; Individual Action, National and Local; Nongovernmental Organizations; Social Action, National and Local

Further Readings

Fowler, C. W. *Systemic Management: Sustainable Human Interactions With Ecosystems and the Biosphere*. New York: Oxford University Press, 2009.

Ross, B. and S. Amter. *The Polluters: The Making of our Chemically Altered Environment.* New York: Oxford University Press, 2010.
Schusler, T. M. and M. E. Krasny. "Environmental Action as Context for Youth Development." *The Journal of Environmental Education*, 41/4 (2010).
Short, P. C. "Responsible Environmental Action: Its Role and Status in Environmental Education and Environmental Quality." *The Journal of Environmental Education*, 41/1 (2009).
van Santen, R., D. Khoe, and B. Vermeer. *2030: Technology That Will Change the World.* New York: Oxford University Press, 2010.

Stephen T. Schroth
Jason A. Helfer
Logan Willits
Jordan W. Willits
Knox College

International Law and Treaties

Effective environmental law often must be international in jurisdiction, because many environmental problems are global in scope. Pollution from one source becomes everyone's problem when it reaches the atmosphere or the oceans, and ongoing climate change is a global phenomenon. However, international law has limits bound by the sovereignty of nations, cultural and value differences, and the varying economic needs of nations, particularly between the developed and developing worlds.

Principles of International Law

There are several important principles guiding the international legal environment.

Sovereignty

First and foremost, countries have the right of self-determination, to establish their own identity in the international community and their own rule of law domestically. Sovereignty is not the same as democracy, which derives from the idea of popular democracy—the idea that sovereignty resides with the people, not with their rulers. Self-determination and equal rights are spoken of in the same breath as the United Nations Charter, and the UN Universal Declaration of Human Rights refers to a "right to a nationality." This is not always as clear-cut a right as it seems, as there are often nations—peoples—within states that claim that the majority community has interfered with their right to self-determination. Movements developing from these complaints are typically described as nationalist; for instance, Basque nationalism has been a vital political movement since the 19th century in France and Spain, and the ongoing Israeli-Palestinian conflict invokes the right to self-determination on both sides. In the United States, such movements are typically secessionist, and there have been a number of unsuccessful secessionist movements in its recent history, including a Hawaii sovereignty movement, an attempt to form an independent Lakota republic, neo-Confederate movements, and recurring attempts to stir up secessionist sentiment in Texas. Internal conflicts over self-determination have impacts on international law because sovereignty is essentially established when it is recognized by outsiders.

Some members of the internationalist political movement see national sovereignty as an obstacle to achieving international goals, an outdated notion in an age when more and more activity transpires at the global scale rather than the national. They call for international law that supersedes national law and consistent worldwide standards of law, justice, and trade relations. The tension between internationalist aims and sovereignty is analogous to the conflict between federalists and anti-federalists in the early days of the United States.

Reciprocity

In foreign relations, the principle of reciprocity states that benefits or penalties granted by one state to the citizens of another should be returned. Foreign authors, musicians, and artists are able to copyright their work, for instance, and nations where piracy is rampant and is not prosecuted by the local authorities come under criticism for failing to uphold the institutionalized sense of trust that underlies international reciprocity.

The Polluter Pays Principle

The polluter pays principle says that the party responsible for producing pollution is responsible for remedying the damage done to the environment. This is a common notion in international environmental law, but is not universal. It is mentioned in the Rio Declaration on Environment and Development and is enacted in the European Union and the Organisation for Economic Co-operation and Development (OECD) countries, having been described by the Swedish government in 1975 as "extended polluter responsibility." Ideally, this is true not only of environmental accidents but of all environmental consequences: the OECD describes extended polluter responsibility as "a concept [whereby] manufacturers . . . should bear a significant degree of responsibility for the environmental impacts of their products throughout the product life-cycle, including upstream impacts inherent in the selection of materials for the products, impacts from manufacturers' production process itself, and downstream impacts from the use and disposal of the products." It is not enough to avoid or remediate pollution caused by the manufacture of a product, if that product winds up in a landfill where it leaches toxins into the soil.

Much of the motive behind enforcing the polluter pays principle comes from the desire not only to attach consequences to environmentally harmful behavior but to shift the cost of consequences away from taxpayers. In the United States, polluter-pays laws include the Superfund law enacted in 1980, which requires responsible parties to clean up environmental disasters and the gas-guzzler taxes established in the 1978 Energy Act to penalize fuel-inefficient vehicles. However, domestic polluter-pays laws are significantly more limited than international ones and are generally more favorable to large corporations than are the European analogues. Ecotaxes like the gas-guzzler tax are generally rather disproportionate in their ratio of environmental impact to revenue collected, for instance, and there are limits in place on how much a polluter can be penalized, especially if no negligence can be proved. Many cases of pollution and mishandling of toxins are resolved in civil cases like class-action lawsuits, where again the amount collected from the polluter may be wildly out of proportion to the actual cost of repairing damage. Some of these differences simply reflect the differences between U.S. and European legal traditions; others reflect the influence of lobbyists in the United States.

The Precautionary Principle

The precautionary principle states that when considering an action suspected of causing public or environmental harm with no scientific consensus affirming that harm, the burden of proof falls on the proponents. In the strongest formulation of this principle, only those actions known to not be harmful should be taken: there is no benefit of the doubt when there is a possibility of harmful consequence, and the economic costs of regulation are irrelevant. The history of DDT, PCBs, aerosols, and anthropogenic greenhouse gases, all of which had unforeseen consequences, informs the precautionary principle, particularly its strong formulation. The weaker formulation says that precautions should be taken even when the scientific evidence for the risk or harmful consequences is not conclusive. The weak formulation is often compared to the use of seatbelts: a passenger does not know that he will be in an accident, but knows that if he is, using a seatbelt will reduce the severity of the consequences. The fact that there is uncertainty about the possibility of the accident does not lead to the decision to stay out of the car, only to mitigate the consequences of the risk.

Consequences that should be considered under the precautionary principle include contributions to climate change, the introduction of harmful substances into the environment, threats to public health, species extinction, new biosafety issues (such as the creation of new bioactive substances), pollution, and the depletion of important resources. In U.S. politics, it is also a principle that has been invoked in opposition to social change.

For most of the world, the precautionary principle is simply a guiding principle in the formation or advocacy of policy. Typically, a cost-benefit analysis of possible risks and consequences against the possible costs associated with guarding against those risks, or avoiding the action entirely, informs the final decision. For a political body, those costs may be political (the possibility of failing to be reelected, political consequences from other entities) rather than financial. In recent years, the principle has particularly been advocated in the regulation of nanotechnology, the environmental and health consequences of which are not yet fully understood. In the European Union (EU), the precautionary principle is a statutory requirement. The European Commission adopted it in 2000, and it has impacted European Union law in international environmental policy, food safety, consumer protection, and technological development. In many of these areas, the EU has been criticized—especially by the industries in countries most affected by an abundance of precaution—for erring too far on the side of caution.

Intergenerational Equity

The idea of intergenerational equity is especially key in international environmental law, as it refers to a need for fairness in relationships between generations—and the need, thus, for the present generation to avoid depleting resources that will be needed by future generations. Putting a value on intergenerational equity avoids situations in which the present generation permits environmental harm so long as its effects are slow enough to be felt that they will never experience the consequences—which history would suggest is the normal human tendency. The concept of seven-generation sustainability is a strong formulation of intergenerational equity: inspired by Iroquois law, seven-generation sustainability calls for considering the impact of an action on the seventh generation to come.

The Common Heritage of Mankind

As international law pertaining to outer space was formulated in the 1960s, the phrase *the common heritage of mankind* was coined in the 1967 United Nations Outer Space Treaty, in reference to those natural territories and cultural elements that are the common property of all mankind and are in need of protection from exploitation. "Outer space shall be free for exploration and use by all states without discrimination of any kind," the treaty stipulates in its first article, "on a basis of equality and in accordance with international law, and there shall be free access to all areas of celestial bodies. There shall be freedom of scientific investigation in outer space." The later Moon Treaty specifically states, "The Moon and its natural resources are the common heritage of mankind." The common heritage of mankind further includes the atmosphere and seas, the continuing environmental health of Earth, and the human genome, and acting in accordance with the principle is a remedy to the dilemma of the tragedy of the commons described in Garrett Hardin's 1968 *Science* article. In the tragedy of the commons, individuals acting from rational self-interest can destroy a shared resource even when each knows that it is to no one's long-term benefit to do so.

Sources of International Environmental Law

Sources of international environmental law include treaties, customary law, and judicial decisions. Customary international law derives from national customs. Judicial decisions are those made by international courts like the International Court of Justice, the European Court of Justice, and the World Trade Organization's Dispute Settlement Board; they are infrequent but important in the development of international law. In environmental matters, nuclear weapons testing cases have been especially formative.

Treaties can be multilateral (many parties) or bilateral (two parties) and typically deal with specific environmental issues such as the treatment of the Arctic or of particular endangered species. Protocols are supplements to existing treaties, adding provisions or adjusting previous agreements because of later developments.

The Kyoto Protocol

The United Nations Framework Convention on Climate Change, produced at the Earth Summit in Rio de Janeiro in 1992, set forth a framework to develop binding limits on greenhouse gas emissions. The subsequent Kyoto Protocol was adopted in 1997 in Kyoto, Japan, and entered into force in February 2005.

Under the protocol, the "Annex I countries" (the European Union and 39 industrialized countries) have committed to a reduction of greenhouse gas emissions of 5.2 percent from the 1990 levels, not including those produced by international aviation and shipping. Relevant gases to the protocol are carbon dioxide, methane, nitrous oxide, sulfur hexafluoride, hydrofluorocarbons, and perfluorocarbons; chlorofluorocarbons are dealt with separately under the earlier ozone-protecting 1987 Montreal Protocol.

The Kyoto Protocol is the key element shaping the current legal and business environment of greenhouse gas emissions, having set up the emissions trading market, the Clean

Development Mechanism, and Joint Implementation as "flexible mechanisms" by which Annex I countries can reduce their effective emissions without reducing their actual emissions. All three flexible mechanisms work on the same essential principle: that the need to reduce emissions by X amount represents a debt of X that can be transferred to and paid by another entity. For instance, as a result of the protocol, emissions limits have been placed on emitters (mostly corporations) in the various participating countries. Under various emissions trading schemes, an emitter that will exceed its limit by X can purchase credit in the amount of X from an emitter that is below its limit by X. The total amount of emissions is therefore the same. It becomes more complicated when carbon offsets are included. Allowing countries and emitters to count carbon offsets like reforestation as "negative emissions" is controversial (and disallowed by the European Union, for the time being) because it is unclear whether the long-term benefits are truly equivalent.

See Also: Antiglobalization Movement; Antiwar Actions/Movement; Biodiversity Loss/Species Extinction; Communication, Global and Regional; Ecotourism; Energy; Environmental Law, U.S.; Global Warming and Culture; Human Geography; Individual Action, Global and Regional; Kyoto Protocol; Law and Culture; Social Action, National and Local.

Further Readings

Bodansky, Daniel. *The Art and Craft of International Environmental Law*. Cambridge, MA: Harvard University Press, 2009.

Houck, Oliver A. *Taking Back Eden: Eight Environmental Cases That Changed the World*. Washington, DC: Island Press, 2009.

Kubasek, Nancy K. and Gary S. Silverman. *Environmental Law*. Upper Saddle River, NJ: Prentice Hall, 2010.

Bill Kte'pi
Independent Scholar

Internet, Advertising, and Marketing

In 2006, a Gallup poll reported that 77 percent of Americans were concerned about the environment and felt a need to get involved. Many look for environmentally friendly goods and services. Properly educated consumers will pay more for ecologically friendly products; a recent survey indicated that about 10 percent were willing to pay a bit more for "green." Consumer education is commonly a part of green marketing strategies.

After the United Nations (UN) Climate Change Conference in Copenhagen in December 2009, businesses and corporations sought methods to continue marketing and advertising while reducing their carbon footprint in an effort to become more energy efficient and pollute less. After production, traditional advertising and marketing were the most energy-consuming activities, so a shift is being made to use sustainable sources or digital methods for advertising and marketing.

Green marketing reduces reliance on non-ecosensitive media and maximizes organic marketing, pay-per-click searches, mobile marketing, and social media marketing. Internet marketing uses a currently existing medium to reach large numbers of consumers without creating disposable, hard waste products. It reduces carbon emissions, energy consumption, and garbage. Green marketing may include promoting a company's pollution reduction initiatives, recycling programs, use of new technology for greater energy efficiency or conservation, fair trade and labor practices, virtual meetings, purchase of carbon offsets, and use of a paperless office. Green marketing may also include using soy-based ink and recycled paper for packaging and marketing materials.

Traditional print marketing uses large amounts of paper, requires heavy machinery to print and transport, and generates trash. Televisions require a heavy use of energy during production and transmission, and creating television advertising consumes energy as well. Transmitters and other television equipment also require a good deal of electricity and fossil fuels. Traditional consumers also burn energy traveling to shops and showrooms.

Green marketers sell their product or service in ways that take into consideration the impact on the environment. Whenever possible, these marketers use Internet-based marketing and advertising. Strategies that minimize ecological damage include reducing the use of billboards, magazines, newspapers, flyers, business and greeting cards, and the like.

Green marketing online saves money and environmental damage by eliminating flyers, CDs, videos, presentations, and catalogs. Electronic marketing allows frequent updates without the cost of printing, and is postage free. Green catalogs are available for viewing and downloading whenever needed and at the consumer's expense if printed.

Rather than advertising in print magazines, green marketers often prefer placing ads on search engines. Using Google AdWords, green businesses can reach a targeted market. Online billing via e-mail saves on printing, mailing, and faxing. Internet advertising is affordable because charges are based on actual consumer clicks on the advertisement. Search engines have tools that allow marketers to select keywords that home in on desired customers. Banner ads in online versions of print magazines are also cost effective. There are also online press release distribution services that offer fast, postage- and paper-free marketing, and allow press kits to be directed quickly to publishers and journalists. E-mail also is effective for faster business-to-consumer communication, and is capable of verifying that a customer has received information and when.

Green Internet marketing is time independent and silent, and chat rooms and social media allow for the instant spread of word-of-mouth advertising. Newsletters on Facebook and other social networking sites offer timely interaction and instant customer feedback.

Rules of Internet Advertising and Marketing

Green industries include recyclers, builders, manufacturers of ecofriendly products (such as organic cosmetics), green energy suppliers, and manufacturers of green technologies. Green marketing makes consumers aware of the availability and positive aspects of these industries and the ways that they impact the environment and the future of humanity.

A responsible green marketer learns the business, products, or services necessary to create an ethical and accurate marketing campaign matching the individual firm's needs and goals. This analysis includes research on the business; its website, competition, and niche; and the business-specific rules, regulations, and licensing requirements. The market strategy that arises from this effort includes Internet marketing for those who want a stronger online presence and have a desire to make the public more aware of what they offer and the green movement in general.

A green business complies with specific rules, guidelines, regulations, and licensing. These rules are available at http://www.business.gov and cover regulations regarding environmental claims, marketing and advertising, organic versus made with organic ingredients, labeling rules, and so on. Companies need to be aware of these rules and abide by them, sometimes assisted by green marketing advisers.

Electronic marketing is covered by the same rules that apply to other forms of advertising and marketing. Above all, the information must be truthful, substantiated, and not misleading to consumers, particularly if health and safety or performance are involved. A deception may be one of representation, omission, or practice, and deception rests on a tendency to mislead or affect decision making, or if it may cause substantial harm greater than other benefits.

Guides published by the Federal Trade Commission (FTC) cover labeling, advertising, promotions, and all other marketing, either direct or implied, whether by word, logo, symbol, brand name or any other means, including digital and electronic means (e.g., Internet and e-mail). FTC regulations cover claims about the environmental attributes of a marketed item in the context of advertising, marketing, selling, or offering for personal, family, commercial, institutional, or industrial use. The FTC guides are not enforceable, and federal, state, and local agencies have precedence in legal action, but the FTC can bring action in cases of deceptive acts or practices.

Liability for claims lies with the seller, but liability for dissemination of the claims may rest with the ad agency or web designer if they help to develop the claims or are aware that they are deceptive. Agencies and designers have an obligation to check the claims themselves; it is not sufficient to rely on the client's claims. They should ask for proof and, if the proof is missing or weak, proceed with caution in creating the marketing copy or site. Marketers are responsible for clear and conspicuous disclaimers, even though disclaimers alone are normally insufficient to remedy false claims. Section 260.5 states that any claim, express or implied, to environmental attributes must have a reasonable basis, defined as "competent and reliable evidence." This evidence may include research, test results, and objective studies by professionals in the relevant area using professional standards. The criteria include clarity of language, including font size, indication of which part(s) of the product or packaging are environmentally friendly (if it is the package only, the marketing must indicate this clearly), that claims cannot be exaggerated, and that comparisons must be verifiable and accurate.

The FTC specifies what is permissible and provides examples to aid in interpreting its guidelines. It does not offer a complete set of what is legitimate or not, merely representative samples. General environmental benefit claims are impermissible because they are ambiguous, hard to document or interpret, and generally misleading. For instance, "eco-safe" may be a misleading brand name if consumers interpret it as a product with environmental benefits that are not necessarily there. The manufacturer needs to clarify that no environmental claim is being made. Further, containers may be environmentally friendly while their contents are not. Similarly, compostable, recyclable, and degradable (including bio- and photodegradable) require qualification. These rules generally apply to hard products rather than to the Internet, but Internet marketing must conform to ensure that claims about hard products are truthful and not misleading.

Other regulated marketing practices applicable to the Internet as well as to other media include 10-day-advance disclosure by franchise and business opportunity sellers, mid-level marketing and pyramid schemes, credit and financial disclosure and billing laws, and electronic fund transfer. Environmental misrepresentation is also included, and marketers are cautioned to qualify or delete broad claims, to avoid claims that are in fact environmentally insignificant, such as labeling a trash bag recyclable while knowing that the bag will end up, along with its contents, at the local landfill.

Internet marketers need to be aware that advertising to children needs to be done with special care because children do not have the ability to evaluate claims and the purposes of advertising. The Children's Online Privacy Protection Act took effect in 2000. It required that general websites or those targeting children under 13 may not knowingly collect information from children without parental consent. At http://www.onguardonline.gov/topics/kids-privacy.aspx, parents and website operators can find an explanation of the law.

One deterrent to broader acceptance of the Internet as a medium for advertising and marketing is the concern about identity theft and other consumer risks. Advertisers in violation of FTC rules can face cease and desist orders or court injunctions with fines up to $16,000 per violation and potential consumer litigation. The Consumer Sentinel Network is a secure online database of complaints available to civil and criminal law enforcement agencies in the United States and elsewhere.

Even so, net-based marketers are slow to provide what consumers want. There are rules about consumer privacy, given the incredible and unprecedented capacity of the Internet for sharing and collecting information. Marketing privacy is good business because it reassures consumers unwilling to purchase online because of doubts about the security of their personal information. As early as 1998, the FTC reported that 85 percent of sites sampled collected personal information, but only 14 percent notified consumers. A 2000 report showed progress in disclosing privacy policies; only 20 percent of the sites used the four fairness practices—notice, choice, access, and security. Only 41 percent provided notice and choice.

Green Servers

Although an improvement over traditional paper marketing, Internet marketing does have a carbon footprint. One estimate reports that 868 billion kilowatt hours (kwh) of energy a year are required to run the Internet and its related infrastructure. Servers use 112.5 billion of these hours. At least one company is offering a green alternative to servers burning electricity and indirectly contributing to carbon dioxide pollution.

Solar Energy Host advertises a carbon-free server, which is an even better option than being carbon-neutral, which entails buying carbon credits to neutralize a site's pollution. If the Internet contained 108.8 million sites as of 2007, then each needs credits for 1,020 kwh each year. Solar Energy Host claims to prevent 20,000 pounds of pollutants a year. Burnishing its green image, it notes its Gardens of Hope Project, where $5 from each account is given toward planting one tree and linking five communities in Lesotho, Africa, into a network of projects pooling ideas, resources, and skills while aiding orphan education. Solar Energy Host says it is perhaps the only commercial carbon-free data center in the world. The company states that although it has batteries, propane, and ties to the grid, it has never needed to use the grid.

Conclusion

Traditional marketing sources are energy inefficient and ecologically harmful. Green marketing is a solution that allows marketing to continue but in the context of ecological virtue. Wendy Roltgen stresses that green marketing has to be real, engaging, and informative if it is going to work. It has to be clear but not overhyped. A company website can provide customers with up-to-date information about new initiatives and products. Trying to actively involve consumers can also be effective, including educating consumers on how to recycle packaging, how to get carbon offsets, and the like. The bottom line is that a consumer concerned about the environment will receive the message that the business is equally concerned and is doing something about it.

In December 2010, the FTC once again proposed that Congress enact a do-not-track law to prevent marketers from using technology to track online purchases and other activities, and from tracking the physical locations of phones and other electronic devices. Almost immediately, opponents warned that such a law would diminish the web experience, and potentially damage the $12 billion online advertising industry. As long as privacy remains an important issue and marketers fail to address consumer concerns, Internet marketing, green or not, will fall short of its potential.

See Also: Advertising; Green Consumerism; Print Media, Advertising; Shopping.

Further Readings

Federal Trade Commission. "Advertising and Marketing on the Internet: Rules of the Road." December 2000. http://business.ftc.gov/documents/bus28-advertising-and-marketing-internet-rules-road (Accessed December 2010).

Federal Trade Commission. "Guides for the Use of Environmental Marketing Claims." http://www.ftc.gov/bcp/grnrule/guides980427.htm (Accessed December 2010).

Raspaile, Sanjeev. "Internet Marketing for Eco-Sustenance." http://www.tutorialspoint.com/white-papers/149.pdf (Accessed December 2010).

Roltgen, Wendy. "How to Benefit From Green Marketing: Highlight Green Business Practices to Attract Customers." Suite 101 (April 5, 2009). http://www.suite101.com/content/how-to-utilize-green-marketing-for-your-business-a107468 (Accessed December 2010).

Roltgen, Wendy. "Online Green Marketing Saves Money and Resources. Sustainable Businesses Use Technology to Communicate With Customers." Suite 101 (March 15, 2010). http://www.suite101.com/content/online-green-marketing-saves-money-and-resources-a213825 (Accessed December 2010).

Solarenergyhost. "How Green Is the Internet?" http://www.solarenergyhost.com/Why-Choose-Solar-Energy-Host/how-green-is-the-internet.html (Accessed December 2010).

Tessler, Joelle. "FTC Proposes 'Do Not Track' Tool for Web Marketing." Associated Press (December 1, 2010). http://news.yahoo.com/s/ap/20101201/ap_on_hi_te/us_tec_ftc_do_not_track (Accessed December 2010).

John H. Barnhill
Independent Scholar

K

Kyoto Protocol

Though many special interest corporations and trade associations opposed the Kyoto Protocol, which targeted reduction of six greenhouse gases, others like BP and Dutch Royal/Shell left the Global Climate Coalition and began vocally supporting greenhouse gas regulation.

Source: Wikimedia Commons

Following an intense heat wave and drought in North America in summer 1988, the United Nations (UN) and World Meteorological Organization (WMO) created the Intergovernmental Panel on Climate Change (IPCC) to evaluate scientific evidence that humans were causing the world climate to become warmer. Acting on behalf of the IPCC's recommendations, policy makers from the world's nations worked for several years on a treaty that would require industrialized nations to implement binding commitments curbing the greenhouse gases associated with climate change. The call to reduce global warming was driven by the growing scientific consensus that the Earth was warming due to human industrial activity that could eventually cause glaciers to melt, sea levels to rise, displacement of people, and numerous other problems. Although the consequences of inaction are dire, regulating greenhouse gases faced fierce opposition from many corporations, think tanks, and front groups. Nevertheless, in December 1997, an agreement, albeit limited, known as the Kyoto Protocol, was reached in Kyoto, Japan.

Opposition to Kyoto

The global consensus surrounding ending global warming has not been without its detractors. Opposition to greenhouse gas regulations coalesced after the IPCC was formed, when several corporations and trade associations formed the Global Climate Coalition (GCC) in 1989. Fossil fuel energy and fossil fuel–intensive corporations mobilized in response to a perceived threat that a societal shift from their cheap sources of energy to one based on sustainable energy would affect their profit margins. Other corporations joined the GCC in solidarity against government regulations more generally, especially after witnessing the agreement reached by international governments to regulate chlorofluorocarbons (CFCs), the cause of the ozone hole. GCC members even created political front groups with environmentally friendly–sounding names to help in the efforts to prevent regulations of greenhouse gases. For instance, the Western Fuels Association created a front group called the Greening Earth Society, which used publications to help spread climate change skepticism. Sharon Beder has also noted that several GCC members, such as General Motors, Exxon, and Mobil, were major funders of the Advancement of Sound Science Coalition (TASSC) front group, which lobbied against climate change regulations. Conservative think tanks have also been critical in the movement against climate change regulations by spreading skepticism regarding scientific evidence for climate change. According to Aaron McCright and Riley Dunlap, think tanks have spread their skepticism during congressional hearings and in their own publications, as well as in newspaper articles that used think tank representatives as sources in the years leading up to Kyoto.

The efforts of this countermovement paid off when in 1997, the U.S. Senate unanimously passed Senate Resolution 98 notifying the Clinton administration that it would not ratify any treaty that might be reached that December in Kyoto, Japan. But news was not all good for the opposition either, as BP and Dutch Royal/Shell left the GCC and vocally supported regulating greenhouse gases. And despite Senate opposition and the influence of the countermovement, the Clinton administration joined 160 countries in adopting legally binding greenhouse gas reductions on December 11, 1997, in Kyoto.

The Agreement

The Kyoto Protocol created emission reduction targets of six greenhouse gases for each participating industrialized nation for the 2008–12 time frame, most notably carbon dioxide (CO_2). The European Union had the highest required emission reductions (8 percent) under this treaty, followed by the United States (7 percent), Japan (6 percent), and Canada (6 percent). Russia negotiated a 0 percent reduction, in other words, no increase in emissions, and some other industrialized countries such as Australia were allowed an increase in emissions. Since the treaty specifies emission regulations through 2012, it remains to be seen if the ratifying countries will meet their obligations. However, if they do, it is projected that the industrialized countries will reduce emissions a little over 5 percent below 1990 greenhouse gas emission levels.

Additionally, it wasn't until 2005 (after Russia ratified the Kyoto Protocol) that the treaty went into effect. This was due to the stipulation that ratification must include enough of the industrialized countries, otherwise known as Annex I Parties, responsible for over 55 percent of the total 1990 greenhouse gas emissions. Although an agreement was reached and has gone into effect, the Kyoto Protocol has some limitations.

Limitations of the Kyoto Protocol

The most glaring limitation of the Kyoto Protocol is that it exempts developing countries from making meaningful greenhouse gas reduction commitments. The vast majority of developing nations have ratified the treaty, but it was easy to do as it only required industrialized nations to reduce greenhouse gas emissions. While this wasn't as problematic at the signing of the treaty in 1997, now that China is the largest emitter of greenhouse gases, and other developing countries like India and Brazil contribute significant amounts of greenhouse gases, it is now imperative that a treaty intended to reduce global warming include developing nations.

Weaknesses in the Kyoto Protocol are not just due to developing nations, but also to the lack of participation from the second-largest emitter of greenhouse gases, the United States. While the Clinton administration did sign the Kyoto Protocol, due to the Senate's resolution unanimously opposing the treaty, it was never brought to the Senate for ratification. Furthermore, shortly after George W. Bush became president, he reversed his campaign pledge to comply with the Kyoto Protocol and pulled the United States completely from participating.

Finally, it was not specified how countries would meet their commitments in the Kyoto Protocol, but was rather left for later negotiations. Further climate negotiations during the three years following the signing of the Kyoto Protocol therefore centered on emission credits and financing of emission reductions. Global warming is also a problem that requires action for decades to come; the Kyoto Protocol is only mandated for the management of emissions through 2012.

Conclusion

In the face of intense opposition, a treaty at Kyoto in the winter of 1997 was agreed upon; however, further work remains. The Kyoto Protocol has reduced greenhouse gas emissions in some countries; however, without the participation of the United States and developing nations, the net total of greenhouse gas emissions will continue to rise. The countermovement to greenhouse gas reductions has caused a culture of global warming skepticism in the United States that, if not overcome, may result in irreversible consequences that threaten human life as we know it.

See Also: Corporate Green Culture; Global Warming and Culture; Nongovernmental Organizations.

Further Readings

Beder, Sharon. *Global Spin: The Corporate Assault on Environmentalism*. Devon, UK: Green Books, 2002.

Dessler, Andrew E. and Edward A. Parson. *The Science and Politics of Global Climate Change: A Guide to the Debate*. New York: Cambridge University Press, 2006.

Fisher, Dana R. *National Governance and the Global Climate Change Regime*. Lanham, MD: Rowman & Littlefield, 2004.

Layzer, Judith A. "Deep Freeze: How Business Has Shaped the Global Warming Debate in Congress." In *Business and Environmental Policy: Corporate Interests in the American Political System*, Michael E. Kraft and Sheldon Kamieniecki, eds. Cambridge, MA: MIT Press, 2007.

McCright, Aaron M. and Riley E. Dunlap. "Challenging Global Warming as a Social Problem: An Analysis of the Conservative Movement's Counter-Claims." *Social Problems*, 47/4 (2000).

McCright, Aaron M. and Riley E. Dunlap. "Defeating Kyoto: The Conservative Movement's Impact on U.S. Climate Change Policy." *Social Problems*, 50/3 (2003).

Newell, Peter. *Climate for Change: Non-State Actors and the Global Politics of the Greenhouse*. New York: Cambridge University Press, 2000.

James Everett Hein
The Ohio State University

Law and Culture

The history of public environmental awareness and the passage of environmental legislation and regulations reveal a strong correlation. Many countries have passed increasingly comprehensive environmental legislation since the rise of the environmental movement and green culture in the late 20th century. A variety of environmental activists have pushed for environmental legislation at the local, national, and international levels over the years. Results have included increased environmental laws and compliance at the national level and the emergence of international treaties and agencies to work at the global level. These same groups have aided the implementation of such legislation and monitoring for compliance, even suing corporations and government agencies when necessary.

Environmental Legislation

Social and cultural environmental awareness increased in the post–World War II period due to a number of factors. Continued population growth and urbanization led to the realization that natural areas and natural resources, including clean air and potable water, were not inexhaustible. Continued industrialization, the growth of disposable products and throwaway packaging, and fertilizer and pesticide use were accompanied by increased pollution levels, health risks, and garbage problems. Many people became aware that their modern lifestyles were threatening the environment, which in turn threatened their health and future. The United States held its first Earth Day on April 22, 1970, galvanizing public and governmental support for environmental legislation. Books such as Rachel Carson's *Silent Spring* further increased public awareness of environmental concerns and their potential impact on the sustainability of life. The result of this social and cultural shift toward environmental preservation included increased demand for environmental legislation.

Individual nations' environmental legislation often features multiple layers of protection at different governmental levels. For example, in the United States, the federal Environmental Protection Agency (EPA), established in 1970, oversees environmental legislation and regulation, but states and municipalities also have environmental agencies and regulations. State legislation, however, must meet or exceed federal legislation. Environmental legislation is also amended over time to strengthen or reduce regulation or add new provisions and address new environmental threats and sustainability needs. Countries without strong green culture movements, such as Mexico and Russia, often lag behind in terms of environmental legislation, regulation, and enforcement.

The U.S. Wilderness Act of 1964 created the National Wilderness Preservation System to preserve publicly owned wilderness areas and to establish a system of evaluation for the addition of future congressionally designated wilderness lands. Wilderness areas included national forests, parks, and wildlife refuges, and were for the most part to be a minimum of 5,000 acres. The administering agency for each designated wilderness area is responsible for its care and the maintenance of its wilderness nature. Provisions prevented commercial enterprises, permanent roads, and vehicles, aircraft, or motorboats, with certain exceptions for administration, safety, or previous allowance of such activity. Information on minerals and natural resources may be obtained as long as the area's wilderness character is preserved. The National Wilderness Preservation System came to consist of over 100 million acres and set a precedent for the permanent preservation of natural areas.

The U.S. federal government created the Environmental Protection Agency (EPA) in 1970 through the reorganization and consolidation of various environmental departments to create a comprehensive national approach to environmental issues. The U.S. government also passed the International Environmental Protection Act of 1983, which was amended in 1986 and 1989 to authorize U.S. aid to other countries in conservation and environmental protection though the Agency for International Development (AID) in accordance with the World Conservation Strategy. In addition, the act fosters participation by local, national, regional, and international nongovernmental organizations (NGOs). The act is designed to ensure that U.S. international development assistance fosters sustainable development through the prevention of habitat and species loss, pollution, and other forms of environmental destruction. The International Environmental Protection Act demonstrated an increasing realization that environmental destruction within any country had global consequences. Many other developed countries have passed similar laws.

Many nations have also committed to international treaties, agreements, and agencies focusing on sustainability and other global environmental issues. Both environmental activists and governments expanded earlier localized approaches as the global impact of environmental problems became increasingly apparent. Some countries, however, balk at surrendering their individual autonomy to international authorities or treaties. Examples of international agreements include the Basel Convention on the Control of Transboundary Movements of Hazardous Wastes and their Disposal, the Environment and Security (ENVSEC) Initiative, the Intergovernmental Panel on Climate Change, the Convention on Wetlands of International Importance (Ramsar Convention), the Vienna Convention for the Protection of the Ozone Layer, and the Convention on Biological Diversity (CBD).

The United Nations (UN) maintains the UN Environment Programme (UNEP) and the UN Framework Convention on Climate Change (UNFCCC) and its added Kyoto Protocol. European nations work together through the European Commission's Environment Directorate-General and the European Commissioner for the Environment as well as abide by European Union environmental laws as outlined in its Environment 2010: Our Future, Our Choice action program and an Environment and Health Action Plan for 2004–2010. Canada, Mexico, and the United States work together through the North American Agreement on Environmental Cooperation (NAAEC), a side agreement of the North American Free Trade Agreement (NAFTA).

Nations and environmental activists also work through public and private nongovernmental organizations (NGOs) at the local, national, and international levels. Examples of such agencies include the Canadian Environmental Law Association (CELA), the Environmental Law Alliance Worldwide, the Environmental Law Foundation (ELF), and the Center for International Environmental Law (CIEL). Goals of international legislation, agreements, and agencies include using and reforming existing laws and passing new legislation in the interest of environmental protection and sustainability; the provision of expert legal advice and resources to governments, the public, and other agencies working toward environmental reform; increased awareness of the global implications of environmental issues; and the establishment of target accomplishments in areas such as carbon emission reductions.

A key area of environmental legislation and the green culture movement has been energy production and consumption. Since the development of nuclear technology and the United States' use of the first atomic weapons against Japan at the end of World War II in 1945, legislation to prevent or minimize the environmental dangers of nuclear energy production and weapons use has been a primary goal of many environmental activists. Goals include the development and regulation of nuclear facilities; the provision of government licensing requirements for civilian use of nuclear materials and facilities; the establishment of government agencies to monitor compliance with government regulations; the outlining of legal uses for atomic energy; the proper and regulated disposal of nuclear waste; the cleanup of old nuclear production, testing, and disposal sites; and the promotion of world peace.

Environmentally friendly energy laws have also focused on the reduction of human dependence on fossil fuels and limiting their damage to the environment. Key concerns include the damage to waters and wildlife due to massive oil spills; the encouragement of energy independence; energy security and supply issues; the development of alternative, clean, and renewable energy sources; the improvement of vehicle fuel economy; the development of alternate and mass transportation; increased energy efficiency; green building technologies; and the debate over whether drilling rights should be expanded into environmentally sensitive areas such as Alaska and the Arctic National Wildlife Refuge and along Florida and the coastlines of other U.S. Gulf states.

Many nations have enacted clear air, water, and soil legislation designed to reduce pollution and contamination. Clean air, water, and soil legislation often establishes quality-level standards and targets for improvement as well as acceptable thresholds for the emissions of polluting substances into the environment. Targets have included industrial and vehicle emissions; the use of products harmful to the ozone layer; greenhouse gas emissions; the burial or other disposal of toxic wastes on land and in rivers, oceans, and other waterways; pesticide use; chemical runoffs; acid rain; smog; mining damages; industrial discharges into the water; and potential climate change and global warming.

Environmental changes advocated by green culture activists and enforced through environmental legislation have included new production and operation methods, changes in product packaging, and changes in the use of raw materials. Other examples include the requirement of obtaining government permits for disposal or emissions and government limits on acceptable methods and amounts. Environmental protection, conservation, and sustainability are key goals of such legislative changes. Environmental activists and governments also realize that environmental legislation must be cost effective to appeal to corporations. These diverse groups can differ, however, on which factors to include and how to interpret the results of such cost-benefit analyses.

Some groups argue that stringent environmental legislation can hinder corporate free market activity and can impinge on private property rights. Others argue that environmental legislation forces businesses to adapt to meet the new regulations and maximize profitability and can actually result in technological advances, sometimes in partnership with governments. Activists also debate whether environmental regulations should be waived for certain groups such as the military if national defense is at stake. Environmental regulations and considerations are frequently ignored during times of war or national emergency. Another key argument is centered on the question of who pays the cost of environmental policy enforcement. For example, will businesses simply pass on higher costs of production and distribution to consumers?

Environmental legislation protects nature, from threatened or endangered plants and animal species to entire threatened habitats and ecosystems. Regulations include limiting or forbidding the hunting or commercial use of endangered species, the requirement of environmental impact studies before development projects are approved, and restrictions on land use and development. In areas where hunting or fishing is allowed, government licenses are often required, with portions of the accompanying fees used to support environmental protection and conservation efforts.

Role of Green Activists in Legislation

Environmental activists' crucial roles in the fight to enact, implement, and enforce environmental legislation have become increasingly visible with the rise of the green culture movement in the mid- to late 20th century. Early advocates of environmental legislation included John James Audubon, Sierra Club founder John Muir, Gifford Pinchot, President Theodore Roosevelt, Aldo Leopold, and Rachel Carson. Many environmental historians mark the 1962 publication of Rachel Carson's *Silent Spring*, a landmark warning about the effects of pesticides such as DDT, as the beginning of the modern environmental law movement. Public reaction helped spur the passage of U.S. environmental legislation, including the Clean Air Act, Clean Water Act, and Safe Drinking Water Act.

Major environmental disasters also spur public demand for environmental legislation. Key examples have included the 1934 Dust Bowl in the Midwestern United States; the smog tragedies in Donora, Pennsylvania, in 1948, and London, England, in 1952; large oil spills such as the *Torrey Canyon* off the British coast, the *Exxon Valdez* off the Alaska coast, and the BP oil spill off the U.S. Gulf Coast; the toxic contamination and health impacts found in New York's Love Canal neighborhood in the 1960s and 1970s; the accidental release of toxic gases by Union Carbide in Bhopal, India; and the nuclear accidents at Three Mile Island in the United States in 1979 and Chernobyl, Russia, in 1986.

International environmental groups include the Bellona Foundation, Biofuelwatch, Conservation International, the Conservation Law Foundation, the Earth Charter Initiative, the Earth Policy Institute, the Environmental Foundation for Africa, the Environmental Investigation Agency, the Forests and the European Union Resource Network, Friends of Nature, Friends of the Earth International, Greenpeace International, the International Environmental House, the International Institute for Sustainable Development, the International Union for Conservation of Nature, The Nature Conservancy, NatureServe, the Nicodemus Wilderness Project, the Stockholm Environment Institute, the World Business Council for Sustainable Development, the World Resources Institute, and the Worldwatch Institute.

By the start of the 21st century, many green culture activists believed that government legislation was still not effective enough in the areas of environmental protection and sustainable development. They also protested the lessened strength or enforcement of environmental legislation that sometimes occurred as the result of administration changes, economic troubles, periods of war or conflict, or other issues that deflect attention from environmental issues. Some green activists have taken political participation a step further, creating political parties such as the European Green Party to elect leaders dedicated to the environment.

Tactics used by green culture activists to seek legislative changes include targeting governmental officials and bodies as well as public awareness. They organize letter-writing campaigns, testify before governmental committees, meet with legislators, and hire government lobbyists to represent their interests against those of the corporations or other groups that often oppose environmental legislation. Green activists are also encouraging the recent movement among many corporate investors and stockholders to ensure that the corporations in which they are invested are in compliance with environmental legislation and maintain sustainable practices. Green investors also support government policies that allow the buying and selling of environmental resources such as energy certificates.

Public relations campaigns include seeking the greatest amount of media exposure possible or the organization of marches and rallies targeting local, national, and international governments, agencies, or conventions and summit meetings to help put the weight of public support behind their campaigns. Environmental groups and activists also work with and provide guidelines for individuals or communities that wish to become involved in the legislative process, often as a result of an individual environmental issue within their community. They encourage individuals to practice environmental citizenship and stewardship, in part through the legislative process.

Local movements to prevent environmental threats have become so commonplace they have earned the popular nickname NIMBY, which stands for "not in my backyard." For example, citizens of the U.S. city of Jackson Hole, Wyoming, successfully launched a nationwide campaign to prevent the U.S. Department of Energy from establishing a nuclear waste incinerator 90 miles outside their city. The group helped raise approximately $1 million for the organization Keep Yellowstone Nuclear Free, forcing the Department of Energy to abandon its plans to avoid potential lengthy legal battles and delays.

Recent environmental issues that have attracted the attention of environmental activists in the 21st century have included possible climate change such as global warming, reliance on fossil fuels, the emergence of new diseases as man encroaches on new habitats, massive deforestation and desertification, growing lists of endangered species, the need to develop a green energy economy and lifestyle, and the harmful impacts of conflicts such as Iraq and Afghanistan on the environment. Green culture activists seek legislation that forces reductions in carbon emissions or encourages the development of public transportation and is used as a key tool in their efforts to reduce these threats to the environment's sustainability.

Green culture activists also play a prominent role in the implementation and enforcement of environmental legislation once it is enacted. Many activists believe that aggressive monitoring is necessary to ensure that laws are being correctly enforced and that violators are punished. Activists realize that national agencies such as the EPA are often underfunded and understaffed and overwhelmed by excessive workloads and deadlines. They also battle against the tendency to push environmental issues to the background in times of economic or political crisis or conflict.

Green activist groups can play a key role through the preparation of guidelines for legal compliance for corporations and the general public and through monitoring compliance and looking for and reporting violations. Green culture activists also frequently use the law as one of their most effective tools in their movement for environmental protection and sustainability, sometimes suing corporations or government agencies to force compliance.

For example, a 2001 case involved the four environmental groups Friends of the Earth, the Environmental Working Group, the Pesticide Action Network, and Pesticide Watch, which sued the state of California's Department of Pesticide Regulation. The four groups successfully claimed that the state agency failed to regulate use of a toxic gas, which can potentially cause nerve damage and birth defects, in violation of the law. Another case occurred at the federal level. The environmental groups Earthjustice, the Center for Biological Diversity, and the Bluewater Network sued 18 U.S. government agencies, including the Departments of the Interior, Commerce, and Energy, for failing to purchase enough alternative-fuel vehicles by the year 2000 to meet the requirements mandated in the 1992 Energy Policy Act.

See Also: Environmental Law, U.S.; Environmental Protection Agency; International Law and Treaties; Kyoto Protocol; Nongovernmental Organizations; Organizations and Unions; United Farm Workers (UFW) and Antipesticide Activities.

Further Readings

Barros, James and Douglas M. Johnston. *The International Law of Pollution*. New York: Free Press 1974.

Cudahy, Richard. "Coming of Age in the Environment." *Environmental Law* (Winter 2000).

Diamond, Jared. *Collapse: How Societies Choose to Fail or Succeed*. New York: Viking, 2005.

Earth Council. http://www.ecouncil.ac.cr (Accessed October 2010).

Earth Day Network. http://www.earthday.net (Accessed October 2010).

"Ecosystems and Human Well-Being: Synthesis." *United Nations Millennium Ecosystem Assessment Report*. Washington, DC: Island Press, 2005.

Environmental Law Institute. http://www.eli.org/About/index.cfm (Accessed October 2010).

Foster, Kenneth R. and David E. Bernstein. *Phantom Risk: Scientific Inference and the Law*. Cambridge, MA: MIT Press, 1993.

Freeman, A. Myrick, Robert H. Haveman, and Allen V. Kneese. *The Economics of Environmental Policy*. Hoboken, NJ: Wiley, 1973.

Gilpin, Alan. *Dictionary of Environmental Law*. Northampton, MA: Edward Elgar, 2000.

Green Party of the United States. http://www.gp.org (Accessed October 2010).

HG.org Worldwide Legal Directories. "Environmental Law: Guide to Environmental and Natural Resources Law." http://www.hg.org/environ.html (Accessed October 2010).

Hurrell, Andrew and Benedict Kingsbury. *The International Politics of the Environment: Actors, Interests, and Institutions*. New York: Oxford University Press, 1992.

International Institute for Sustainable Development. http://www.iisd.org/default.asp (Accessed October 2010).

McNeill, J. R. *Something New Under the Sun: An Environmental History of the Twentieth-Century World*. New York: W. W. Norton, 2000.

Mowrey, Marc and Tim Redmond. *Not in Our Backyard: The People and Events That Shaped America's Modern Environmental Movement*. New York: W. Morrow, 1993.

Natural Resources Defense Council. http://www.nrdc.org (Accessed October 2010).

Patton-Hulce, Vicki. *Environment and the Law: A Dictionary*. Santa Barbara, CA: ABC-CLIO, 1995.

Portney, Paul R. and Roger C. Dowger. *Public Policies for Environmental Protection.* Washington, DC: Resources for the Future, 1990.
Sagoff, Mark. *The Economy of the Earth: Philosophy, Law, and the Environment.* New York: Cambridge University Press, 1988.
Sax, Joseph L. *Defending the Environment: A Strategy for Citizen Action.* New York: Knopf, 1971.
Stone, Christopher D. *The Gnat Is Older Than Man: Global Environment and Human Agenda.* Princeton, NJ: Princeton University Press, 1993.
Stone, Christopher D. and Garret James Hardin. *Should Trees Have Standing? Toward Legal Rights for Natural Objects.* Los Altos, CA: W. Kaufmann, 1974.
United Nations Environment Programme. http://www.unep.org (Accessed October 2010).
U.S. Department of Energy. "Energy Technology Engineering Center Closure Project." http://www.etec.energy.gov/ (Accessed October 2010).
U.S. Environmental Protection Agency. http://www.epa.gov/ (Accessed October 2010).
Vig, Norman J. and Regina S. Axelrod. *The Global Environment: Institutions, Law, and Policy.* Washington, DC: CQ Press, 1999.
Vogel, David. *Trading Up: Consumer and Environmental Regulation in a Global Economy.* Cambridge, MA: Harvard University Press, 1995.
Wargo, John. *Our Children's Toxic Legacy: How Science and Law Fail to Protect Us From Pesticides.* New Haven, CT: Yale University Press, 1996.
Weinberg, Philip. *Environmental Law: Cases and Materials*, 3rd ed. Lanham, MD: University Press of America, 2006.
Wenz, Peter S. *Environmental Justice.* Albany: State University of New York Press, 1988.
Wilkinson, David. *Environment and Law.* London: Routledge, 2002.

Marcella Bush Trevino
Barry University

Leisure

Leisure can be thought of as time spent as one wishes (i.e., time not spent working or undertaking duties or obligations). At various points during modern history, the concept of leisure has been held up as the prize that would be awarded to members of society as a result of "advances" in such things as technology, work habits, and international relations. Certainly, this promise of increased leisure was true of the Industrial Revolution that took place from roughly the 18th to 19th centuries.

One of the great promises of the Industrial Revolution was that with hard work—and great increases in average wages—the newly minted urban workforce would eventually experience greater quantities of leisure time, or in other words, greater quantities of time when they did not have to work. The disposable income that was increasingly in the pockets of the workers could thus be spent in order to make the most of this leisure time.

Similarly, after World War II, the Allied troops going home were promised that the lives to which they were returning would be increasingly filled with leisure. The military technological boom that had occurred in the name of winning the war would be turned to domestic pursuits and people would be freed from domestic drudgery.

The promise of leisure, highlighted in the historical examples above, has been also been prevalent, more generally, in the discourse of "progress," especially "technological progress." What technology was going to increasingly bring to us, all of us, perhaps more than anything else, was the freeing of time from the tyranny of undesirable domestic tasks such as cooking and cleaning, as well as making general day-to-day tasks such as transportation that much more efficient. And from the mid-19th to the mid-20th centuries, according to Harvard University economist Juliet Schor, there was a decrease in average weekly hours worked (at least in wealthier capitalist countries such as the United States), but more recently, the trend has been reversed.

Schor, author of *The Overworked American: The Unexpected Decline of Leisure,* is the most well-known and respected researcher on the modern-day phenomenon of increasing work and decreasing leisure. Schor's research has shown that while there was a period between the mid-19th and the mid-20th centuries in which average working hours decreased, since then, working hours have increased (e.g., Schor says that between 1969 and 1987, work increased by the equivalent of a month per year) while average leisure time (vacation, holidays, paid leaves) decreased by the equivalent of a month per year.

Schor's work has also highlighted that women have borne the brunt of the increase in work and decrease in leisure, such that while women have increasingly entered the paid workforce, their domestic workload has not changed much (i.e., from just more than twice the men's quantity of domestic work to just less than twice the men's quantity of domestic work).

Within this brief look at how leisure has been steadily decreasing while work has been steadily increasing is the question, "Why have these changes occurred?" Schor's analysis gives some weight to employer demands for increased hours that are placed on employees, but she gives the greatest explanatory credence to what she calls the cycle, or culture, of "work and spend." This cycle, or culture, is based on the encouragement that people receive everywhere in modern, mediated society to "buy more" and "have more" and, although not as explicitly articulated in the messages, "spend more." The cycle, explored by Schor using the analogy of getting stuck in a squirrel's cage, is discussed throughout *The Overworked American.* The promise that lures us into the squirrel's cage is that by having more we will be happier, more content. The reality as shown by numerous studies—research that is nicely overviewed by Tim Kasser in his 2002 book *The High Price of Materialism*—is that the opposite is true. The more intent we are on consumption, and the more we believe that through consumption we will be happier, the less happy we actually are.

So the question we might ask next is, "Why does it matter if people are working more and having less leisure time?" If people are convinced that their lives will be made better through consumption and, as such, they pursue a "work and spend" lifestyle, then why should we, at a societal—or global level—care? One part of the answer is that, in general, happier members of society make society, and the planet, a better place. More specifically, as Schor highlights in *The Overworked American*, "our continuing confinement in the 'squirrel cage' of work . . . holds the potential for ecological disaster." But is ecological disaster actually a possibility?

In 2001, then–United Nations Secretary-General Kofi Annan wanted to assess "consequence of ecosystem change for human well-being and the scientific basis for actions needed to enhance the conservation and sustainable use of those systems and their contributions to human well-being." In other words, Annan was interested in determining if the planet could continue to sustain "business as usual."

In order to do this, Annan convened over 1,300 expert scientists and social scientists—the "largest group ever assembled to assess knowledge in this area"—and asked them to reach a consensus about the state of the world's ecosystems. Based on the extensive analysis conducted by this large group of scientific experts over four years (2001–05), and with a $24 million budget, the main conclusion of the Millennium Ecosystem Report assessment was that "human activity is putting such a strain on the natural function of Earth that the ability of the planet's ecosystems to sustain future generations can no longer be taken for granted."

There was a time in human history when concerns about increasing work and decreasing leisure could have been dismissed as fanciful, idealistic—and unrealistic—academic conjecturing. As we now are much more aware of the impact that our endlessly demanding consumptive lifestyles are having on a planet with finite resources (and out-of-balance ecosystems), calls for critical thought about work and leisure seem much less fanciful and much more critical.

So, while the origins of this discussion about leisure may go back to the Scientific Revolution (when we first began, in earnest, to believe we could "conquer" the mysteries and limitations of the material world), and while some of the roots certainly come from the Industrial Revolution (when products spilled off the assembly line and workers actually had to be taught how to also be consumers), perhaps most of all we should focus on the time since the end of World War II. It is in these past 65 or so years that we have been encouraged—everywhere and endlessly—to work and spend. Leisure and, as it turns out, our well-being, be damned. But in this era of degraded and dying ecosystems, climate change, and general environmental alarm, the embracing of leisure may contain a significant piece of the puzzle for how we can find our way to happier and more sustainable living.

And, as Ronald Inglehart has proposed in his post-materialism thesis, one may very well have leisure to thank for what environmental awareness does exist on the planet today. In other words, without fundamental material needs being met, people would not have leisure, and without leisure, people would not have been able to turn their attention to environmental issues. More leisure, less consumption, and more time and energy for the planet: This would, indeed, be a potent mix.

See Also: Fashion; Garbage; Home; Malls; Nature Experiences; Shopping; Vacation; Weekends; Work.

Further Readings

Kasser, Tim. *The High Price of Materialism*. Cambridge, MA: MIT Press, 2002.
Pieper, Josef. *Leisure: The Basis of Culture*. Alexander Dru, trans. San Francisco, CA: Ignatius Press, 2009 [1952].
Schor, Juliet. *The Overworked American: The Unexpected Decline of Leisure*. New York: Basic Books, 1993.

Jennifer Good
Brock University

Locavores

Locavores (also called localvores) are people who limit the food they eat to that which is grown or produced in a restricted area. The term *local* is typically used to describe foods grown within 50 to 400 miles of the buyer's home. Choosing to eat mainly or only local

Farmers markets, one of the most popular sources for local food, give shoppers access to a wide variety of produce while also allowing the shoppers to interact with the farmers. There are over 4,000 markets across the United States.

Source: USDA Agricultural Marketing Service

foods has become increasingly popular over the past decade and has become a national movement. Locavores cite many reasons for choosing to eat foods grown nearby, including the desire for fresh nutritious food, energy conservation, and environmental protection.

The Rise of the Locavore Movement

In 2001, ecologist Gary Nabhan published *Coming Home to Eat*, which chronicled his year-long attempt to eat only foods grown, caught, or gathered within 250 miles of his Arizona home. Inspired by Nabhan, a group of four women in northern California began calling themselves "the Locavores" in 2005. The Locavores established the Eat Local Challenge, which calls on Americans to define what local means to them (a 100-mile radius of one's home is often suggested) and eat only foods from this area for a month.

Other locavore chronicles have increased the popularity of this concept. One notable example is Alisa Smith and James Mackinnon's blog and subsequent book *Plenty: One Man, One Woman, and a Raucous Year of Eating Locally*, which describes their attempt to eat only foods grown and produced within a 100-mile radius of their home in British Columbia. Another is author Barbara Kingsolver's *Animal, Vegetable, Miracle: A Year of Food Life*, written with her husband and daughter, which depicts the family's attempts to grow, buy, preserve, and eat local food in rural Kentucky.

Today, thousands of Americans have pledged to do their best to eat locally. To support this effort, many grocery stores feature local products. Furthermore, over 1,000 school districts in 35 states are participating in farm-to-cafeteria programs, which provide students with locally grown food.

Why Eat Locally?

Locavores choose locally produced food for a number of reasons. One reason is food quality. Many locavores believe food picked when fresh and eaten shortly after tastes better and is more nutritious than food that has been transported hundreds or thousands of miles before it is consumed. Furthermore, because locally grown produce is often organic and eaten or preserved shortly after it is picked, it is often free of pesticides and chemical preservatives.

Many people also adhere to a locavore diet to decrease their carbon footprint and conserve energy. Food sold in U.S. supermarkets today travels an average of 1,500 miles before it is consumed. This is a 25 percent increase in distance traveled since 1980. As a result, about a fifth of the petroleum used in the United States is related to agricultural needs.

Buying locally grown food significantly decreases energy costs associated with food transportation and storage.

Furthermore, buying from local producers, typically small-scale, family-owned farms, reduces reliance on large factory farms. Locavores emphasize that smaller-scale, local farms are less likely to negatively impact the environment through pesticide contamination of soil, water, and air or soil erosion. In addition, eating locally produced foods also benefits local or regional economies and increases regional self-sufficiency.

The Locavore Diet

The main restriction to a locavore diet is location. Many locavores try to limit themselves to foods grown and produced within a 50-, 100-, or 150-mile radius of their homes. Others choose foods primarily grown within the borders of their state. While some locavores choose to be vegetarian or vegan, many others eat locally raised meat. Because of the location restrictions, many North American locavores must forgo coffee, chocolate, wheat, tropical fruits, spices, beer, and olive oil. While some locavores strictly adhere to eating only local foods, others incorporate dried spices, chocolate, or coffee into a mainly local diet. More recently, many people have decided to eat strictly local diets for a week or a month annually and then do their best to eat locally for the rest of the year.

Locavores get food from many sources, such as farmers markets, farm stands, food cooperatives, or community supported agriculture groups (CSAs), and even supermarkets. Many also grow some food themselves. When eating locally produced food is not an option, many locavores choose food that is organic or produced by a family farm.

Farmers markets are one of the most popular sources for local food, with over 4,000 markets across the United States. Farmers markets give shoppers access to a wide variety of local foods, including produce, meat, poultry, fish, dairy, and eggs. These markets also allow shoppers to interact with the farmers who grow the food and ask questions about farming methods and pesticide use. Food co-ops and CSAs are also becoming more popular and appealing to locavores. While food co-ops are similar to traditional grocery stores in many ways, co-ops emphasize supporting local farmers and the local economy. CSAs allow locavores to receive weekly baskets of seasonal farm produce in return for investment in a community farm.

See Also: "Agri-Culture"; Alternative Communities; Cooking; Diet/Nutrition; Green Consumerism; Individual Action, National and Local; Organic Foods; Shopping.

Further Readings

Kingsolver, Barbara, Camille Kingsolver, and Steven L. Hopp. *Kingsolver's Animal, Vegetable, Miracle: A Year of Food Life*. New York: HarperCollins, 2007.

Nabhan, Gary Paul. *Coming Home to Eat: The Pleasure and Politics of Local Foods*. New York: W. W. Norton, 2002.

Smith, Alisa and J. B. Mackinnon. *Plenty: One Man, One Woman, and a Raucous Year of Eating Locally.* Toronto: Random House, 2007.

Alexa J. Trumpy
The Ohio State University

Love Canal and Lois Gibbs

The story of residents in the Love Canal neighborhood near Buffalo, New York, becoming ill as a result of exposure to hazardous waste buried around their homes exploded on the national conscience in 1978. What made this instance of exposure unique was not that residents of a neighborhood were exposed to dangerous industrial waste but, rather, the response engendered. Led by Lois Gibbs, lower-middle-class residents, mostly women, mobilized through the auspices of the Love Canal Homeowners Association (LCHA) to try to force the government to purchase their homes and thus free them to move elsewhere. Coming on the heels of an explosion in federal governmental attention to environmental issues in the 1970s, this local story garnered tremendous national media attention. Lois Gibbs soon found herself advising and linking together local anti-toxics groups across the country—and at the vanguard of a national grassroots movement.

Love Canal became a symbol of the worst excesses of big business, polluting the environment to the detriment of human health. And events at Love Canal served as the major impetus for the passage of the Comprehensive Environmental Response, Compensation, and Liability Act (CERCLA), more commonly known as "Superfund." Lois Gibbs moved to the Washington, D.C., metro area and became a lifelong activist, renaming the LCHA the Citizens Clearinghouse for Hazardous Waste (CCHW; later renamed the Center for Health, Environment and Justice), a small organization that continues to serve as a leading group in the anti-toxics movement in the United States and a forerunner to a broader environmental justice movement.

What Is Love Canal?

Constructed by entrepreneur William T. Love in the 1890s, Love Canal was purchased by the Hooker Electric Chemical Company (now part of Occidental Chemical Corp.) in 1947. Utilized as a dumpsite by Hooker from 1942 to 1954 and by the city of Niagara Falls until 1953, the canal was estimated to contain 25,000 tons of chemical waste, including hazardous substances such as hexachloride, chlorobenzenes, and dioxin. After being approached by the Niagara Falls Board of Education, the entire canal plot was transferred to the school board by Hooker in 1953 for the price of $1. By 1960, the 99th Street School and surrounding homes had been constructed, with development of the surrounding area continuing through the mid-1970s.

It was through informal discussions among housewives that Gibbs and her neighbors came to realize there were unusually high incidences of miscarriages, birth defects, leukemia, and other illnesses in the community. Looking through old newspapers, Gibbs began to uncover the toxic legacy of the neighborhood. Finding government officials unresponsive to their concerns, residents organized for redress, with Gibbs emerging as one of many leaders. In 1978, led by Gibbs, residents formed the LCHA with the aim of forcing government officials to acknowledge the patterns of illness they had discovered. Their efforts were wildly successful in bringing significant attention to the cause, including that of the national media, which devoted significant news space to Love Canal and hazardous waste issues in 1978–80. The group also conducted their own scientific research, developing the "swale theory," which proposed that chemicals had traveled through the soil via natural drainage tracts located throughout the community.

As a result of their organizing efforts and activism, and the national media attention it garnered, the state and federal government agreed to purchase 240 homes from the "inner ring," those closest to the canal, in August 1978. Yet the surrounding homes, known as the "outer ring," the area where Gibbs and other leaders lived, were not purchased, and the struggle to have the entire area evacuated continued. On May 19, 1980, results of a chromosome test performed by the Environmental Protection Agency (EPA) finding that 11 of 36 residents examined displayed rare chromosomal damage were released. Two days later, tensions culminated in two EPA officials who attended a meeting of Love Canal residents being held hostage for approximately five hours. In an attempt to ameliorate the situation, President Jimmy Carter declared a state of emergency at Love Canal on May 21, 1980. By the end of 1980, the federal and state governments had agreed to purchase the outer ring homes surrounding the canal, allowing residents to move to safer locations.

As Love Canal held the spotlight in the national news, citizens across the country, concerned about the safety of their own communities, realized their situations were not unique. Beginning in the early 1970s but intensified by news of Love Canal, many communities began to oppose incidents of toxic waste dumping. Gibbs soon found herself inundated with requests for assistance from groups battling instances of local toxic contamination. She provided advice and served an important function, stitching these local groups together into a national grassroots movement. As a result of events at Love Canal and the emergence of a broader toxics movement, Superfund legislation was enacted in December 1980. Gibbs had been transformed from "ordinary" housewife to political activist, using her settlement monies to move to the Washington, D.C., area and to establish the small but influential CCHW. Acting as an information clearinghouse and mutual support organization, the CCHW served a critically important networking function that fostered the organization of distinctly local groups into state and regional level coalitions. The "toxics movement" continued to develop throughout the 1980s as both new national anti-toxics organizations (such as the National Toxics Campaign) were founded and many venerable environmental organizations (such as Greenpeace and the Sierra Club) began to focus attention on the toxic waste issue.

Significance Today

Ultimately, Love Canal became a cultural icon, synonymous with corporate malfeasance and governmental disregard for citizen health impacts of industrial pollution. The emergence of Love Canal on the national stage signaled the development of a national anti-toxics movement that fundamentally changed the U.S. environmental movement and federal environmental policy. In Love Canal, as in many instances of local contamination, it was housewives—those in charge of family health and who extensively networked with neighbors—who first noticed the signs of contamination, uncovered patterns, and led efforts at redress. Gibbs has estimated that 80 to 85 of the local activists assisted by the CCHW were women. Events at Love Canal also symbolize the development of a distinct wing of the national environmental movement, where the elite-driven national movement was paired with a more broadly encompassing grassroots movement. Neither Gibbs nor the majority of people in her neighborhood who participated in protests would consider themselves environmentalists. Indeed, members of local anti-toxics movements and Gibbs herself were highly critical throughout the 1980s of national environmental organizations, which they saw as more concerned with wildlife than human health and often beholden to large corporate donors. It is this legacy that leads to the modern image of a dual environmental movement—one elite driven and nationally based, and one led by working-class folks and concerned primarily with issues of human health. And so, in addition to being at the vanguard of what grew into a national anti-toxics movement focused on "plugging the pipe" of industrial hazardous waste production, the CCHW and toxic waste movement, more generally, have played a central role in the development of a broader environmental justice movement.

See Also: "Agri-Culture"; Environmental "Bads"; Environmental Justice Movements; Grassroots Organizations; Toxic/Hazardous Waste.

Further Readings

Gibbs, Lois. *Love Canal: My Story*. Albany: State University of New York Press, 1982.

Mazur, Alan. *A Hazardous Inquiry: The Rashomon Effect at Love Canal*. Cambridge, MA: Harvard University Press, 1998.

Szasz, Andrew. *EcoPopulism: Toxic Waste and the Movement for Environmental Justice*. Minneapolis: University of Minnesota Press, 1994.

Erik W. Johnson
Joseph Kremer
Washington State University

MAATHAI, WANGARI

Wangari Maathai is an internationally recognized environmental, human rights, and democracy activist and the first African woman to receive the Nobel Peace Prize (2004). She is best known for founding the Green Belt Movement (GBM), a widespread grassroots organization that enables people to plant trees where they live to provide themselves with food and fuel in such a way as to prevent soil erosion and desertification.

Born in Nyeri, Kenya, in 1940, Maathai began her university studies in the United States, earning her bachelor's in Biological Sciences from Mount St. Scholastic College in Atchison, Kansas, in 1964. In 1966, she obtained her Master of Science degree from the University of Pittsburgh. During that same year she met, Mwangi Mathai, another Kenyan who had studied in the United States. He would later become her husband. One year later, at the urging of one of her professors, she traveled to Germany, where she undertook doctoral studies at the University of Giessen and the University of Munich. In the spring of 1969, she returned to Nairobi, where she continued her studies at the University College of Nairobi while working as an assistant lecturer. Shortly afterward, she married Mwangi Mathai and gave birth to her first child. In 1971, she gave birth to the second of her three children and obtained her Ph.D. from the University of Nairobi, becoming the first woman in east and central Africa to have earned a doctoral degree. In the years that followed, she continued to teach at the university, where she became a senior lecturer in Anatomy in 1974, chair of the Department of Veterinary Anatomy in 1976, and associate professor in 1977.

In the early 1970s, Maathai also got involved in a number of civic organizations outside the university. For example, following the establishment of the Environment Liaison Centre in 1974, established in Nairobi to promote the participation of nongovernmental organizations (NGOs) in the work of the United Nations Environment Programme (UNEP), Maathai was asked to be a member of the local board. Around this same time, Maathai also joined the National Council of Women of Kenya (NCWK) and was active in this organization from 1976 to 1987. Through her work in various volunteer associations, it became evident to her that the root of most of Kenya's problems stemmed from environmental degradation. Thus, in 1976, she introduced the idea of a broad-based, grassroots organization that later became known as the Green Belt Movement.

When her husband won a seat in Parliament in 1974, his promise to create jobs to limit the rising unemployment in Kenya led Maathai to connect her ideas of environmental restoration to providing jobs for the unemployed. This led to the founding of Envirocare, Ltd., a business that involved ordinary people in the planting of trees to conserve the environment. Envirocare ran into multiple problems, however, chief among them funding. Although the project failed, UNEP made it possible for Maathai to attend the first United Nations conference on human settlements, known as Habitat I, in June 1976.

In 1977, Maathai spoke to the NCWK concerning her attendance at Habitat I. She proposed further tree planting, which the council supported, and which led to Save the Land Harambee. On June 5, 1977, the NCWK marched in a procession from Kenyatta International Conference Centre in downtown Nairobi to Kamukunji Park on the outskirts of the city to mark World Environment Day. There, they planted seven trees in honor of historical community leaders. This was the first "Green Belt" planted by the members of the movement. Maathai encouraged the women of Kenya to plant stands of trees throughout the country by searching nearby forests for seeds to grow trees native to the region. She also agreed to pay women a small stipend for each seedling that was later planted in other areas.

In 1986, the movement established a Pan African Green Belt Network (an environmental nongovernmental organization focused on the planting of trees, environmental conservation, and women's rights) that has introduced over 40 individuals from other African countries to this approach. Thus far, countries such as Tanzania, Uganda, Malawi, Lesotho, Ethiopia, and Zimbabwe have successfully launched similar initiatives throughout Africa.

Maathai and the Green Belt Movement have received numerous awards, most notably the 2004 Nobel Peace Prize. In June 1997, Maathai was elected by *Earth Times* as one of 100 persons in the world who have made a difference in the environmental arena. She has served on the boards of several organizations, including the Jane Goodall Institute, Women and Environment Development Organization (WEDO), and Green Cross International. In December 2002, Maathai was elected to parliament with an overwhelming 98 percent of the vote. Between January 2003 and November 2005, she also served as assistant minister for Environment, Natural Resources and Wildlife in the government of President Mwai Kibaki. In her most recent book, *The Challenge for Africa* (2009), a reflection on the problems currently facing Africa, she emphasized the necessity for Africans to come up with and implement their own solutions to these dilemmas.

See Also: Nongovernmental Organizations; Social Action, Global and Regional; Tree Planting Movement.

Further Readings

Maathai, Wangari. *The Challenge for Africa*. New York: Pantheon Books, 2009.

Maathai, Wangari. *The Green Belt Movement: Sharing the Approach and the Experience*. New York: Lantern Books, 2003.

Maathai, Wangari. *Unbowed: A Memoir*. New York: Knopf, 2006.

Danielle Roth-Johnson
University of Nevada, Las Vegas

MALLS

A far cry from the first open-air strip malls with only a common parking lot, enclosed shopping malls like the Mall of America in Minneapolis (shown is the amusement park at its center) established themselves as shopping destinations.

Source: Jeremy Noble/Wikimedia Commons

The concept of placing multiple stores in one place dates to the marketplaces, town squares, souks, and bazaars of the Middle Ages. Modern shopping malls developed in the 1920s as strip malls consisting of stores arranged in a row with a common sidewalk, driveway, and parking lot. The first malls were built for the new generation of car owners and were placed on major thoroughfares for customer convenience. Later, shopping malls diverted consumers away from downtown, toward the empty spaces between the city and the suburbs, and developed simultaneously with the suburbanization of America and the population migrations that filled out the Sun Belt and converted former farmland in the Midwest and New England into housing developments. The fully enclosed mall did not become the standard until the 1970s; even the first retail complex to use the word *mall*, the Bergen Mall in Paramus, New Jersey, opened as an open-air shopping center before rebuilding as an enclosed space in 1973. Enclosed malls were introduced in the late 1950s and soon added the hallmark benefits of shopping malls: air conditioning in the summer, heat in the winter, and an open food court with seating that underscored the idea of the mall space as a place in and of itself, not just the space between stores.

Rather than depending on attracting motorists on their way home from work, enclosed shopping malls established themselves as shopping destinations. Much like other shared business spaces, vendors contribute a fee that goes toward mall expenses like utilities, land, security, and advertising. One of the benefits for both businesses and consumers is the ease of introducing a business with a very narrow focus that would have more difficulty attracting sufficient business in a stand-alone setting: Mrs. Fields Cookies and Herbert's Potato World (selling cookies and French fries, respectively) are two food court examples of special-interest stores that flourish when they can depend on a customer base that is already physically present, and among their retail analogues in the main body of the mall are stores selling (nothing but) baseball caps, posters, greeting cards, sunglasses, or candles. Furthermore, a mall provides a good location for stores that exist only seasonally, selling seasonal needs (gift wrapping and other holiday supplies at Christmas, costumes and accessories at Halloween, beads and throws at Mardi Gras) and for stores selling new categories of products. The enclosed shopping mall was well established when video game consoles and personal computers were introduced in the 1970s and 1980s, and many curious shoppers examined the products and watched product demonstrations in stores when they were at the mall for some other shopping purpose. The same was true for the VCR, and malls have long been sites for dedicated kiosks selling first pagers and later mobile phones.

Environmental Pluses and Minuses

There are some environmental advantages to shopping malls. The consumer arrives by car (presumably) and shops on foot, rather than having to make separate trips to different stores—even if we assume that consumers would not replicate their exact shopping if forced to do it at different stores instead of in the mall (eliminating impulse or convenience purchases made only because of the consumers' proximity, purchases "not worth" a separate trip), this is clearly less driving and fewer carbon emissions, provided the mall is within a distance from the consumer's starting point equal to or less than the sum of the travel it would take to reach those other stores. For suburban consumers, this is a fair assumption: Their home town is not likely to have an extensive downtown retail area, as more and more shopping has been diverted to the outskirts, be it at shopping malls or at "big box stores" like Target and Walmart. With urban consumers, the determination is more difficult to make. In a small city like Nashua, New Hampshire, stores may be spread out around the city and surrounding towns, necessitating getting on and off the highway several times to replicate the purchases of a trip to the shopping mall. Even a larger city like New Orleans has effectively "outsourced" much of its shopping to nearby Metairie, Kenner, and Slidell, which makes shopping by public transportation effectively impossible because bus routes do not cross from one town to the next. But in the biggest cities like New York, car ownership can be more inconvenience than convenience, and public transportation has developed to get locals from virtually any point to any other point, making the emissions cost of these separate store trips negligible.

Essentially, it is difficult to talk about the environmental impact of shopping malls without talking about the environmental ramifications of the settlement patterns they have evolved in and that have evolved around them. The emissions cost of leaving town to go to the mall may be lower than leaving town to go to four separate stores, but if downtowns had never declined and most living and shopping still centered around the center of one's own town, this would not be the case. The shopping mall does not exist in a vacuum; it came about as part of other ongoing trends in land use and settlement.

But the above speaks only to the environmental impacts of each consumer's actions. There is additionally the environmental impact of the mall itself—the physical facility. The aggregate impact is made up of the impacts of the preparation of the land (and removal of whatever used to be there, often including fields or forest and animal habitats), the construction of the mall, the electricity used to heat or cool the mall and light both the interior and the parking lot (with parking lot lights often remaining on all night), and the transportation costs of stocking inventories. Pavement covers about 10 percent of all arable land in the United States, some 60,000 square miles. Parking lots alone, the largest of which are typically found at shopping malls, occupy nearly 2 million acres. Large-scale paving puts water quality at risk because increasing the acreage of impervious surfaces increases the volume of water runoff, which contributes to larger and more frequent floods as well as longer periods of low stream levels. Water temperatures rise, as does water acidity, and the water quality of some rivers and lakes degrades to a level unsuited for drinking water.

Those narrow-focused stores in malls arguably contribute to material waste in a significant way. Seasonal decorations are too often discarded instead of saved for the next year, especially when they are inexpensive. The social pressure to purchase gifts on holidays leads to many purchases of content-free, utility-free presents that may be briefly amusing but soon wind up as plastic hunks in landfills: dancing plastic flowers, stuffed dogs that sing Christmas carols, and so forth. Small tokens just to fulfill the need to give a gift may be thoughtful but are far more resource intensive than their low prices suggest.

Real Versus Online Stores

Further, the better comparison to make may not be between a shopping mall trip and trips to individual stores. Many suggest that online shopping is much more environmentally friendly than shopping at the mall. There are significant energy savings in the physical facilities because a store like Amazon.com consists only of inventory space and office space—no display space, no expensively air conditioned food court, or three-story-high interior. The Center for Energy and Climate Solutions in a 2007 comparison of Amazon .com with a traditional large chain bookstore found that Amazon.com's energy costs per square foot were almost half that of the bookstore: $0.56 instead of $1.10. But the big difference was in the energy cost per $100 in sales: $0.03 for Amazon, $0.44—more than 13 times as much—for the bookstore. Internet vendors turn over their inventory faster and are responsible for much less new construction (one of the most energy-intensive economic activities). Even when new warehouses are built, they are more energy-efficient space than retail space, and in certain cases there is no storage space necessary at all, as when digital media sales replace physical media, be they mp3s instead of CDs, digital movie or streaming movie files instead of DVDs, or e-books instead of physical books.

Online shopping is already having an impact on shopping malls. In the first decade of the 21st century, strip malls and regional malls lost stores at a rapid and increasing rate, with more commercial space in malls vacated in the first quarter of 2009 than in the entirety of 2008 (though some of this was due to the worst of the recession hitting at that time).

See Also: Advertising; Auto-Philia/Auto-Nomy; Green Consumerism; Shopping.

Further Readings

Farrell, James J. *One Nation Under Goods: Malls and the Seduction of American Shopping.* Washington, DC: Smithsonian, 2003.

Redclift, Michael. *Sustainable Development.* New York: Routledge, 1987.

Rogers, Elizabeth and Thomas Kostigen. *The Green Book.* New York: Three Rivers Press, 2007.

Torgerson, Douglas. *The Promise of Green Politics: Environmentalism and the Public Sphere.* Durham, NC: Duke University Press, 1999.

Bill Kte'pi
Independent Scholar

Media Greenwashing

Media greenwashing is a derogatory term that originated in the 1980s to criticize companies for making false or misleading claims of environmental responsibility about the companies or their products to the public through advertising or public relations campaigns in the mass media. The term combines the notion of *green*, meaning ecological, and the word *wash* from the word *whitewash*, which means to mislead or cover up the truth. Its legitimate counterpart is generally called corporate social responsibility, a marketing concept that advises verifiably portraying businesses and products as good ecological stewards. Environmental groups, consumer watchdogs, and the federal government have raised concerns about greenwashing and made concerted efforts to expose companies engaged in greenwashing, to educate consumers about detecting greenwashing, and to regulate greenwashing under the truth-in-advertising regulations of the Federal Trade Commission (FTC).

The term *green* itself has been traced to two environmental political parties founded in Germany in the late 1960s, the Green Campaign for the Future and the Green Lists. A book titled *The Greening of America* in 1970 infused the term with an overarching cultural context of change. Applying "green" to consumerism possibly emerged approaching the Earth Day anniversary in 1990. Critics of this concept point to the implicit assumption that environmental problems are the result of choosing among products, rather than with the mass production and consumption of products. In green consumerism, the question of the threat to the environment posed by consumer society itself is left unasked. Consumers themselves have demonstrated increasing interest in making purchases based on "green" product criteria, and manufacturers and businesses have responded by marketing products for environmental values instead of promoting decreased consumption.

The devastating chemical explosion in Bhopal, India, in 1984 has been identified as one tragedy that led to a growing awareness among both industry and environmentalists of the need for environmentally safe business practices. The Bhopal disaster, which has been called the worst-ever industrial accident, resulted the next year in a safety code for the chemical industry. Accompanying responsible reactions to the need for greater environmental safety was the growth of greenwashing. For example, while the international business community set up a program to implement standards for sustainable development in 1990, the environmental group Greenpeace published a book on greenwashing to distribute at the 1992 Rio Earth Summit in Brazil.

In 2007, TerraChoice Environmental Marketing conducted a study that found widespread misleading or false environmental claims among more than a thousand products reviewed using FTC guides. The study warned that if greenwashing succeeds, its success is short-lived, and in the long term more damaging, especially as business competitors, environmental and consumer groups, and bloggers are all eager to unmask transgressors. TerraChoice identified seven marketing "sins" of greenwashing. These include (1) the "hidden trade-off," which highlights one good environmental practice but neglects to mention another practice that damages the environment; (2) offering "no proof," which makes a claim that is not substantiated, a particularly risky practice since many groups exist to certify environmental claims; (3) "vagueness," which involves a claim, such as "all-natural," that is meaningless; (4) "worshipping false labels," which falsely suggest a product is endorsed by a third party; (5) "irrelevance," which associates an environmental benefit that is true but not relevant to the product; (6) "lesser of two evils," in which an environmentally harmful product is claimed to be environmentally sound, such as "organic cigarettes"; and (7) "fibbing," which claims certification that does not exist.

Fighting Greenwashing

To combat greenwashing, independent groups have emerged that certify products, such as the Chlorine Free Products Association (CFPA), Forest Stewardship Council, Green Guard, and Green Seal. Such groups provide the names of certified products to the public and may expose greenwashers who claim that their products are certified. Green Seal, for example, is a nonprofit organization that started in 1989, developed standards, and certified its first products in 1992. It has certified hundreds of products in almost 200 categories. Another independent nonprofit group, Greenguard Environmental Institute, started in 2001 and developed standards to certify building and related materials for emissions that affect indoor air quality. Also independent, the CFPA promotes advanced technologies that are free of chlorine, seeks to inform the public of alternatives, and develops international markets for certified products. In response to the 1992 Rio Earth Summit, the Forest Stewardship Council was formed in 1993 to set standards and policies for sustainable forest practices worldwide. The U.S. chapter, formed in 1995, oversees forest management standards nationally, conducts public outreach, and accredits certifiers to apply its criteria.

In addition to these are groups that monitor media greenwashing. Prominent among these is Greenpeace's StopGreenwash.org, which exposes greenwashing and provides the public, environmental groups, and government agencies information to counteract the practice. CorpWatch issues Greenwash Awards to criticize corporations for creating environmental public images in contradiction to their records. One CorpWatch target is British Petroleum, which CorpWatch argues is spending $45 million for a solar energy corporation, while spending $5 billion to explore for oil in Alaska. Other targets include Shell, Ford Motor Co., Monsanto, Dow, and DuPont. The Greenwashing Index is a watchdog website involving the University of Oregon School of Journalism and Communication and EnviroMedia Social Marketing, an advertising and public relations agency. Consumers post and rank advertisements in this online forum. Another group, Greenwash Guerrillas, uses street theater to draw attention to greenwashing. One event included a group wearing biohazard suits that arrived at the National Portrait Gallery in London to protest British Petroleum's Portrait Award ceremony. News media also have begun covering greenwashing with a variety of columns, blogs, and articles. The *Guardian* newspaper in England publishes a weekly column in a series called "Greenwash: Exposing False Environmental Claims." American Public Media's *Marketplace* radio program began the Greenwash Brigade in 2007 to cover green businesses through the efforts of four bloggers. Since 2009, the *New York Times* has published a weekly "Green Inc." column, complemented by a Green Blog about energy and the environment.

Interest in green products has grown, as has consumer skepticism about environmental marketing claims. Environmental marketing surveys evidence the widespread interest consumers have in green products. In 2008, 90 percent of consumers surveyed predicted increasing their environmentally based purchases in the following two years. Seventy percent thought environmental labels improved decision making. And about a third of respondents indicated almost half of their purchases are affected by environment considerations. Boston College's 2008 Green Gap Survey found that about 40 percent of Americans are seeking and purchasing products because they are environmentally beneficial, although respondents are unsure what being an "environmental product" means. Also, nearly half believed a product marketed as environmental was fully so, while only a quarter understood that environmental marketing terms are relative to previous or other products. Almost half also trusted companies to be truthful in environmental messages and accepted that the information about the product's impact on the environment is accurate. More than 60 percent said they comprehended the terms used in the environmental marketing. At the same time, public cynicism about energy company greenwashing was evident in the 2008 presidential campaign, when environmental groups produced parody ads, such as the Alliance for Climate Protection's satire presenting a fictitious coal company claiming that coal is clean, and smells good, too. In December 2008, major oil companies British Petroleum's and Shell's advertising of their green investments was offset by the knowledge that their main products are damaging to the environment and minuscule resources are applied to alternative energy research and development. Automakers also have advertised their environmentalism, including the highly acclaimed Toyota Prius, which an advertising authority in England criticized for being misleading.

Regulatory Response to Media Greenwashing

The regulatory response to environmental marketing has occurred through the Federal Trade Commission's Bureau of Consumer Protection. The FTC regulates misleading or false claims about the environmental attributes of products and services under its legal authority to enforce laws against fraudulent claims. The FTC first issued Guides for the Use of Environmental Marketing Claims, also known as Green Guides, in 1992, and last revised them in 1998. In early 2010, the FTC was in the process of updating the guides and was expected to address the increasing problem of greenwashing. The FTC has focused on expanding and more strictly enforcing the guidelines. Since the guidelines were introduced, the FTC has fielded about 45 complaints, although none were filed during the presidency of George W. Bush. Seven have been filed during the administration of President Barack Obama. In 2009, for example, the FTC successfully brought a complaint against Kmart Corp. to stop misleadingly advertising paper plates as biodegradable. Four companies were charged with misleading claims that bamboo clothing was ecological because stringent chemicals were used in producing the clothing. The Green Guides have developed criteria for measuring such concepts as "recyclable" and "biodegradable," so that businesses have benchmarks for assessing the accuracy of their environmental claims. The guidelines enacted in early 2010 address newer terms such as *sustainable*, *carbon neutral*, and *renewable energy credits*. The guides do not require companies to meet environmental standards. Enforcement of the guidelines falls under the Federal Trade Act prohibition against companies making any fraudulent claims. The Green Guides also have served as models for state laws in California and Indiana.

In its consumer guide for environmental marketing, the FTC responds to the use of terms such as *environmentally safe*, *recyclable*, *degradable*, and *ozone friendly*. Its six recommendations for consumers include the following:

- The claim should be specific. If the product claims to be recycled, the label must include information about what percentage of the produce or package is recycled. More labels include information about the source of the recycled material in the product. If it contains used or reconditioned parts, the label must specify that fact. Comparisons about recycled material should include a specific competitor's product or a previous product. If a product is labeled "nontoxic," it indicates the manufacturer believes it will not pose a risk to people or the environment.
- Vague claims are not meaningful, such as "environmentally friendly . . . safe . . . preferable," or "ecosafe." All products impact the environment, and the specific information required to assess the products' environmental qualities is missing.
- Recyclable claims mean the product is proved to be able to be collected and used again. A claim of "Please recycle" has no relevance unless the product is recycled in a given community.
- A claim that a product is degradable is meaningful only if the product is taken to a landfill, where degradation occurs extremely slowly. "Biodegradable" means the material will

decompose when exposed to air, water, and bacteria. Composting does allow "degradable" products to degrade. A label claiming the product is "compostable" means it is safe to compost at home.

- Claiming a product is "CFC-free" or "ozone-friendly" should be based on evidence that it does not harm the atmosphere. Chlorofluorocarbons, or CFCs, have been banned as propellants since 1978 and are being phased out in all products.
- Symbols showing that a product is recyclable and what type of plastic it is made of also can be helpful.

See Also: Advertising; Blogs; Communication, National and Local; Corporate Social Responsibility; Demonstrations and Events; Environmental Communication, Public Participation; Environmental Media Association; Green Consumerism; Greenwashing; Popular Green Culture; Print Media, Advertising; Shopping; Television, Advertising.

Further Readings

Clapp, Jennifer and Peter Dauvergne. *Paths to a Green World: The Political Economy of the Global Environment*. Cambridge, MA: MIT Press, 2005.

CQ Researcher. *Issues for Debate in Corporate Social Responsibility*. Thousand Oaks, CA: Sage, 2010.

Federal Trade Commission. "Facts for Consumers: Sorting Out 'Green' Advertising Claims." http://www.ftc.gov/bcp/edu/pubs/consumer/general/gen02.shtm (Accessed September 2010).

Hart, Ted, Adrienne D. Capps, and Matthew Bauer. *Nonprofit Guide to Going Green*. Hoboken, NJ: John Wiley & Sons, 2010.

Jedlicka, Wendy. *Sustainable Graphic Design: Tools, Systems and Strategies for Innovative Print Design*. Hoboken, NJ: John Wiley & Sons, 2010.

Nelson, Gabriel. "FTC Moves May Signal Start of 'Greenwashing' Crackdown." *New York Times* (February 3, 2010).

TerraChoice Environmental Marketing. *The Six Sins of Greenwashing: A Study of Environmental Claims in North American Consumer Markets*. Ottawa, Canada: TerraChoice Environmental Marketing, 2007.

Tokar, Brian. *Earth for Sale: Reclaiming Ecology in the Age of Corporate Greenwash*. Cambridge, MA: South End Press, 1997.

Paul Grosswiler
University of Maine

Music

Environmental themes have been woven throughout the long history of popular music. From early folk, bluegrass, and old-time music to modern-day rock, alternative, and country, popular musicians have voiced environmental concerns both on and off the stage. It is increasingly common to see both direct and symbolic references to environmental concerns like global warming, pollution, urban growth and development, modernization, strip mining, deforestation, nuclear energy, natural disaster, water rights, peak oil, and general environmental decline in the lyrics of contemporary popular music in the United States and abroad. While some artists make passing references to environmental issues or themes, others use their music as a platform to promote critical inquiry among their listening audience and to promote environmental stewardship or action. More recently, this "greening" of popular music has spread into various dimensions within the circuit of culture, including the production, consumption, distribution, performance, and marketing. Though most prominent in the folk music genres, environmental themes are becoming increasingly prominent as environmental consciousness seeps into everyday discourse and other realms of popular culture.

Place and Displacement in Appalachian Music

In the 1930s and 1940s, bluegrass, old-time, and American folk music genres represented the widespread community change encountered through commercial and natural resource development of the Appalachian region. Entire mountain communities were displaced and mountaintops laid bare to extract coal, natural gas, timber, and other natural resources. The songs from this era expressed the precarious relationship between social life in these communities and the natural environment that surrounded them. While the landscape was experiencing rapid change, bluegrass musicians voiced their reactions to these changes and articulated their attachment to these vanishing landscapes in the song titles and lyrics of these early ballads and fiddle tunes. Dominant themes frequently reference the hometown or "old home place" as a site of interpersonal warmth and relational stability that is threatened by the forces of time and economic modernization. "Home" for these early musicians and their audiences was defined by the landscapes that they longed for after being displaced from their farms, communities, and livelihoods through economic development.

Early bluegrass artists like Bill Monroe (Kentucky), Lester Flatt and Earl Scruggs (Tennessee/North Carolina), and the Stanley Brothers (Virginia) reference the difficulties of residential displacement and life on the road in songs like "I'm Going Back to Old Kentucky," "I'm Going Back to the Old Home," "I'm Coming Back but I Don't Know When," "Blue Ridge Cabin Home," and "Homestead on the Farm." Place names also abound in their songs, which are replete with references to familiar towns, counties, roads, ridges, mountain peaks, hollows, rivers, creeks, and valleys that characterized rural life in the Appalachian region. These early songs also memorialized the histories of environmental events like floods and storms including Monroe's "Muddy Water" and "Wasn't That a Mighty Storm" about the catastrophic Galveston flood of 1900.

Songs like John Prine's "Paradise," which became a standard in the bluegrass genre, articulate the environmental destruction encountered by those living in mining towns (in this case, Paradise, Kentucky). The chorus describes a child asking his father to take him back to the place of his youth (Muhlenberg County, Kentucky) and the father lamenting "*Sorry my son, but you're too late in askin'/Mr. Peabody's coal train has hauled it away.*" The song goes on to explain the environmental devastation encountered by this mountain town: "*Then the coal company came with the world's largest shovel/and they tortured the timber and stripped all the land/well, they dug for their coal 'til the land was forsaken, then they wrote it all down as the progress of man.*" A metaphor for the development of landscapes throughout the Appalachian region and beyond, this song captures a distinct link between lived human experience and environmental decline.

Environmentalism in the 1960s Folk Revival

In the folk revival in the 1960s and early 1970s, artists like Pete Seeger, the Weavers, and Woody Guthrie tapped into this long, lyrical folk tradition to voice growing concerns about the state of the environment. In Seeger's "Sailing up My Dirty Stream," he sings about the polluted Hudson River that he hopes "*someday, though maybe not this year*" will run clear. Throughout the song, he traces the river from its origins "*high in the mountains*" to its departure into the Atlantic Ocean. Along the way, however, pollution abounds, starting with "*floating wrappers of chewing gum*" to more egregious industrial dumping of "*five million gallons of waste a day*" from the Consolidated Paper Plant. He continues to ponder: "*Down the valley one million toilet chains/Find my Hudson so convenient place to drain/ And each little city says, 'Who, me?/Do you think that sewage plants come free?'*" Other artists such as Joni Mitchell ("Big Yellow Taxi"), Simon and Garfunkel ("The Sun Is Burning"), Tom Paxton ("When It's Gone, It's Gone," "Whose Garden Is This"), and John Denver ("Rocky Mountain High," "Earth Day Every Day," "To the Wild Country," "Children of the Universe") carried these themes into the 1970s.

One of the most notable environmentally themed songs to emerge from this era is Mitchell's "Big Yellow Taxi," which she wrote from her hotel room on her first trip to Hawaii. Inspired by the beauty of the Pacific Islands, she looked down from her hotel window to see a parking lot as far as she could see, which inspired the chorus "*They paved paradise and put up a parking lot.*" She goes on to decry themes of deforestation: "*They took all the trees and put 'em in a tree museum/And then they charged all the people 25 bucks just to see 'em*"; and environmental pollution: "*Hey farmer, farmer, put away your DDT now. Give me spots on my apples but leave me the birds and the bees, please.*"

Notable folk musicians have also engaged in various forms of environmental advocacy by partnering with environmental organizations for stewardship campaigns, participating in green-themed benefit concerts or tours, using their public stature to press forward on climate legislation, or creating their own nonprofit organizations to address climate change. One of the earliest examples of such work was John Denver's Windstar Foundation. Cofounded by Denver in 1976, this foundation raises awareness about environmental issues through educational programming. Denver continued this work through a 1995 benefit concert for the Wildlife Conservation Society to raise awareness of wildlife and the conservation of wilderness areas and remained a staunch environmental advocate until his death in 1997.

Modern Rock and Environmental Advocacy

It is not only folk musicians who wear their environmental ethos on their sleeves. A growing number of modern rock musicians have incorporated explicit environmental messages in their music. For example, Chrissie Hynde, singer and lyricist from the 1980s-era progressive rock band The Pretenders, penned "My City Was Gone," a song lamenting the growth, development, and pollution of her childhood home near Akron, Ohio. In the song, Hynde expresses her dismay that her favorite places had been "pulled down" to make way for parking spaces and shopping malls: "*my pretty countryside had been paved down the middle by a government that had no pride/The farms of Ohio had been replaced by shopping malls/And Muzak filled the air from Seneca to Cuyahoga Falls.*" Similar themes are reflected in the Talking Heads' song "(Nothing but) Flowers," which finds a disoriented singing subject reacting to an ironic and inverted picture of the development expressed in the Pretenders song: "*There was a factory, now there are mountains and rivers . . . /There was a shopping mall, now it's all covered with flowers . . . /Once there were parking lots, now it's a peaceful oasis . . . /As things fell apart, nobody paid much attention.*" Despite the unexpected transformation of the landscape back to its previous state, the picture is a familiar one—rapid and undesirable development and a longing for the way things were prior to human (or in this case, nature's) intervention.

Other examples of environmental themes in rock and popular music include bands as diverse as Dave Matthews ("One Sweet World"), Bad Religion ("Kyoto Now"), Miley Cyrus ("Wake Up America"), Michael Jackson ("Earth Song"), John Cougar Mellencamp ("Rain on the Scarecrow"), and Jack Johnson ("The 3Rs [Reduce, Reuse, and Recycle]").

Increasingly, popular musicians are using their stature and public visibility to draw attention to the environment. For example, Peter Gabriel, prolific solo artist and cofounder of the rock band Genesis, has also had an active career advocating human rights and environmental concerns. In 1988, he traveled to the former Soviet Union to help establish Greenpeace, a global environmental action nonprofit organization. He also participated in the 1990 "One World, One Voice" album, which brought together artists from around the world to create one continuous "chain tape" blending various world genres and styles of music with the objective of raising awareness of global environmental issues. The album gave rise to a concert performance featuring artists from the album (including Sting, The Chieftans, Laurie Anderson, Gypsy Kings, Afrika Bambaataa, David Gilmour, and various others) as well as a live DVD recording of the event. Other artists such as Sting, Elton John, and Bono (U2) have taken a more active role in environmental politics.

Coming off the Live 8 concert tour in the summer of 2005, which aimed to address issues of extreme poverty and environmental decline, both Sting and Elton John appeared as panelists in a televised debate about the G8 that included United Nations Secretary-General Kofi Annan, British Chancellor Gordon Brown, President Benjamin Mkapa of Tanzania, and Chris Martin from the rock band Coldplay. Sting has also been involved with celebrity fundraisers for the Rainforest Foundation and Prince Charles's Rainforest Project. For his latter collaboration, on the anniversary of his Police hit "Message in a Bottle," in September 2009, Sting sent his public "SOS to the World" to raise awareness about deforestation and CO_2 accumulation in the atmosphere in a brief video released before the Copenhagen round of climate negotiations. U2's "front man" Bono has also used his public stature to support the organization Greenpeace and protest the construction of a nuclear facility in Sellafield, England. Other bands, like Coldplay, have been vocal proponents of environmental causes, including the support of the Future Forests initiative. Hip-hop artists and staunch vegetarians The Roots have also demonstrated their environmental ethic through their meat-free diet and public support of People for the Ethical Treatment of Animals (PETA). These artists see their position as famous rock musicians adding an element of responsibility to their work and use their position of public prominence as a platform to raise awareness about these issues.

Green Production and Touring

The music industry has an enormous, though often overlooked, impact on the environment. Beyond the musicians and music itself, there has been a recent movement to "green" musical production, consumption, and performance. The music industry itself is catching on to the market niche of green consumerism with bands packaging their latest CDs in recycled plastic or paperboard sleeves and hitting the road in biodiesel tour buses.

At the production level, Warner Brothers Music celebrated Earth Day 2008 by fully offsetting the carbon emissions at their corporate headquarters. Additionally, they have made a commitment to reduce waste and increase recycling practices at corporate-sponsored events. While some may consider such campaigns a form of corporate green-washing, others see this as a necessary step in protecting our fragile planetary balance. Record companies like the alternative record label Sub Pop have worked to reduce their CO_2 output to zero through the purchase of "green tag" vouchers that subsidize the use of renewable energy. Other record companies, mostly through the leadership of individual artists, have sought to reduce the environmental impact of their CDs and DVDs by using recycled or biodegradable paper, cardboard, or plastic jewel cases; soy-based inks; packaging that uses a greater percentage of post-consumer waste; or distributing music exclusively through digital formats.

While touring bands use considerable amounts of fuel and energy going from show to show with their equipment, some are discovering ways of reducing their carbon footprint. For example, the rock giants Pearl Jam introduced a carbon portfolio strategy to offset the carbon emissions from their tour of the United States, Europe, and Australia through direct financial support to multiple environmental advocacy organizations. Other notable artists like Willie Nelson, Sheryl Crow, The Foo Fighters, and Alanis Morissette have dedicated their efforts to reduce their CO_2 emissions through the use of biodiesel transport for their personal motor coaches and touring fleet.

San Francisco, California's bluegrass-inspired Hot Buttered Rum have taken this initiative to the next level. Their band website features a "green" menu link that addresses their environmental "philosophy and mission," which explains their efforts to tour and make music in environmentally responsible ways. Included are a "green resume" and a section devoted to the "Green Machine," their tour bus that has been fueled by recycled vegetable oil and biodiesel since 2003. The website includes detailed diagrams and technical information about the diesel engine and modifications used on the bus. In fact, one of their songs, "Well-Oiled Machine," references the bus directly: "*what the restaurant kitchens dump/I want to pump/To fill the tank of my well-oiled machine*." This is one of a dozen or more environmentally themed and politically driven original songs that they feature on their website (including both lyrics and recordings of live performances) for free download.

Green Festivals and Concerts

The venues of musical performance are also experiencing an environmental revolution, with some music festivals promoting "carbon-neutral" events through the incorporation of alternative energy carbon offsets into ticketing prices or through the promotion of recyclable or compostable cups at the vending booths. The practice of greening the festival or concert site is more than simply good business practice—it serves a pedagogical function that educates concertgoers about sustainability and waste management.

Clean Vibes (www.cleanvibes.com) is a North Carolina–based company dedicated to managing concert waste in an environmentally responsible manner. Ecoconscious festival promoters, including those for Bonnaroo, High Sierra, Lollapalooza, and the band Phish have hired Clean Vibes to enable concert attendees to separate their waste into landfill, recycling, and compost through the use of color-coded bins placed throughout festival sites. Some festivals and bands also have designated volunteer "green teams" who stay long after the festival or concert has ended to ensure the site is left free of trash and other waste. Their objective is to send a "clean vibe" to other concertgoers and to promote environmental stewardship with the objective of influencing the habits and behaviors of their fellow scene members.

To further encourage environmental stewardship, other festivals have incorporated the use of educational workshops and kiosks. For example, the Oregon Country Fair is a yearly three-day countercultural arts and music festival outside Eugene, Oregon, that features an entire environmentally themed "energy park" and "community village" that provide visitors with educational opportunities through hands-on demonstrations and information kiosks about organic gardening, saving seeds, harvesting and reusing rainwater, planting native residential plants, and other ecofriendly household practices. These sites also provide a host of exhibition tables about green and renewable energy and innovatively applied technologies including biodiesel engines, solar cookers, windmills, hydroelectric generators, human-powered blenders, and other technologies that could be adopted for home use. The festival also features a site-wide composting, recycling, and waste management system that seeks to considerably reduce the amount and costs for waste disposal for the 40,000 daily attendees.

Another Oregon event, the Pickathon Festival—a benefit festival for a local community radio station in Portland featuring folk and American roots music—has created an initiative to eliminate all plastic from their event. All food served at the venue is packaged in compostable or biodegradable containers and served with compostable, corn-based utensils. Additionally, to eliminate the waste of plastic beer cups, the festival partnered with Kleen Kanteen, which sold reusable stainless steel pint glasses to those visiting the beer garden and included a station where the cups could be washed, dried, and stored while not in use. Rather than sell plastic water bottles, they also supplied free drinking water to all attendees, distributed through multiple water tanks throughout the venue. Planet Bluegrass, a Colorado event production company that hosts the Rocky Grass and Telluride Bluegrass Festivals, takes its sustainability efforts one step further by offsetting all energy consumed by and at the festival, including travel for all attendees.

See Also: Art as Activism; Green Consumerism; Greenwashing; Media Greenwashing; Musicians; Popular Green Culture.

Further Readings

Brickington, Dan. *Celebrity and the Environment: Fame, Wealth and Power in Conservation.* London: Zed Books. 2008.

Dibben, N. "Nature and Nation: National Identity and Environmentalism in Icelandic Popular Music Video and Music Documentary." *Ethnomusicology Forum*, 18/1 (2009).

Ingram, David. "'My Dirty Stream': Pete Seeger, American Folk Music, and Environmental Protest." *Popular Music and Society*, 31/1 (2008).

White, Ryan. "Pickathon Music Festival Takes Sustainability Further This Year With Steel Beer Cups." *The Oregonian* (August 6, 2010).

Wignall, Alice. "When the Music's Over." *The Guardian* (August 21, 2008).

Robert Owen Gardner
Linfield College

Musicians

Musicians have worked both individually and collectively to raise environmental awareness and promote a variety of green issues and practices. Music-based organizations include Musicians United to Sustain the Environment, Reverb, and the Green Music Alliance. Green practices promoted by musicians have included ecofriendly touring, raising audience environmental awareness, and funneling music sale and concert profits to environmental organizations or initiatives.

Musicians have long been involved in writing and singing about the environment, helping to promote social awareness and inspire environmental activism among their fans. Although social protest music is often associated with certain genres such as folk, other popular musicians have also added songs with social messages to their musical catalogues. Classic songs with an environmental message have included Joni Mitchell's "Big Yellow Taxi," Kansas's "Death of Mother Nature Suite," John Cougar Mellencamp's "Rain on the Scarecrow," The Pretenders' "My City Was Gone," Rush's "The Trees," and James Taylor's "Traffic Jam," just to name a few.

Musicians have also promoted environmental awareness and green culture through involvement with music-based and environmentally based nongovernmental organizations (NGOs). One of the leading music-based volunteer organizations is Musicians United to Sustain the Environment (MUSE). MUSE is composed of professional touring musicians who wish to be actively involved in environmental efforts to save the planet. This organization has two goals: they want to create grassroots environmental projects and a heightened awareness of environmentalism in general. MUSE members direct the profits of their music sales toward conservation organizations. A for-profit organization called MusicMatters, a marketing company, works with musicians to make their tours green.

Reverb is a nonprofit organization in Maine that helps musicians incorporate green practices into their touring operations. Environmental activist Lauren Sullivan, who worked for the Rainforest Action Network, and her musician husband Adam Gardner of the indie rock band Guster founded Reverb in 2004. They have helped over 80 musical tours go "green." Some of their well-known clients have included Alanis Morissette, Sheryl Crow, the Red Hot Chili Peppers, and Bonnie Raitt. Musicians who use Reverb want not only to make their tours ecofriendly but also to spread the message of the environmental movement to their fans.

Reverb-coordinated music tours feature Eco-Villages, which feature interactive and informative tents that teach concertgoers about the serious ecological issues of the day. They also enlist volunteers to sort out recyclable garbage from other refuse after the concerts are over. On most Reverb-coordinated tours, the performers contribute a portion of ticket sales or pay toward the purchase of renewable energy credits from a Vermont company called NativeEnergy. These renewable energy credits provide for the construction of wind, solar, and biomass electricity generators to offset the amount of carbon (carbon footprint) produced by each tour. Willie Nelson, Sting, and Don Henley are just a few artists who have raised millions of dollars for ecofriendly charities.

A number of musicians have worked with the international environmental organization Greenpeace, which has utilized concerts as opportunities for mass activism. Greenpeace asked attendees of the Rock in Rio concert held in Lisbon, Portugal, in 2008 to go home and record the first few minutes of Beethoven's 5th Symphony on any instrument they had at their disposal. Studies have shown that the 5th Symphony improves brain functions and promotes creativity. Greenpeace then played an amalgamation of the best recordings for the political leaders that met at the G8 summit in Japan that year. Greenpeace's musical message was clear: the people want solutions to global warming. The band Green Day has participated in Greenpeace's MusicWood campaign, which is attempting to convert the guitar industry to the use of sustainable wood sources.

Concerts With a Message

Musicians have participated in group concerts with environmental messages. The city of Chicago has a Green Music Festival, combining attendees' passions for music and environmental preservation. In addition to the music, there are retail areas, arts and crafts vendors, and food. Beverages are served in biodegradable corn-based cups. Some musicians have also made individual efforts to promote green culture, both within and outside the music business. Musicians such as Bonnie Raitt are also themselves environmental activists. Raitt is a founding member of the organization Musicians United for Safe Energy. Raitt has also held benefit concerts that promote alternative energy solutions. She dedicated a humanitarian award to Julia Butterfly Hill, who sat in a tree for a year in an effort to save the California redwoods, even hiking a mountain trail and using a pulley to give Hill the award as she sat in the tree.

Musicians like Dave Matthews and Willie Nelson often travel in buses fueled by biodiesel made from vegetable oils. Neil Young was one of the first musicians to use biodiesel fuel. This type of fuel emits about 75 percent less pollution than standard diesel fuel. Other artists utilizing biodiesel on tour include KT Tunstall, the Indigo Girls, and Keith Urban. The band Coldplay plants trees in third world countries. Some musicians have also lobbied national and international governments to implement climate change policies. For example, the band Green Day has advocated that the United States move away from its dependence on oil. Global warming is at the forefront of the Dave Matthews Band's social activism interests. In their hometown of Charlottesville, Virginia, they help scatter free bikes around town to promote ecofriendly travel.

Canadian singer Sarah Harmer and her band hiked the Niagara escarpment, a wild area that includes Niagara Falls, to draw public awareness of the environmental threats posed by the area's commercialization as a tourist attraction. She and her band also raise awareness for the preservation of wild lands, cofounding an environmental group called Protecting Escarpment Rural Land, or PERL. The organization helps protect wild lands from human destruction. The band Pearl Jam has donated thousands of dollars to nine different environment organizations to offset the carbon emitted from their tour buses and shows. They also seek to educate children about the environment and help protect forests near their hometown of Seattle, Washington.

The band Barenaked Ladies incorporates a number of green culture practices into their daily lives and tours, including the consumption of organic food, the use of plates and forks made of biodegradable potato starch, and reusable drinking canteens as opposed to plastic bottles. The band also purchased renewable energy credits from concert profits to offset the electricity that was used during their tours. Before each gig, they play a slideshow on global warming for their audiences to raise awareness. The band even began a unique program to retrieve broken and used instrument strings from the stage after concerts; these were then recycled into jewelry,

Some other notable green touring acts include John Legend, John Mayer, Bon Jovi, Incubus, The Fray, The Beastie Boys, Avril Lavigne, and Jack Johnson. Incubus, for instance, sells promotional items that include organic t-shirts and posters printed on recycled paper using soy ink. The Fray use vintage guitars and turn their used guitar strings into jewelry. The Beastie Boys give a free Beastie Boys item to anyone who drops off cell phones and other accessories at their on-site recycling booths. There is even a female folk duo called the Ditty Bops who traveled 4,700 miles on bicycles to promote their album, playing shows along the way. Neil Young, Bonnie Raitt, Dave Matthews, and Willie Nelson all have tours that are carbon neutral.

The CD titled "Power Up the Planet," released by the music company Planetwize, helps support positive changes in the environment. Every download or physical CD sold helps fight climate change and poverty, in part through the funding of SolarAid, an organization that trains people in poor communities to harness solar power. SolarAid brings items such as solar lanterns and mobile phone and battery chargers to remote locations in Africa. SolarAid projects have been successful in countries such as Tanzania, Malawi, Zambia, and Kenya.

See Also: Art as Activism; Individual Action, National and Local; Music; Nelson, Willie.

Further Readings

Brown, Steven and Ulrik Volgsten. *Music and Manipulation: On the Social Uses and Social Control of Music*. New York: Berghahn Books, 2006.
Crawford, Richard. *America's Musical Life: A History*. New York: Norton, 2001.
Green Music Alliance. http://www.greenmusicalliance.org (Accessed September 2010).
Grossman, Elizabeth. *High Tech Trash: Digital Devices, Hidden Toxics, and Human Health*. Washington, DC: Island Press, 2006.
Krull, Kathleen and Kathryn Hewitt. *Lives of the Musicians: Good Times, Bad Times (and What the Neighbors Thought)*. New York: Harcourt Brace Jovanovich, 1993.
MacDonald, Raymond A. R. and David J. Hargreaves. *Musical Identities*. New York: Oxford University Press, 2002.
Musicians United to Sustain the Environment. http://www.musemusic.org/index.php?pr=Home_Page (Accessed September 2010).
Peddie, Ian. *The Resisting Muse: Popular Music and Social Protest*. Aldershot, UK: Ashgate 2007.
Wilkinson, Alec. *The Protest Singer: An Intimate Portrait of Pete Seeger*. New York: Alfred A. Knopf, 2009.

Marcella Bush Trevino
Barry University

National Park Service

The National Park Service promotes development for tourism and scenic enhancement, like this trail in Apostle Islands National Lakeshore, as long as most of the land stays untouched. Today, over 285 million people visit the areas protected by the National Park Service yearly.

Source: National Park Service

On August 25, 1916, President Woodrow Wilson signed the National Park Service Organic Act and created the National Park Service. This federal bureau is regulated by the Department of the Interior and is responsible for protecting national parks, monuments, seashores, parkways, and other important areas. The National Park Service has undergone a myriad of changes over the past century, but remains one of the most popular federal agencies. Today, there are over 390 diverse areas that comprise the national park system, from the White House in Washington, D.C., to Volcanoes National Park in Hawaii.

Founding of the National Park Service

The origin of the National Park Service can be traced to the founding of Yellowstone National Park in March 1872, the United States' first national park. Although Congress had reserved several million acres of land to create national parks throughout the country since that event, there was no uniform national park policy. As a result, the park service was run very differently than it is today.

The need for a comprehensive and coherent national park policy became apparent in the early 20th century when San Francisco Mayor James Phelan proposed damming

Hetch Hetchy Valley, a remote area of Yellowstone National Park, in order to build a pipeline to bring a steady source of water from the valley to San Francisco. Although President Woodrow Wilson signed the Raker Act in 1913, which authorized the construction of the dam, the debate over the proper use of Hetch Hetchy Valley exposed many of the inconsistencies in the then-current national park management and philosophy. For example, while hotels and roads were constructed in Yellowstone, Yosemite, and Glacier National Parks, other parks remained largely undeveloped. Similarly, while hunting was authorized in Yellowstone, it was prohibited in many other national parks. The Hetch Hetchy debate also highlighted tensions between two prevalent groups with clashing views of natural resources: conservationists and preservationists. Conservationists argued national parks and other public land should be used to produce the greatest good for the greatest number; therefore, they believed well-developed roads and hotels in national parks would increase accessibility and use. On the other hand, preservationists argued national parks should be preserved, protected, and remain relatively unaltered by human development for the public's enjoyment and appreciation of nature.

According to the language of the Organic Act, the National Park Service was designed to regulate and protect national parks and conserve natural scenery and wildlife, leaving them unaltered for future generations. This language, particularly the mandate to leave the parks unimpaired, supported the preservationists' philosophy and vision. Yet although the first director of the National Park Service, Stephen Mather, supported this philosophy, in practice the National Park Service did not follow a policy of strict non-interference and preservation. Instead, development and construction were promoted in the parks to increase popularity and promote tourism, thus advancing a conservationist viewpoint. Overall, Stephen Mather set the precedent that national parks should be developed to promote tourism and enhance the natural scenery, but remain otherwise untouched. Mather referred to this practice as a double mandate of public use and preservation. This double mandate continues to influence national park policy and use today.

Expansion of the National Park Service

Until the mid-1920s, all of the national parks except Acadia National Park (in Maine) were in the western United States. In 1926, many areas in the eastern Appalachian mountain region were also designated as national parks, including Shenandoah and the Great Smokey Mountains. In 1933, the national park system also incorporated parks and monuments previously managed by the War Department, national monuments maintained by the Forest Service, and the national capital parks, including the White House, the Washington Monument, and the Lincoln Memorial. Parkways designed for scenic driving, such as the Blue Ridge Parkway between Great Smokey Mountains and Shenandoah National Parks, were also added to the National Park Service in the 1930s.

New types of parks continued to join the National Park Service. Cape Hatteras in North Carolina became the first National Seashore to be protected by the Park Service in 1937. National Lakeshores on the Great Lakes were added in 1966. Furthermore, under the National Trails System Act of 1968, the Park Service became responsible for the Appalachian Trail, which runs over 2,000 miles from Georgia to Maine. The Alaskan National Interest Lands Conservation Act of 1980 added over 47 million acres of land to the National Park Service, more than doubling its size.

National Park Service Today

Today, over 285 million people yearly visit the 392 areas protected by the National Park Service. Areas protected by the Park Service exist in every state except Delaware. Protected areas are also found in the U.S. Virgin Islands, Guam, American Samoa, and Puerto Rico. The Park Service employs over 20,000 people and has over 145,000 volunteers.

There is great diversity in use of the park system. Popular areas such as Great Smokey Mountains National Park, the Blue Ridge Parkway, and the Grand Canyon consistently have millions or tens of millions of visitors each year. On the other end of the spectrum, many isolated protected areas such as Noatak National Preserve and Kobuk Valley National Park (both in Alaska) receive only a few thousand visitors a year. In fact, Aniakchak National Monument and Preservation Area in Alaska had only 60 visitors in 2006. Yet despite the vast diversity in use, Americans continue to praise and enjoy the Park Service. Public opinion polls have consistently ranked the National Park Service as one of the most popular federal agencies, and it is likely to remain popular in years to come.

See Also: Ecotourism; Environmental "Goods"; Leisure; Nature Experiences.

Further Readings

Albright, Horace M. and Marian Albright Schenck. *Creating the National Park Service: The Missing Years*. Norman: University of Oklahoma Press, 1999.

Runte, Alfred. *National Parks, the American Experience*. Lincoln: University of Nebraska Press, 1987.

Sellars, Richard West. *Preserving Nature in the National Parks: A History*. New Haven, CT: Yale University Press, 1997.

Alexa J. Trumpy
The Ohio State University

National People of Color Environmental Leadership Summit

The First National People of Color Environmental Leadership Summit in 1991 and the Second National People of Color Environmental Leadership Summit in 2002 were conferences in which representatives of the environmental justice movement in the United States and around the world came together to define their movement and develop common strategies to address environmental problems, with a specific focus on environmental racism (policies and practices that negatively affect the environment of communities of color at a higher rate or scale than white communities). Major outcomes of the summits include a formal coalescing of the environmental justice movement, the establishment of interorganizational networks, a united platform and many policy papers detailing the contours of specific environmental problems facing communities of color, and potential solutions. Environmental justice scholars, such as Robert D. Bullard, chair of the Environmental

Justice Resource Center at Clark Atlanta University, who served as a key organizer for both summits, consider the first summit in particular to be the single most important event in the movement's history.

Out of initially disparate local struggles, the environmental justice movement grew into a national phenomenon during the 1980s. A major moment of cohesion occurred when the United Church of Christ's Commission for Racial Justice, led by environmental justice activist and scholar Benjamin Chavis, helped convene the First National People of Color Environmental Leadership Summit October 24–27, 1991, in Washington, D.C., at the Washington Court Hotel. The conference was attended by over 650 delegates representing over 300 organizations from around the world, including all 50 U.S. states, Puerto Rico, Chile, Mexico, and the Marshall Islands. They came together to share their experiences of environmental racism and their action strategies for addressing these problems. Bullard, a key organizer of the conference, maintains that the summit served to broaden the environmental justice movement beyond its early anti-toxics focus to encompass issues as varied as public health, housing, transportation, and community empowerment. The summit also confirmed the environmental justice movement as a multi-issue, multi-ethnic, and multinational movement.

A major outcome of the First National People of Color Environmental Leadership Summit was the 17 Principles of Environmental Justice. These principles were to act as a defining and guiding document for the movement and announced to the world the existence of a cohesive environmental justice movement. The document affirms the sacredness of and humanity's dependence on Mother Earth; pledges to fight the destruction of the communities and lands of people of color; promotes the use of environmentally sound economic alternatives; and seeks to secure the political, economic, and cultural liberation for people of color collectively facing oppression and histories of colonization. The summit also produced the 9 Principles of Working Together, guiding organizational partnerships within the movement and addressing accountability.

An early victory claimed by the environmental justice movement, invigorated by the summit, was the increased level of public concern garnered for environmental justice issues, prompting President Bill Clinton to issue Executive Order 12898, Federal Actions to Address Environmental Justice in Minority Populations and Low-Income Populations, on February 11, 1994.

Out of a need to assess accomplishments, galvanize support, share information, develop strategy, and reinvigorate the work of the environmental justice movement, the Second National People of Color Environmental Leadership Summit was convened October 23–27, 2002. Held once again in Washington, D.C., this time at the Hyatt Regency Hotel, Summit II, as it is colloquially known, drew over 1,200 delegates and another 200 nonvoting participants representing environmental justice organizations from every inhabited continent of the world. The attendees were composed of academics, activists, students, and scientists representing grassroots organizations, faith-based groups, unions, civil rights groups, youth organizations, and academic institutes.

Chaired by Beverly Wright, director of the Deep South Center for Environmental Justice at Xavier University, Summit II issued a national call for papers designed for use at the conference's workshops and discussions. In all, 26 papers were submitted, encompassing such topics as suburban sprawl, transportation equity, smart growth, public health and the environment, land use, land rights, sustainable development, worker and occupational safety, environmental education, energy, climate justice, and globalization. From these resource papers, the delegates were able to analyze issues of environmental racism and advocate environmental justice solutions. Summit II concluded with the delegates reaffirming their commitment to environmental justice.

See Also: Clinton, William J. (Executive Order 12898); Demonstrations and Events; EcoPopulism; Environmental Justice Movements; United Church of Christ.

Further Readings

Bullard, Robert D. "Environmental Justice in the 21st Century." http://www.ejrc.cau.edu/ejinthe21century.htm (Accessed August 2010).
Bullard, Robert D., ed. "Second National People of Color Environmental Leadership Summit Resource Papers: A Synthesis." http://www.ejrc.cau.edu/SummitPolicyExSumm.html (Accessed August 2010).
Dorsey, Michael K. "Toward an Idea of International Environmental Justice." In *World Resources 1998–99: Environmental Change and Human Health*. Washington, DC: World Resources Institute, 1998.
"Environmental Justice for People of Color Summit Draws 1,200 Delegates to Washington." *The Black Commentator*, 16 (November 12, 2002).
Environmental Justice Resource Center. "People of Color Environmental Summit II." http://www.ejrc.cau.edu/EJSUMMITwlecome.html (Accessed August 2010).
Second National People of Color Environmental Leadership Summit. *Environmental Justice Timelines—Milestones*. Washington, DC: Summit II National Office, 2002.
Voices From the Earth. "The Second National People of Color Environmental Leadership Summit." *Voices From the Earth*, 4/1 (2003).

Robert Connell
University of California, Berkeley

Native American Culture and Practices

Contemporary scholarship shows that Native Americans factor into the environmental movement in important and sometimes ironic ways. A full picture requires the integration of many diverse sets of information, including an accurate understanding of prehistoric and historic lifeways; the impacts of colonial encounters; Euro-American characterizations of indigenous Americans; and the cultural, economic, and political struggles over land use and dominion that characterize historic and recent interactions.

Prehistoric Practices for Managing Natural Resources

As humans migrated into and across the Americas between 30,000 and 15,000 years ago, small nomadic groups subsisted on a wide range of resources. The migrants and their new environments adjusted to each other, varying across space and time. Once, human activity was faulted for the mass extinctions of Pleistocene animals, an issue politicized by mid-20th century anti-treaty-rights groups to undercut Native claims for ethical stewardship. Today,

The image of the Native Americans as one with nature created the idea of both the environmentally friendly "noble savage" as well as the bloodthirsty or animal-like "ignoble savage."

Source: iStockphoto

scientific study has reduced the human role in extinctions to a very small factor.

Over time, populations increased, and many became sedentary. The development of horticulture facilitated the emergence of complex societies such as at Caral in Peru 4,600 years ago or at Poverty Point, Louisiana, 3,600 years ago. In the ensuing millennia, numerous urban civilizations that engaged in large-scale agriculture flourished and diminished. Each encountered and attempted to manage the problems associated with aggregated populations at economic, political, or religious centers: soil degradation, resource depletion, waste management, transportation, and interregional competition.

Other groups maintained more flexible human–environment relations in smaller-scale village societies practicing intensive horticulture, often with great surpluses. Many groups located along the Pacific coast regularly harvested abundant wild Pacific coast resources and developed large socially stratified societies capable of conspicuous consumption without animal husbandry and domestication. Other groups continued hunter-gatherer lifeways. While this lifeway was once regarded as environmentally ephemeral, recent research contradicts this idea. In South, Central, and North America, evidence shows that over time the daily practices of foragers shaped old-growth forest composition.

Overall, regions described by Europeans as "unbroken wildernesses" are now known to have consisted of networks of highly managed resources. These landscapes act as historical narratives of human-environment relations and allow for the reinterpretation of the writings of early European explorers and missionaries. These authors were often unable to recognize or unwilling to credit indigenous people with such achievements. Indeed, their goals of imperialism or conversion encouraged them to emphasize that Native Americans were uncivilized.

Historic Trajectories

Much of this evidence was overlooked until recently due to the aftermath of one of the most tragic episodes in human history: the arrival of Old World diseases among New World peoples. While paleodemographic research into this episode is problematic, some estimates put precontact populations at 40 to 100 million and mortality rates as high as 80 to 90 percent. Compounded by wars that raged over trade goods and land, most survivors suffered through serious socioeconomic and political disruptions. This included forced removal of entire groups to reservations far from their traditional homelands. The resumption of traditional environmental relationships and subsistence practices in new areas proved problematic for these groups.

Myths and Realities of Native American Environmental Relationships

From these early encounters, two ecological stereotypes of Native Americans emerged among Europeans, the "ignoble" and "noble savage." Both are rooted in perceptions of Native American peoples' relations with the environment. The "noble savage" appears in the earliest European writings about the New World, characterizing Native Americans as honest, carefree, peace-loving, wise, and "natural"—one with nature. Louis Armand, Baron de Lahontan, even critiqued Old World culture through his fictive native character Adario, who stood for the noble "other" to contrast Europeans. Jean-Jacques Rousseau similarly used Native Americans to argue that human nature is essentially good. These stereotypes were carried into recent centuries by romanticists, including Henry David Thoreau and Henry Wadsworth Longfellow.

The ignoble savage was born out of ethnic hatreds resulting from the violent conflicts of conquest, some religious ideologies, and the negative "other" of racism. It is exemplified in the writings of Lord Jeffrey Amherst, who infamously distributed blankets infested with smallpox. Proponents of this stereotype used the same "natural man" concept to portray Native Americans as "animal-like" or inhuman, irrational, and bloodthirsty.

By the mid-20th century, Americans were increasingly apprehensive about environmental degradation. Rachael Carson's 1962 book *Silent Spring* captured and ignited that sentiment, and the first Earth Day in 1970 inaugurated the "environmental decade." A year later, its spokesperson, played by Native American actor Iron Eyes Cody, made his first appearance. "The Crying Indian" debuted in a televised commercial sponsored by Keep America Beautiful, Inc. Cody instantly became the new archetype for the "noble savage" stereotype as it transformed into the "ecological Indian." Amid the political and social uncertainties of the day, many Native Americans embraced the stereotype for its positive connotations, and civil rights organizations incorporated ecological concerns into social justice campaigns.

Despite this perception, Native and Euro-Americans are equally capable of both environmental preservation and degradation. Shepard Krech critiqued blanket characterizations in his controversial 1999 book *The Ecological Indian*, which Native American groups reviled and anti-treaty rights groups used as anti-Native American propaganda. The root issue lies in the use of overgeneralizations concerning any group's talent for sustainable environmental stewardship. Preservation and degradation are more closely linked to a society's scale and economy than ethnic identity. Scientific study has shown that degradation linked to state-level agriculture in both ancient Europe and indigenous America more often occurred due to abandonment, war, disease, or climate change rather than mismanagement.

Modern Conflicts, Compromises, and Solutions

Recent environmental conflicts between corporations or governments and Native Americans have been resolved in a myriad of ways and with mixed results. The episodes described below illustrate that different natures of tribal identities, non-Native agents, issues, resources, and laws lead to diverse outcomes of cases fought across the continent. On occasion, common concerns have stimulated alliances between Native Americans and non-Native Americans against environmentally destructive mining, milling, and logging interests. Among the many mainstream Native American enterprises, some include a specific environmental focus, for example, modern ecological forestry, bison production, and ecotourism.

The U.S. Forest Service approved construction of ski facilities on northern California's Mount Shasta, sacred to many California tribes. However, the development was blocked through the skilled use of federal environmental regulations.

Disreputable practices secured corporate and government officials the profitable rights to uranium on Navajo reservations, essentially without compensation. Extreme poverty led Navajo to low-paid, unprotected mining work, resulting in high incidences of cancer. The Radiation Exposure Compensation Act of 1990 was passed, but required culturally inappropriate documentation of work histories and relationships, excluding most families from payments.

Dam construction deprived the Wisconsin Menominee of the culturally and nutritionally significant lake sturgeon, a species susceptible to overharvesting. Menominee attempts to take fish from off-reservation spurred violent conflict with impoverished non-Native Americans whose livelihood relied on sport-fishing tourism. In 1993, a cooperative agreement between tribal, federal, and state governments led to successful reintroduction of sturgeon to the Menominee reservation.

Conclusion

Native American societies are as diverse as any on Earth. They have experienced the egalitarianism of minimalist foraging lifeways and the excesses of socially stratified state-level societies, yielding sequences of ecological resilience and degradation. Only ongoing multidisciplinary study can help to begin eliminating persistent positive and negative environmental stereotypes and facilitate a future of cooperative environmental relationships.

See Also: Environmental Law, U.S.; Human Geography; Political Persuasion; Social Action, National and Local.

Further Readings

Cronon, William. *Changes in the Land: Indians, Colonists, and the Ecology of New England.* New York: Hill and Wang, 1983.

Deloria, Vine, Jr. *Custer Died for Your Sins: An Indian Manifesto*. New York: Macmillan, 1969.

Gutiérrez, Ramón. *When Jesus Came, the Corn Mothers Went Away: Marriage, Sexuality, and Power in New Mexico, 1500–1846*. Stanford, CA: Stanford University Press, 1991.

Krech, S. *The Ecological Indian: Myth and History*. New York: W. W. Norton, 2000.

Mann, C. *1491: New Revelations of the Americas Before Columbus*. New York: Vintage Books, 2005.

White, Richard. *The Roots of Dependency: Subsistence, Environment, and Social Change among the Choctaws, Pawnees, and Navajo*. Lincoln: University of Nebraska Press, 1983.

Daniel Cadzow
State University of New York, Buffalo

Nature Experiences

Nature experiences are leisure activities outdoors, including camping, hiking, backpacking, mountain climbing, riding off-road vehicles like all-terrain vehicles or snowmobiles, and gardening. The environmental impact of all these activities varies, particularly according to the individual's practices and habits. In many ways, these activities can be seen as falling into two categories: gardening, which is discussed in more depth in its own entry, brings nature to the home; other activities take one away from the home and into "the great outdoors," a place of ritualistic escape.

Hiking, for instance, is of relatively low impact as long as one is careful not to disturb the natural environment one is walking through, but littering, trampling plants, cutting down foliage to make a path, or bringing along loud portable radios are all actions that would drastically increase the negative impact.

While the use of trails originated for the sake of convenience and to avoid getting lost, they have the additional benefit of keeping hikers concentrated and thus limiting their impact. The footprints of a single hiker or a small group may not have much effect, but the cumulative effect of many hikers traveling through an area leads to soil compaction, which can have a number of deleterious effects on the ecosystem. Even this, though, is a mild effect compared to the soil compaction caused by heavy vehicles.

The effects of camping are much the same as hiking, but campers are more likely to produce waste, in addition to lighting campfires. While the emissions of a campfire are of relatively little concern in the big picture, green campers should bring firewood with them rather than gathering it from their campsite. The fallen wood in a forest decomposes and nourishes the soil. Furthermore, campers concerned with the environment should not strip branches from trees for use as tent poles, marshmallow or hot dog sticks, or other purposes. Living wood should almost always remain on the tree; an experienced forest guide may know which trees can benefit from pruning. Dishes should be washed in a container, with the used dishwater discarded away from freshwater sources and plants—dishes should never be washed directly in a body of water, nor should wastewater be dumped anyplace where plants will wind up drinking it. Truly green campers can use solar panels and crank-powered devices like flashlights and radios in lieu of batteries; in any event, batteries should never, under any circumstances, be disposed of in the woods. Cosmetics, chemical deodorants, and other creature comforts of suburban life should be left at home rather than contaminating water sources with them as a result of bathing or swimming. The basic tenet of green camping is "Leave No Trace": when one leaves, the forest should appear entirely undisturbed except for trail markers and other man-made artifacts that are part of the permanent campground (if applicable). Though a number of green camping products are manufactured, green habits are ultimately more important and have a greater effect than buying a better backpack.

Vehicles like all-terrain vehicles (ATVs), off-road vehicles (ORVs: jeeps, motorcycles, dune buggies, campers, pickup trucks, when used off-road), and snowmobiles are associated with "enjoying the great outdoors" as much as hiking is, but their environmental impact is severe and often overlooked. The body of research on the air quality impacts of these vehicles is incomplete, focusing principally on snowmobiles, but show that emissions exceed human health standards. Better understood is the effect of soil compaction and the shear forces of motor vehicles' turning wheels, which in off-road environments lead to mud holes and gullies, changing local hydrologic patterns and accelerating erosion, in addition to any ecosystem damage. The noise pollution caused by all these vehicles, often significantly worse than automobiles and, of course, much closer to wildlife habitats, can put stress on local animals. During mating season, stress can have a severe effect on the life cycle of the animals. One reason the body of research is greater for snowmobiles is because of the number of studies conducted to challenge the notion that snowmobiles have relatively little environmental impact because they ride on top of snow and thus usually do not disrupt soil. In actuality, snow compaction jeopardizes small mammals that live below the snow layer and depend on the loosely packed snow for insulation. Hard-packed snow is more thermally conductive and causes the ground below it to freeze to a greater depth, jeopardizing subterranean animal life, possibly interfering with plant life, and contributing to erosion during the spring thaw. Similarly, on beaches, compacted sand can disrupt the ability of turtles to nest.

Particularly when used outside designated areas, off-road vehicles destroy vegetation and fish-spawning areas, damage water supply sources, disrupt already-threatened wetland habitats, and increase the siltation of freshwater bodies. Conservation agencies regularly petition for ATV/ORV/snowmobile activity to be strictly regulated, limited to specific areas. Even in jurisdictions where it is, this is often not enforced unless a complaint is made. Many ATV riders think that they are safe and responsible so long as they avoid private property, but it is public lands and wilderness that are most likely to be home to the most vulnerable animal and plant ecosystems.

Until recently, there were few restrictions on snowmobiling in national parks, apart from wildlife reserves and other protected spaces. Because of their hydrocarbon emissions, however, snowmobiles account for 80 percent of total hydrocarbon emissions and 50 percent of carbon monoxide emissions at Yellowstone National Park in the winter, and for this reason, only the least-polluting style of snowmobile—the four-stroke engine, considerably less common than the two-stroke engine—is now allowed in Yellowstone in an effort to curb wintertime emissions levels. Consumer demand is driving an attempt to build a cleaner snowmobile, but even the use of the four-stroke engine is a fairly recent innovation, and anything cleaner is likely to be years away from market. Noise pollution, on the other hand, has been considerably improved; the average 21st-century snowmobile is only 10 percent as loud as the average 1960s snowmobile, provided it is not using an aftermarket exhaust system (which increases engine power but raises noise levels, despite sometimes being called a "silencer").

See Also: Environmental "Goods"; Forest Service; Gardening and Lawns; Leisure; Vacation.

Further Readings

Lodico, N. J. *The Environmental Effects of Off-Road Vehicles: A Review of the Literature.* Washington, DC: U.S. Department of the Interior, 1973.

Redclift, Michael. *Sustainable Development.* New York: Routledge, 1987.

Rogers, Elizabeth and Thomas Kostigen. *The Green Book.* New York: Three Rivers Press, 2007.

Vancini, F. W. *Policy and Management Considerations for Off Road Vehicles: Environmental and Social Impacts.* Ithaca, NY: Cornell University, 1989.

Bill Kte'pi
Independent Scholar

Nelson, Willie

Willie Hugh Nelson is an American singer, songwriter, actor, author, and social activist. Much of his advocacy work involves green initiatives, particularly with his ventures in biodiesel and agricultural reform.

Willie Nelson was born in 1933 in Abbott, Texas, to Ira Nelson and Myrle Greenhaw (née) after his parents left the isolation of north central Arkansas in 1929. Soon after his birth, Myrle returned to her family and Ira left Willie and his sister Bobbie Lee to the care his parents, Alfred and Nancy Nelson. Willie's childhood was set against a backdrop of poverty, and he worked from an early age, like most other Abbott children, picking cotton to supplement the family income. Music was central to the Nelson family's life and spirituality. Alfred and Nancy had been influential music teachers in Arkansas, and the family often ended a day of hard labor by playing and singing together. Nelson credits much of his subsequent political advocacy to his childhood experiences.

Music

Though active musically since the 1950s, Nelson is most recognized for his contributions to country music's "Outlaw Movement" during the late 1960s and 1970s. In reaction to the formulaic "Nashville sound," self-professed "Outlaws" including Nelson, Johnny Cash, and Waylon Jennings left Nashville for Texas and revived country's popularity. Increasingly disillusioned with the inauthenticity of corporate country, Outlaws encouraged creative control over material and production. Cementing Outlaws' popularity was the critically acclaimed *Red Headed Stranger* (1975), an album in which Nelson defied the conventions of the Nashville sound by writing a sparsely produced concept album.

Activism: Biofuels

Much of Nelson's environmental advocacy work is tied to the biodiesel industry, using biofuel based on vegetable oil to power his fleet of tour vehicles. His first experience with renewable fuels occurred during his sponsorship of Gatewood Galbraith's run in the Democratic primary of Kentucky's 1991 gubernatorial election. Both vocal advocates of the legalization of cannabis, Nelson and Galbraith toured Kentucky in a Mercedes fueled by hemp oil.

It wasn't until his fourth and current wife, Ann-Marie "Annie" D'Angelo, returned to their Maui home in a Volkswagen Jetta a decade later with a diesel engine converted to run on recycled cooking oil that Nelson's commercial interest in biofuels was sparked. In 2004, Willie and Annie partnered with Robert and Kelly King of Pacific Biodiesel, Inc., (PacBio) to build two biodiesel plants in Salem, Oregon, and Carl's Corner, Texas. Originally conceived in 1995 in response to environmental and health concerns around the volume of cooking grease ending up at the Central Maui Landfill, Pacific Biodiesel got its start when King built a plant to recycle the restaurant waste. Pacific Biodiesel currently operates 10 plants in the United States and diverts a combined 1,298 tons of waste per month at its Maui and Oahu plants.

In 2005, Nelson lent his name to "BioWillie" (Willie Nelson Biodiesel) in association with Peter J. Bell, chief executive officer of biofuel distributor Distribution Drive, to market the fuel to truck stops. BioWillie was exclusively licensed to Dallas's Earth Biofuels, Inc., the publicly traded parent of Distribution Drive active in the production, distribution, and sale of renewable fuels. Though creditors filed for involuntary bankruptcy of Earth Biofuels in 2007 after it amassed significant corporate debt, the company has since reorganized as Evolution Fuels with plans to build a chain of renewable fuel stations and convenience stores in the southern United States modeled after the Willie's Place Truck Stop at Carl's Corner, Texas.

In 2006, Nelson backed Richard S. "Kinky" Friedman's Texas gubernatorial campaign in endorsement of Friedman's support of alternative energy production and decriminalization of marijuana.

Activism: Farm Aid, Agricultural Reform

Tied to Nelson's biodiesel projects are his concerns about the plight of the American family farmer, particularly through the work he has done with Farm Aid. Nelson, along with Neil Young and John Mellencamp, organized the first Farm Aid concert on September 22, 1985, after Bob Dylan asked to reserve some of the proceeds of the Live Aid concert earlier that year to assist farmers in financial crisis. Dave Matthews subsequently joined the board of directors in 2001. In 2007, Nelson partnered with Ben & Jerry's to release "Willie Nelson's Country Peach Cobbler" ice cream, with the proceeds of sales donated to Farm Aid.

Initially intended to prevent foreclosures and provide disaster/emergency relief by funding farm and rural service organizations, Farm Aid has expanded to fight corporate agribusiness factory farms and plummeting crop prices. Outreach efforts through the Good Food Movement encourage Americans to buy local and family farmed food. Campaigns for long-term policy initiatives challenge Congress to enact farm bills supporting family farm interests rather than those of industrial agriculture. Nelson advocates a wide view of the farm bill by framing it around issues of nutrition, environmental stewardship, and renewable energy. Defending both family farms and biodiesel, Nelson encourages family farm production of biofuels, expanding markets for farmers while driving sales of clean-burning fuels. As a proponent of domestic energy independence, Nelson envisions the combination of biodiesel farming as a means to reduce America's reliance on Big Oil and foreign nonrenewable energy while stimulating the national economy.

Nelson's support of biodiesel and local agrarianism has been invaluable in raising public awareness of the issues. Nelson also works with the National Organization for the Reform of Marijuana Laws for the legalization of marijuana and has recently become involved with animal welfare campaigns, working with Habitat for Horses and the Animal Legal Defense Fund to campaign for the humane treatment of horses and improve the living conditions of animals within the dairy industry.

See Also: "Agri-Culture"; Hybrid Cars; Music; Musicians; Popular Green Culture; Social Action, National and Local.

Further Readings

Associated Press. "Willie Nelson Bets on Biodiesel." *Wired* (January 14, 2005). http://www.wired.com/science/discoveries/news/2005/01/66288 (Accessed May 2010).

"Corporate Report: Earth Biofuels, Inc." *Herald Tribune* (July 13, 2007). http://www.heraldtribune.com/article/20070713/BUSINESS/707130592/1007?p=1&tc=pg (Accessed May 2010).
Cusick, Richard. "The Pope of Austin, Texas." *High Times* (January 2008 http://hightimes.com/lounge/rick/6419?utm_source=rss_home (Accessed May 2010).
Essex, Andrew. "Farm Aid at Fifteen: Willie Nelson Keeps Up the Fight." *Rolling Stone* (October 26, 2000).
Hakim, Danny. "His Car Smelling Like French Fries, Willie Nelson Sells Biodiesel." *New York Times* (December 30, 2005. http://www.nytimes.com/2005/12/30/business/30biowillie.html (Accessed May 2010).
Nelson, Willie. "Take Action: Support a Better Farm Bill." *Mother Earth News* (June/July 2007). http://www.motherearthnews.com/Sustainable-Farming/2007-06-01/Take-Action-Support-a-Better-Farm-Bill.aspx (Accessed May 2010).
Nelson, Willie and Bud Drake. *Willie: An Autobiography.* New York: Simon & Schuster, 1988.
Nelson, Willie and Turk Pipkin. *The Tao of Willie: A Guide to the Happiness in Your Heart.* New York: Gotham Books, 2006.
Patoski, Joe Nick. *Willie Nelson: An Epic Life.* New York: Little, Brown, 2008.

Krista Weger
York University

Nongovernmental Organizations

Nongovernmental organizations (NGOs) have been described as associations formed on personal initiative by a few committed people dedicated to the design, study, and implementation of development projects at the grassroots level. They work outside the government structures but function within the legal framework of a country. The rise of NGOs as a strong institutional alternative reflects the growing recognition that the central government and private sector lack sufficient capacity to respond to the challenges of poverty alleviation and other forms of developments.

NGOs are more popular than ever in official circles these days. Just a few decades ago, their popularity lay largely in their supposed efficiency in meeting the basic needs of people at the grassroots level, especially in tackling poverty; today NGOs are being heralded, according to the United Nations Development Programme (UNDP) Human Development Report (1993), as representative par excellence of civil societies in the global south. In the post–Cold War era, international institutions and donor agencies have turned their attention increasingly to concerns about democratization and popular participation. According to the UNDP report, greater people's participation is no longer a vague ideology based on the wishful thinking of new idealists but has become an imperative—a condition of survival.

NGOs and People's Participation

The ideas and institutions that traditionally governed aid to the global south have entered into crisis. There is growing disenchantment with the results of billions of dollars in official development assistance from state donors and with the very idea of the state as the main actor in promoting development. In the search for alternatives, NGOs have emerged as the main candidates to promote just and democratic forms of development. NGOs working in the area of development assistance are hardly a new phenomenon. They trace their origins back to the work of Western missionaries in the colonized world and adopted their present form in the post–World War II period when they began to provide relief to the peoples of war-torn Europe. For decades, NGOs were perceived as peripheral actors, providing humanitarian assistance and organizing small, grassroots development projects on the margins of the global south. Recently, however, NGOs themselves, as well as their observers, have begun to focus their role on supporting civil society. As long as supporting civil society is a main objective, popular participation becomes an essential element of a successful strategy. They have basically three strategies for promoting participation: instrumental, localized, and political.

Instrumental participation is valued primarily for its contribution to the efficient implementation of a project. Beneficiaries' participation is limited to involvement in the work and benefits of the project, and they do not have real control over the design and evaluation. Control is in the hands of an outside actor, which may be the state, a political party, or the NGO itself. Localized participation attempts to increase the abilities of participants to run their own affairs; it involves a real degree of control by local organizations but does not promote broader participation in processes of social change. Although some believe that increased participation at the local level and empowerment of individuals are necessary elements in the transformation of broader power structures, the connection between localized participation and national democratization is unclear. Political empowerment aims not only at greater control by community members at the local level but also their involvement in the broader social movement seeking increased political participation by excluded groups in national decision-making processes. Ideally, all these activities and missions of NGOs have been well perceived by, among others, various donors; however, criticisms about the role of NGOs are also widespread. As NGOs move from "development partner" to "development alternative," tensions also erupt between NGOs and states. NGOs' involvement in development projects and influencing national politics raises numerous questions and skepticisms. Northern NGOs have been subjected to considerable criticism by observers from the south for their arrogant and interventionist approaches. NGOs in Bangladesh are a good example, which will be examined later in this article.

Empowering Civil Society and Promoting Good Governance

The proliferation of NGOs in the developing countries is a common phenomenon these days. NGOs have been credited for empowering civil society and promoting good governance in various parts of the world. However, some critics claim that for NGOs to operate, for markets to penetrate and to hold authority, for private organizations to take hold of the societies' power, the first thing that should be done is to diminish the power and authority of the state by curtailing its role in providing services to its citizens and by reducing its control of resources. This is a neoliberal agenda; it advocates the rhetoric of "good governance," which is paradoxical in meaning and operation. G. Wood calls this scenario "franchise state" (state franchising its responsibility to NGOs).

In the global north, good governance is generally explained as a democratic process with strong accountability between state and people, removing the prospects of dictatorial oppressive governments and underpinning, therefore, the protection of fundamental human rights. Wood calls it "hypocrisy" embodied in the Western preoccupation of the theme "good governance." He argues that "good governance" represents a revival of ethnocentric, modernizing ideology, attempting to make the myths of one society reality in another. Giving the example of the United Kingdom, he claims that good governance is more possible elsewhere than in those countries that purport to be the keepers of the discourse.

When one talks about "good governance," there arise many questions and problems, especially the problem of accountability. First of all, "good" is not universal, rather relative, and contingent upon cultural expectations and distributional outcomes. The paradoxes in the notion of good governance, as Wood puts it, include the following: First, the thrust of policy is to undermine the monopoly of the state in service provision and allocation of resources, thereby creating more opportunity for exit choices and thus reducing the necessity for government to be good. Second, the preoccupations with privatizations and markets on the one hand, and good governance on the other, do not easily sit side by side. Third, adherence to neoliberal views about the efficacy and the responsiveness of the market as an allocator of public goods crucially slides over the issue of responsibility. However, markets tend to ignore responsibility and have been proved to be failures in distributing resources. Markets, rather, serve the capitalists for accumulation and legitimation. Fourth, good governance is geared to improve participation. It is very contradictory, as most NGOs are operated in an authoritative manner. Finally, good governance undermines and limits the capacity and power of the state, but the state remains responsible for defining, guaranteeing, and regulating entitlements on the one hand and delivery on the other. NGOs that are operating to improve "good governance" are basically working to break the state monopolies in both service and goods delivery and to remove regulations and licensing to allow the market to breathe.

If Wood is right, one can argue that the "franchise model" cannot be alternative to the state and the market because markets have been proved to be inefficient allocators. Despite criticisms and skepticism, there are, however, empirical proofs that NGOs play a vital role in empowering local communities, creating vigorous civil societies, ensuring participation of the local community in development activities, and in making development more meaningful and accepted for them. Various environmental and social justice NGOs have become successful in creating environmental and social justice awareness and policies that were lacking before.

NGOs in Bangladesh as a Case Study

Following the war of independence (1971), Bangladesh witnessed a large scale of foreign aid from the developed countries. This was the starting point of NGO activities in Bangladesh. Now, Bangladesh has an NGO density of about four per square kilometer. However, with the growth of the enormous number of NGOs in Bangladesh, many people have been critical of them. One of the reasons is NGOs' involvement in national politics, including their participation in the national elections. The critics claim that the NGOs are running an invisible government, which is a threat to the national security and an assault on the culture of the country.

The origin of NGOs in Bangladesh can be traced back to Christian missionary activities during colonization. No one is certain about the exact time that NGO activities started in Bangladesh. Bengal (present-day Bangladesh) was one of the richest lands in the world during medieval times. According to some scholars, the first NGO activities were started by Christian missionary groups. Like the present rulers of Bangladesh, the Great

Moghal rulers of Delhi (the present capital city of India) opened the doors of the province of Bengal to Christian missionaries in 1517. By 1793, the famous British missionary William Carey had arrived in Calcutta (the capital city of West Bengal in India) and started missionary as well as social welfare activities through the missionary-run schools established all over Bengal.

Until the Liberation of Bangladesh in 1971, there were only a few NGOs funded by donor agencies or foreign NGOs working in the region. In the post-1971 era, however, growth of NGOs has mushroomed in Bangladesh. Initially, NGOs helped in rebuilding and rehabilitating the war-ravaged country and then in trying to eradicate poverty, unemployment, malnutrition, illiteracy, and disease. The 1980s and 1990s witnessed a new phenomenon as the number of international and national NGOs operating in Bangladesh increased sharply. Many took unprecedented decisions to establish field offices, especially those traditionally operating programs from headquarter offices. The main reason is that Bangladesh is a densely populated country where the government has not been able to provide food, clothes, and shelter for its poor population. Another reason for the emergence of the hundreds of NGOs is the enormous violations of human rights in the country. Donor agencies encourage NGOs to protect and promote human rights. Similarly, the government neglect of women and children has given rise to many organizing programs for the betterment of women and children. Donor agencies have shifted their programs from the government to the NGOs because of the failure of the government in providing sufficient services and input. Presently, there are about 20,000 local and international NGOs registered in Bangladesh, providing services to nearly one-fifth of the population of 150 million.

Despite the activities of NGOs in Bangladesh with their main agenda of poverty alleviation, poverty still persists as a major problem in Bangladesh. According to an Asian Development Bank (ADB) report, Bangladesh remains one of the poorest countries in the world with a per capita income of approximately $337 in 1998 and nearly one-half of the population living below the poverty line. ADB also reported that, although poverty is the central thrust of the development agenda of the government and all the major funding agencies in Bangladesh and world's active NGOs and microcredit movement, the pace of poverty reduction has been very slow. One research study on the success and expenditures of NGOs in Bangladesh found that only about 5 percent goes to the target group, and the huge amount of money is spent for other purposes. The Bangladesh NGO Bureau identified that more than 52 NGOs are directly involved in Christian missionary activities under the banner of development activities.

NGOs have established thousands of schools all over the country. For instance, Bangladesh Rural Advancement Committee (BRAC) has established more than 5,000 primary schools and kindergartens where about half a million students are enrolled. It is very interesting that government schools have 15 percent attendance whereas NGOs have about 90 percent attendance. Consequently, the literacy rate has been increased. Latest statistics show that the adult literacy rate has increased over the past few years, especially among females, from 16 percent to 38.1 percent. The literacy rate has increased from 23.8 percent in 1981 to 47.3 percent in 1996. This is, of course, a success for NGOs, though the quality of education is highly questioned.

Although NGOs are nonprofit organizations, research shows that the majority of the NGOs are exploiting the rural people, especially women, through microcredit programs. An analysis of microfinance data of 495 NGOs in the country indicates that about 95 percent of NGOs are directly involved in such suppression. It has been noted that 95 percent of NGOs charge interest rates in the range of 12–25 percent. More than 70 percent of NGOs commonly use an interest rate of 15 percent per annum, whereas the government-initiated private company that provides loans to the NGOs charges interest at 3–5 percent. The interest rates of the top 20 NGOs (out of 495) range between 12 and 15 percent, and these NGOs have lent money to 85 percent of the total borrowers in the sector. The largest NGOs like BRAC, ASA, Proshika, and Grameen Bank charge 15 percent, 12.5 percent, 18 percent, and 20 percent, respectively. Most of the NGOs' loan products are underpriced, effectively at 25 percent approximately, just keeping roughly 5 percent as margin.

Despite being nonpolitical organizations, many believe that NGOs are involved in the politics of the country. In 1994, the government NGO Bureau identified 16 NGOs that were politically active. The NGOs also publish magazines and newspapers and take political stands. They provide financial assistance to political parties of their choice. Even after the 1996 election, some opposition political parties claimed that the NGOs seriously campaigned for the then–ruling party candidates. Some claim that NGOs are increasingly assuming the role of an invisible government, having little regard for the history, culture, customs of the people, and rules and regulations of the government. NGOs run a very powerful parallel government, and they can undo any order of the government anytime they like.

Some NGOs with their northern backing sometimes humiliate the government of Bangladesh. When the NGO Bureau of the government took action against two powerful NGOs—ADAB (Association of Development Agencies of Bangladesh) and SEBA (Society for Economic and Basic Administration)—canceling their registration on the grounds of defalcation of funds and receiving money from a foreign embassy without the permission or even the knowledge of the government, the foreign embassies allegedly compelled the government to withdraw the cancellation order within three hours of the issuance of same. Since this humiliation, the democratic government of the country has refrained from taking action against the corrupt NGOs and their executives involved in undesirable activities, including violating government rules and indulging in political activities. Some harsh critics term some foreign NGOs as "Neo–East India Company" in Bangladesh. Different newspapers highlighted the helplessness of the government with three- to four-column headlines such as "The government yields to dishonest and vicious NGO circle," "Plot to snatch away our independence through the tactics of East India Company," and "The government fails to control NGOs," and so forth.

Khaleda Zia's government (1991–96) indicated that it would like to enact a law to regulate the activities of the NGOs and to ensure that they operate within the ethics and discipline of the country's social and legal framework. But the ADAB coordinator warned that it would create a suffocating atmosphere and make it impossible for the NGOs to conduct development work. ADAB also rejected the proposed law as incompatible with the government's stated policy of democratization and liberalization. However, the Awami League (AL) government (1996–2001) was soft-minded toward the NGOs and missionaries, since the party got tremendous help from them during the parliamentary election in 1996.

The public of Bangladesh has launched a massive protest against missionary NGOs. The simmering crisis spilled into the streets in 1994 when large demonstrations against the NGOs took place in various towns and cities of Bangladesh. They claimed that the government must act before the country lost its sovereignty and the nation its identity. The main reason for this protest was that great majorities of the NGOs were allegedly engaged in missionary activities. During the cyclone in 1991, hundreds of people stormed the office of an NGO in Kutubdia in Chittagong to protest against being asked to change their faith if they wanted to receive the relief supplies. Muslims were given a meager share of the relief aid. The Church of Bangladesh argued it had come from "Christian" countries.

In Bangladesh, tension between NGOs and religious groups always persists. NGOs' allegation against the Islamic groups is that they (the Islamic groups) have challenged NGOs involved in the rural development, female education, and income-generation activities by women. For instance, the Bangladesh Rural Advancement Committee (BRAC) complained that 1,400 of its 20,000 schools are vandalized and a good number of them burned. The Grameen Bank, another well-known NGO, has complained about the fundamentalists. In 1994, when the United Action Council (UAC), a coalition of 13 religious parties and supporting groups, pushed the government to bring NGO activities under its control, it was seen as a collision between liberals and fundamentalists.

What we see is that despite having tremendous success in alleviating poverty, providing basic needs to the needy population, and creating environmental and social awareness, many NGOs are in acute tensions with governments and sometimes with local people. NGOs still remain a very powerful development partner, if not alternative, and therefore, cooperation between the state and the NGOs needs to be fostered for a greater result.

See Also: Ecotourism; Environmental Justice Movements; Law and Culture; Organizations and Unions.

Further Readings

Holloway, R. *Supporting Citizens' Initiatives: Bangladesh's NGOs and Society.* London: International Technology Publications, 1998.
Hossain, F. "Mullahs Escalate NGO-Bashing." *Holiday International* (April 1, 1994).
Islam, S. S. "Impact of Technology and NGOs on Social Development: The Case of Bangladesh." *Journal of South Asian and Middle Eastern Studies*, 21/2 (Winter 1998).
Keck, M. E. and K. Sikkink. *Activists Beyond Borders: Advocacy Networks in International Politics*. Ithaca, NY: Cornell University Press, 1998.
Khan, A. K. *Christian Missions in Bangladesh: A Survey.* London: Islamic Foundation, 1981.
Nuruzzaman, M. "NGOs: The Web of New Colonialism: Aid Merchant Buying Up Sonar Bangla." *Impact International* (July 1994).
Rashiduzzaman, M. "Islam, Muslim Identity and Nationalism in Bangladesh." *Journal of South Asian and Middle Eastern Studies*, 18/1 (Fall 1994).
Wood, G. "States Without Citizens: The Problem of Franchise State." In *NGOs, States and Donors: Too Close for Comfort?* David Hulme and Michael Edwards, eds. New York: St. Martin's Press, 1997.

Md Saidul Islam
Nanyang Technological University

Novels and Nonfiction

American writer Nathaniel Hawthorne (1804–64) once wrote that while words are powerless on their own, they become a "potent for good and evil . . . in the hands of one who knows how to combine them." This has certainly been true of words written by environmentally conscious writers who have done so much to inform the public about environmental threats that include global warming, devastation to valuable and limited resources, toxic waste dumping, and vanishing species of animals, birds, sea life, and insects. Rachel Carson (1907–64), who was a zoologist and ecologist as well as a writer, understood the need to protect the environment long before it became an issue of public concern, and one of the most powerful pieces of environmentalist literature is *Silent Spring*, published in 1962. It quickly became the most controversial book of the year. For her efforts in promoting environmental issues, Carson has been credited with launching the modern environmentalist movement in the United States.

Rachel Carson was a major influence in shaping the environmentalist views of President John F. Kennedy. In 1963, she testified before the U.S. Senate about the dangers inherent in using pesticides such as DDT, which was later banned. Carson is often credited with being the main influence in the creation of the Environmental Protection Agency in 1970. She was repeatedly honored for her efforts to educate the public about the role that animals, birds, and sea life play in the ecology of the planet. The anger that Carson faced from the chemical industry and antienvironmentalists has not faded since her death from breast cancer in 1964. In honor of her 100th birthday, in 2007 Congress debated a bill acknowledging her contributions to the American environmentalist movement. In response, Senator Tom Coburn (R-OK) lashed out at Carson, calling her a "junk scientist" and charging her with the responsibility for allowing malaria victims around the world to die because of the banning of DDT.

After the stage was set by Rachel Carson, a score of writers of both novels and nonfiction began to take on the subject of environmentalism, raising public awareness of the issue and influencing the founding of an international environmental movement, the passage of laws and regulations, and the establishment of both public and private agencies devoted to sustainability.

One of the most important lessons that environmentalist writers have to offer is that the subject is not confined to a single discipline. It encompasses politics, literature, economics, engineering, biology, physics, business, philosophy, architecture, nutrition, and a host of other fields.

Reading Lists

Based on the notion that great books serve as motivators for change, a number of individuals and groups have developed "green" reading lists. Some of the books included on the Green Warrior Reading List, for instance, are *Asphalt Nation: How the Automobile Took Over the Nation and How We Can Take It Back* (1998), Jane Holtz Kay; *A Better Place to Live, Reshaping the American Suburb* (1997), Philip Langdon; *Geography of Nowhere: The Rise and Decline of America's Man-Made Landscape* (1994), James Howard Kunstler; *My Ishmael* (1998), Daniel Quinn; *The Next American Metropolis: Ecology, Community, and the American Dream* (1993), Peter Calthorpe; *Downsize This!* (1997), Michael Moore; *Common Courage* (1994), Noam Chomsky; *The Tao of Pooh* (1983), Benjamin Hoff; *Cool Companies: How the Best Businesses Boost Profits and Productivity by Cutting Greenhouse Emissions* (1999), Joseph J. Romm; and *The Sand Dollar and the Slide-Rule: Drawing Blueprints From Nature* (1997), Delta Willis.

In its list of books that have had a significant impact on the environment, the *New World Encyclopedia* identifies 14 nonfiction books that span a period of 152 years: *High Tide: The Truth About Our Climate Crisis* (2003), Mark Lynas; *Crimes Against Nature* (2004), Robert F. Kennedy, Jr.; *A Sand Country Almanac* (originally published in

1949 and reprinted in 1966), Aldo Leopold; *Desert Solitaire* (1968), Edward Abbey; *Silent Spring* (1962), Rachel Carson; *Walden* (1854), Henry David Thoreau; *The Everglades: River of Grass* (1978), Marjory Stoneman Douglas; *The Global Environmental Movement* (1995), John McCormick; *Encounters With the Archdruid* (1971), John McPhee; *Man and Nature* (1864), George Perkins Marsh; *The Consumer's Guide to Effective Environmental Choices: Practical Advice From the Union of Concerned Scientists* (1999), Michael Brower and Warren Leon; *The World According to Pimm* (2001), Stuart L. Pimm; *An Inconvenient Truth* (2006), Al Gore; and *The Revenge of Gaia* (2006), James Lovelock. Only three fiction books are considered significant enough to be included in the list: *The Monkey Wrench Gang* (1975) by Edward Abbey, *The Lorax* (1971) by Dr. Seuss, and *Hoot* (2002) by Carl Hiaasen. It is important to note that both the second and third books on this list of fiction are children's books.

Many academic journals also frequently offer reading lists of current books that are potentially important to the environmentally conscious reader. One such list compiled by *Bioscience* in 2009 included *Avian Invasions: The Ecology and Evolution of Exotic Birds*, Tim M. Blackburn et al; *Cave Biology: Life in Darkness*, Aldemaro Romero; *Chasing Molecules: Poisonous Products, Human Health, and the Promise of Green Chemistry*, Elizabeth Grossman; *Dinosaur Odyssey: Fossil Threads in the Web of Life,* Scott D. Sampson; *Mr. Jefferson and the Giant Moose: Natural History in Early America,* Lee Alan Dugatkin; and *Sexy Orchids Make Lousy Lovers: And Other Unusual Relationships,* Marty Crump.

General Novels

One of the first novels to combine environmentalism with fiction was Edward Abbey's *The Monkey Wrench Gang* (1975), which tells the tale of a group of misfits that includes a Vietnam veteran (George Hayduke), a New York–born feminist saboteur, a wilderness guide, and a libertarian who sets fire to billboards. This unlikely group joins forces to save the desert of the American Southwest from encroaching development. Abbey continues to promote environmentalism in the sequel, *Hayduke Lives!* (1990), in which the protagonist sets out to stop the progress of earth-moving machines through his beloved desert. In Steve Amick's *The Lake, the River, and the Other Lake* (2005), another Vietnam veteran (Roger Drinkwater) is determined to save Michigan's Lake Meenigeesis and Oh-John-Ninny River from jet skiers and polluters. Tom Barbash sets *The Last Good Chance* (2002) in Lakeland, New York, where hometown-boy-made-good Jack Lambeau becomes involved in the effort to revive a dying port town. In *The Lives of Rocks: Stories* (2006), Rick Bass turns to the short story to deal with a wide range of environmental issues. The story told by Don DeLillo in *Underworld* (1997) covers a span of 50 years, allowing the reader to see the evolution of the environmental movement. Environmentalism also takes center stage in Jonathan Franzen's *Strong Motion* (1992), when a seismologist uncovers human causes for a devastating earthquake in Boston.

Genre Fiction

Many novels of mystery and suspense also deal with environmentalism, including Liz Adair's *Snakewater Affair* (2004), part of the Spider Latham mystery series, and Grace Alexander's *Hegemon: A Novel* (1996), which deals with efforts to poison the Czech water supply. In Lindsay G. Arthur's *The Litigators* (2005), a young lawyer takes on the case of a gravely ill mother who has developed neurological problems as the result of living near an abandoned gas station, which served as the site for an experimental cleanup product. Alaska salmon runs provide the focal point for Michael T. Barbour's *The Kenai Catastrophe* (2002). Barbour turns to Hawaii in *Blue Water, Blue Island* (2004), dealing with threats to the state's vulnerable coral reefs. Best-selling mystery writer Nevada Barr sends her protagonist Anna Pigeon to the ruins of the Mesa Verde National Park in Colorado in *Ill Wind* (1995). *Silent Justice*, the ninth book in William Bernhardt's Ben Kincaid series, deals with a class-action suit involving the pollution of a local water supply by toxic chemicals. Michael Black also deals with illegal disposal of toxic waste in *A Killing Frost* (2002).

Known for his compelling medical thrillers, Robin Cook has also used environmental themes in his books. In *Fever* (1982), a research physicist bucks the medical establishment in his quest to find a cure for his daughter's leukemia. Donna Cousins's *Landscape* (2005) has a similar plotline, but the main character is a business owner involved in a hazardous waste exposé. Hazardous waste is also the theme of *The Wise Pelican* (2000) by Rubin Douglas, *Dead and Buried* (2001) by Howard Engel, and *Countdown in Alaska* (2005) by Lidia Llamas-LoPinto and Charles LoPinto. David Michael Donovan chooses the area of Louisiana known as "Cancer Alley" as the setting of *Evil Down in the Alley* (1999). Linda H. Kistler also focuses on environmental causes of cancer in *Cause for Concern* (2003). British writer William Haggard adds the mafia to his tale of illegal waste dumping in *The Vendettists* (1993). Threats to biodiversity shape the story lines of Janet Dawson's *Don't Turn Your Back on the Ocean* (1994) as well as Ken Goddard's *Prey* (1992), *Wildfire* (1994), and *Double Blind* (1997); and *Greenwar* (1997) by Steven Gould and Laura J. Mixon. John Grisham's *The Pelican Brief* (1992) on the same subject was made into a blockbuster movie starring Julia Roberts and Denzel Washington. Sara Paretsky's V. I. Warshawski, who was featured in a 1991 movie starring Kathleen Turner, also deals with environmental themes in *Blood Shot* (1988). Environmental threats coupled with ethnic controversies make up the plot of John Hockenberry's *A River out of Eden* (2001).

The genres of women's fiction, chic lit, and romance have also dealt with environmental issues to a lesser extent. In the English romance *A Fabulous Fling* (2000) by Geraldine Bedell, the protagonist finds love with a television environmentalist. In Betsy Burke's Canadian novel *Hardly Working* (2005), most of the action takes place in an environmental protection agency called Green World International. Pollution at a paper mill is the focus of Barbara Delinsky's *Looking for Peyton Place: A Novel* (2005). Jill Giencke also deals with environmental issues in *Still Waters* (1991).

Despite the use of environmentalism as a common theme in other genres, it is the science fiction and fantasy genre that is most closely associated with incorporating environmental issues into story lines. British writer and philosopher Olaf Stapledon (1886–1950), who is best known for *Last and First Men* (1930) and *Star Maker* (1937), was one of the first to use works of science fiction to call public attention to environmental issues. California-born writer David Brin, who was influenced by Stapledon and by environmentally conscious movies such as *Soylent Green* and *Silent Running* during the 1960s, admits to being an environmental activist and acknowledges that this influences his writing, motivating him to repeatedly use words such as *sustainability* and *diversity* in works that include the *Uplift* novels, for which he is best known. Brin names Kim Stanley Robinson, Greg Bear, Octavia Butler, and Nancy Kress as other sci-fi writers who consciously deal with environmentalism in their works.

An art critic exposed to radioactive waste becomes a superhero in Michael Bishop's *Count Geiger's Blues* (1992). Unregulated waste products from swine and illegal animal research shape the plot of Bryan Eytcheson's *Host* (2005). Magic and toxic materials form the joint plot of Alan Dean Foster's *Slipt* (1987) and Harry Turtledove's *The Case of the Toxic Spell Dump* (1993). The search for alternative sources of power is the subject of Stewart Farrar's *A Novel of Eco-magick* (2007). A male with environmentally motivated super strength is the focus of Gary Goshgarian's *Rough Beast* (1995). Toxic waste imbues a nerdy janitor with superpowers in Lloyd Kaufman's *The Toxic Avenger* (2006). Protecting the planet is the theme of both Judith Moffett's *The Ragged World: A Novel of the Hefn on Earth* (1992) and Charles Sheffield's *How to Save the World* (1995).

Juvenile and Young Adult Books

Teaching children about protecting the environment is essential to creating a concerned citizenry of future writers, politicians, scientists, and business owners. Lists of recommended children's books about the environment are available through a number of resources such as Teachers.net, the Center for Environmental Education, the National Association for Humane and Environmental Education, and Ecobusiness. Many of the children's books on "green" lists convey the message that environmentalism should be an essential element of daily life. At the same time, these books encourage children to develop a sense of environmental responsibility. Books written for young children often teach environmentalism through pictures as well as words, including those by Australian writer and illustrator Jeannie Baker. Simple messages, such as those presented in Ruth Lercher Bornstein's *Rabbit's Good News* and Marcia Bowden's *Nature for the Very Young: A Handbook of Indoor and Outdoor Activities* keep the message simple enough so that even young children can learn to appreciate nature.

The children's book most often considered to have had a significant influence on developing environmental responsibility in young readers is *The Lorax* by Dr. Seuss. The Lorax clearly proclaims, "I speak for the trees, for the trees have no tongues." While many other books have not received the recognition of *The Lorax*, they have exerted considerable influence in shaping children's viewpoints. In *Alejandro's Gift*, which was a Reading Rainbow selection, Richard E. Albert teaches children about the importance of gardening. In Chris Van Allsburg's *Just a Dream*, the protagonist discovers that the threat of a neighborhood landfill makes environmentalism a personal issue. Likewise, Sheri Amsel and Jim Arnosky pack strong environmentalist messages into *A Wetland Walk* (1993) and *Guide to Knowing Animal Habitats* (1997). Harry Behn's *Trees* not only conveys an environmentalist message, it puts environmentalism into practice by being printed on recycled paper.

One popular theme in children's environmentalist literature is the combining of significant historical figures from various fields with lessons about the environment. In *Linnea in Monet's Garden*, for instance, Swedish writer Christina Bjork uses reproductions and actual photographs to describe a fictional journey to the garden of French artist Claude Monet. At times, historical names are used only to catch the public's attention. In *Winston of Churchill: One Bear's Battle Against Global Warming,* author Jean Davies Okimoto tells the story of a polar bear named Winston who lives in Churchill in Manitoba, Canada, to explain how carbon dioxide, methane, and nitrous emissions are destroying the planet along with polar bear habitats. Poetry and biography also are used to promote sustainability. In *The Earth Is Painted Green: A Garden of Poems About Our Planet*, Barbara Brenner includes the poems of writers as diverse as Carl Sandburg and Shel Silverstein; and biographies such as Leila M. Foster's *The Story of Rachel Carson and the Environmental Movement* give young readers a context for the entire range of environmental issues.

Children's classics such as Rachel Carson's *The Sense of Wonder* (1965) are being reprinted to teach environmentalism.

Protecting the rainforests is an ongoing theme in children's books, and Lynn Cherry's *The Great Kapok Tree* packs a powerful message about the loss of animal habitats in response to deforestation. *The Big Book for the Planet*, edited by established children's writers Ann Durrell, Jean Craighead George, and Katherine Paterson, covers a range of environmental issues that include overpopulation, pollution, and waste disposal. A number of books, including Pamela Friedman's *Earth Day Activities* and Cait Johnson's *Celebrating the Great Mother: A Handbook of Earth-Honoring Activities for Parents and Children,* provide young people with hands-on projects that give them opportunities to protect the environment around them. Environmental protection is also the theme and British Columbia is the setting in Rosemary Allison and Ann Powell's *Mrs. Beaver Goes West* (1993). Laurence Anholt and Arthur Robbins offer a twist on a familiar children's story in *Eco-Wolf and the Three Pigs* (2004).

Young adults tend to be concerned about the environment, and a number of books written for this audience address that concern. Protecting biodiversity is the issue in Chester Aaron's *Spill*; the rainforest adventures of Larry Bischof et al.'s Widget, an alien hero, in *Amazon Adventures* (1992); and Joe Cardillo's *Pulse: A Novel* (1996). Dina Anastasio's *Flipper* (1996) is about a boy and a dolphin threatened by environmental hazards. The familiar theme of illegal toxic dumps forms the plotline of Karen Ball's *The Hazardous Homestead* (1992) and Sigmund Brouer's *Cobra Threat* (1998). Intrigue surrounds the disappearance of an environmentalist concerned about health issues connected to the use of DDT in Christopher Barry's *Mosquito Point* (1996). The Girl Talk series examines environmental issues in L. E. Blair's *Earth Alert!* (1991) and *Queens* (1991). In *Green Boy* (2002), Susan Cooper explores environmental issues in the Bahamas and in the fictional land of Pagnaia. Protecting the Everglades and the Native American lands of the American Southwest are the main themes of Cynthia C. DeFelice's *Lostman's River* (1994) and Bev Doolittle and Elise Maclay's *The Earth Is My Mother* (1999), respectively. Endangered forests take center stage in Anthony Dorame's *Peril at Thunder Ridge* (1993). Journalist Linda Ellerbee turns her talents to young adult fiction in *Girl Reporter Blows Lid off Town!* (2000) about an 11-year-old who stumbles onto an environmental cover-up.

In addition to the sampling of books touched on in this article, there are scores of others. The ubiquitous nature of environmental themes in both novels and nonfiction suggests that writers are well aware of the importance of the subject to their audiences. Many also see the written word as a way to promote sustainability in a nonthreatening way.

See Also: Biodiversity Loss/Species Extinction; Popular Green Culture; Toxic/Hazardous Waste.

Further Readings

Anderson, Nancy. "An Annotated Bibliography of Children's Literature With Environmental Themes." http://teachers.net/archive/envirobks.html (Accessed September 2010).

Bordson, Lauren and Laura L. Barnes. "Environmental Novels: An Annotated Bibliography." http://www.istc.illinois.edu/info/library_docs/other_pubs/Environmental-Novels.pdf (Accessed September 2010).

Carson, Rachel. *The Sea Around Us*. New York: Oxford University Press, 1951.

Carson, Rachel. *Silent Spring*. New York: Houghton Mifflin, 1962.

"Carson's 'Silent Spring' Still Making Noise." *Weekend All Things Considered*. NPR. May 27, 2007.
Foster, John Bellamy. *The Ecological Revolution: Making Peace With the Planet*. New York: Monthly Review Press, 2009.
MacGregor, Sherilyn. *Beyond Mothering Earth: Ecological Citizenship and the Politics of Care*. Vancouver: UBC Press, 2006.
McLennan, Jason. "Appendix A: The Green Warrior Reading List." In *Philosophy of Sustainable Design*. Bainbridge Island, WA: Ecotone Publishing, 2004.
"New Titles." *Bioscience*, 59/11 (December 2009).
Schor, Juliet B. and Betsy Taylor. *Sustainable Planet: Solutions for the Twenty-First Century*. Boston: Beacon Press, 2002.
Starbridge, Saren. "An Interview About Science Fiction and the Environment." *Living Planet* (Winter 2000).
Steffen, Alex. *Worldchanging: A User's Guide for the 21st Century*. New York: Abrams, 2006.
Torgerson, Douglas. *The Promise of Green Politics: Environmentalism and the Public Sphere*. Durham, NC: Duke University Press, 1999.

Elizabeth Rholetter Purdy
Independent Scholar

Obesity and Health

Convenience foods, fast-food meals, and restaurant meals are all likely to be higher in calories and fat than home-cooked meals. Fast-food restaurants' increasing presence is seen globally; for instance, McDonald's operates in over 100 countries and has over 800 restaurants in China alone, including this one in Beijing.

Source: Wikimedia Commons

Obesity is becoming an increasing health concern worldwide and in developing as well as industrialized nations. Obesity is associated with increased risk for many serious health conditions as well as reduced quality of life, discrimination, and social stigma, resulting not only in shorter life span, fewer years of healthy life, and lower quality of life for the obese individual but also increased burden on the healthcare system. For instance, in the United States in 2006, it was estimated that medical costs for obese individuals were $1,429 higher (per person) than for normal-weight individuals and that overall medical costs associated with obesity were $147 billion. For the years 1987–2001, one study found that more than a quarter (27 percent) of increases in medical costs were associated with diseases related to obesity.

Obesity is customarily defined as increased body weight caused by excess accumulation of body fat (adiposity), but because direct measures of the percent of body fat are cumbersome, for broad-based studies such as population surveys, obesity for adults is usually defined using the Body Mass Index (BMI; also known as the Quetelet Index). The BMI is calculated by dividing body weight in kilograms by squared height in meters. According to the World Health Organization (WHO), a person with a BMI under 18.5 is underweight, a person with a BMI in the range 18.5 to 24.9 is normal weight, one with a BMI from 25.0 to 29.9 is overweight, and a person with a BMI over 30.0 is obese. Some also classify levels of obesity so that a BMI of 30.0 to 34.9 is level I, a BMI from 35.0 to 39.9 is level II, and a BMI of 40.0 or higher is level III or extreme obesity.

For children, overweight (the preferred terminology due to the stigma attached to obesity) is determined by percentile range using national growth charts (for instance, the Centers for Disease Control and Prevention [CDC] produces charts for use in the United States and the WHO produces them for international use). The percentile range method enables a physician to rank a child in relation to other children of the same age and gender, with the following categories currently in use: underweight = less than the 5th percentile; healthy weight = 5th percentile to less than the 85th percentile; at risk of overweight = 85th to less than the 95th percentile; overweight = 95th percentile or higher.

Overweight and obesity are major public health concerns because they are associated with increased morbidity and mortality: for instance, some studies have identified obesity as second only to tobacco smoking as a preventable cause of illness and premature death. Obesity is associated with the metabolic syndrome (defined by factors that include increased waist circumference, abnormal triglyceride levels, high blood pressure, decreased HDL cholesterol, and insulin resistance), which is a major risk factor for several diseases and health conditions, including stroke, type II diabetes, and coronary artery disease. Even if the metabolic syndrome is not present, obesity is a risk factor for many diseases, including osteoarthritis, sleep apnea, type II diabetes, stroke, coronary heart disease, and various cancers, including those of the endometrium, colon, gall bladder, prostate, kidney, and breast. Overweight children and adolescents have a higher probability of becoming obese adults than do normal-weight children, and serious conditions and diseases previously associated with adults are increasingly being diagnosed in obese children and adolescents as well. These include type II diabetes (previously called "adult-onset diabetes" because it rarely occurred in children), high blood pressure, and high cholesterol.

Obesity Prevalence

Obesity has been a major public health issue in the United States for at least two decades and, despite wide publicity and the implementation of many programs to combat the problem, a trend toward increasing or flat levels of obesity continues. For instance, the national health promotion initiative Healthy People 2010, which sets a series of health objectives for the nation, included as one of those objectives that obesity prevalence among adults be reduced to 15 percent. As of 2009, no state is meeting this goal: according to data from the annual Behavioral Risk Factor Surveillance System (BRFSS), 26.7 percent of American adults are obese, with rates by state ranging from 18.6 percent in Colorado to 34.4 percent in Mississippi. Nine states had an obesity prevalence rate over 30 percent, while only Colorado and Washington, D.C., had rates below 20 percent. Because the BRFSS uses self-reported data on height and weight to calculate obesity, it is likely the actual prevalence is even higher. Looking at children in the United States in 2007, 10.5 percent of children ages 2–5 were obese, as were 19.6 percent of those ages 6–11, and 18.1 percent of adolescents ages 12–19.

Perhaps even more disturbing than the high overall rates of obesity is the fact that obesity is more common among the more disadvantaged members of society, suggesting that poverty and other social factors play a role in the probability of an individual's becoming obese. For instance, for U.S. adults who did not graduate from high school, the obesity rate was 32.9 percent (versus 26.9 percent overall), and African Americans and Hispanics also had rates of obesity above the national average (36.8 percent and 30.7 percent, respectively). The same trends are noticeable among infants and children: in 2007–08, Hispanic boys were significantly more likely to be overweight than white boys, and African American girls were significantly more likely to be overweight than white girls.

Globally, the WHO estimates that in 2005, 1.6 billion adults (age 15 or older) were overweight and 400 million were obese; by 2010, if current trends continue, these numbers will increase to 2.3 billion overweight adults and 700 million obese adults. In addition, WHO estimates that in 2005, at least 20 million children under the age of 5 were overweight.

Obesity and the Environment

The main cause of overweight and obesity for most people is an energy imbalance: they consume more calories than they expend in physical activity. Changes in daily life are considered by many scholars to play an important role in global increases in the number of people who are overweight and obese. The major factors identified are (1) changes in diet toward foods away from traditional plant-based foods that are high in fiber and micronutrients (such as vitamins and minerals) toward processed foods and animal products that are high in calories, often from sugar and fat; and (2) decreased physical activity due to the changing demands of the workplace, increasing urbanization, and a shift to motorized rather than human-powered transportation (e.g., driving a car or riding a bus rather than walking or riding a bicycle). The pervasiveness of these changes worldwide is one reason that the problem of obesity has remained fairly intractable despite many government and private programs that have attempted to reverse the trend toward increasing obesity.

Many of the changes are related to social trends. For instance, the fact that in the United States it is common for both parents in a family to hold full-time jobs means that there is less time available for cooking, and thus a higher proportion of convenience foods are consumed within the home and a higher proportion of meals are taken outside the home (and convenience foods, fast-food meals, and restaurant meals are all likely to be higher in calories and fat than home-cooked meals). These trends are seen globally, including in developing countries: for instance, the fast-food chain McDonald's currently operates in over 100 countries and has over 800 restaurants in China alone. In addition, the process of food distribution has been seen as best handled by the free market, which has left some consumers at a disadvantage. For instance, the economic segregation seen in many communities has left some low-income areas as virtual "food deserts" without supermarkets offering healthy foods such as fruits and vegetables, whole grains, and low-fat milk but with multiple fast-food outlets offering high-calorie, low-nutrition foods. Although the term *food desert* was first applied to urban areas, a similar problem exists in some rural areas where the closing of local food markets has left people who do not have access to an automobile with limited opportunity to purchase healthy foods. In addition, more nutritious foods may be priced outside the range of low-income consumers, while fast foods offer a high number of calories for a low price.

Reduced physical activity is also a trend in the world as in the United States. Some of this is related to the increasing availability of motorized transportation and evolution of the workplace to more sedentary professions. Although in one sense these changes are beneficial (it is unlikely that many readers of this encyclopedia would care to spend their days as physical laborers or commit to lifelong use of only walking and bicycling for their transportation needs, for instance), when coupled with changes in food consumption, the result is often increased levels of obesity. Leisure-time physical activity has also decreased, with many factors playing a potential role, including the cost of joining a gym or a team, fears about exercising in parks or other public areas, and increased time devoted to work and commuting.

Even schoolchildren exhibit less physical activity than those of a few generations ago, in part because the move to consolidated school districts has made it less common for children to walk or bike to school, and increased emphasis on academics and/or cutbacks in the school budget have decreased or eliminated time devoted to physical education during the school day. While some children participate in organized sports leagues or after-school programs that include a high degree of physical activity, the availability of such programs is often related to family income (i.e., they are more likely to be available in wealthier areas). In addition, concerns about crime may motivate parents to keep their children indoors when not in school, so they are more likely to watch television or engage in other sedentary activities than to take part in active outdoor play.

Although traditionally behaviors as food selection and physical activity have been considered individual choices or characteristics that reflect the values and preferences of the individual making the choice, more recently scholars have turned their attention toward looking at social forces and circumstances that may influence those choices. Some such as Kelly Brownell have gone so far as to claim that the United States has a "toxic food environment" that promotes unhealthy eating habits and obesity. One of the implications of such a shift is that efforts to reduce obesity should focus not on motivating individuals to make different choices but on changing the choices available to them or the incentives offered for making different choices. To take the example of food deserts cited above, some would argue that it makes little sense to expect individuals to begin shaping their diet around whole grains, low-fat milk products, and fruits and vegetables if those products are not readily available to them at a price they can afford. Similarly, exhortations to increase physical activity levels make little sense if opportunities for safe, convenient, and low-cost physical activity are not available to the people receiving the message.

Some scholars believe that modern building and development patterns promote obesity. In particular, they have highlighted trends toward increasing urban sprawl, defined as an approach to community design that separates areas designated for living, working, and recreation and is characterized by a single-use, low-density design approach. Such design limits individuals' potential to walk or bike rather than use an automobile for transportation, not only due to safety issues (for instance, some suburbs have been built without sidewalks and with roads too narrow to safely accommodate bicycles and automobiles simultaneously), but also because they present an uninviting environment for physical activity (noise and pollution from roadways, lack of any facilities such as grocery stores within walking or biking distance). Studies have found that people living in environments characterized by urban sprawl tend to drive more and exercise less than people living in denser environments, although such studies have been complicated by the fact that individual preferences for activity and for an activity-friendly environment are not independent.

Criticisms of Obesity Research

Not everyone agrees that overweight and obesity are as great a public health risk as the CDC and other public health agencies claim. Some point to the fact that most statistics about obesity at the population level are calculated using the BMI, which does not directly measure an individual's proportion of body fat but is used in many public health studies because information about height and weight is more readily available than direct measures of proportion of body fat (such as underwater weighing), which require special equipment and trained technicians. A few highly publicized cases of professional athletes whose BMI indicated they were technically overweight or obese (for instance, the baseball player Mark McGwire) because they were extremely muscular (and thus had a high body weight relative to their height) have highlighted this weakness. In addition, some have found that sufficient physical activity can outweigh the negative health effects of obesity and charge that obese people have been unfairly targeted due to a societal prejudice against fat people.

See Also: Auto-Philia/Auto-Nomy; Diet/Nutrition; Gardening and Lawns; Locavores; Veganism/Vegetarianism.

Further Readings

Brownell, Kelly D. *Food Fight: The Inside Story of the Food Industry, America's Obesity Crisis & What We Can Do About It.* New York: McGraw-Hill, 2003.

Cardello, Hank and Doug Carr. *Stuffed: An Insider's Look at Who's (Really) Making America Fat.* New York: HarperCollins, 2009.

Centers for Disease Control and Prevention. "Childhood Overweight and Obesity." http://www.cdc.gov/obesity/childhood/index.html (Accessed September 2010).

Centers for Disease Control and Prevention. "Vital Signs: State-Specific Obesity Prevalence Among Adults—United States, 2009." *Morbidity and Mortality Weekly Report* (August 3, 2010). http://www.cdc.gov/mmwr/preview/mmwrhtml/mm59e0803a1.htm?s_cid=mm59e0803a1_e%0D%0A (Accessed September 2010).

Danei, Goodarz, Eric L. Ding, Darlush Mozaffarian, Ben Taylor, Jurgen Rehm, Christopher J. L. Murray, and Majig Ezzati. "The Preventable Causes of Death in the United States: Comparative Risk Assessment of Dietary, Lifestyle and Metabolic Risk Factors." *PLOS Medicine*, 6/4 (2009).

Office of Disease Prevention and Health Promotion, U.S. Department of Health and Human Services. "Healthy People" http://www.healthypeople.gov/default.htm (Accessed September 2010).

Ogden, Cynthia L., Margaret D. Carroll, Lester R. Curtin, Molly M. Lamb, and Katherine M. Flegal. "Prevalence of High Body Mass Index in US Children and Adolescents, 2007–2008." *Journal of the American Medical Association*, 303/3 (January 2010).

Wehung, Jennifer. "The Food Desert." *Chicago Magazine* (July 2009).

World Health Organization and the Food and Agriculture Organization of the United Nations. "Diet, Nutrition, and the Prevention of Chronic Diseases." Technical Report Series 916. Geneva: World Health Organization, 2003.

Sarah Boslaugh
Washington University in St. Louis

Organic Clothing/Fabrics

Growing a t-shirt takes a third of a pound of chemical fertilizer and pesticides. Organic fabrics do not use chemicals. As the public becomes more aware that resources are finite and decreasing, a shift is on to healthy and sustainable ways of living. Even Walmart touts its sustainable business practices and sells organic foods and clothing. Sustainable clothes and organic clothing are eco-fashionable.

"Sustainable" and "organic" do not mean the same thing. Organic clothing is a product of the organic farming movement, while sustainable clothing is a creation of the environmental movement. Organic is farm based; sustainable comes from the lab. Sustainable is more concerned with reuse and recycling of manufactured products to reduce landfill demand and provide consumer goods at lower prices. Although recycling is important, sustainability requires more—environmentally friendly and energy-efficient facilities, shipping, and products. Companies touting their environmental awareness boast of their voluntary efforts to recycle, plant trees, replace natural gas with landfill methane, and support sustainability.

Sustainability includes the Design for the Environment program that involves the Environmental Protection Agency (EPA) and businesses and the Green Building Council. Sustainability justifies itself by reduced cost due to reclaiming, recycling, and reducing energy. Organics are less concerned with economic sustainability. And sustainability may include synthetics and emit chemicals such as formaldehyde at low levels as long as they are manufactured in environmentally sound ways.

"Organic" is not the same as "synthetic organic." There is also a synthetic organic fiber industry that creates noncellulosic fibers for yarn, monofilament, stable, and tow (the coarse fiber prepared for spinning) for use in knitting machines, looms, and other textile-creation equipment. Synthetics date from 1913 and peaked in the 1970s and early 1980s due to economic problems and environmental regulation. Experiments with extruded filaments date to the 18th and 19th centuries, but the industry became viable only after World War I, and the first commercial organic synthetic dates to the 1930s. Man-made fibers were stronger and cheaper than natural cotton, wool, silk, and such. Cellulosic synthetics come from treated liquefied wood pulp. Examples are polyester, spandex, olefin, and nylon.

Organic clothing appeared initially in specialty stores, either the high-end boutiques or the hippie-style stores. By 2009, organic was mainstream and almost common. In 2009, Organic Monitor touted organics due to increased ethical consumerism. Investors were putting their money in new product lines and new producers and retailers.

Organics do not use toxic or persistent pesticides or artificial fertilizers or genetically modified seeds. Federal regulations govern the manner in which organics are grown. Organic certification requires inspection by an accredited certification organization under national or international standards. Organic farmers safeguard the soil and their animals from toxic chemicals.

Organic Fabrics

Organic fabrics are those created from organic raw materials. The benefits are a better environment and better (healthier) people.

Hemp was grown in China as early as 4500 B.C.E. Linen weaving and spinning in Egypt dates to 3400 B.C.E.; silk spinning, a precursor to creation of artificial fibers, dates to 2640 B.C.E.; and buckwheat hull pillows have been used for centuries in Asia. Agricultural chemicals came into widespread use only after World War II, harming workers, the environment, and consumers of tainted clothing and other textiles.

Growing cotton the traditional way is chemical intensive. Using less than 3 percent of agricultural land, cotton consumes 20 percent of all pesticides and 22 percent of insecticides. Cotton uses 46 chemicals; 5 are extremely hazardous, 8 are highly hazardous, and 20 are moderately hazardous. Because fields are commonly sprayed, drift is likely to affect neighboring wetlands and soil.

Irrigation of cotton is usually by controlled flooding, and three-fourths of cotton plantations are irrigated. Cotton is usually irrigated by water taken from nearby rivers and lakes, then returned after rinsing pesticides from the acreage into the water supply. Local water suffers from salinization, pollution, and eutrophication, and local wildlife and people suffer. Chemicals are also used in processing, dyeing, fixing the dye, and the rest of the manufacture—up to 8,000 chemicals for a t-shirt. Organics are chemical free in the field and through the process. Early organics were slightly rougher and off-white because they had no chemical treatments, including bleaching. Natural enzymes serve in organic processing, but dyeing is less effective as a result.

Traditional cotton farming is a multibillion-dollar industry and the source of 48 percent of all fabrics. However, organic farming is growing. The Soil Association reports 25,000 organic cotton farmers in Africa and organic cotton growing in 22 countries as farmers find they can save money by using organic pest and weed control and growing other crops with their cotton. Organic cotton can be grown with color and can be colored with eco-safe dyes. Conversion of all cotton growing to organic agriculture would cut the use of insecticides by a fourth.

Hemp and bamboo have the potential to rival cotton because both grow extremely fast on even marginal land. Their fibers are versatile, but neither is grown on the scale of cotton. Organic hemp fibers, seed, seed meal, and seed oil are usable. Organic hemp lacks the psychoactive ingredient that makes marijuana a drug. It weaves into a linen-like fabric suitable for clothes, home furnishings, and carpet. It also is antimicrobial and anti-mildew.

Bamboo is a fast-growing plant—as much as several inches each day. Pesticides are unnecessary. Bamboo pulp has a texture similar to that of silk. It is antimicrobial, breathable, and resistant to UV rays.

Organic wool holds a temperature comfortable for the wearer regardless of time of year. It is hypoallergenic and resistant to bacteria, mold, mildew, and flame. Wool-producing animals include sheep, llamas, and alpaca.

Organic kapok is a product of the Ceiba tree found in the Asian wilderness. It is extremely lightweight (30 times lighter than cotton), silky in texture, and capable of supporting 30 times its weight in water. It cannot be spun because it is too delicate. It is nontoxic, odorless, rot-resistant, and nonallergenic, unsuitable for clothing but useful for water safety equipment and pillows.

Fast growing in wet areas, naturally pest-resistant organic millet and buckwheat are used in pillows in Asia because they conform easily to head and shoulders. They are good for temperature maintenance, keeping a person warm in winter and cool in summer.

Organic silk is a product of Asian silkworms. It is soft and insulating, strong and long-lasting, and luxurious for pillows, linens, and clothing.

Production and Sales

In 2003, organic fiber sales were $85 million, up 22.7 percent from the previous year. In 2009, organic fiber sales were up 10.4 percent over the previous year at $521 million.

In 2009, world organic cotton production rose 20 percent to over 800,000 bales on 625,000 acres. That totaled just over three-fourths of 1 percent of the world total. Turkey led, with the United States fifth (10,731 acres in 2009, up from 8,539 in 2008, but the peak year was 2001 with 11,586 acres). Organic cotton farmers numbered 220,000. Sales of cotton apparel and home textiles totaled about $4.3 billion in 2009, a rise of 35 percent from $3.2 billion in 2008.

Organic cotton costs more to grow. Conventional cotton requires genetically modified seed, petroleum-based fertilizer, and pesticides and herbicides. Organic cotton has the same problems of weeds and insects, and nonchemical methods cost more than chemical methods (which is why conventional cotton growers use chemicals).

Organic cotton is harvested without use of chemical defoliants and desiccants that kill unwanted leaves and weeds. Chemical cotton also uses ethephon-containing chemicals to speed opening of the bolls. Organic harvesting is more labor intensive.

Manufacturing costs are higher for organics due to smaller quantities and more costly per-pound ginning, cleaning, and manufacture. Because it is produced in facilities that also create conventional cotton items, the ginning, weaving, and knitting machines have to be cleaned before organic production occurs, and the costs of downtime for cleaning and conversion are added to the costs of organic clothing. Smaller runs on shared sewing machines also add to the cost of garments.

The major added cost is that of labor. Organics take pride in not using near-sweatshop labor. Rather than paying labor costs of pennies a day in developing and poor countries, organics such as Earth Creations and Blue Canoe support fair trade and fair pay and manufacture their items in the United States. Other higher costs include shipping, advertising, and marketing. To reduce this handicap as well as lower the costs of buying supplies, organics have formed trade organizations such as Green People, Soil Association, Co-op America, and the Organic Trade Association. Increasingly, nonprofits are promoting organic clothing.

The conditions that apply to cotton also pertain to other natural fibers, including corn, soya, wool, bamboo, and hemp.

Organic products are more expensive than their conventional counterparts. For instance, a Faded Glory tank top of ordinary cotton costs about $4 at Walmart, including shipping, while an organic site, LotusOrganics, charges $24 with free shipping for an Earth Creations natural-dye tank top that is 55 percent hemp and 45 percent organic cotton. Organic clothing is more expensive because the business is new and dominated by small firms or "mom-and-pop" stores and thus lacks economies of scale. The argument for organics is that price is secondary to sustainability, ethics, and doing the right thing for consumers, workers, and the planet.

See Also: Fair Trade; Fashion; Green Consumerism; Individual Action, National and Local; Organic Foods; Organics; Popular Green Culture; Shopping; Social Action, National and Local.

Further Readings

Chamberlin, Christine. "Introduction to Organic Fibers." http://www.organic.org/articles/showarticle/article-224 (Accessed September 2010).

Hemp Industries Association. http://www.thehia.org (Accessed September 2010).

Highbeam Business. "Industry Report: Manmade Organic Fibers (Excluding Cellulosic)" (2010). http://business.highbeam.com/industry-reports/chemicals/manmade-organic-fibers-except-cellulosic (Accessed September 2010).

LotusOrganics.com. "The High Cost of Organic Clothing" (2006). http://organicclothing.blogs.com/my_weblog/2006/12/the_high_cost_o.html (Accessed September 2010).

LotusOrganics.com. "Sustainable Clothing—Emerging Standards Purely Beautiful & Sustainable Clothing" (2006). http://organicclothing.blogs.com/my_weblog/2006/05/sustainable_clo.html (Accessed September 2010).
Organicbabywearhouse.com. "What Is Organic Fiber?" (2008). http://www.organicbabywearhouse.com/what-is-organic-fiber.html (Accessed September 2010).
Organic Trade Association. "The National Organic Standards Board Definition of 'Organic'" (2007). http://www.ota.com/organic/definition.html (Accessed September 2010).
Organic Trade Association. "Organic Cotton Facts" (June 2010). http://www.ota.com/organic/mt/organic_cotton.html (Accessed September 2010).
Sustainable Cotton Project. http://www.sustainablecotton.org (Accessed September 2010).
Williams, Jeremy. "The Fabric of Progress: Why Organic Clothing Matters." Celsias. News and Opinion. http://www.celsias.com/article/fabric-progress-why-organic-clothing-matters (Accessed September 2010).

John H. Barnhill
Independent Scholar

Organic Foods

"Natural" foods have a limited amount of additives and preservatives, while the term *organic* is the product of an entire system of agricultural production, distribution, and marketing. These plants, being sprayed with pesticides, are no longer considered organic.

Source: iStockphoto

Organic foods have been promoted over time via diverse interest groups seeking to advance social change. The premise and philosophies underlying organic agriculture and foods serve to magnify not only the problems inherent in industrial agriculture but in the larger, dominant social systems that create it. Organic foods are one aspect of a multifaceted social movement (including local foods, "health" foods, vegetarian, and vegan foods) that strives for the betterment of the world through changing how foods are produced, valued, and consumed. As opposed to "natural" foods, which are limited in additives and preservatives, "organic" foods are the product of an entire system of agricultural production, distribution, and marketing, thereby having wider, systemic impacts on the food system.

Prior to the advent of chemically synthesized fertilizers, pesticides, and herbicides that facilitated industrial agriculture, farmers relied on organic matter such as manure, guano, fish meal, leguminous crops, or cover crops to enrich soil with nutrients, enhance its water retention and tilth, and minimize erosion. The earliest players in what is collectively referred to as the "organic food movement" were farmers, including Thomas Jefferson, who in his 1793 diaries described his use of manure and cover crops to build the soil at his Monticello estate to the "best quality." He wrote this in response to the predominant cotton and tobacco farming methods of the day, which left soil depleted and nonproductive, necessitating the clearing of forest to grow crops.

In the 1840s, development of chemical fertilizers revolutionized agriculture. German chemist Justus von Liebig (1803–73), a "father of the fertilizer industry," discovered that it was the inorganic mineral salts contained in organic matter that nourished plants. He advocated a scientific approach to agriculture to increase efficiency through the replacement of compost with inorganic fertilizers. Organic agricultural methods developed in response to this reductionist approach and growing concerns about its ecological and soil quality effects. The first documented use of the term *organic* as applied to agriculture comes from Baron Walter James Northbourne (1896–1982) in his book *Look to the Land* (1940). Here, he described the farm as an organism to be managed as an integrated system. Other advocates also promoted practices to nourish soil, microbes, and humus beyond simply providing nutrients to plants, thereby attempting to mimic natural forest processes of soil development on forest floors. Early leaders in this area included Rudolph Steiner (1861–1925), whose lectures on farming practices, including cosmological and spiritual aspects of planting and harvest, became the foundation of contemporary biodynamic farming. Steiner's work inspired the development of the first organic labeling and certification system in 1924. Sir Albert Howard (1873–1947) was a pioneer in organic farming and composting. His Indore process of composting, developed in the Indore Province of India, continues to be foundational to contemporary practices. He stressed that soil quality produced wholesome foods that would promote human health.

Organic Food as a Social Movement

Another group to forward the movement of organic food was the dietary reformers who, partly in response to the advent of processed, industrially produced foods, stressed the importance of whole, unadulterated foods as part of a holistic practice for physical, mental, and spiritual "hygiene." In the mid-19th and early 20th centuries, many of these health crusaders couched their messages about food and health in Christian doctrine. They advocated dietary temperance, often including vegetarianism, to suppress licentious urges and promote health. Leaders in this movement included Reverend Sylvester Graham (1794–1851), Seventh-Day Adventist medical doctor John Harvey Kellogg (1852–1943), and Mary Baker Eddy (1821–1910), founder of the Christian Science Church. The spiritual-based idealism of these evangelists would reemerge among the 1960s counterculture, with a less Christian tone, to focus on food as a means for reunifying body, spirit, and Earth.

Other seminal events occurred near this time, including the publication of Upton Sinclair's *The Jungle* (1906); the ensuing passage of the federal Pure Food and Drug Act (1906); and the creation of the Bureau of Chemistry, which would later become the Food and Drug Agency (1930). Food contamination scares such as lead arsenate (insecticide) in apples spurred wider public concern over the wholesomeness of industrially produced foods. In 1940, J. I. Rodale (1898–1971) established a farm in Emmaus, Pennsylvania, where he practiced organic farming methods that complemented his interest in disease prevention through diet. He was among the first in the 20th-century United States to promote concepts from earlier European organic agriculturalists, linking organic methods of farming to human health. His magazine *Organic Gardening and Farming* (OGF) was started in 1942 with little popularity. In 1939, dichlorodiphenyltrichloroethane (DDT) was found to be an effective pesticide. It was used by the U.S. Army during World War II to reduce malaria and typhus among soldiers fighting in the Southern Hemisphere. After the war, questions about its safety were routinely disregarded by agricultural and scientific experts. Pronouncements of its safety by the National Research Council's Food Protection Committee and the Public Health Service intended to assuage the public's concerns about its use as an agricultural pesticide and trace amounts found in many foods. In 1962, Rachel Carson's *Silent Spring* alarmed the public further to the systemic effects of DDT and other organic chemical fertilizers such as malathion via their ubiquity in groundwater, foods, and accumulation in human body fat and breast milk. Other environmental concerns such as the contamination of the Great Lakes with industrial waste, particularly mercury, continued to raise the public's circumspection of industrial agriculture. In 1969, a survey conducted by the National Technical Information Service reported that approximately 60 percent of those surveyed believed agricultural chemicals were dangerous to health.

At this time, with increased public interest in ecology and the growing "back to the land" counterculture, organic foods, which promised to be free of chemical residues, grew in popularity, as did readership of *OGF* magazine. Even though it did not have the same countercultural style of other similar magazines at the time (e.g., *Mother Earth News*), the "hippie bible" *Whole Earth Catalog* described the magazine as one of the "most subversive" publications of the day through its challenge to the foundations of the industrial state. It reached its apex of popularity in 1971 with a readership of 700,000 people.

In 1969, public polls showed that most Americans believed "naturally" or "organically" grown foods were more nutritious than conventionally grown food. Rodale knew that there had been no conclusive evidence to support the idea that organic food was more nutritious than conventionally grown. In fact, since that time, most studies have shown organic food to offer little if any benefit over conventional food in terms of the nutrients it contains.

Other contributors to the organic food movement who have carried the movement into the 21st century are those with interests based in politics. In 1960, a group of young former communist activists formed the New Left, whose messages of social reform in an era marked by the civil rights movement and outrage about the Vietnam War had strong appeal on U.S. college campuses.

The perspective of the New Left regarding the shortcomings of capitalism and corporate structures overlapped with those of many American Liberals, including attorney and political activist Ralph Nader. Nader and activists within the environmental movement argued that corporate food could be blamed for much of the nation's pollution and ill health. With this influence, the New Left began to adopt the issue of food as an important entry point for raising public awareness of the imperialism of corporate America.

In the 1960s, young adults began to break out of the family-centered structures of the 1950s that were often viewed as oppressive. This "counterculture" of "hippies" and "dropouts" explored other forms of community, with interests generally not as grounded in critical theory and politics as those of the New Left, yet equally as condemning of the destructive effects of American social structures and institutions (such as capitalism, government, organized religion) on human and planetary well-being. The counterculture was also influenced by Eastern philosophies, including Zen and macrobiotics, which stressed eating foods that were detoxifying and free of chemical contaminants. The response was to return to the land and the simple rules of the natural world, a central means of which was the consumption of wholesome, unprocessed, often organic foods.

Several mainstream publications of the 1970s alerted the general public to the effects of industrial food on the nation's health, particularly degenerative diseases. Leading titles include *Consumer Beware! Your Food and What's Been Done to It* (1970) by Beatrice Hunter, James Turner's *The Chemical Feast* (1970), Gene Marine's *Food Pollution: The Violation of Our Inner Ecology* (1971), and *Diet for a Small Planet* (1971) by Frances Moore Lappé, which uniquely linked diet not only to health but to ecological and social justice concerns as well.

Throughout the 1960s and 1970s, growing consumer demand led to increasing numbers of farms claiming to produce organic foods. With the backing of scientific inquiry such as the work of biologist John Todd and the New Alchemy Institute beginning in 1969, organic foods were gaining a new respectability. The movement was presented by mainstream media such as the *New York Times* and *Newsweek* with a new orthodoxy that challenged earlier counterculture associations. More health food and cooperative markets appeared in response to heightened public interest in organic and whole foods. Given an increasingly affluent, educated clientele, these markets began to look more like upscale, mainstream supermarkets.

By the early 1970s, in California alone there were more than 300 health food stores and 22 restaurants selling organic foods. One of these was Alice Water's Chez Panisse in Berkeley, which she founded on the premise of providing wholesome, fresh, locally produced (with later emphasis on organic) foods. Her work spurred the "California" style of cuisine, marked by fresh, whole, and local foods, that has since become popular throughout the United States.

Rise of Big Organics: Organic Foods Since the 1970s

During the 1970s to the 1990s, many New Left advocates and counterculturalists became professionals in academia, politics, government, or industry. As the organic food industry moved into the late 20th and early 21st centuries, a growing conflict arose between those who viewed organic food as an "industry" relying on business models, profit seeking, and market share and those who remained loyal to the philosophical underpinnings of organics as a "movement" emphasizing stewardship, social justice, and small, community-based agrarian farming. Much of this conflict arose from the evolution of large-scale, centralized organic farms and distribution networks in the 1980s and 1990s that, to an extent, modeled industrial agriculture systems. Proponents of organics as industry have argued that business can be used to promote many of the ideals of the organic movement on a large scale if practiced correctly, emphasizing change from within the corporate system. These organic businesses demonstrated the profitability and potential of organic foods to mainstream agribusiness, eventually leading to buyouts of organic companies by larger parent companies with or without strong commitments to organic principles. Early examples of this trend include Fresh Express bagged lettuce and salad greens, which grew to annual sales of $1 billion and was sold to Chiquita for $855 million in 2005. Stonyfield Farm, manufacturer of the number one–selling brand of organic yogurt in the United States, was sold to the French company Groupe Danone in 2001. Stonyfield Farm continues to source milk through the Coulee Region Organic Produce Pool Cooperative (CROPP), today known as Organic Valley, the nation's largest cooperative of independent, organic family farmers. White Wave Foods, one of the first large-scale organic soy milk producers, was procured by Dean Foods in 2002. Dean Foods has since been accused of transitioning its products away from organics without notifying retailers or consumers. By 2004, eight of the largest conventional food corporations owned the 38 largest organic businesses, raising concerns among loyalists of compromised integrity of organic standards by agribusiness to increase profitability. Government regulations became a focal point for maintaining the rectitude and control of organic foods.

Large-scale production of organic foods has both facilitated and been advanced by the appearance of organic foods among large-scale ("big box") retailers. Most notable of these is Walmart, which began selling organic foods in the summer of 2006 in an attempt to refashion its appeal to less penny-conscious consumers. Walmart's decision, in part, catalyzed the development of organic products by some of its large, conventional food suppliers, such as Kellogg's, Kraft Foods, and General Mills. Other large-scale retailers have developed their own private label lines of organic food, such as Safeway's "O organics." The combination of large-scale production and retail involvement has resulted in availability and affordability not previously seen in organics. Walmart, for example, announced that organic products would be priced only moderately (approximately 10 percent) above conventional comparable items.

As with large-scale production of organic foods, promotion by "big box" retailers has been met with divergent responses from organic advocates, with some supporters stressing the ecological benefits of expanded organic agriculture, while skeptics view the involvement of big business as potentially damaging to the integrity of the social justice and environmental stewardship values that underpin the movement.

Regulation of Organic Foods

A key role for government in the regulation of organic farming and foods has been to create policies that define what it is and build mechanisms for enforcement. On a global scale, the International Federation of Organic Agriculture Movements (IFOAM) established the IFOAM Accreditation Program (IAP) in 1992, which defines existing standards in many countries. The Codex Alimentarius of the United Nations Food and Agriculture Organization (UN-FAO) and the World Health Organization (WHO) include guidelines for organic that emphasize consumer protection and trade promotion. The most basic international standards for certification require avoidance of synthetic chemical inputs, use of land that is free of synthetic inputs for three or more years, documentation of production and sales, physical separation of organic and conventionally produced foods, and periodic on-site inspection.

In the United States, private, mostly nonprofit organizations began developing regulation standards for organics in the 1970s to protect consumers from fraud and to support organic producers. Accreditation was offered by some states in the 1980s, but standards were not uniform across these institutions. This led the U.S. Department of Agriculture (USDA) to begin developing uniform standards via the 1990 congressional passage of the Organic Food Production Act. The USDA's National Organic Program (NOP) was implemented in 2002 to provide a framework for regulating these national organic standards. With the exception of producers with annual sales under $5,000, all claims of "organic" must first meet USDA organic certification criteria. The NOP regulates all aspects of production, processing, packaging, distribution, and sale of organic foods. A USDA Organic seal may be applied to products containing at least 95 percent organic ingredients.

Today, organic food is one of the fastest-growing food sectors, with an estimated 20 percent annual growth and consumer sales of $14.6 billion in 2005 (2.5 percent of total U.S. food sales). Organic products are now widely available across the mainstream food system via more than 20,000 natural foods stores and 73 percent of conventional grocers as well as through direct producer-to-consumer channels, including farmers markets and community supported agriculture (CSA) programs. The outlook for organic foods is strong, with expanding efforts to promote direct marketing by producers to maintain economic viability and increase accessibility to organic foods across diverse ethnic, economic, and social subpopulations.

See Also: Cooking; Diet/Nutrition; Locavores; Organics; Veganism/Vegetarianism; Veganism/ Vegetarianism as Social Action.

Further Readings

Belasco, Warren. *Appetite for Change: How the Counterculture Took on the Food Industry.* New York: Pantheon Books, 1989.

Canavari, Maurizio and Kent D. Olson. *Organic Food: Consumer's Choices and Farmer's Opportunities.* New York: Springer, 2007.

Fromartz, Samuel. *Organic, Inc.: Natural Foods and How They Grew.* New York: Hartcourt, 2006.

Kristiansen, Paul, Acram Taji, and John Reganold, eds. *Organic Agriculture: A Global Perspective.* Ithaca, NY: Comstock Publishing, 2006.

Levenstein, Harvey. *Paradox of Plenty: A Social History of Eating in Modern America.* Berkeley: University of California Press, 2003.

Julia L. Lapp
Ithaca College

Organics

"Organic" refers to agricultural production without the use of synthetic chemicals. Organic agriculture is based on principles of sustainability—the ability to be continued in the long term without negative environmental effects—and designed to improve and support soil health. It prohibits using human-made chemicals such as pesticides, herbicides, and synthetic fertilizers. For livestock farming, organic agriculture prohibits the use of antibiotics, hormones such as bovine growth hormone, or synthetic medications. Organic agriculture also prohibits the use of genetically modified plants or animals. Instead, organic agriculture relies on more labor-intensive management such as using a chipper to remove weeds and the use of natural inputs such as composts and homeopathic animal health treatments. Within the United States, the Organic Food Production Act of 1990 and the U.S. Department of Agriculture (USDA) National Organic Program govern use of the term *organic.*

Organics have become a significant, though small, part of the agrifood system. Global organic sales are estimated to have doubled from $25 billion in 2003 to $50.9 billion in 2008. In the United States, organic food sales were about $24.8 billion in 2009 and have continued to grow despite recent economic declines. There were more than 4.8 million acres of certified organic land in the United States in 2008, although this is still less than 1 percent of total agricultural land in the nation. Globally, there are over 35 million hectares of organic agricultural land. The countries with the largest amounts of certified organic land are Australia, Argentina, and Brazil. California is the largest producer of organic fruits and vegetables, which are the largest portion of organic sales. While growing rapidly, organic sales were still less than 3 percent of food sales in the United States in 2009.

History

Contemporary organic farming, especially in North America, is largely based on the writings of F. H. King in the 1890s and early 1900s. King was a professor of agricultural physics at the University of Wisconsin–Madison who had concerns that modern agriculture was rapidly depleting soil fertility and would be unsustainable. King studied techniques to protect the fertility, aeration, and health of soils. From 1902 to 1904, King worked for the USDA Department of Soils but was forced to resign after two years because his ideas about sustainability were considered too radical. King emphasized the importance of nutrient balance in the soil and wrote about sustainable practices used in Asian agricultural communities. His writings taught the practices of crop rotation, recycling soil nutrients through manure, and the ability of a largely vegetarian diet to provide more food at lower cost than meat-rich diets.

Sir Albert Howard was also a significant innovator in the branch of organics that emphasized soil fertility and sustainability. Howard was a British botanist who went to India to teach Western styles of agriculture but instead became interested in local sustainable agricultural practices that protected the soil and supported communities despite few resources. He espoused the benefits of composting, and his 1940 book *An Agricultural Testament* is an influential organic textbook. King and Howard both emphasized the importance of caring for the soil and creating organic agriculture that is sustainable in the long term.

A second major strand of organics, more common throughout Europe than North America, is biodynamic agriculture. Biodynamic agriculture is based on the teachings of Austrian philosopher Rudolf Steiner, specifically those taught in his agriculture course in Koberwitz, Poland, in 1924. Steiner coined a social philosophy he called "Anthroposophy" that was a combination of science and mysticism. Biodynamic agriculture brought this spiritual scientific perspective to farming. Steiner emphasized the importance of a resilient farm system that was self-reliant. He also introduced specific practices that are still central to biodynamic agriculture, particularly the use of animal manure and different types of compost to support soil health. Steiner also emphasized seasonality and connecting the land to the cosmos through timing with the lunar calendar.

Organic agriculture is one part of a larger sustainable agriculture movement. *Sustainable agriculture* is an umbrella term that encompasses a variety of alternative agricultural practices designed to avoid the environmental and social problems of conventional chemical agriculture. Community supported agriculture (CSAs), integrated pest management, local food systems, and low-till agriculture are other types of sustainable agriculture. Consumer trends such as "buy local" campaigns, slow food, and fair trade have emphasized the consumption side of sustainable agriculture.

Reasons for Organics

The organic movement is diverse and is driven by both producers and consumers. Producers—farmers and gardeners—have developed organics out of environmental concerns and fears for their own health and well-being. Consumers have turned to organics to provide safe, healthy, and environmentally sound food. Overall, modern organic agriculture has grown from concerns about the negative environmental and social effects of conventional agriculture.

During the 1930s and 1940s, the use of synthetic chemical fertilizers, pesticides, and herbicides on farms became widespread during the so-called green revolution. Environmental and social concerns about the heavy use of chemicals rose quickly along with the modern environmental movement. Early practitioners of organic agriculture were ridiculed by agricultural scientists and farmers for being anti-modern. After concerns about the heavy use of agricultural chemicals became more widespread, however, modern organic agriculture became more popular by the 1960s and 1970s.

Scientists and environmentalists who were concerned about water pollution, air pollution, and health problems from chemical exposure criticized the early use of agricultural chemicals. Rachel Carson's 1962 book *Silent Spring* documented the devastation of bird populations caused by heavy use of DDT, a highly toxic and long-lasting pesticide. It raised public awareness of the problems of chemical agriculture and is often credited with reigniting the contemporary environmental movement. Socially, high oil prices of the 1970s raised concerns about whether farms could continue to rely on chemical inputs derived from petroleum. The farm crises of the 1970s and 1980s, with high rates of bankruptcy, foreclosure, and rural poverty, also shook the confidence of many farmers and scientists in conventional agriculture.

Consumer interest in organic agriculture has largely focused on health concerns from chemical exposure. In the 1980s, media coverage of health problems associated with Alar, a growth-regulating chemical sprayed on fruits and vegetables, significantly increased consumer interest in organic agriculture. Studies continue to demonstrate the significant health concerns for consumers from chemical residue on food. The "dirty dozen" are foods that have received particularly heavy criticism for their high rates of chemical residue: apples, cherries, grapes, nectarines, peaches, pears, raspberries, strawberries, bell peppers, celery, potatoes, and spinach. Concerns about the potential impact of bovine growth hormone on breast cancer and reproductive health have spurred consumer interest in organic milk, as have fears of antibiotic resistance from heavy use of antibiotics in cattle. Studies of organic consumers show that health concerns are a primary motivation for buying organic produce.

There are several reasons why farmers typically consider conversion to organics and pursuing organic certification, including (1) environmentalism, (2) health concerns, (3) financial incentives, and (4) some combination of these motivations. Most organic farmers have serious environmental concerns about conventional farming and a desire to better care for the environment. Also, a health crisis such as nerve damage or cancer from chemical exposure often breaks farmers' trust in chemical agriculture and motivates them to consider organics. A serious financial crisis such as bankruptcy can also motivate farmers to consider organic production. Organic agriculture has much lower input costs because it relies on labor-intensive management instead of chemical inputs, and organic produce is also usually sold at a premium. The desire or need to capture financial premiums is a motivation for organic certification.

Certified Organic

"Certified organic" refers to a farm or product that meets a specific set of organic production standards. A label on the product usually indicates certification; it allows consumers to verify organic practices and producers to charge premium prices for organic goods. Organic certification typically requires a three-year conversion period during which any nonorganic inputs such as synthetic fertilizers, herbicides, or pesticides are phased out of use. At the start of organic certification, each farm must have a comprehensive management plan that outlines strategies for mitigating problems without prohibited inputs. Plans must address strategies for maintaining soil fertility, pest control, weed control, and animal health if the farm has livestock.

Certification relies on auditing to guarantee that a farm meets organic standards. First-party auditing is when producers audit themselves; second-party audits are when the customer or buyer audits producers; third-party audits are when a neutral party audits producers. Most organic certification relies on third-party audits. First, auditors examine the management plan to make sure that it is adequate. After beginning certification, organic farms are audited annually to ensure that they continue to meet the standards and are not using prohibited inputs. Additional unannounced audits are possible at any time. Any input, such as fertilizer, that is used on the farm must be reported in the audit with records to verify where the inputs were purchased and that the inputs are organic. For most certified organic systems, inputs must be audited and certified independently. This means that every product used on a certified organic farm must also be certified organic.

Certification can be expensive and time consuming. Producers must pay to register and audit their farms and must maintain careful management plans and records.

Organic Standards

Organic standards outline specific practices and inputs that are prohibited or encouraged in organic agriculture. The goal of organic standards is to ensure that only legitimate organic producers can use the organic label. In this way, standards protect both organic consumers and producers. Organic standards also help to make the meaning of the term *organic* consistent.

Local and regional organic standards were in use as early as the 1940s, but formal organic standards became more widely used in the 1970s. Early organic standards were created by organic farmers and gardeners and environmentalists themselves, and utilized first-party audits. Most early standards were very simple and short. They broadly prohibited chemical inputs, but they also focused on what producers do to ensure sustainability such as diversifying their farms and protecting soil fertility. Early organic standards were more prescriptive than current standards, meaning that they gave equal importance to what producers should do as well as to what they should not do. The Rodale Press organic standards established in 1972 were influential early organic standards, created by the Rodale Institute to codify the agricultural practices taught by Sir Albert Howard. The Rodale Institute created these standards and an experimental organic farm to demonstrate the economic and environmental viability of organic agriculture.

The International Federation of Organic Agriculture Movements (IFOAM) founded in 1972 is an important organic standard system. IFOAM is an umbrella organization that represents the worldwide organic movement; it is composed of members from 750 member organizations across 108 countries. IFOAM creates the International Basic Standards for Organic Production and Processing. Accredited member organizations can certify farms and processors to this IFOAM standard to receive international recognition of their organic produce. Instead, however, most member organizations participate in IFOAM's equivalency program. Member organizations submit their own organic standards to IFOAM for auditing; the standards are evaluated to determine whether they are functionally equivalent to IFOAM's Basic Standards. This equivalency program allows some variation in organic standards while guaranteeing that basic requirements are met.

Demeter International standards, created in 1928 and still more popular in western Europe than in North America, formalized biodynamic farming principles into another important organic standard and the oldest ecological label for food products. Similar to IFOAM, Demeter International uses an international equivalency program for biodynamic standards. Demeter International has a specific biodynamic standard, but it also recognizes 44 national and regional Demeter standards through its equivalency program. Each national or regional biodynamic association creates its own Demeter standard, which is then audited by the International Accreditation Council for equivalency with Demeter International standards.

Many governments also have regulations for organic standards. The European Union, the United States, China, Japan, Canada, New Zealand, and Australia all have government regulations that set and enforce organic standards. All of these government organic standards also rely on third-party auditing through independent certifiers to ensure compliance. These government organic standards restrict the import and export of organic products across nations.

USDA National Organic Program

By the 1980s, a large variety of organic standards were in use throughout the United States. Many local and regional groups of activists and farmers had created their own standards and were using their own auditing practices to claim the label "organic." Several states such as California and Washington had created state laws governing the use of the term *organic*. As the consumer market for organic food grew, it became a problem to have multiple organic standards. All of the standards used the term *organic*, but they each had different recommendations and prohibitions. As some farmers and food companies began to sell organic produce in different states and communities, the variation of standards became a concern.

In 1990, the U.S. Congress included the Organic Foods Production Act as part of the 1990 Farm Bill under pressure from a coalition of organic farming groups, consumers, and environmental groups. The Organic Foods Production Act was meant to establish a national set of standards governing use of the term *organic*. The goal of the regulation was to guarantee a consistent meaning of the term *organic* and to create a system to ensure compliance with the standard. Large organic producers and food companies particularly supported the creation of the bill. The Organic Trade Association, created in 1985 to represent large organic food businesses, was a significant supporter of the Organic Foods Production Act.

In 1992, the USDA created the National Organic Standards Board (NOSB) to advise on the national organic standard. The NOSB is composed of four farmers, two handlers/processors, one retailer, one scientist, three consumer/public-interest advocates, three environmentalists, and a certifying agent. The Secretary of Agriculture appoints the NOSB. From 1992 to 1997, the NOSB collected information and opinions on a national organic standard through public meetings and input from stakeholders. In 1997, the NOSB made its first proposal for the USDA National Organic Program (NOP) standards and procedures. However, this first proposal allowed the use of many inputs and practices traditionally excluded from organic agriculture such as genetically modified organisms. Public and industry groups widely rejected the standard, and the USDA received a record number of complaints. A new proposal that incorporated almost all of the recommendations from activists and farmers was enacted on October 22, 2002. The NOSB continues to advise on revisions to the standards and procedures.

Any product sold as organic in the United States must be certified to meet the USDA NOP and may display the USDA Organic label. Processed foods must contain 95 percent USDA NOP certified ingredients in order to carry the USDA Organic label. The USDA NOP does not govern the use of the term *organic* on health and beauty products. The USDA NOP relies on third-party auditing to enforce the organic standard. Certifying agencies must become accredited by the USDA NOP and are subject to auditing and oversight from the USDA. These agencies then audit organic farms and processors to guarantee compliance with the USDA NOP standards. At the time of this writing, there are 57 accredited certifiers within the United States and 43 international accredited certifiers. The USDA can either directly accredit international certifiers or the USDA can allow government agencies within those nations to accredit certifiers. Organic farms or companies that wish to sell organic produce in the United States must be certified by one of these accredited certifiers. They are subject to the three-year conversion period, comprehensive management plans, and annual and unannounced audits. Any significant noncompliance with the standards such as the use of a prohibited input can result in the loss or suspension of certification.

Controversies and Concerns

The growth of the organic industry has not been without controversy. Many early organic activists see problems with the current industry, arguing that it is too corporate and too focused on profit. There are concerns that financial incentives for organic production have encouraged agribusiness firms and corporate farmers to enter the industry, "watering down" the radical social and environmental goals of organics. Contemporary organics does not address, for example, labor rights, local production, and greenhouse gas emissions. At the same time, many people interested in development issues have disparaged high certification costs and the technicality of organic certification. They argue that these costs keep small family farms and third world farmers from gaining the benefits of certification.

Proponents of conventional agriculture have argued that organic agriculture cannot produce the yields needed to feed a growing population, that conventional agriculture can produce much less expensive food to feed the global poor, and that organics has exaggerated the risks of agricultural chemicals.

Organics still has many critics within the movement itself and within conventional agriculture, but it continues to thrive as part of our agrifood system. Organics was born out of a recognition and rejection of the problems of chemical agriculture and offers an alternative, more sustainable approach to feeding the world.

See Also: "Agri-Culture"; Green Consumerism; Locavores; Organic Clothing/Fabrics; Organic Foods; Permaculture; Veganism/Vegetarianism.

Further Readings

Duram, Leslie A. *Good Growing: Why Organic Farming Works*. Lincoln: University of Nebraska Press, 2005.

Guthman, Julie. *Agrarian Dreams: The Paradox of Organic Farming in California*. Berkeley: University of California Press, 2004.

Kristiansen, Paul, Akram Taji, and John Reganold, eds. *Organic Agriculture: A Global Perspective*. Ithaca, NY: Comstock, 2006.

Raynolds, Laura T. "The Globalization of Organic Agro-Food Networks." *World Development*, 32 (2004).

U.S. Department of Agriculture. "National Organic Program, Regulatory Text." http://www.ams.usda.gov/AMSv1.0/ams.fetchTemplateData.do?template=TemplateF&navID=RegulationsNOPNationalOrganicProgramHome&rightNav1=RegulationsNOPNationalOrganicProgramHome&topNav=&leftNav=&page=NOPRegulations&resultType=&acct=noprulemaking (Accessed June 2010).

Rebecca L. Schewe
University of Wisconsin–Madison

Organizations and Unions

Unions and environmental organizations sometimes work at cross purposes, as was the case with environmental regulations resulting in a dramatic decrease in the production of coal and the debates over car mileage regulations and logging rights. Over time, however, the two groups have begun forming alliances for the purpose of working toward the twin goals of protecting the environment and the rights and health of workers around the world. Miners and factory workers, for instance, are particularly vulnerable to the presence of particulate matter in the air, and environmental groups have been instrumental in the passage of regulations that mandate the use of equipment that cleans the air and protects the health of all who work in mines and factories. Even though some jobs have been lost to environmental improvements, their loss has been offset to some extent through the addition of jobs created to promote sustainability. Despite the successes, the alliance remains fragile because of the tendency toward clashing interests between what is called the blue-green coalition in the United States and the red-green coalition in Europe. The designation of "red" in the European alliance refers to the preponderance of socialists and left-wingers who join such coalitions to affect change. In Denmark, for instance, the Red Green Alliance (*Enhedlisten*) won six parliamentary seats in 2010. In Australia, the two groups come together as the Green Union Caucus.

One of the most effective ways in which the partnership between environmental organizations and unions works is the lobbying of governmental bodies to make sure their united voices are heard and that the alliances are being considered for available grant monies, a move that benefits governmental bodies as well as union-environmentalist coalitions. In general, alliances between unions and environmentalists tend to be stronger in Europe than in the United States. European unions have endorsed environmental practices such as creating more sustainable industries and organic farming. The strength that comes from unity makes it easier for European alliances to take a more active role in shaping governmental policies. For instance, the European Commission, the executive arm of the European Union, maintains close ties to unions and nongovernmental organizations (NGOS) that include environmental groups, consulting them regularly before preparing legislation that impacts their interests and encouraging their participation at meetings where the EU agenda is being established. For example, representatives from the union-environmentalist alliance were asked to take part in a pre-conference meeting leading up to the 2000 World Summit on Sustainable Development. Coalitions of the two groups give them added strength in such proceedings.

Many trade unions now have on-site environmental representatives. Their roles include encouraging employers and workers to be more environmentally responsible by taking small steps such as recycling and turning off lights when not in use. Such efforts save money as well as protect the environment. Education is a key factor in creating green working environments, and some unions offer courses to teach representatives how to accomplish this goal. Some companies have also joined the alliance between unions and environmentalists. Ethix Merch, for instance, provides many of the goods handed out free at industry trade shows, and provides its clients with ecofriendly union-made products that do not add to a company's carbon footprint.

César Chávez, Delores Huerta, and the UFW

Union organizers César Chávez and Delores Huerta, the founders of the United Farm Workers of America, were in the vanguard of the environmental movement of the 1960s. Although their focus remained on winning the right to unionize and to engage in collective bargaining, they were quick to realize that the constant exposure to pesticides used in agricultural fields across America was responsible for a host of conditions among farmworkers that included cancer, birth defects, miscarriages, sterility, respiratory illnesses, and increased mortality rates. Chavez contended that farmworkers, like canaries in mines, served as an early warning system for the rest of society by identifying environmental dangers before they spread to the greater population. Winning greater access to healthcare for farmworkers remained a high priority for the UFW.

John Sweeney and the AFL-CIO

In the United States, the initial motivating force behind blue-green alliances was John Sweeney, president of the AFL-CIO from 1995 to 2009. Sweeney recognized that both groups had much to gain by working together and much to lose if they did not. In 1997, he instigated the first blue-green discussions. That same year, the group joined with anti-globalist Republicans to block President Bill Clinton's efforts to gain fast-trade negotiating authority. Internationally, members of union-environmentalist groups and other opponents of such efforts turned to the Internet to launch a campaign to stall talks on the Multilateral Agreement on Investments. One of Sweeney's most successful strategies was the appointment of Jane Perkins, a former union leader who had become the president of Friends of the Earth, to lead discussions on coalition strategies and give them cohesion.

The always-fragile coalition is constantly being threatened by disagreements concerning the issue of global warming policy, particularly the role of the World Trade Organization (WTO) in designing those policies. While environmentalists believe the reach of the WTO should be somewhat curtailed, unions believe the WTO can be a powerful force in the international battle for social justice, including workers' rights.

Health Issues

Since the early days of industrialization, trade unions around the world have been involved in the fight to protect the health of workers. The fact that certain workers are more often exposed to toxins and hazards than are members of the general public is generally accepted. It is also common knowledge that major accidents in the workplace frequently spread to the surrounding community. These disasters may even cross national borders. Such extensive threats to the environment become the immediate concern of both unions and environmental groups, heightening the need for alliances. This need was made abundantly clear during three separate disasters that took place in the 1980s: the 1984 methyl-iscocyanate spill at the Union Carbide Plant in Bhopal, India, where several thousand people were killed; the 1986 explosion of a nuclear reactor at Chernobyl in the Soviet Union that killed 20 and forced the evacuation of thousands; and a fire that same year at the Sandoz warehouse in Basel, Switzerland, which led to the leaking of tons of agricultural chemicals and mercury into the Rhine River. The drinking water of four other countries was also polluted by the Basel disaster.

In 1989, the case for cooperation between unions and environmental organizations again reached international proportions after an American journalist reported that a British company, Thor Chemicals, Inc., had been poisoning workers and the greater community by seepage from mercury that had been stored after being shipped in from plants around the world, including two American companies, Borden Chemicals and Plastics, and American Cyanamid Company. It was revealed that mercury levels in the Mngeweni River, which flowed into several other bodies of water, were 1,000 times the acceptable mercury-contamination level for U.S. companies. The World Health Organization identified a number of deaths and illnesses directly related to the mercury exposure. Despite being charged with culpable homicide and other workplace violations in a British court, Thor Chemicals received a negligible fine and a "slap on the wrist."

Potentially fatal illnesses related to the workplace have been documented for centuries, particularly among agricultural workers, miners, and factory workers. Other industries have received less attention but the dangers are no less serious. For instance, in the trucking industry, workers are constantly exposed to fine particulates in the air around them, especially those emitted from vehicle exhaust systems. As a result, there is a well-documented connection between working in the industry and the contracting of cardiovascular diseases and lung cancer. One 2007 study examined the aggregate records of more than 54,000 male workers who had been employed at one of four national trucking companies between 1985 and 2000. As expected, trucking company employees were more likely than the general public to suffer from lung cancer and ischemic heart disease. The risk was greatest for those employees who worked as truck drivers.

Blue Green Alliance

In the United States, the Blue Green Alliance (BGA) operates on the principle that unions and environmentalists share a mutual interest in promoting a green economy, renewable and efficient energy sources, and mass transit and rail systems that help to slash levels of carbon dioxide in the atmosphere. The desire to put union workers into new sustainable jobs continues to fuel the partnership between unions and environmental organizations. The formal entity known as the Blue Green Alliance was established in 2006 when the United Steel Workers and the Sierra Club joined together to promote the greening of American business and industry. Other unions, including the Communications Workers of America (CWA), the Natural Resources Defense Council (NRDC), the Service Employees International Union (SEIU), the Laborers' International Union of North America (LIUNA), the Utility Workers Union of America (UWUA), the American Federation of Teachers (AFT), the Amalgamated Transit Union (ATU), and the Sheet Metal Workers' International Association, later joined the alliance, bringing total membership to more than 8 million individuals.

BGA maintains that its emphasis continues to be on lobbying for legislation dealing with clean energy and climate change that is capable of achieving optimum reduction targets based on scientific evidence while protecting existing jobs and creating new green ones for American workers; restoring the rights of unions to organize and engage in collective bargaining; promoting trade policies that reflect a 21st-century focus on labor, environment, and human rights at the global level; and protecting workers from improper use of toxic chemicals. Lobbying policy makers at all levels of government is considered essential to accomplishing BGA's goals, and the alliance has expressed public support for the Employee Free Choice Act, which is designed to expand the rights of workers, and the Transportation Act Reauthorization Bill, which promises to create additional green jobs for workers in the transportation industry. BGA also conducts an education campaign, aimed at teaching both workers and the general public about the ways in which green jobs promote solutions to the global warming crisis. Alliances with groups such as Al Gore's Alliance for Climate Protection and the Twin Cities' Mayors Green Manufacturing Initiative extend the reach of the Alliance.

The Blue Green Alliance has often enabled unions and environmental groups to present a united front when dealing with government officials, and it has been more successful at obtaining grant money than either group would be on its own.

Maxxam Corporation and the Union-Environmental Alliance

Union-environmentalist alliances have generally been most successful when targeting individual companies that have records of denying workers' rights while disregarding the environment. In 1999, the Blue Green Alliance issued a list of the 10 worst companies in the United States from an environmental perspective. Since both Walmart and Cintas were in the group, many saw the list as a not-so-subtle attack on nonunionized companies. Some of the most effective of these campaigns have involved union rank-and-file efforts. One example that epitomizes the need for the Blue Green Alliance is the case of steelworkers employed by Maxxam Corporation. In 1998, at the same time that Maxxam was recklessly cutting down ancient redwood trees in northern California, hostilities came to a head between management and workers who were disgusted with constant givebacks. When the Steelworkers Union ordered a strike, Maxxam brought in scabs, including a number of the loggers who had been cutting down the redwoods while employed by Maxxam-owned Pacific Lumber. Maxxam rejected the offer of the steelworkers to return to their jobs. In February 2000, the locked-out steelworkers aligned themselves with environmentalists to launch a weeklong protest against Maxxam before the opening session of the World Trade Organization.

Alliances in Practice

Despite the need for unions to cooperate with environmental groups, they often form uneasy alliances that stem from the need to balance jobs gained or lost with environmental concerns. The alliances are particularly heavily strained in some countries, including South Africa, where the economy has always been dependent to some extent on the export of precious gems and minerals. In the early 21st century, gold, diamonds, and platinum made up the bulk of those imports. Because of the importance of mining, South African workers have always been vulnerable to health and safety issues. Consequently, the unions took on the major responsibility for their protection. In response to environmental concerns, unions broadened their focus, often aligning themselves with environmental groups to accomplish specific goals. These efforts were also heavily influenced by the Thor Chemical case discussed above. However, certain events have stretched the union-environmental alliance almost to the breaking point. For instance, when Richards Bay Minerals announced that it would open a titanium mine in a protected wetlands area, environmentalists were up in arms, but the union was forced to consider the potential for much-needed jobs. In 1995, in response to public debate, the government vetoed the opening of the mine. That same year, another battle ensued when environmentalists objected to plans for Iscor Steel to commission a project near a protected lagoon. Environmentalists lost that battle when Iscor convinced authorities and unions that the economy would suffer both job and foreign exchange losses if the project were canceled.

Alliances between unions and environmentalists sometimes result when the two groups join together to fight government policies that threaten both the rights of workers and the environment. In summer 2010, trade unions in Papua New Guinea (PNG), led by the PNG Trade Union Congress (PNGTUC), angrily attacked new amendments to the Environment and Conservation Act, insisting that they were passed to promote the interests of multinational corporations rather than of the people of Papua New Guinea. Trade unions were particularly concerned about the possibility that the laws left no avenues for challenges since no lawsuits could be filed against sanctioned resource projects, even those that damaged the local environment and threatened the livelihood of the people. Proclaiming that the new legislation presented major threats to the "pristine and priceless" environment, union officials claimed that the government had sold out the people of Papua New Guinea.

While the outlook for a successful alliance between environmentalists and unions may seem fragile, the need to present a united front is likely to become more insistent in the face of globalization and the global warming crisis.

See Also: Chávez, César (and United Farm Workers); Environmental Justice Movements; Huerta, Dolores; United Farm Workers (UFW) and Antipesticide Activities.

Further Readings

"About the Blue Green Alliance." http://www.bluegreenalliance.org/about_us?id=0001 (Accessed September 2010).

Bombay, Peter. "The Role of NGOs in Shaping Community Positions in International Environment Fora." *Review of European Community and International Environment Law*, 20/2 (2001).

Chivian, Eric, et al., eds. *Critical Condition: Human Health and the Environment.* Cambridge, MA: MIT Press, 1993.

Dahl, Judy. "True Green." *Credit Union Magazine*, 74/7 (2008).

Ellis, Ryan. "Big Labor Partnering With Green Groups to Increase Political Power." *Human Events*, 62/37 (October 30, 2006).

Foster, John Bellamy. *The Ecological Revolution: Making Peace With the Planet.* New York: Monthly Review Press, 2009.
"A Global Green Union." *Earth Island Journal*, 13/4 (1998).
Jarman, Holly. "The Other Side of the Coin: Knowledge, NGOs, and EU Trade Policy."*Politics*, 28/1 (2008).
Kaye, Leon. "Ethix Merch Pairs Unions With the Environmental Movement." http://www.triplepundit.com/2010/08/ethix-merch-pairs-unions-with-the-environmental-movement (Accessed September 2010).
Laden, Francine, et al. "Cause-Specific Mortality in the Unionized U.S. Trucking Industry." *Environmental Health Perspectives*, 115/8 (2007).
MacGregor, Sherilyn. *Beyond Mothering Earth: Ecological Citizenship and the Politics of Care.* Vancouver: UBC Press, 2006.
Megane, Pelelo, et al. "Chapter 7: Unions and Environment: Life, Health, and the Pursuit of Employment." http://www.idrc.ca/en/ev-138116-201-1-DO_TOPIC.html (Accessed September 2010).
Miodonski, Bob. "Labor Group Donates to Green Building Movement." *Contractor Magazine*, 54/2 (2007).
"PNG (Papua, New Guinea) Unions Slam Environment Law." *Post Courier* (June 2, 2010).
Rootes, Christopher. "The Transformation of Environmental Activism, Organizations, and Policy-Making Innovation." *The European Journal of Social Sciences*, 12/2 (1999).
Schor, Juliet B. and Betsy Taylor. *Sustainable Planet: Solutions for the Twenty-First Century.* Boston: Beacon Press, 2002.
Steffen, Alex. *Worldchanging: A User's Guide for the 21st Century.* New York: Abrams, 2006.
Torgerson, Douglas. *The Promise of Green Politics: Environmentalism and the Public Sphere.* Durham, NC: Duke University Press, 1999.
"Trade Unions and the Environment." Union Learn. http://www.unionlearn.org.uk/education/learn-1912-f0.cfm (Accessed September 2010).

Elizabeth Rholetter Purdy
Independent Scholar

P

Permaculture

Permaculture is an ecological design approach conceived by Bill Mollison and his student David Holmgren in the 1970s. When they originally began using the word *permaculture*, it represented a contraction of "permanent agriculture" but over the years it has been broadened to also encompass "permanent culture." The practice and movement associated with permaculture originated in Australia and has since evolved to become global in scope. Permaculture has been applied in rural and urban contexts and has been used as a framework not only to design food systems and landscapes but also to develop balcony gardens as well as ecovillages. The key to all permaculture strategies is finding creative solutions to living a more sustainable life.

Permaculture is first and foremost a design system, but it has evolved into a land use and community building movement and a philosophy. The permaculture approach is one of observing natural systems as a means of learning how to live in ways that emulate the diversity, stability, and resilience of natural ecosystems. Most often, this science and form of design is applied to developing food production systems and ecological human habitats that restore, preserve, and extend natural systems while simultaneously yielding an abundance of food, fiber, and energy for the provision of local needs in a sustainable way.

Most permaculture practitioners and proponents borrow liberally from traditional wisdom and contemporary knowledge and practices from around the world. They utilize and combine diverse ideas, skills, and ways of living such as indigenous knowledge of edible and medicinal native plants, with more green ideas from systems ecology, ecoforestry, bio-intensive organic gardening, and sustainable community design. Hence, the specific techniques and components of permaculture are not original, but it is the overarching and interdisciplinary framework that many find new and enticing.

The main vehicle for permaculture training has been through a permaculture design course based on a curriculum that was codified by many of the original advocates of permaculture but with evolved regional divergences in the content of the courses. The organized system of design that permaculture emphasizes is not dogmatic; it includes broad principles, not detailed prescriptions, and is intended to be used in combination with local knowledge. In promoting ecological design for the regeneration of people and place, permaculture is purposely site and context specific, responding to the needs, resources, and potential of each individual locale and its inhabitants. Subtle differences of microclimate, soil and vegetation, water availability, and plant requirements are taken into account, as are the preferences and lifestyles of the human and other inhabitants.

While dynamic diversity lies at the heart of permaculture, nonetheless those undertaking it are guided by broad ethical principles. The ethical foundations of permaculture design include care for the Earth (protection and regeneration of soil, water, and forests); care for people (looking after oneself, kin, and community); and fair share (setting limits on consumption and reproduction and redistributing any surplus created). Yet these general precepts are not meant to replace the teachings and philosophies of great thinkers and religious leaders or to displace the insights of scientific discovery, but rather to act as constraints and guides for what are appropriate and desirable outcomes for living in the world.

Permaculture Design Principles

Permaculture also includes a set of design principles that act as a toolkit of strategies and techniques used by practitioners to create urban and rural habitats that are resilient, regenerative, productive, and sustainable. They are largely drawn from ecology, particularly systems ecology, but also from other sources like ethnobiology, geography, and appropriate technology design. Depending on whom one turns to as a source of permaculture inspiration and guidance, they can vary in number and emphasis but, in general, include most of the following principles:

- *Observe and interact.* Permaculture is not a passive pursuit but instead involves active learning by doing. Careful observation and interactive thought are the preconditions for generating models of land use and living that work with nature rather than against it. As a part of this strategy for being in the world, restraint should be exercised so as to ensure the activities undertaken create the most effect with the least amount of disturbance. Designs should be organized around the basic premise that it is possible to substitute ecological information and human stewardship for today's dependence on capital, hardware, chemicals, machines, genetically engineered organisms, and destructive technologies.
- *Catch and store energy.* Most of our current systems rely on the unsustainable use of fossil fuel energy and other resources and consequently draw down the reserves of natural capital. Through the creation of plant-based ecological systems, permaculture designs capture solar energy, along with carbon and nitrogen, in plant biomass and cycle these elements repeatedly through the system. While advocating a healthy dose of skepticism regarding technological innovation as primary in addressing sustainability, this principle draws attention to the need to employ technologies and knowledge to consume less fossil fuels, convert to a greater reliance on renewable energy, and rebuild reserves of natural capital.
- *Obtain a yield.* This principle emphasizes the need to design systems that provide for self-reliance at all levels, including the needs of humans. The long-term success of permaculture depends on humans being treated as an integral component of the system and solutions and building and managing functional and productive systems that provide immediate and useful yields in energy, food, water, or income.
- *Apply self-regulation and accept feedback.* This principle points to the need to limit or discourage inappropriate growth and behaviors by paying attention to consequences of actions and designs. Using feedback loops as a guide, systems should be designed to be self-regulating and thereby reduce the need for repeated and harsh corrective management.

- *Use and value renewable resources and services.* Permaculture designs make the best use of nonconsuming natural services as possible and emphasize harmonious interactions between humans and nature.
- *Produce no waste.* Permaculture encourages frugality and care so that resources are used wisely and pollution is not created. Nature does not create waste, and so systems should be designed in such a way that all outputs are used effectively through closed loops that mimic natural systems.
- *Design from patterns to details.* Finding the appropriate pattern for a permaculture design is more important than understanding the details of every element in the system. In practical terms, this can be accomplished by determining zones of intensity of use around an activity center, such as a house, as a guide for the placement of design elements and considering the impacts of the sun, wind, flood, and fire on various sectors of the system. At the same time, as many failed international development projects illustrate, the larger social and community context can determine if a particular design solution for a particular site will be achievable and successful in the long term.
- *Integrate rather than segregate.* Permaculture design and practice is a system of assembling conceptual, material, and strategic components in a pattern that functions to benefit all life forms, with more of a focus on the relationships and interconnections between the components than on the components themselves. In permaculture design, every component should function in many ways, and every essential function should be supported by many components.
- *Use small and slow solutions.* Systems should be designed to perform functions at the smallest scale possible that is practical and energy efficient. The use of appropriate technology, starting small, and purchasing from local businesses are all ways to practically integrate this principle into a permaculture plan.
- *Use and value diversity.* Diversity lies at the heart of permaculture's allowing for systems of complexity and resilience to emerge. Permaculture practitioners often make use of polycultures, or the growth of multiple crops at once, but diversity can also be enhanced through the use of different cultivation techniques, incorporating a diversity of structures and materials on a site, and conservation and nurturing of not only a wide variety of plants and animals but also languages and cultures.
- *Use edges and value the marginal.* Marginal elements can sometimes be the most influential components of a system, just as the edges of an ecosystem are often the most productive and biodiverse.
- *Creatively use and respond to change.* Permaculture designs should allow for and accommodate the system in demonstrating its own evolution.

Conclusion

Sometimes explicitly, but more often implicitly, permaculture advocates call into question conventional agriculture, forestry, landscape design, and gardening that depend on chemical inputs, monoculture planting patterns, and fossil fuel consumption. Instead, they advocate a holistic approach that looks for the connections between the different parts of the system and promotes a strategy of how to enhance these connections so they can work more harmoniously. At its heart, permaculture is about living more sustainably and learning how to reduce humanity's ecological impact, or more precisely, to turn our negative ecological impacts into positive ones. Hence, if this movement as a whole advocates a politics, it is that real change should take place at the bottom and move up, rather than waiting for change at the top to filter down.

See Also: "Agri-Culture"; Alternative Communities; Green Consumerism; Social Action, National and Local.

Further Readings

Hemenway, Toby. *Gaia's Garden—A Guide to Home-Scale Permaculture*. White River Junction, VT: Chelsea Green, 2001.

Holmgren, David. *Permaculture: Principles and Pathways Beyond Sustainability.* Hepburn, Australia: Holmgren Design Services, 2002.

Mollison, Bill. *Permaculture—A Designer's Manual*. Sisters Creek, Australia: Tagari Publications, 1988.

Permaculture Cairns. "World Changing Permaculture Resources." http://www.permaculturecairns.com (Accessed June 2010).

Permaculture Institute. "Permaculture—Key Concepts." http://www.permaculture.org/nm/index.php/site/classroom (Accessed June 2010).

Permaculture Principles. "Introduction." http://www.permacultureprinciples.com (Accessed June 2010).

Lorelei Hanson
Athabasca University

Plastic Bags

Paper or plastic? What was once a routine question at grocery stores throughout the world has created much controversy over the past two decades. In 1977, the single-use plastic bag was introduced to American grocery stores, but their use and availability was limited. The emergence of the plastic bag in stores did not occur nationwide in the United States until the early 1980s when new technologies made them cheaper to produce than traditional paper shopping bags. Today, four out of five bags handed out at grocery stores are made of plastic. For nearly 20 years, these bags were used without hesitation, until environmental concerns began to surface in the mid-1990s.

Environmental groups estimate that between 500 billion and 1 trillion plastic bags are used worldwide each year. Over 12 billion gallons of oil are used in the production of these bags, which, after use, proceed to be recycled, sent to landfills, or littered into the environment. While over 100 million pounds of plastic bags are recycled in the United States each year, this is only a percentage of the bags produced. Most plastic bags end up in landfills where they can take thousands of years to decompose. Many bags end up in our streets, trees, bushes, creeks, rivers, streams, and oceans. According to California Assemblyman Lloyd Levine, not only do communities endure a high financial cost for cleaning up this litter (California spends roughly $300 million annually), but the impacts on wildlife and humans can be significant.

Tens of thousands of birds and marine animals die each year from ingesting plastic products such as plastic bags. When the bags degrade in the ocean, they break into smaller pieces and concentrate in an area known as the Great Pacific Garbage Patch, where the semi-degraded plastics are consumed by animals mistaking it for food. The indigestible

Tens of thousands of birds and marine animals die each year from ingesting plastic products like plastic bags, which here litter a tree in South Africa. Various strategies are being put in place to reduce, minimize, or eliminate plastic bag usage.

Source: Wikimedia

plastic remains in the animals' stomachs. The more plastic that is consumed, the less space for food, which may eventually result in death due to lack of room for food. Plastic bags have also been linked to animal deaths by asphyxia.

Plastic bags have economical and social consequences as well. Littered bags have clogged storm drains in Mumbai, causing massive flooding that produced civilian casualties and left many others homeless. While this is disheartening, according to the plastics industry, it is the third-largest manufacturer in the United States, with more than 1.1 million employees and $379 billion in annual revenue in the U.S. economy.

Presently, several strategies are being employed to reduce, minimize, or eliminate plastic bag usage, ranging from banning to education. In the United States, a state-mandated ban of plastic bags was passed in 2009 to protect the delicate coastal ecosystem of the North Carolina Outer Banks. In California, bill AB 1998 to ban single-use plastic bags was proposed in 2010; however, the bill was not passed by the California State Senate. Single-use plastic bags have been banned in Bangladesh, Taiwan, South Africa, and other countries. Some countries have implemented taxes or levies, which proved successful in the case of Ireland, where the 2002 tax on plastic bags resulted in a 95 percent reduction in their use. In order to encourage voluntary reduction in the use of plastic bags, Scotland and the United Kingdom have worked with retailers to reduce the use of plastic bags. Many stores offer a minimal monetary incentive (e.g., five cents as a credit or charity donation) to entice shoppers to bring their own reusable bags. Additionally, many stores provide recycling bins for plastic bags. What are the alternatives for consumers who choose not to use single-use plastic bags and how sustainable are those choices?

Alternatives to plastic bags include single-use paper bags, single-use biodegradable bags, and reusable shopping bags. Paper bags are made from trees, which are a renewable resource. They can be manufactured using recycled materials, can be recycled after use, have a higher recycling rate than plastic bags (21 percent compared to less than 10 percent of plastic bags, with some studies estimating plastic bag recycling rates as low as 1 percent), and are frequently reused. However, paper bags generate more air and water pollutants in manufacturing, weigh more, are bulkier, and require more energy and water to produce and recycle than plastic bags. Additionally, paper bags do not break down in today's landfills due to lack of air, light, water, oxygen, and other elements required for decomposition.

Single-use, biodegradable bags are viewed as a sustainable alternative to plastic bags because they will decompose at the end of their life. However, they take months or years to decompose and will not decay in the modern landfills or other noncomposting environments. Care must be taken that these bags not go into recycling because they can contaminate the recycling process. Biodegradable bags are about the same size and weight, but cost 12 to 30 cents per bag, compared to one cent per bag for single-use plastic bags.

Reusable shopping bags, another alternative to plastic bags, may be used hundreds of times, can be laundered, and are made from a variety of materials including virgin or recycled plastic, canvas, cotton, nylon, or other kinds of cloth. While their larger size and weight may consume more materials in manufacturing than single-use plastic bags, this could be offset by their reuse.

Considering the impacts of single-use plastic bags and the alternatives, one common solution that environmentalists, manufacturers, and policy makers can agree on is education. Whether recycling, litter reduction, or reuse is the end goal, the most effective tool to achieve it is education of the public.

See Also: Environmental "Bads"; Environmental Protection Agency; Garbage; Recycling; Shopping.

Further Readings

Environmental Literacy Council. "Paper or Plastic?" http://www.enviroliteracy.org/article.php/1268.html (Accessed September 2010).

ICF International. "Master Environmental Assessment on Single-Use and Reusable Bags." San Francisco, CA: Green Cities California, 2010.

Ritch, Elaine, Carol Brennan, and Calum MacLeod. "Plastic Bag Politics: Modifying Consumer Behaviour for Sustainable Development." *International Journal of Consumer Studies*, 33 (2009).

Roach, John. "Are Plastic Grocery Bags Sacking the Environment?" *National Geograhic News* (2003).

U.S. Environmental Protection Agency. "Questions About Your Community: Shopping Bags: Paper or Plastic or . . .?" http://web.archive.org/web/20060426235724/http://www.epa.gov/region1/communities/shopbags.html (Accessed September 2010).

LaDona Knigge
Eli Goodsell
California State University, Chico

Poetry

Green poetry, more commonly "ecopoetry," is poetry that incorporates a strong ecological emphasis or message. The ecological message is not new, it may well be ageless in fact, but the recent ecopoetic movement is stronger than ever before. In little more than two decades, ecopoetry has achieved a status as a subgenre written by English-speaking poets.

Origins

The Club of Rome warned of in the 1960s and formalized in *The Limits of Growth* (1972) the dangers of the culturally based misunderstanding of nature as inexhaustible and available for arbitrary consumption that induced the ecological crisis that threatened to destroy not only the ecosystem but mankind. Earth Day, April 22, 1970, was the official birthday of ecological awareness, but political action only began in the 1990s. The old anthropocentric worldview disappeared, to be replaced by a circular man–nature view. The search for a new ethic—new values—became a literary effort, initially sporadic, but with time, more organized.

Ecopoetry is poetry that ties humanity to the world with an implicit responsibility. It shares with other engaged models (feminism and Marxism, for instance) an envelopment by ethical concerns. That said, there is no solid definition of ecopoetry or ecopoetic projects or poiesis. There is a general acceptance that such projects include a commitment to ecology for both the person and the society. It is a mainstream subgenre and recognized as such.

Terminology

Practitioners and advocates seemingly enjoy playing word games that attempt to explain the origins of the term *ecopoetry*. "Eco" is the planet, the house shared by millions of species. *Ecology* is a term dating from 1873, and *ecocriticism* arose a century later as a term for the different ways of imagining and portraying the environmental-human connection. Coincident with ecocriticism was the use of "ethnopoetics," which looks to the past and away from the modern and experimental. A subset of ethnopoetics is ecopoetics.

The biological term for the area between adjacent ecosystems is *ecotone*, so ecopoetics is the boundary touching several disciplines, the cutting edge if done right. Ecopoetics is the zone of contact between ecology, poetry, and ethnopoetics. The journal *Ecopoetics* explores the points where writing (mostly poetry) touches ecology, preferring the term *poiesis*, making or writing, rather than poetry.

Eco- or Nature Poetry

In *Ecopoetry*, the strongest anthology currently available, J. Scott Bryson surveys nature poetry as a dominant genre from Beowulf to Blake, a declining force in the Darwinian era, subject to abuse by anti-Romantics such as Frost, and reemergent with the Beats and Gary Snyder. Even John Clare, the 19th-century poet, understood nature as in decline as men's fences erected unnatural barriers, imprisoning their flocks and small minds on small lots.

As a subset of nature poetry, ecopoetry acknowledges that all creatures are interdependent, that the relationships among living creatures, including humans, should be approached with humility rather than anthropocentric hubris, and advocates caution and skepticism about the technological and rational culture that dominates the current era. Some regard ecopoetry as similar to nature poetry, but others argue that ecopoetics is clearly not nature poetry. The latter traditionally is a human meditation on a nonhuman, perhaps nonanimate natural work of nature as a means of understanding the human. Nature appears in ecopoetry as the dominant worldview of the work.

Nature has always been closely tied to poetry, but as humanity has drawn away from its roots, a gap has opened between humanity and nature. Pastoral poetry, the songs of the simple and natural life, is dead, replaced by a new understanding of nature as humanity's damaged victim. Ecopoetry is one of the efforts to make people aware of the crisis and push them to take necessary action.

Content

The technological, self-interested, economic-prioritizing current world has isolated itself from nature and is poorer for it even as the Earth is endangered and more dangerous. Nature poetry tends to be passive, while ecopoetry is issue oriented, a reminder that all life is interdependent, nature is other and wild, and human efforts to plunder and tame nature are irresponsible. The ecopoet is more deeply engaged and more aware of the environmental negatives of the human presence. Thus, ecopoetics takes on a political coloration, incorporates a subtle undercurrent of political criticism, the same as ecocriticism, Marxism, and feminism. Ecopoetry is engaged with the world, enveloped by ethical concern and responsibility for the environment. Ecopoetry tends to be innovative rather than mainstream, because mainstream poetic values and practices tend to reflect the larger society, whereas ecopoetry is critical of the larger society's values and practices.

Lawrence Buell says that environmental orientation includes four elements. First, the nonhuman environment must be more than a framing device, must in fact be a presence sufficient to at least hint that human history is tied to natural history. Also, human interest may be important but it is not the single interest. The text must implicate humans as accountable for the environment. The environment must be understood as process, not constant.

James Engelhardt explains that ecopoetry is a means of connecting from the core to the world through language. He notes that it sometimes borrows approaches and strategies from postmodernism and related offshoots, but that it is not postmodern. To an extent it shares scientific concerns, specifically the concern with nonhuman nature (and ecocriticism shares this value). But most poetry throughout history has shared this concern, too, the question of how humans can connect to a nonhuman nature that is so much larger than a human. Prior poetry has attempted to give unrealistic praise or to applaud human dominance over nature. This style of poetry is pastoral or georgic.

Ecopoetry recognizes with science that nature is neutral and will never become human. Ecopoetry seeks to approach nature, even as nature slips away, by naming it scientifically. Engelhardt rejects attempts to approach nature spiritually because humans are not spirits but bodies that have to deal with nature from what they are. And humans are of families, broader than the nuclear, more at the level of culture, and they have to understand how this affects their linking to nature. And the ecopoet, like the feminist or the Marxist, must decide if the poem is to be simply esthetic or if it is to have a broader meaning, an activist element. And an ecopoem must incorporate the human pleasure of play, including wordplay.

Ecopoets

Green poetry tries to answer the question of whether poetry can be ecological. Forrest Gander of the Poetry Foundation explains that poetry can have inherent ecological values, and can express the interrelationship between environment and humans. Gander feels that poetry is deft enough to capture the fragments of natural processes already disrupted by human observation, and events that have no beginning or end and that are multilayered.

Clearly green, but not labeled as such, are works such as Jay Ramsay and Carole Bruce's *The White Poem* (1988), *Bosco* (1999, 2001), and *Heavy Water: A Poem for Chernobyl* (2004). New anthologies and books are overt in using the idea of ecopoetry; among them are *The Thunder Mutters,* edited by Alice Oswald (2005) and *Ecopoems*, edited by Neil Astley (2007). In 2002, J. Scott Bryson edited *Ecopoetry: A Critical Introduction,* which includes essays on the classics from Ralph Waldo Emerson and William Butler Yeats to Robinson Jeffers and William Carlos Williams, and it also deals with the major current ecopoets—W. S. Merwin, Mary Oliver, Wendell Berry, and Gary Snyder—as well as lesser current practitioners. Neil Astley's *Earth Shattering* is a collection of more than 200 poems about ecological balance and environmental degradation. Subjects include global warming, species extinction, and other whole-Earth themes as well as smaller local or regional worries such as pollution of rivers or seas or air and destroying forests. Ecopoetic themes also include the now lost or nearly lost natural world, with either lament for its passing or hope that recycling, conservation, and rethinking might recover it.

Broadening of the Movement

Global warming poetry by slam poets was introduced to Sundance in 2008 in collaboration with Youth Speaks. Robert Redford wanted to involve young people, particularly those who had a high environmental priority. Many participants were first-time voters, and organizers wanted their ideas to move the audience to be more active in working toward a future compatible with climate change.

In March 2010 in New York, the Poets House group began a series on poetry and the environment, titled Ecopoetic Futures. It later began a program that sought to raise environmental awareness by installing poetry in zoos in New Orleans, Little Rock, Jacksonville, and Brookfield (Illinois). The installations include poetry and nature books and programs already offered at the local public libraries, as well as poets in residence. Lodged in a "green" building, Poets House helps raise awareness of environmental issues through its programs.

A Persian leader in poetry was Forough Farrokhzad, an estranged woman writing in the 1940s and 1950s. Her poetry, particularly "I Pity the Garden," predated the messages of Al Gore, Jr., and others warning of imminent environmental danger, if not collapse. Current Persian musicians incorporating her poetic view warn that the Western divide (dating from the enlightenment) between nature and culture is driving a crisis. Contemporary biologists are becoming aware of how blurred the line can be between an individual and the environment, something that Persians understood a thousand years ago.

The University of East Anglia has begun Ecopoetry in Schools, an outreach program in conjunction with the Norfolk Wildlife Trust and the Norfolk County Council. The program seeks to give children aged 5 and 6 engagement with nature to increase their creativity and bring them closer to nature. Children interact and then write poems about their experiences.

Stylistically, ecopoetry rejects traditional forms, preferring a simple structure to clearly spread the message to a wide audience, even in translation, as called for by the United Nations Educational, Scientific and Cultural Organization (UNESCO) during the 2007 World Day of Poetry.

The poet laureate for the United States in 2010 was W. S. Merwin, the most prominent green poet of the time, with over 40 books and two Pulitzer Prizes to his credit. An excerpt from Merwin's 1977 "The Last One" (*The Lice*, 1977) appeared in the *Los Angeles Times* with the announcement of his appointment: "Well they cut everything because why not/ Everything was theirs because they thought so/It fell into its shadows and they took both away."

See Also: Art as Activism; Novels and Nonfiction; Redford, Robert.

Further Readings

Arigo, Christopher. "Notes Toward an Ecopoetics: Revising the Postmodern Sublime and Juliana Spahr's 'This Connection of Everyone With Lungs.'" *How2*, 3/2 (2007).

Bryson, J. Scott. *Ecopoetry: A Critical Introduction.* Salt Lake City: University of Utah Press, 2002.

Ecorazzi. "Robert Redford Brings Green Poetry Slam to Sundance" (July 17, 2008). http://www.ecorazzi.com/2008/07/17/robert-redford-brings-green-poetry-slam-to-sundance (Accessed September 2010).

Engelhardt, James. "The Language Habitat: An Ecopoetry Manifesto." *Octopus Magazine* (Winter 2006–07).

Gander, Catherine. "Ecopoetry in Schools." http://www.uea.ac.uk/ams/ecopoetry (Accessed September 2010).

Hawkinns, David. "Earth Shattering: *Ecopoems* Edited by Neil Astley." *The Ecologist Reviews* (June 1, 2008).

Kuipers, Dean. "Merwin Is Green as U.S. Poet Laureate." *Los Angeles Times* (August 29, 2010).

Lila Sound Productions. "Persian Poets, Environmental Devastation, and Electronic Music: Green Memories Is an Ambient Voice of the Earth." *Payvand Iran News* (December 8, 2008). http://www.payvand.com/news/08/dec/1084.html (Accessed September 2010).

Merwin Conservancy. http://www.merwinconservancy.org (Accessed September 2010).

Merwin, W. S. "The Last One." In *The Lice,* 1977.

Mickey Z. "Green Glossary: Eco-Poetry Mickey Z" (December 11, 2008). http://planetgreen.discovery.com/fashion-beauty/green-glossary-eco-poetry.html (Accessed September 2010).

Peacework Magazine (October 2003). http://www.peaceworkmagazine.org/pwork/0310/031013.htm (Accessed September 2010).

Poetshouse.org. "News: Current." http://poetshouse.org/news.htm#g (Accessed September 2010).

Porritt, Jonathan. "Review of *Earth Shattering.*" Bloodaxe Books. http://www.bloodaxebooks.com/titlepage.asp?isbn=1852247746 (Accessed September 2010).

Timm, Lenora A. "Brittany's Eco Warrior: The Environmental Poetry of Anjela Duval." http://www.breizh.net/anjela/saozneg/environmental_poetry1.php (Accessed September 2010).

John H. Barnhill
Independent Scholar

Political Persuasion

Political persuasion has played a key role in the growing influence of the green movement. Although political persuasion is a significant factor in many policy decisions, it works slightly differently with regard to sustainability and environmental issues. Political persuasion has uniquely affected green issues insofar that influence has been exerted both from a governmental level that has changed consumer and industrial behavior, and from a grassroots base that has swayed legislation and practices.

The modern environmental movement took root during the 1960s as part of the larger counterculture questioning of traditional practices. While embraced by certain politicians, government officials, and corporate leaders early on, the green movement's early successes were largely the result of a grassroots demand for change at the local, statewide, and national levels. As interest in sustainability and environmentally friendly policies became more mature, legislation was enacted that affected both citizens and corporations. Political persuasion continues to shape and mold the green movement, and greatly influences government policy as well as individual behavior. The unique grassroots movement has been evidenced by citizen suits that have allowed individuals to bring suits against companies that have committed environmental offenses. The effect that the green movement has had on politics is further reflected through the protection of citizen suits by Congress and others, who protect them as part of a democratic process that allows citizens to address social problems and issues, such as environmental pollution, the destruction of natural resources, or other key tenets of the green movement.

Beginnings of the Green Movement

The modern American green movement has its roots within the conservation movement of the early 20th century. Concurrent with this movement, political control over environmental issues began to shift from local government control to that of state and federal governments. Proponents of the conservation movement advocated for a long-term plan that would allow economic development that was controlled in a manner so as to maximize the benefits of natural resources. Conservationists were opposed by owners of private property, such as lumber and mining companies, who favored a more laissez-faire approach that allowed owners of property to do whatever they wished with their land. Throughout this period of initial understanding of the importance of the environment, legislative and popular focus centered on the establishment of state and national parks, forests, wildlife refuges, and other areas designed to preserve natural features and areas of natural beauty. The political will that supported early conservation was deeply intertwined with the progressive movement. As a result, conservation's proponents focused on general expansion of their political power as a key issue, acting in response to the considerable influence of industrialization and urbanization. Early conversation advocates believed strongly, much like progressives, in notions of efficiency with resources and land development. Taking this path, the conservationists enjoyed legislative victories under the administration of Theodore Roosevelt.

Legislative examples of the conservationists' successes include introduction of the Reclamation Act of 1902, also known as the Lowlands Reclamation Act. The Reclamation Act was concerned with irrigation projects in the western United States, and sought to increase efficiency in the allocation of natural resources by permitting irrigation while also carefully monitoring power generation and flood control. The Reclamation Act was considered highly successful at the time because it was able to permit settlement of large previously uninhabitable areas of the west while also making attempts to limit erosion. The Reclamation Act led to the founding of the precursor to the U.S. Bureau of Reclamation. In addition to the Reclamation Act, early legislative victories included the creation of the U.S. Forest Service, as well as the creation of five national parks, 51 bird reserves, four game preserves, and 150 national forests. Gifford Pinchot, a close ally of Roosevelt, was named the first head of the U.S. Forest Service. Through their concentrated efforts to resist much harmful expansion, the conservationists laid the framework for the oncoming green movements.

Although the conservation movement sought more responsible stewardship of natural resources, it did not favor changes like those advocated for by environmentalists, such as John Muir. Environmentalists took a much more doctrinaire approach to natural resources. Holding natural resources to be an almost sacred trust, many environmentalists opposed the harvesting of timber on public lands or the damming of rivers to supply water for urban settings. The environmentalists were especially opposed to the O'Shaughnessy Dam that created the Hetch Hetchy Reservoir in Yosemite National Park, which Roosevelt approved, as it greatly altered the terrain of that area. Environmentalists feared that the conservationist movement permitted too much use of lands that were, in the environmentalists' view, best left in as pure a state as possible. Political leaders and legislative bodies were often challenged by advocates of diametrically opposed positions. Despite these challenges, support for the preservation of natural spaces would grow and expand, although not without setbacks.

For much of the ensuing 50 years, green political issues received little attention, although during the 1930s, the administration of Franklin D. Roosevelt expanded the National Forest System through the purchase of privately held lands. During the 1960s and 1970s, however, popular interest in environmental issues again increased. Part of this growth came from increased interest in the counterculture and antiwar movements, with their distrust of many establishment figures and practices. Political support for environmentally friendly policies and laws grew, and politicians from both the Democratic and Republican Parties embraced legislation that promised a cleaner Earth. Indeed, Republican Richard Nixon supported the founding of the Environmental Protection Agency (EPA), and signed legislation such as the Clean Air Act and the Clean Water Act that greatly expanded federal oversight of pollution.

After this burst of activity, public commitment to protecting the quality of the environment remained strong, despite predictions of public disinterest that often follows landmark legislation. Indeed, supporters of green policies strongly opposed many of the anti-regulatory positions that were supported by Ronald Reagan's administration during the 1980s. Political battles were fought over issues such as offshore oil drilling, clear-cut timber harvesting, and measurement of carbon dioxide emissions. Citizen involvement often forced a retreat by government agencies that advocated for pro-development policies. These struggles created a general mistrust among many regarding the government's ability to manage environmental problems. This mistrust led to the rise of the grassroots green movement, as environmentalists began to view the government as both an ally and hindrance to green goals. After the movement away from government control of the process, environmentalists sought alliances with scientists, lawyers, and other individuals who possessed technical knowledge that would further their goals. Green issues continued to be contentious through the administrations of Bill Clinton, George W. Bush, and Barack

Obama, with many advocates for sustainability and environmentally friendly policies feeling stymied by government policies that attempted to take a middle road that accommodated business interests. Although many in the green movement in the United States identified with the Democratic Party, many who favored environmental change began to explore alternative avenues for their political involvement.

Alternative Advocacy Groups

Although many alternative political groups advocate for environmentally friendly policies, the Green Party has consistently pushed for more sustainable action. Its guiding principles have also been influential with both mainstream and alternative groups. The current American Green Party began as the Green Committees of Correspondence (GCC) in 1984. The GCC was charged with the task of organizing local green groups in order to consolidate power and provide further support to the grassroots movement that had taken hold in the United States as a resistance to governmental agencies. The GCC was initially constituted of regional confederations that met three times a year. The regions were self-defined, and based around natural regions rather than upon state boundaries, as a further repudiation of the rigid political boundaries that many in the movement opposed.

The GCC favored "bottom up" administration, with key values and ideas being generated from the membership, not leaders or other experts. The GCC, and later the Green Party, developed 10 key values that have proved influential to other organizations attempting to use political persuasion to affect sustainable development and environmental protection. These 10 key values included the following:

- Grassroots democracy
- Ecological wisdom
- Social justice and equal opportunity
- Nonviolence
- Decentralization
- Community-based economics
- Feminism
- Respect for diversity
- Personal and global responsibility
- Future focus and sustainability

The 10 key values were meant to provide a moral and philosophical underpinning for green advocacy. As a result, many involved in the green movement have sought their group's own purpose, structure, processes, and actions. Because of the decentralized nature of many green advocacy groups, the political activities of their members are varied, including such diverse pursuits as electoral politics, public education, media and publications, and the formation of citizen watchdog groups.

Voters and activists who favor sustainability and environmentally friendly policies have been able to wield influence at the state and local level, especially with regard to issues that were difficult to build a consensus for at the national level. California citizens, for example, were able to use political persuasion to allow passage by the legislature of the Global Warming Solutions Act of 2006, also known as California AB 32. California AB 32 seeks to bring that state into compliance with the provisions of the Kyoto Protocol. Local demonstrations, events, and celebrations also influence local and statewide public opinion, practices, and voting behaviors.

Conclusion

As interest in the green movement grows, advocates for sustainable practices and environmentally friendly policies are placing more focus on increasing their political power. While this move is in some ways a repudiation of the nonelectoral movement that was caused by disillusion with government policies of the 1980s and 1990s, it recognizes the growing popularity of green issues. As a wider spectrum of voters have become advocates of green issues, it is not surprising that they are looking for candidates in elections who share their views.

See Also: Antiglobalization Movement; Antiwar Actions/Movement; California AB 32; Demonstrations and Events; Grassroots Organizations; Organizations and Unions; Public Opinion; Social Action, National and Local.

Further Readings

Aldy, J. E. and R. N. Stavins, eds. *Architectures for Agreement: Addressing Global Climate Change in the Post-Kyoto World*. New York: Cambridge University Press, 2007.

Mutz, K. M., G. C. Bryner, and D. S. Kenney, eds. *Justice and Natural Resources: Concepts, Strategies, and Applications*. Washington, DC: Island Press, 2002.

Ross, B. and S. Amter. *The Polluters: The Making of Our Chemically Altered Environment*. New York: Oxford University Press, 2010.

Stephen T. Schroth
Jason A. Helfer
Jordan K. Lanfair
Knox College

Popular Green Culture

As the environmental movement has grown in visibility and urgency, environmentally friendly behaviors and products have become a significant part of *popular culture*, a term that describes the mainstream cultural preferences of a given community and that includes all aspects of the vernacular experience. Concerns for the environment have gone from being considered obscure or even unpopular by the population at large to being a centerpiece of popular culture and communication in the 21st century. In short, "it's cool to be green" in 21st-century America. The popularity of environmental concerns in popular culture can be evidenced by its centrality to popular American design, art and architecture, food and agriculture, entertainment media (including film, mass media, music, and Internet), education, government, and business. While this popular attention for the environment has helped bring awareness of serious problems to millions of individuals, some do have concerns about the degree to which significant changes are enacted by mere pop culture exposure.

Emergence of Green Culture

The American environmental movement can trace its roots to the mid- and late-19th-century writings of Henry David Thoreau and John Muir, whose sublime depictions of the natural world first brought public attention to the beauty of nature. However, despite the popularity of those authors through the decades, the environmental protection movement began with small activist groups who noted problems with the way in which natural resources were conceived of and used by most people in the United States. While groups like the Sierra Club and the National Wildlife Federation can point to long, active traditions of environmental protection, their membership and friends remained mostly marginal voices in a chorus of contrary messages from corporate America that urged human progress and generous use of natural resources, which were considered to be generally unlimited. Despite moments in the 1970s and 1990s when environmental concerns gained momentum, being green was eccentric rather than "cool."

By the 21st century, Americans took notice of the environment and its limited resources due to a combination of events that included increased costs for fossil fuel energies, growing popularity of mass media messages about environmental preservation, and highly publicized natural disasters like Hurricane Katrina and the severe floods of several Midwestern states in 2007. By this time, concerns about the environmental movement moved from the margins into mainstream society, and the term *green* was adopted to refer to all things that could be considered better for the environment. Green became cool, which is to say that to associate with the green movement brought both individuals and corporations cultural capital in fashion or pop culture trends. Hollywood celebrities like Leonardo DiCaprio and Pamela Anderson associated their voices and images with environmental documentaries and animal rights organizations. As green gained in popularity, individuals became interested in making greener choices as a means of performing this cultural value as a trend. Clothing with recycling logos became popular, as did clothing that was made from organic or natural fibers. In fact, even Walmart stores carry clothing made of organic cotton. Environmental concerns, under the guise of the green movement, can be said to be cool on multiple fronts within popular American culture.

Although the concerns of environmentalists are not necessarily issues embraced by all parts of the ideological spectrum, the environment as object has become a popular component or unit of popular culture, traceable in nearly all facets of the vernacular experience.

Aspects of Green Popular Culture

Evidence that green culture has become "cool" in popular culture can be located in American design, art and architecture, food and agriculture, entertainment media (including film, mass media, music, and Internet), education, government, and business.

The popularity of the green movement is first evident in the creative world of art, architecture, and design. The field of environmental art, or art designed to relate to or comment upon the natural world, has emerged in the fine arts. Notable examples of environmental art include "Oak Bike" and "Green Beatle" by Danish artist Morten Flyverbom, who uses natural materials to create human consumables like bicycles or cars. Home design has become green as well: designing attractive rain-return barrels and water retrieval systems, streamlined corn- or pellet-burning stoves, and using organic and natural materials. Many architecture projects involve gaining certification from an organization called Leadership in Energy and Environmental Design (LEED) to demonstrate their adherence to green building and maintenance practices.

In the areas of food and agriculture, green concerns have impacted food production, distribution, and consumption in the United States. The fastest-growing sector of the food industry is organic and natural foods, and these products have space on the shelves of even the most mainstream groceries, including Walmart Super Centers. To accommodate growing demands for local farm-grown produce, farmers markets have sprouted in communities across the United States. Similarly, a plan of agriculture called "community supported agriculture" (CSA) has emerged, in which a farmer offers a certain number of shares for sale to consumers, who in turn receive a portion of the goods produced on the farm all season. A renewed interest in backyard and community gardening has also emerged, and even residents of crowded cities now participate in the gardening of abandoned lots or construct rooftop gardens. In nearly all facets of food production and consumption, green initiatives have become significant concerns.

Media theorists claim that the entertainment media both reveals and constructs consumer interests, and the topic of environmentalism has been important fodder for many popularly successful films, books, and television programs. One of the most notable popular films about environmental concerns was Davis Guggenheim's documentary *An Inconvenient Truth*, about Al Gore's campaign to educate the public about global warming; the film won the 2006 Academy Award for Best Documentary. Other documentary movies with environmental themes that garnered popular and/or critical popularity included Robert Kenner's *Food Inc.*, Aaron Woolf's *King Corn*, and Louie Psihoyos's *The Cove*. Films and television shows depicting nature and the environment enjoy public popularity, with notable examples being Disney's *Earth* film or the National Geographic cable channel. Print publishing also reflects the public appetite for green material with books including *The Omnivore's Dilemma* by Michael Pollan, *Exposed: The Toxic Chemistry of Everyday Products* by Mark Schapiro, and *Deep Economy* by Bill McKibben earning significant public interest. There are also countless titles to inform and instruct the consumer about leading a more environmentally friendly lifestyle. Finally, the Internet has provided limitless space for promoting environmental concerns, and all prominent environmental activist groups maintain an Internet presence; citizens can discover important information about many different environmental concerns with a quick search. Green media is a significant source of popular culture production, reflected in all these varied ways.

More evidence for the prominence of environmental concerns in popular culture emerges from its growing presence in government debates and legislation. Federal and state governments have enacted green initiatives that promote environmentally friendly lifestyle choices. The federal government has developed the Energy Star program, which is certification and labeling for products meeting certain energy-efficiency standards, and the Internal Revenue Service (IRS) offers tax credits on federal income tax filing for the purchase of Energy Star products. An additional federal initiative was the popular Cash for Clunkers program, which helped consumers finance fuel-efficient cars while simultaneously removing thousands of fuel-inefficient cars from regular use. Other examples of federal initiatives for green lifestyle choices include IRS tax credits for improving the energy efficiency of one's home or for purchasing a hybrid car, and the Appliance Rebate Program, which offered cash rebates for homeowners who replaced their inefficient household appliances with Energy Star appliances. Politically, no contemporary U.S. presidential candidate can avoid having a clear platform that includes positions on significant environmental debates like oil drilling and cap-and-trade regulations for limiting pollution.

In higher education, environmental studies has emerged as a discipline of study that prepares students for careers in environmental protection and advocacy. Environmental topics are also studied by scholars in communication, sociology, English literature, history, and most of the sciences, among others. College campuses have embraced green lifestyle changes, and more than 650 American university presidents have signed the American College and University Presidents Climate Change Commitment, which means their campuses are committed to reducing their overall carbon footprint. In secondary education, many public and private schools are following the lead of universities and are increasingly switching to environmentally friendly lighting, heating, and energy consumption.

Finally, the popularity of green concerns has impacted business in multiple ways, including the products and services offered and the advertising and public relations messages disseminated. With the emergence of a green consumer consciousness, American industry has introduced numerous new environmentally safe or conscientious products to the market. In addition to the organic and natural food market described above, companies have marketed environmentally safe cleansers and laundry detergents, paper towels and tissues made from post-consumer waste, and other recycled products for nearly every imaginable human need. In the service industry, hotels request that patrons reuse towels and sheets in an effort to help them attain a smaller carbon footprint, and restaurants promote their use of locally grown products. In addition to providing products that assist a consumer in leading a green lifestyle, businesses also market products in ways that help associate themselves with the green movement or with environmentally friendly choices. Greenwashing is a technique used by advertisers to help make even the least environmentally friendly businesses appear green, often through the use of deceptive euphemisms.

Potential Benefits and Costs of Environmental Pop Culture

The environment is clearly a "hot" topic in all aspects of American popular culture as discussed above; however, whether this popular exposure actually helps or harms efforts to protect and conserve the environment remains a debated topic. Seemingly, one of the most significant positive impacts on the environment from increased exposure in popular culture has been increased awareness to the environment and its problems. More Americans are aware of the melting glaciers from climate change and of the plights of dolphins in a hidden cove in Japan as a result of the popular culture status of green concerns, and increased exposure typically results in increased monetary support for popular causes. Another possible benefit to the popularity of green concerns is the greater ease with which consumers can locate and purchase more environmentally friendly products and services.

While there are obvious benefits to the increased status of green concerns in popular culture, some critics are concerned that this exposure comes with risks. First, some critics are concerned that environmental problems that are deemed attractive by audiences are given more attention than those considered unattractive or uninteresting. However, these critics argue, even though the destruction of Midwestern wetlands is devastating for waterfowl and other animals dependent upon them, there is less public interest in this because the areas are not deemed attractive or interesting. Furthermore, critics note that there can be a certain emptiness to popular culture messages that disallow real education to happen through these means. Lastly, some critics argue that having the environmental concerns so constantly present serves as a replacement for actual changes in behavior, since participants in popular culture often feel as if they are participating in the culture they consume.

Green is "cool." Although this has not always been true, the presence of environmentally friendly messages in so many facets of popular culture suggests that the environment is an important part of early 21st century American culture. Whether the popularity of the green movement will benefit the environment is a debate over which cultural theorists vary in their overall opinion.

See Also: Artists' Materials; Corporate Green Culture; Education (Climate Literacy); Fashion; Green Consumerism; Law and Culture; Print Media, Advertising; Print Media, Magazines; Print Media, Newspapers; Television, Advertising; Television News; Television Programming.

Further Readings

Herndl, Carl. *Green Culture: Environmental Rhetoric in Contemporary America.* Madison: University of Wisconsin Press, 1996.

Kirk, Andrew G. *Counterculture Green: The Whole Earth Catalog and American Environmentalism.* Lawrence: University Press of Kansas, 2007.

Sturgeon, Noel. *Environmentalism in Popular Culture: Gender, Race, Sexuality and the Politics of the Natural.* Tucson: University of Arizona Press, 2008.

Jennifer Adams
DePauw University

Population/Overpopulation

The human population is currently expanding by about 1.5 million people per week and is growing more rapidly than expected, while at the same time, global lifespan also increases.

Source: iStockphoto

Numerous population and environmental theorists characterize human population growth as an unsustainable pandemic accountable for a variety of ecological problems. However, regional consumption patterns amplify the environmental impact of a population, making the two factors (consumption and population) difficult to evaluate separately. For instance, a billion subsistence farmers may instigate less environmental impact than a much smaller number of rich consumers. Furthermore, controversy surrounds calls for population reduction. Many environmentalists advocate wider distribution of family planning services, contraception, and sex education to prevent population growth.

Meanwhile, some rights advocates insist that population growth is the symptom of larger cultural injustices and that contraceptives are inappropriate tools to address these underlying inequities.

In 1970, the global population stood at 3.7 billion people. Today it is nearly twice that sum. The population of humans is currently expanding by about 1.5 million per week—roughly equivalent to adding the population of San Francisco to the world's numbers every 86 hours. The U.S. Census Bureau expects the world population to peak at about 9 billion by 2043. The impacts of this population size will vary depending upon consumption habits in different societies. Most rich-world inhabitants each consume much more than a subsistence farmer in terms of products, services, and energy. These lofty consumption practices intensify a population's impact on the biosphere. How much, precisely, is a matter of contention.

Environmental theorists debate the impact of population and consumption on the global ecosystem. Cornucopian theorists believe that technological innovation will allow for continuing growth despite a growing population with high levels of consumption. However, most environmentalists criticize population growth and high consumption patterns as problematic over the long term. Some argue that human population will surpass the maximum carrying capacity—the Earth's ability to provide sustenance to all inhabitants. Others argue that it already has. Presumably, in order for all humans on the planet to enjoy middle-class American levels of consumption, multiple planets would be required to provide the requisite natural resources. But we have only one planet. So, would a smaller population be ideal?

Considering an Optimum Population

For hundreds of years, population theorists have posited optimum population levels that are ideal for Earth to support over many generations. Their estimates range dramatically from fewer than one billion to over 1,000 billion. The nonprofit Optimum Population Trust advocates shrinking the population over time to between one and two billion people, a population they argue Earth can support, yet still afford the potential for every inhabitant to live at a high standard. Activists frequently call upon such optimum population estimates to justify campaigns for either shrinking or expanding the population.

Mainstream politicians, journalists, and academics frequently avoid discussing population issues because notions about population are often seen as being politically or ideologically motivated. As a result, population debates are generally argued from the political margins. Some groups characterize population growth as a Ponzi scheme, whereby increasing numbers of youth are constantly required to support older generations. Anti-immigration activists mobilize overpopulation fears in an attempt to justify legislative actions against immigration. Environmental organizations generally cite population growth as an unsustainable stress on natural ecosystems and resources. Some argue that population growth is a problem for poor nations to address, while others point out that rich-world populations instigate the bulk of environmental harms.

Population Dynamics

The United States' population is growing more rapidly than expected. In 1984, the U.S. Census Bureau predicted that the nation would comprise 309 million residents by 2050.

But the nation's population surpassed that estimate 40 years early. Revised population estimates for 2050 range from 420 to 500 million.

Shrinking the human population over time is not as straightforward as it might seem. Even if the global birth rate were to drop from the current average of 2.6 children per couple to the replacement rate of about 2.1 children per couple, the world population would continue to expand for 70 years before stabilizing at about 13 billion inhabitants due to population momentum (the differential between reproductive-age and older cohorts). A disproportionately large proportion of the global population is young—40 percent of the world's population is under 20 years old, and natural reproduction rates will guarantee population increases as this cohort moves through childbearing years. Humans also now live longer than their ancestors. In 1900, humans survived an average of just 30 years. Today, the global average lifespan is 67 years. Residents in wealthy nations live an average of 78 years. Medical and longevity advancements will likely extend life spans further.

Populations are aging in several European nations and Japan, leaving fewer young workers to support a sizable number of elderly individuals. Declining tax income combined with increasing eldercare costs threaten funding for established social welfare systems. Economists are closely monitoring these countries to determine if and how they can maintain their high standards of living during this impending demographic transformation. Meanwhile, social scientists point to several benefits afforded to nations as their populations age. While these nations must pay more for eldercare, they also have fewer children to birth, clothe, bathe, house, and educate. Demographers maintain that crime rates tend to fall as populations age, since younger people perpetrate most crimes. As a result, aging regions can spend less on policing, crime investigations, and jails.

Contemporary Population Debates

Past attempts to control population levels for various political, economic, social, and environmental endeavors form a long and contentious history. In the early 19th century, Thomas Robert Malthus argued in his book *An Essay on the Principle of Population* that while population can expand exponentially, food supply is limited to grow at a slower linear rate. He observed that large swells in population were typically followed by famine as a result of limited food supplies. He therefore argued against benefits or charity for the poor, citing that such generosity removed natural population checks. Numerous political leaders evoked his theories in order to justify cutting social welfare programs. Charles Dickens modeled the character of Scrooge after Malthus.

With the advent of modern agriculture, fertilizers, and the green revolution, Malthus's theory no longer held, though over the years, population scholars refashioned some of his arguments in efforts to justify population reduction programs. During the 1960s and 1970s, rich nations prescribed and funded population control initiatives in the developing world. Human rights advocates criticized these top-down programs for framing women as wombs. In the most disturbing cases, women were sterilized without their consent.

Today, contraception proponents argue that preventing population growth does not have to be coercive. They maintain that contraception has long formed an effective separation between sex and conception. In a variety of contexts, popular demand for contraceptive devices increases as they become more easily available, especially when accompanied by fertility education. Contraception advocates point to Costa Rica, Iran, Sri Lanka, and Thailand, countries that have cut their overall fertility rates in half by providing basic fertility education and contraception choices. Still, throughout the world, many women are denied access to contraception by their husbands, mothers-in-law, religious authorities, and even medical providers. Contraception proponents advocate not only greater access to contraceptive devices worldwide but also education to correct misinformation regarding their use.

Even though most rights advocates agree that contraception access is an important issue, many argue that focusing on contraception obscures the much broader struggle for comprehensive women's rights. They envision high fertility rates as a by-product of broader economic and gender inequities. Rights advocates maintain that premising family planning initiatives on reproductive rights, human rights, and social justice rather than simply birth rate reduction will yield the greatest benefits to women, their families, and the environment.

Population concerns are broadening to include potential risks of climate change and related impacts on human civilizations. In 1994, rich nations pledged to fund climate initiatives chosen by poor nations. Over subsequent decades, poor nations identified the need to address local population growth. Grassroots organizations emphasize three central concerns. First, they maintain that population growth endangers important natural resources such as forests and topsoil. Second, a growing population increases demand for limited resources such as food and water. Third, large populations amplify human vulnerability to natural disaster. Local governments identify both reproductive rights and human rights initiatives as central to addressing these extended challenges.

See Also: Environmental Justice Movements; Social Action, Global and Regional; Theoretical Perspectives.

Further Readings

Connelly, Matthew. *Fatal Misconception: The Struggle to Control World Population.* Cambridge, MA: Harvard University Press, 2008.

Ehrlich, Paul. *The Population Bomb.* New York: Ballantine Books, 1968.

Whitty, Julia. "The Last Taboo." *Mother Jones* (May–June 2010).

Zehner, Ozzie. *Coming Clean.* Lincoln: University of Nebraska Press, 2011.

Ozzie Zehner
University of California, Berkeley

Print Media, Advertising

The late 1980s were marked by a wave of increasing consumer interest in environmentally sound products in the United States, with businesses responding to this demand through the introduction of environmentally safer products. Jacquelyn Ottman has noted that the count of newly introduced products claiming environmental benefits increased over 13 times from 1986 to 1991, making up 13.4 percent of the total of all new products introduced in 1991 from 1.1 percent in 1986. Print media advertisements making environmental claims to sell these new products were relatively new in advertising, marking a major shift in a corporate culture that now incorporates a green message and has therefore become part of the green culture in America.

What Is Green Advertising?

According to Subhabrata Banerjee, Charles S. Gulas, and Easwar Iyer, green advertising markets (1) an image of green responsibility, (2) an environmentally sound lifestyle, or (3) the relationship between a product or service and the natural environment. Through advertising of greener products, like 100 percent recycled paper, companies try to reach environmentally conscious consumers by underscoring the environmental credentials of the company and product. The hope of these companies is that environmentalists will respond by purchasing their products. What is interesting is that the environmental concerns that first influenced corporate culture to create products have now become artifacts of green culture. The increasing availability of such products and following advertising has also resulted in growth of green culture through increased consumption of green products by environmentalists, creating a symbiotic relationship between green culture and corporate culture.

Problems With Green Advertising

The picture of the early growth of green advertising isn't completely rosy. The attitudes of environmentally conscious consumers have largely been skeptical of print advertisements making green claims, especially if their sources are corporations. Critics have labeled misleading and false environmental claims in some of these advertisements "greenwashing," with some state attorneys suing businesses for making false claims, and numerous studies exposing deceptive statements. Skepticism of green claims being made in advertisements is in part due to many environmentalists blaming corporations for various environmental problems, from air pollution to destruction of the rainforests.

An additional problem is that advertisements proclaiming a product to be environmentally friendly do not address what might be a larger problem: a dominant culture that emphasizes constant consumption over reducing one's desire for needless products. New consumer products mean more garbage produced once the lifetime of a product runs out, and print media advertisements end up as waste as well. A chemical-free organic t-shirt is certainly better than a t-shirt made with toxic dyes, but one does not need a 20th t-shirt to begin with. Other lifestyle choices can better address environmental problems than consuming cleaner products, such as living within walking distance of work instead of purchasing a hybrid vehicle and commuting from the suburbs. Such examples aren't meant to depict one action as environmental and the other as not, but rather that actions come in varying shades of green, with some more environmentally sound than others.

Print Advertisements as a Political Tool

Advertising in newspapers has also been used by some corporations to combat the environmental movement. According to Peter Newell, advertising was used by the Western Fuels Association as a means to discredit climate change science. Mobil also advertised in the *New York Times* as a means to influence policy makers and the public on the Kyoto Protocol issue. Due to greater control over an organization's message without having the interference of a journalist or an editor repackaging the message, advertising became an attractive tactic for Mobil. Additionally, *New York Times* journalists did not rely upon Mobil as a source when covering climate change in 1997, which led Mobil to purchase 10 ads in the *New York Times* in 1997 in order to spread its opposition to climate change.

Mobil's ad insertions dramatically increased in the two months prior to the Kyoto conference in December. The ad issued in March used the findings of a Charles River Associates (CRA) study released at that time that warned against the economic costs on consumers if climate change regulations were enacted. Ads in June and July were placed in conjunction with a United Nations climate change conference in New York City in June and Senate Resolution 98, which received a 95–0 vote in July against any climate change protocol in Kyoto. Finally, an ad condemning the Clinton administration's decision to sign the Kyoto Protocol was placed in December after the conference. In an acknowledgment of the power of the growing green culture, these ads relied upon claims that used environmental language, although they are misleading claims.

Conclusion

Businesses trying to meet the rise in demand of environmentally safer products through the creation of new green products in the late 1980s led to the introduction of green claims in the advertising world. This resulted in a symbiotic relationship between green culture and corporate culture that has not always been met with open arms. Some environmentalists have been critical of green advertisements from the very corporations that are involved in accelerating the severity of numerous environmental problems. This suspicion is certainly not eased by the use of print advertisements to discredit the environmental movement in political battles, such as the climate change battle. What cannot be denied, however, is that print media advertisements are a part of a culture in America that is growing greener.

See Also: Environmental Law, U.S.; Garbage; Global Warming and Culture; Kyoto Protocol; Nongovernmental Organizations; Print Media, Newspapers; Public Opinion; Recycling.

Further Readings

Banerjee, Subhabrata, Charles S. Gulas, and Easwar Iyer. "Shades of Green: A Multidimensional Analysis of Environmental Advertising." *Journal of Advertising*, 24/2 (1995).

Carlson, Les, Norman Kangun, and Stephen J. Grove. "A Classification Schema for Environmental Advertising Claims: Implications for Marketers and Policy Makers." In *Environmental Marketing: Strategies, Practice, Theory, and Research*, Michael Jay Polonsky and Alma T. Mintu-Wimsatt, eds. London: Haworth Press, 1995.

Newell, Peter. *Climate for Change: Non-State Actors and the Global Politics of the Greenhouse*. New York: Cambridge University Press, 2000.

Ottman, Jacquelyn. *Green Marketing: Challenges and Opportunities for the New Marketing Age*. Lincolnwood, IL: NTC Business Books, 1993.

James Everett Hein
The Ohio State University

Print Media, Magazines

In many ways, magazines act as an adjunct to other media outlets. For example, they may publish fiction or book excerpts; they cover the arts, music, and movies, without competing with them; they provide a forum for an ongoing conversation through articles and letters (and peer review, in the case of academic and scientific work); and while they cannot present news that is as up to date as television, radio, or the Internet, their publishing schedule allows them the time to conduct longer interviews and present more in-depth coverage in the form of feature articles. The magazine format of presenting various articles and ongoing features of different lengths and subject matter is one that is flexible and successful enough to have been replicated first by radio and later by television.

Like television, magazines are financed by a combination of advertising and subscriptions (magazines are also available for purchase at a wide variety of outlets, but the price per issue is usually significantly greater than the subscription cost). The first general-interest magazine was published in 1771 in London, while newspapers and special-interest magazines became popular in the previous century. A golden age of magazines followed: in the 19th and early 20th century, when there were no real-time outlets like radio and television to compete with, the magazine was the most relied-upon form of media for American entertainment, news, and culture. The muckrakers who deliberately stirred public opinion against social ills and corporate misdeeds in the Progressive Era were as present in magazines as they were in newspapers, and the second incarnation of *Life* magazine was one of the most important presences in the history of photojournalism. This incarnation began in 1936 and was published until 2000, although its influence waned after 1972 when it abandoned its original weekly schedule.

Time publisher Henry Luce had purchased the original *Life* magazine—a light entertainment magazine—purely for the sake of using its name, and published *Life*, the first all-photographic American news magazine, as a visually striking complement to the incisive and text-heavy *Time*. Its sales exceeded 13 million a week at its peak, comparable to the highest-rated TV shows today, and the magazine developed a reputation for generating and showcasing iconic photographs. As environmentalism became a public concern in the wake of Rachel Carson's 1962 *Silent Spring*, *Life* arguably became the most popular magazine to discuss the effects of pollution, with photographs highlighting wildlife and natural beauty in jeopardy, the impacts of air and water pollution on animals and plants, and other issues. An early 1970 *Life* cover declares, "Ecology Becomes Everybody's Issue." Interestingly, *Life* had long been considered a conservative magazine; its (often positive) coverage of hallucinogenic drugs came before the illegalization of same. The years just before the abandonment of the weekly format saw a small shift toward the left, with increasing attention paid to environmental matters, and coverage of the Vietnam War, which, while generally apolitical, would be difficult to frame as supportive.

Magazines like *Life*—and later the *Vanity Fair* Green Issue (now defunct)—have long been both an indicator of and an influence on public opinion. Photos of the war in Vietnam put a face and an image on events in a country so little known in North America that news anchors early in the war did not know how to pronounce the country's name. Photos of polluted waterways, the effects of acid rain, and smog-filled skies made vivid the realities of man's effect on the environment in ways that essays couldn't bring to life and opposing rhetoric could not erase. In the United States, the golden age of magazines was a pivotal force in the creation of a nationwide culture of Americans rather than a confederacy of statewide cultures, and from the progressive muckrakers on, magazines provided a forum in which opinions on national matters could be aired.

Of course, any environmental message in a print magazine is undercut somewhat by the large amounts of waste and consumption associated with the magazine industry. Major magazines, in particular, are full of advertisement pages, multiple subscription and gift card offers, fragrance samples, and other advertising gimmicks, all of which add considerably to the environmental impact of the magazine in order to generate more revenue for the publisher. More than 2 million tons of paper per year are used in the printing of American magazines, and nearly all of the paper used is from virgin (nonrecycled) fiber. Even magazines that use some recycled content for their paper typically contain only 10 to 20 percent recycled content, and virtually all magazine paper has been bleached with chlorine or chlorine compounds. Unlike books, magazines are usually discarded after reading, and nearly every magazine is produced with the assumption that it will be thrown away. Studies show that fewer than 25 percent of them are recycled. Worst of all, 3 billion magazines a year are delivered to newsstands; if not sold, they are destroyed, rarely through recycling.

See Also: Advertising; Blogs; Environmental Law, U.S.; Garbage; Media Greenwashing; Print Media, Advertising; Print Media, Newspapers; Public Opinion; Recycling; Television News; *Vanity Fair* Green Issue.

Further Readings

Redclift, Michael. *Sustainable Development*. London: Routledge, 1987.

Rogers, Elizabeth and Thomas Kostigen. *The Green Book*. New York: Three Rivers Press, 2007.

Torgerson, Douglas. *The Promise of Green Politics: Environmentalism and the Public Sphere*. Durham, NC: Duke University Press, 1999.

Bill Kte'pi
Independent Scholar

Print Media, Newspapers

In the past, an adage about newspapers was that today's copy is tomorrow's fish wrap. Not anymore. These days, you are much more likely to find yesterday's news inside the walls of your home, in your fireplace, or your garden, or even in the lining of your dog's bed. Common throughout North America and Europe, newspaper recycling has become both good business and big business. Almost three-quarters of newspapers today wind up in recycling bins across America, and more than a quarter of that ends up making fresh newsprint. Today, newspaper recycling has become a leading component of the green movement, which promotes resourcefulness, profitability, and creativity in reusing precious natural resources. In addition to doing its part to recycle its own product, newspapers have been on the forefront of environmental coverage during the past decade. Since 2006, the amount of green-related business stories in newspapers has increased 700 percent, according to Yale's *Environment 360* online magazine. This is a stark contrast to the 1990s, when a study of environmental themes on mainstream broadcast television showed a mere two and a half hours worth of total coverage from 1991 to 1997, an amount less than the average Monday Night Football broadcast.

Media Gatekeeping and the Environment

As producers of news content, the media has considerable power in shaping what information the public consumes. When environmental issues like contaminated drinking water or chemical spills affect the general public, the media frames the debate in ways the public can understand clearly. Newspapers, with their in-depth approach to newsgathering and nuanced reporting, are particularly suited to this public service. It also doesn't hurt that at the heart of most environmental issues lies the one thing newspapers thrive on: conflict. Whether it is lumberjacks versus "tree-huggers," corporations versus townships, or scientists versus climate change skeptics, environmental issues are tailor-made for the news cycle. Such instances are also highly emotional, which increases reader interest. In short, newspapers have a large role in shaping public opinion about pressing environmental issues, one of which, ironically, is what to do with all of those used, wasted newspapers that clog up landfills. That's where "greening" comes in.

Newsprint Recycling

Because printing newspapers from new materials is more expensive than using recycled paper, businesses—both the companies that recycle it and the media outlets that use it—prefer to go the "green" route. From an environmental standpoint, using recycled paper also eliminates unnecessary deforestation. Even as wood pulp is increasingly coming from trees planted especially for that purpose, companies are finding it cheaper still to manufacture newspapers from recycled materials. This is because it is easier and faster to collect used newspapers and run them through recycling plants than to wait for trees to reach their required length. Also, it takes two-thirds more wood pulp than recycled paper to produce newsprint, and companies save about a quarter of the amount of energy recycling newspapers than creating new materials.

History

Over the past four decades, as more and more newspapers wound up in landfills across North America, crowding valuable space and going to utter waste, governments and industries began to reevaluate newspapers' plight. The newsprint recycling boom started in earnest in the late 1980s as the industry began demanding more recycled materials from their paper suppliers. To keep up with the demand from newspapers, suppliers began to invest more money in recycling plants at their facilities. The demand and supply model paid off as newspaper recycling rates shot up from 35 percent in the late 1990s to more than 75 percent in 2001. Ten years ago, 9 million tons of old newspapers in the United States alone were recycled, and almost 40 percent were refurbished for fresh newsprint. The rest found its way into a variety of consumer paper products—everything from bathroom tissue and coffee cup sleeves to garden mulch and particleboard.

Economic and Political Benefits

Newsprint recycling is but one component of a larger recycling effort that began almost 40 years ago. For what started out as a minor environmental movement in the 1960s, recycling in general has come a long way. In the United States alone, recycling revenue tops $200 billion. As consumers in nations around the globe jockey for an increasingly finite amount of natural resources, those companies willing to invest in recycling technology obviously stand to make plenty of money. Companies profit from the economic logic of recycling: it costs much less to produce goods from refurbished materials like metals, wood, and glass than to assemble them from raw materials. Increasing recycling saves energy, too. In the beverage containment arena, companies save as much as 95 percent of their energy costs by using recycled aluminum. These savings are passed on to consumers, who may ultimately benefit the most from their own recycling efforts. Governments, too, benefit from newsprint recycling efforts because land-use issues and waste disposal are vital functions of city and county decision making. Newspapers in landfills create problems, filling up precious space quickly and possibly contaminating the soil by leaching chemicals and inks used in the printing process. In contrast, using recycled paper reduces the amount of bleaches and chemicals needed to produce newsprint. As such, when newsprint gets recycled, governments get a breather. Consequently, as the federal government turns to "green jobs" to help in the recycling process, citizens benefit a second time as new hiring opportunities are created and tax bases are expanded. In the end, recycling newsprint helps everyone involved.

Newsprint Shelf Life

Recycling newsprint holds many advantages over making newspapers from new material. But it must be noted that paper initially must be made from raw materials—the first time—before it can ever get recycled. And newsprint cannot be recycled indefinitely. There are only enough fibers in newspapers to last about a half dozen times before the paper is useless. At that point, new paper must be created from virgin materials before the recycling process can continue. In the long run, however, it is apparent that the process pays: with one ton of recycled paper, 700,000 new newspapers can be printed.

Emerging Issues

Two other issues about recycling newsprint should be considered. Although more and more consumers are turning online to get their news content, one should be cautious about inferring that the environment is directly benefiting. While trees do get saved when people read their papers electronically, it still requires a great deal of carbon expenditure to power the computers and other digital devices needed to view it. More research must be done to study the competing costs—both environmental and economic—of traditional print journalism and its online counterpart.

Also, consider that as the United States, Canada, Europe, and Australia lead the way in newsprint recycling, developing nations are consuming more and more resources in relation to their industrial global counterparts. Take, for example, Asia. It has resisted the trend common among American newspapers, which are seeing their readers flock to their online versions. In Asia, young consumers are clamoring for newsprint, and Asian media outlets are turning to recycled newsprint to keep up with the current demand. How long this trend will hold remains to be seen. Regardless, newsprint recycling is and will remain an essential part of the green revolution.

See Also: Media Greenwashing; Print Media, Advertising; Print Media, Magazines; Recycling.

Further Readings

Green Energy News. "Recycling Comes of Age." http://www.green-energy-news.com/arch/nrgs2010/20100058.html (Accessed September 2010).

Newspaper Association of America. "Newspapers and the Environment." http://www.naa.org/Sustainability/Newspapers-and-the-Environment/Newspapers-and-the-Environment.aspx (Accessed September 2010).

Christopher Mapp
University of Louisiana at Monroe

Public Opinion

Every public decision-making process, as German philosopher Jürgen Habermas believed in his early years, should be based on thorough, complete, and equal-accessed public discourse. He later joined many other scholars and believed that the unavoidable complexity of some issues, such as energy and environment policy, makes most citizens unqualified to participate in those decision-making processes. Common citizens, these scholars believe, can still assist in making scientific decisions by participating in discourses with experts. Studies show that public opinion is actually an important factor shaping environmental policy in the United States.

Elected politicians generally follow their constituents' preference on environmental issues. For example, in 1966, when the media coverage of the smog crisis in New York City on Thanksgiving Day and the deteriorating air quality in other cities attracted Americans' attention, the U.S. federal government immediately took over the smog matter from the state governments, put it on the congressional agenda, and passed the Air Quality Act of 1967, which required the states to establish air-quality and automobile-emission standards. Later, Congress passed the National Environmental Policy Act of 1970 and the Clean Air Act of 1970, established the Council on Environmental Quality, and created the Environmental Impact Statement (EIS) process. Presidents are also responsive to the public demand about environmental issues. In the 1960s and early 1970s, the world price of oils rose, and the public wanted it lowered. President Richard Nixon responded with wage and price controls to satisfy voters despite the opposition from most big businesses. President Bill Clinton also gained support to block Congress's low-standard proposal of the Clean Water Act by responding to public opinion.

Public opinion also determined the rapid demise of nuclear power as an energy source in the 1970s and the moratorium on offshore oil drilling business in the 1990s. Although U.S. public opinion has slowly turned negative concerning nuclear power since the late 1960s, until the late 1970s, a majority of Americans still supported nuclear power. The Three Mile Island disaster dramatically decreased public support, which dropped from 50 percent in January 1979 to only 39 percent by April of that year. Congress and state governments then imposed more and more costly regulations and licensing hurdles on nuclear power plants. Despite its wealth and power, the nuclear power industry began to die. In the 1980s and 1990s, as oil prices fell, Americans began to appreciate the coastal beauty more and turned against offshore oil drilling. The oil industry has gained little access to offshore oil fields since then.

Environment-related companies have learned to pay close attention to public opinion. During the Gulf Shore oil spill crisis, for example, BP was reported to have used public relations techniques, such as search engine optimization, to mitigate negative public opinion, which saw the oil spill as a major environmental disaster and was angry with the U.S. government for not doing enough in handling the crisis and regulating BP.

Researchers began to conduct polls on public environmental attitudes in the 1960s, when the environmental movement attracted national attention. Most of the time from then on, only around 2 percent of the public chose the environment as the most important issue facing the country in the polls. However, many people still reported that they were concerned about the environment. Public concern with environmental problems climbed to its highest point around the first Earth Day in 1970. After that, first rapidly and then slowly, the public concern decreased through the 1970s. Public concern over the environment dramatically increased in the 1980s, partly due to the President Ronald Reagan's pro-development policies, and continued to increase when President George H. W. Bush took office. By the spring of 1990, it reached an unprecedented high. More and more people thought that the government was not spending enough time or resources on the environment and were willing to pay higher prices to protect the environment themselves. Recently, public opinion polls show that environmental issues gradually become partisan. A 2009 Pew survey found that 35 percent of Republicans saw solid evidence in global climate change, whereas 75 percent of Democrats saw that evidence.

Public attention on environmental issues usually increases when the president pays less attention to the issue, as in the Reagan and the two Bush administrations, and lowers if the president addresses the issue, as in the Clinton and Obama administrations.

See Also: Clinton, William J. (Executive Order 12898); Energy; Environmental Law, U.S.; Global Warming and Culture.

Further Readings

Dunlap, Riley E. and Rik Scarce. "Poll Trends: Environmental Problems and Protection." *Public Opinion Quarterly*, 55 (1991).

Fischer, Frank. *Citizens, Experts, and the Environment: The Politics of Local Knowledge*. Durham, NC: Duke University Press, 2000.

Gillroy, John M. and Robert Y. Shapiro. "The Polls: Environmental Protection." *Public Opinion Quarterly*, 50 (1986).

Johnson, Renee. "Environment." In *An Encyclopedia of Public Opinion,* S. J. Best and B. Radcliff, eds. Westpoint, CT: Greenwood Press, 2005.

Nisbet, Matthew and Teresa Myers. "Twenty Years of Public Opinion About Global Warming." *Public Opinion Quarterly*, 71 (2007).

Rosa, Eugene A. and Riley E. Dunlap. "Poll Trends: Nuclear Power: Three Decades of Public Opinion." *Public Opinion Quarterly*, 58 (1994).

Smith, Eric. *Energy, the Environment, and Public Opinion*. Lanham, MD: Rowman & Littlefield, 2002.

Qingjiang Yao
Fort Hays State University

R

Rainforest Action Network

The Rainforest Action Network has nonviolently protested actions by companies such as Bank of America, Cargill, Burger King, Massey Coal, Citigroup, and Ford. Here, they protest the expansion of palm oil and soy plantations into critical ecosystems.

Source: Wikimedia Commons

The Rainforest Action Network (RAN) has been in existence since 1985. RAN differs from other rainforest organizations in its preference for grassroots action, its networking, and its ongoing mobilization of indigenous citizen activists in direct and immediate response to threats to forests.

It works on behalf of preservation of rainforests and their inhabitants and ecosystems. The aim is to nonviolently organize and educate with the goal of converting the global marketplace to sustainability while maintaining the human rights of those in and near to the forests. RAN advocates green fuels, wants international banks to fund renewable energy, and opposes old growth logging and rainforest agribusiness while calling for protection of forest habitats for people, plants, and animals. It wants less electricity use and more sustainable green business as well as a halt to global warming and greater protection of human rights.

RAN is headquartered in San Francisco, California, but also has offices in Edmonton, Canada, and Tokyo, Japan. RAN is a member of the Business Ethics Network and is listed in WiserEarth's database. Founders are Randall Hayes and Mike Roselle, and Michael Brune is executive director. Contributions over $30,000 in 2005 came from three anonymous donors as well as the Ecology Trust, the Rockefeller Brothers Fund, Rudolf Steiner Foundation, and Wallace Global Fund. Gifts from $30,000 to $100,000 came from an anonymous donor as well as nine others, including Working Assets. RAN-supporting celebrities include Whoopi Goldberg, Ozzy Osbourne, Bonnie Raitt, and others.

RAN convened the first international conference on rainforests as one of its first measures. The meeting of 35 organizations established an action plan. Subsequent conferences have promoted awareness and growth of the rainforest movement.

RAN maintains close links with other grassroots organizations, and those organizations provide the bulk of local-level education and generate activists in numbers sufficient to pressure for change. RAN produces Action Alerts each month to keep its members informed about where rainforests are under attack and what can be done about it. RAN takes pride in being a major stimulus in making rainforest preservation more than a temporary issue. Publications, conferences, and media campaigns keep the issue alive in hope that someday the rainforests will be safe.

RAN consumer education programs address the harvesting of old growth forests for consumer products—RAN wants a ban on all logging in those forests. RAN encourages a consumer boycott of old-growth products other than those certified as products of socially and ecologically sound logging practices. Informed consumers should confront noncompliant store managers and write letters to noncomplying corporations, lenders, and governments. Nonviolent demonstrations are also in order because governments will not act until pressed by their citizens.

RAN media campaigns include full-page ads in the *New York Times*. Environmental and human rights groups in 60 countries work with RAN to exchange information, coordinate American activities in their countries, and apply local pressure to preserve the forests and inhabitants. RAN provides networking and financial assistance to indigenous groups.

RAN is better known for its activism. Its first direct action campaign was a national boycott of Burger King (BK), the fast food company, which imported beef from tropical areas where cattle pastured on cleared rainforest. In 1987, Burger King sales declined 12 percent, leading BK to cancel $35 million of beef contracts in Central America and to stop importing rainforest beef. This campaign heightened consumer awareness of the public power to influence matters by what they choose to buy.

RAN has used civil disobedience and consumer boycotts in campaigns against Boise Cascade, Citigroup, and Mitsubishi. Another RAN campaign took Ford to task for opposing increased mileage standards. The Citibank effort was an April 1, 2008, Fossil Fools International Day of Action, blockade of Citibank's New York City headquarters by 25 "billionaires for coal," two of whom chained themselves to the door, and others, wearing top hats and tails, publicized Citi's funding of mountaintop removal mining and new coal power plants. The two "billionaires" were arrested. In September 2008, RAN activists replaced Dominion chief executive officer Thomas F. Farrell's presentation at the Bank of America annual investment conference with a slideshow of protests at Dominion's Wise County Coal Plant project in Virginia the same day.

Other actions were the November 14–15, 2009, National Day of Action Against Coal Finance that included RAN as well as Greenpeace, Mountain Justice, and many others. Citibank and Bank of America (BOA) branches and ATMs were targets, and infiltrators once again interfered with the BOA annual conference. In 2010, RAN was among those demonstrating at Environmental Protection Agency (EPA) headquarters against mountaintop removal or mining (the latter term preferred by those who support it). And RAN was part of the coalition against Massey Coal, the mountaintop mining company.

In May 2010, RAN alleged that Cargill was cutting down rainforests in Borneo. Cargill denied that it was violating Indonesian law and the Roundtable on Sustainable

Palm Oil. RAN retorted that it had maps and photos showing Cargill was at least not in compliance with permits and more likely was destroying forests, peatlands, and water supplies while not compensating affected communities as prescribed by local and national law. The report, maps, and photos were placed on the web, the report on RAN's website (http://www.ran.org/cargillreport), and the maps and photos on rainforestactionnetwork .smugmug.com/Palm-Oil/Cargills-Problems-Wi.

RAN successes outside the United States date from 1994. Most recently, in 2008, RAN convinced an Australian bank to stop funding destruction of Tasmanian forests for wood pulp. RAN also achieved forest preservation action against corporations and tougher old-growth protection laws in Canada.

See Also: Demonstrations and Events; Environmental Justice Movements; Grassroots Organizations; Institutional/Organizational Action; Social Action, Global and Regional; Social Action, National and Local.

Further Readings

CommonDreams.org. "Rainforest Action Network Stands by Evidence That Cargill Is Destroying Rainforests" (May 6, 2010). http://www.commondreams.org/newswire/2010/05/06-17 (Accessed September 2010).

RAN.org. "Our Mission and History." http://ran.org/content/our-mission-and-history (Accessed September 2010).

Sourcewatch. "Rainforest Action Network." http://www.sourcewatch.org/index.php?title=Rainforest_Action_Network (Accessed September 2010).

SuperGreenMe. "Rainforest Action Network" (June 8, 2009). http://www.supergreenme.com/RainforestActionNetwork (Accessed September 2010).

John H. Barnhill
Independent Scholar

Recycling

Recycling is an activity that aims to transform used consumer products and packaging into new products rather than disposing them of in landfills or by other means. Recycling is intertwined with production and consumption, and recycling practices have changed over time as production and consumption have changed. The modern understanding of recycling came out of the environmental movement. Concern for the environment is not a reliable predictor of recycling, but the strength of one's belief in a recycling norm and convenience are predictors. Commonly cited benefits of recycling are the reduction of waste sent to landfills and incinerators, prevention of pollution attributable to manufacturing products from virgin materials, energy savings, the decrease of greenhouse gas emissions, conservation of natural resources, and protecting manufacturing jobs in the United States. Although recycling has broad public support and is generally taken for granted to be good for the environment, many criticisms exist.

History of Recycling in the United States

The recycling process is connected to production and consumption; as production and consumption changed over time, so did recycling. Prior to the Industrial Revolution, waste was not a major social problem. Individual households managed the waste they generated in a variety of ways. Some material was saved to put to other uses. What was not saved was burned or thrown into nearby rivers. The rise of industrialization increased household consumption as well as the amount of waste resulting from production processes. At the same time, the demand for raw materials for manufacturing increased. A market for scavenged materials to utilize in manufacturing developed, and waste became a commodity. The increased amounts of waste and the commoditization of waste led to the initiation of organized garbage disposal programs and the growth of the waste recovery and resale trade. This trade was largely perceived as dirty work engaged in by immigrants, but it provided many people lucrative business opportunities.

Another significant period in the history of recycling was during World War II, when the federal government promoted recycling to the public as a patriotic activity. People were asked to save anything that could be reused to contribute to the war effort. The success of the program as a way to collect materials that could actually contribute to the war has been debated, but an important consequence was that the public perception of what waste was changed. While the waste recovery and resale trade had existed long before, the work was no longer seen as dirty, and it became acceptable for all Americans to participate. This change, however, was temporary, as the economic growth and rampant consumerism of the post–World War II period emphasized disposability. Landfills were seen as a cleaner way to dispose of trash and became the dominant model, and recycling was no longer a common household practice.

With the rise of the environmental movement during the 1960s and 1970s, private community recycling centers where people could bring newspapers and cans were established. Some municipal recycling programs were also established during this time. Governments also began to establish cash deposit programs for beverage containers as a financial incentive for recycling. By the 1980s, incineration was seen as problematic, many landfills were designated as Superfund sites, and there was a popular perception that landfills were running out of capacity, all of which provoked a renewed emphasis on recycling. Recycling also came to be understood as a practice that was beneficial for environmental reasons instead of economic reasons.

Recycling Today

Recycling has broad public approval, and a majority of Americans consistently indicate support for recycling programs. Recycling programs are accessible to a great number of people. Approximately 8,600 communities had curbside recycling programs in 2008. However, the number of curbside recycling programs has decreased since 2002, when there were more than 8,800. While there is widespread support for recycling and a large number of programs available, the recycling rate remains somewhat low. About 250 million tons of trash were generated in the United States, and 83 million tons (33 percent) were recovered for recycling or composting. Of the 83 million tons recovered, 61 million tons, roughly 24 percent, were recycled. Some materials are recycled at higher rates than others. For example, about 55 percent of paper and paperboard goods, containers, and packaging were recycled in 2008. Plastic bottles, on the other hand, are the most recycled plastic products, but less than 14 percent of plastic containers and packaging were recycled in 2008.

The recycling rate has increased from less than 7 percent in 1960 to about 33 percent in 2008. However, the amount of waste per capita going to landfills and other disposal channels is about the same as it was in 1960. This is because the amount of waste generated per capita has increased from about 2.68 pounds per person per day in 1960 to 4.5 pounds per person per day in 2008. Since the population has also increased during this period, there has been a net increase in the amount of waste disposed in landfills.

A variety of factors affect participation in recycling programs. One factor is ease or convenience of recycling. For instance, people who have access to curbside recycling programs are more likely to recycle than those who are obligated take their recyclable materials to a drop-off center. Furthermore, people who have access to mixed or commingled recycling bins are more likely to recycle than people who have to sort recyclable materials themselves. Another factor found to correlate with higher recycling rates is the provision of an economic incentive to recycle in the form of bottle deposits. Existing deposit programs, however, cover only a small proportion of the overall waste generated.

Criticisms of Recycling

Some critics of recycling point out that recycling should instead be called "downcycling" because the recycling process often results in degraded materials that are insufficient to reproduce the original product and consequently do not offset the need to use virgin materials in manufacturing new products. In order to be effective and reduce the need to exploit nonrenewable resources, new products must be manufactured from the recycled materials; this is known as "closing the loop."

The potential for recycling to contribute to sustainable communities has been questioned. Critics point out that the prevailing recycling model is dependent on a market for recyclable materials. In order to be profitable, new products must be manufactured using the recycled material and consumed. This process takes emphasis off the need to reduce consumption, the increase in which is a major factor in the waste increase in the first place. Even with increases in recycling rates, without reductions in consumption, the amount of waste sent to landfills and incinerators and the amount of raw materials consumed will continue to increase. The existence of recycling programs is sometimes represented as a defense against the need to reduce consumption. For example, the bottled water industry, which has been criticized on multiple environmental grounds, notes that it encourages the recycling of plastic bottles. The majority of recycled plastic beverage bottles, however, are manufactured into other products, and new plastic bottles are made from virgin materials.

Another criticism of recycling is that people who work in the recycling industry are exposed to toxins and hazards. The recycling industry disproportionately employs people from disadvantaged communities. Early community recycling centers were often businesses that directly benefited these communities, but recycling has become a major industry where large, profitable corporations pay very low wages.

The Future of Recycling

New products and materials create the need for new recycling technologies. Products such as personal electronic devices, televisions, and computers that rapidly become obsolete and are frequently replaced generate large amounts of electronic waste, or e-waste, which is currently a major topic in recycling. Another factor that has the potential to shape the future of recycling is a movement toward designing products and packaging with the end of the product's life in mind at the outset, an approach called "cradle to cradle." The goal of this approach is to plan ahead and eliminate unnecessary waste before it is created, rather than needing to plan a destination later, which is referred to as "cradle to grave."

Recycling is tied to many social processes, and it is unclear what the future holds. Municipalities are moving in divergent directions with regard to recycling. Some cities have abandoned their recycling programs for economic reasons. Others have increased recycling requirements despite economic hardship. San Francisco, California, has adopted a goal of "zero waste" by 2020 and has mandated recycling of all recyclable materials and composting of all organic wastes. In California, there have been recent proposals to include all product packaging, rather than exclusively beverage containers, in deposit programs.

See Also: Corporate Social Responsibility; Environmental "Goods"; Garbage; Social Action, Global and Regional; Social Action, National and Local; Soil Pollution; Theoretical Perspectives; Toxic/Hazardous Waste.

Further Readings

Carlson, Ann E. "Recycling Norms." *California Law Review*, 89/5 (2001).

McDonough, William and Michael Braungart. *Cradle to Cradle: Remaking the Way We Make Things*. New York: North Point Press, 2002.

Pellow, David N. *Garbage Wars: The Struggle for Environmental Justice in Chicago*. Cambridge, MA: MIT Press, 2002.

U.S. Environmental Protection Agency. "Municipal Solid Waste Generation, Recycling, and Disposal in the United States: Facts and Figures for 2008." http://www.epa.gov/osw/nonhaz/municipal/pubs/msw2008rpt.pdf (Accessed January 2010).

Zimring, Carl A. *Cash for Your Trash: Scrap Recycling in America*. New Brunswick, NJ: Rutgers University Press, 2005.

Loran E. Sheley
California State University, Sacramento

Redford, Robert

Robert Redford is an American actor, director, producer, philanthropist, and founder of the Sundance Institute, the Sundance Film Festival, and the Sundance Channel. A firm and unwavering proponent for the environment since the early 1970s, he has devoted much of his energy and resources since that time to advocate preservation and conservation of nature and its resources, Native American rights, and the arts.

Born Charles Robert Redford Jr. on August 18, 1936, in Santa Monica, California, Redford is the son of Martha W. (née Hart) and Charles Robert Redford Sr., a milkman from Pawtucket, Rhode Island, who later become an accountant. After graduating from high school in 1954, Robert Redford Jr. attended the University of Colorado before dropping out to attend the École des Beaux Arts in Paris in pursuit of his ambition to become a painter. Upon returning to the United States, he continued coursework in painting at the Pratt Institute in Brooklyn and theatrical set design at the American Academy of Dramatic Arts in New York.

Redford's career as an actor began in 1959, when he appeared as a guest star on many programs, including *The Untouchables, Perry Mason, Alfred Hitchcock Presents, Route 66, Dr. Kildare*, and *The Twilight Zone*. Four years later, he experienced great success as the uptight newlywed Paul in Neil Simon's play *Barefoot in the Park* (1963), a role that he later took up again opposite Jane Fonda in the 1967 film version. In 1969, he costarred with Paul Newman in George Roy Hill's highly successful classic *Butch Cassidy and the Sundance Kid*. Three years later, his career swung into high gear with the release of the films *Jeremiah Johnson* (1972), *The Candidate* (1972), *The Way We Were* (1973), and *The Sting* (1973), a film for which he received an Oscar nomination. In the following years, he continued these successes by appearing in *The Great Gatsby* (1974), *The Great Waldo Pepper* (1975), *Three Days of the Condor* (1975), and *All the President's Men* (1976), all enormous hits with the general public.

In 1979, he produced *The Solar Film*, a short film about solar energy that was nominated for an Academy Award. Other documentaries he has produced include the award-winning *Yosemite: Fate of Heaven* (1989) and the feature-length documentary *Incident at Oglala* (1992). In the area of feature films, he went on to both direct and produce *The Milagro Beanfield War* (1988) and *A River Runs Through It* (1992), two films with environmental themes.

Although Redford has continued to participate in the making of many other mainstream Hollywood movies from the 1980s to present time, since 1981, he has invested much of his energy in the nonprofit Sundance Institute he founded in Park City, Utah, to provide much-needed support for independent filmmakers. In addition to the institute, Redford also created the Sundance Film Festival and the Sundance Channel, all three of which are located in and around the Park City region north of the Sundance ski area he had purchased several years before creating the institute. Dedicated to the support and development of promising screenwriters and visionary directors, the Sundance Film Festival has been internationally recognized as a vital source of independent cinema. Launched in 1996, the Sundance Channel, a cable network, offers viewers independent feature films, shorts, documentaries, and world cinema.

An ardent conservationist and environmentalist, Redford has served for several decades as a trustee of the board of the Natural Resources Defense Council. Actively engaged in promoting social responsibility and involvement in the political process, he has also advocated many legislative initiatives, including the Clean Air Act (1974–75), the Energy Conservation and Production Act (1974–76), and the National Energy Policy Act (1989).

Well known internationally for his environment work, Redford has received many awards for his activism, including the 1989 Audubon Medal Award, the 1987 United Nations Global 500 Award, the 1993 Earth Day International Award, and the 1994 Nature Conservancy Award. In 2004, the Natural Resources Defense Council (NRDC) bestowed upon him the Forces for Nature Lifetime Achievement Award.

In addition to his work with the NRDC, he is also a board member of the Gaylord A. Nelson Environmental Endowment at the Institute for Environmental Studies at the University of Wisconsin and serves on the National Council of the Smithsonian Institution's National Museum of the American Indian.

In 2007, Redford launched a block of ecothemed television programming titled *The Green* on his Sundance cable channel, an endeavor designed to put the spotlight on individuals and the creative solutions they are proposing and inventing in the face of the current environmental problems facing the planet. In this programming, the channel focuses on a different green theme each week in order to inspire audiences to merge green thinking and practices into all facets of their daily lives.

Redford was one of the few Hollywood celebrities to voice his opinion on the Gulf oil spill of 2010, the worst oil spill in U.S. history. Weighing in on the crisis, he severely criticized efforts by major energy companies to show off their environmental credentials. He also voiced the opinion that Americans need to move away from their dependence on oil in favor of alternative energy sources such as wind and solar power.

See Also: Art as Activism; Communication, Global and Regional; Communication, National and Local; Corporate Green Culture; Corporate Social Responsibility; Education (Climate Literacy); Native American Culture and Practices; Popular Green Culture; Sierra Club, Natural Resources Defense Council, and National Wildlife Federation; Sundance Channel; Television, Cable Networks; Television Programming.

Further Readings

Quirk, Lawrence J. *The Sundance Kid: A Biography of Robert Redford*. Lanham, MD: Taylor Trade, 2006.

Redford Center. http://www.redfordcenter.org (Accessed September 2010).

Sundance Channel. "It's the Green Sundance Channel." http://www.sundancechannel.com/thegreen (Accessed September 2010).

Danielle Roth-Johnson
University of Nevada, Las Vegas

Religion

Religion is a set of beliefs that involves worshipping one god, or gods, or even goddesses, as supernatural being(s); a worldview that provides answers to some philosophical questions around the meaning and existence of life and the universe, and rituals and moral conducts that guide and govern devotional practices and daily life. God or gods are regarded as the sacred, transcendental, divine, and the highest order beyond human capacity, and human fates are contingent upon the favor of that power and authority. Religion provides answers to questions such as: Who created the universe, why, and in what way? What is the purpose of human life in this ephemeral world? What will happen after death? Religion provides sacred rulings and guidance to organize rituals and devotions and codes of conduct that separate sacred from profane, or lawful from unlawful. There are a few thousand religions in the world, notable among them Christianity, Judaism, Islam, Hinduism, Buddhism, Confucianism, and so forth. Christianity, Judaism, and Islam are regarded as the monotheistic religions, as the followers largely believe in one sacred God; these three religions are known as Abrahamic faiths as they originated from the prophet Abraham. While there is considerable diversity, and sometimes tensions, between different religions, most religions provide moral codes of conduct that are quite similar to one another.

The Qur'an, pictured here, and other religious texts advocate harmony between humans and nature. While there is a wide variety of ideology between religions, there is also a surprising unity in the philosophical idea of harmonious living.

Source: Wikimedia Commons

Religion as a Social Force

Founders of sociology such as Émile Durkheim and Max Weber predicted a secularizing trend in industrial/urban societies, and Karl Marx regarded religion as the opiate of the masses. However, empirical evidence does not support their prediction and Marx's judgments. Conversely, we find in contemporary society that religion still persists as, and continues to be, a vigorous force on the sociopolitical scene and empowers individuals in crisis situations.

Religion not only provides personal empowerment through spirituality, it also becomes a great force that leads to social change. Around the world, religion provides answers to perplexing questions about ultimate meanings such as the purpose of life, why people suffer, and the existence of an afterlife. Similarly, religious rituals that enshroud critical events such as birth, marriage, illness, and death provide emotional comfort at times of joy and crisis. Rituals provide a sense of joy, comfort, peace, emotional support, consolation, and social cohesion and may generate a social force. Religion is a motivating factor as well as a source of social movements that give meaning to life for millions of people around the world. Viktor E. Frankl (1905–97), a Jewish psychiatrist and author, drew on his experience as a survivor of the Holocaust to develop the discipline of logotherapy, a form of psychotherapy that, by stressing the need to find meaning even in the most tragic circumstances, offered solace to millions of readers of his classic book, *Man's Search for Meaning: An Introduction to Logotherapy*. Mahatma Gandhi (1869–1948) in India and Martin Luther King, Jr. (1929–68) in the United States initiated great civil rights movements, and they got inspiration from their religions. Both are great religious figures of the world and were the principal leaders of civil rights movements and advocates of nonviolent movements. Another figure is Malcolm X (1925–65), whose criminal life was transformed to a subsequent "Minister of the Nation of Islam" by religion. His charismatic leadership gave meaning to a great number of African Americans. Always charismatic and witty, his words and speeches provided quotes for people with many different social and political goals. His most enduring message remains one of black self-respect and self-help, combined with his uncompromising rejection of racism.

Religion and Green Philosophy

Despite having diverse explanations, one common understanding within the theology of Christianity, for example, is that religion separates humans from nature. In older religious traditions, humans were seen as part of nature, rather than the rulers of nature. In many animistic religions, there is a belief that every tree, mountain, or spring, as well as flora and fauna has a sacred spirit that needs to be respected. According to Lynn White, Jr., in contrast with paganism and Eastern religions, Christianity not only established a dualism of man and nature but also insisted that it is God's will that man exploit nature for his proper ends. He noted that Christianity was a complex faith, and different branches of it differ in their outlooks. But, in general, he proposed that Christianity, and Western civilization as a whole, held a view of nature that separated humans from the rest of the natural world and encouraged exploitation of it for our own ends. Marcia Bunge challenges some of what White had to say and cites biblical passages to show that the Bible affirms the goodness and intrinsic value of all living things; it points out commonalities between human beings and other living things; and it contains the mandate that we treat the natural world with care and respect. For instance, "And God created the great whales, and every living creature that moveth, which the water brought forth abundantly, after their kind, and every winged fowl after his kind: and God saw that it was good. And God blessed them, saying, Be fruitful and multiply, and fill the waters in the seas, and let fowl multiply in the earth" (Genesis 1:21–22).

In Judaism, society and nature has been intertwined. Different elements of nature not only are crucial but also teach humanity the powerful lessons of a profound blending. "But ask the beasts, and they will teach you; the birds of the sky, and they will tell you; or speak to the Earth and it will teach you; the fish of the sea, they will inform you. Who among all these does not know that the hand of the Eternal has done this?" (Job 12:7–9). In Buddhism, a well-noted story tells the profound relations between man and nature. Rajah Koravya had a king banyan tree called Steadfast, and the shade of its widespread branches was cool and lovely. Its shelter broadened to 12 leagues. None guarded its fruit, and none hurt another for its fruit. Now there came a man who ate his fill of fruit, broke down a branch, and went his way. Thought the spirit dwelling in that tree, "How amazing, how astonishing it is, that a man should be so evil as to break off a branch of the tree after eating his fill. Suppose the tree were to bear no more fruit." And the tree bore no more fruit.

A sacred and harmonious nexus between human community and rest of the natural order is evident in Islam: "The creation of the heavens and the Earth is far greater than the creation of mankind. But most of mankind do not know it" (Qur'an, 40:56). There is a growing movement among Muslims—led by thinkers like Seyyed Hossein Nasr—who place an immediate priority in dealing with the intractable problems that the human race is creating for itself by overexploiting and degrading the planet beyond repair. The Islamic call for a responsible and harmonious society–nature relationship has more similarities than differences with other major religious traditions of the world. The Islamic worldview is based on the belief in the existence of an all-powerful creator who is the same God (*Allah* in Arabic) of the other monotheistic faiths, Judaism and Christianity. Muslims learn from the Qur'an that God created the universe and every single atom and molecule it contains, and that the laws of creation include the elements of order, balance, and proportion: "He created everything and determined it most exactly" (Qur'an, 25:2) and "It is He Who appointed the sun to give radiance and the moon to give light, assigning it in phases . . . Allah did not create these things except with truth. We make the signs clear for people who know" (Qur'an, 10:5).

Despite the apocalyptic premise of Samuel Huntington's book *The Clash of Civilizations*, which prophesies an inevitable war between the armies of God and the armies of Allah, Islam and Christianity have much in common. In their view of the natural world, both the Bible and the Qur'an share many of the same stories, heroes, and ethical concepts. But there are some differences. The Qur'an might even be said to be the "greener" of the two holy books. The word "Earth" (*ard*) appears no less than 485 times in the holy book of the Qur'an. Sharia, the word for Islamic Law, literally means "source of water."

Like the New Ecological Paradigm (NEP) developed and proposed by William Catton and Riley Dunlap, Islam guarantees equal rights to other creatures living on the planet to exist and thrive. Not only that, in Islam, human beings are expected to protect the environment since no other creature is able to perform this task. Humans are the only beings that Allah has "entrusted" with the responsibility of looking after the Earth. This trusteeship is seen by Islam to be so onerous and burdensome that no other creature would "accept" it. Islam urges humanity to be kind to nature and not to abuse the trust that has been placed on the shoulders of humans. In fact, to be kind to animals is an integral part of Islam for Muslims. In Islam, the relationship between humankind and the environment is part of social existence, an existence based on the fact that everything on Earth worships the same God. This worship is not merely ritual practice, since rituals are simply the symbolic human manifestation of submission to Allah. The actual devotions are actions, which can be practiced by all the creatures of Earth sharing the planet with the human race. Moreover, humans are responsible for the welfare and sustenance of the other citizens of this global environment.

In Islam, preservation and protection of natural resources for other creatures is a kind of worship, and wasting them is a sin. As befits a faith born in the desert, water is honored in Islam as "the secret of life." Islam forbids the wastage of water and the usage thereof without benefit. According to Gar Smith, the preservation of water for the drinking of mankind, animal life, bird life, and vegetation is a form of worship that gains the pleasure of Allah. While the current economic patterns permit factories to release toxic wastes to water and atmosphere and thereby destroy marine life, other species, and habitats, Islam regards these practices as an act of sin.

Islam, however, endorses the transformation of wilderness into agriculture and cattle pastures. Human beings are instructed in Islam to deal with animals with utmost beneficence and compassion and strive to ensure the preservation of the different species. In Islam, it is forbidden to kill an animal for mere play or amusement. Islam has forbidden wastage of animals and plants in peacetime and in wartime. If someone kills a bird for amusement, the bird will demand justice from that person on the Day of Judgment. In fact, there was no concept of "animal rights" or, for that matter, much civility by the strong toward the weak in the rough Arabian society that Prophet Muhammad was born into more than 1,400 years ago. He also talked of the great rewards of kindness to animals.

While the modern conservation movement in the West started in the mid-19th century, Islam introduced it in the early years of Islam in the 7th century. In the centuries following Muhammad's passing, Islamic scholars introduced the idea of *hima*—a protected zone. Many Islamic countries now set aside certain wild areas that cannot be developed or cultivated. These have become modern wildlife reserves. Islamic jurisprudence contains regulations concerning the conservation and allocation of scarce water resources; it has rules for the conservation of land with special zones of graded use; it has special rules for the establishment of rangelands, wetlands, green belts, and also wildlife protection and conservation. Many of the traditional institutions and laws associated with sound environmental practice in Islam have now fallen into disuse.

Human life is sacred in the view of Islam. No one is permitted to take the life of another person except as life-for-life. Suicide is a crime in Islam. Under Islam, it is incumbent on every Muslim to contribute his or her share in improving greenery. Muslims should be active in growing more trees for the benefit of all people. Even during battle, Muslims are required to avoid cutting trees that are useful to people. The Prophet instructed the faithful that any Muslim who plants a crop that feeds another person, animal, or bird will receive a reward in paradise. Cutting down trees is seen as an abomination. How important is the planting of trees? In the words of the Prophet, "When doomsday comes, if someone has a palm shoot in his hands, he should plant it."

Unlike modernity, which preaches that humans are given domain over the entire Earth to exploit and conquer nature, Islam entrusted man as a "central created being" in the universe with authority and freedom as well as responsibility to preserve society and the environment. While modern environmentalism emerged as a response to the troubled environmental and social legacies of modernity, Islamic tenets for a harmonious society-nature relationship came mainly to maintain and establish a society based on ecological and social justice and principles as well as to warn people of future ecological crises.

Despite considerable diversity and sometimes tension between different religions, there is a surprising unity among most religions of the world when it comes to the philosophical relationship between society and nature. This unity and profound philosophy may even attract the secular ideologies that also seek a harmonious relationship between society and nature.

See Also: Bartholomew I: The Green Patriarch; Biblical Basis for Sustainable Living; Religious Partnership for the Environment; United Church of Christ.

Further Readings

Ali, M. Y. and M. S. Islam. "Methodological Tensions Between Sociology and Islam." *Journal of Islamic Science*, 18/1–2 (2002).

Khalid, F. M. "Islam and the Environment." In *Encyclopedia of Global Environmental Change*, T. Munn, ed. Chichester, UK: John Wiley & Sons, 2002.

Nasr, S. H. *Man and Nature: The Spiritual Crisis in Modern Man*. London: Unqin Hyman, 1990.

Nasr, S. H. *The Need for a Sacred Science*. Surrey, UK: Curzon Press, 1993.

Nasr, S. H. *Religion and the Order of Nature*. New York: Oxford University Press, 1996.

"Religion and Environment." http://daphne.palomar.edu/calenvironment/religion.htm (Accessed June 2010).

"Religious Tolerance." http://www.religioustolerance.org/var_rel.htm (Accessed June 2010).

Smith, G. "Islam and the Environment." *Earth Island Journal*, 17/2 (2002).

White, Lynn, Jr. "The Historical Roots of Our Ecological Crisis." *Science*, 155/3767 (1967).

Md Saidul Islam
Nanyang Technological University

Religious Partnership for the Environment

The National Religious Partnership for the Environment (NRPE) is a coalition of four major religious organizations and alliances in America. These are the Coalition on Environment and Jewish Life (COEJL), the Evangelical Environmental Network (EEN), the National

Council of Churches of Christ (NCCC), and the U.S. Conference of Catholic Bishops. Taken together, these organizations connect in some way to most congregations of Christians and Jews throughout the United States.

The NCCC is composed of 34 highly diverse religious bodies, including mainline Protestant denominations as well as churches in Anglican (Episcopalian), Eastern Orthodox, African American, and other traditions. They formed an Eco-Justice Working Group in 1984 to work together to address issues integrating social and economic justice with ecological well-being. Several of the individual denominations had earlier issued policy statements and begun work on environmental concerns, for example, during the energy crisis of the 1970s. Presbyterians initiated a faith-based response to the toxic crisis at Love Canal in 1979, immediately inviting the involvement of other mainline denominations in the formation of the Ecumenical Task Force at Love Canal, which in time came to engage in support of the neighbors at other toxic waste sites as well. The environmental justice movement can be seen as the confluence of this movement with the civil rights movement.

In his 1990 World Day of Peace message, *The Ecological Crisis: A Common Responsibility*, Pope John Paul II provided the impetus for Catholic engagement in environmental matters. This mandate was taken up by the U.S. Conference of Catholic Bishops in making its 1991 statement *Renewing the Earth* and by creating its Environmental Justice Program in 1993. The program was directed by layman Walter Glazer from its founding through 2007. One of its particular emphases was on the effect of environmental problems on the poor.

The COEJL was created in 1993 by three institutions: the Jewish Council for Public Affairs, the Jewish Theological Seminary of America in New York, and the Religious Action Center of Reform Judaism. It came to include many more Jewish organizations from all four American Jewish movements (Reform, Conservative, Reconstructionist, and Orthodox). Environmentalism had not been a central focus for these organizations prior to a 1992 Consultation on the Environment and Jewish Life convened by Al Gore, Paul Gorman, and Carl Sagan. Initial funding for COEJL came entirely from the NRPE, but the organization eventually developed its own funding from foundations and individuals.

The EEN is a coalition that has come to include 23 evangelical Christian programs and educational institutions. It originated from a 1992 forum at the Au Sable Institute, an environmental studies center in Michigan, to which many Christian colleges send their students and faculty. One of the most prominent evangelical voices was that of Calvin DeWitt, the president of the Au Sable Institute and a biologist/ecologist at the University of Wisconsin. Uniting scientific credentials, environmental ethics, and biblical theology with political astuteness, DeWitt brought to the Republican-controlled U.S. Congress in 1996 the image of the Endangered Species Act as the Noah's Ark of our times and ensured its passage.

Of the four participating faith communities, the evangelicals met the most entrenched resistance to environmentalism in parts of their constituency, particularly in attack articles from the well-funded politically conservative sector and indifference from those who considered the nearness of the Second Coming of Christ to negate environmental concerns.

The NRPE was founded in 1993 by leaders of these four religious communities. Its founding executive director was Paul Gorman. Gorman retired in 2010 from the NRPE and was succeeded by Walter Grazer as interim director during the search for a replacement. During Gorman's 17-year tenure, the NRPE raised over $30 million, mostly from foundations, that it distributed to the four faith communities for projects related to care of the environment. It created materials such as printed curricula, videos, web pages, and newsletters that helped local congregations interpret environmental concerns through the lens of faith.

Each of the participating faith communities has its own culture, and the environmental programs that each develops reflect those differences, though frequently they inspire another religious community to join in or adapt the program for its own membership. For example, in 2006, COEJL promoted the use of compact fluorescent light bulbs to replace incandescent light bulbs in synagogues and homes to conserve energy and raise awareness about global climate change. The campaign was keyed to Hanukkah, the Festival of Lights, celebrated by lighting candles on the menorah that commemorate the miracle by which a day's lamp oil lasted for eight days. The idea was picked up by many Christian churches for their own winter holy days that have a theme of light, including Epiphany.

At the Christian holy day Palm Sunday, congregants wave palm branches reminiscent of the entry of Jesus into Jerusalem at the beginning of Holy Week preceding Easter. Lutherans learning of the ecological damage done by unsustainable palm harvesting developed a relationship with a University of Minnesota–based project for sustainable palm harvesting in Central America. Other congregations joined in purchasing "eco-palms," affording an opportunity for educating their youth and adults in concepts of fair trade and ecological sustainability.

In the culture of evangelical Christianity, many young people wore wristbands or t-shirts with the initials WWJD, asking "What Would Jesus Do?" A clever campaign of the Evangelical Environmental Network played on this by asking, "What Would Jesus Drive?" In 2003, the organization's executive director, the Rev. Jim Ball, toured the Bible Belt in a hybrid Toyota Prius, speaking about environmental stewardship at churches and on Christian radio stations. Ball's work linked to that of other leaders of the NRPE when they went together to meet with Ford Motor Company officials to encourage Ford to build more energy-efficient models.

While most of these environmental initiatives might have occurred within the individual faith communities, the NRPE during the 1990s and 2000s accelerated them through its funding and multiplied their impact in the American policy arena by encouraging them to speak with a united voice.

See Also: Bartholomew I: The Green Patriarch; Biblical Basis for Sustainable Living; Religion; United Church of Christ.

Further Readings

Coalition on Environment and Jewish Life. http://www.coejl.org (Accessed November 2010).
Evangelical Environmental Network. http://creationcare.org (Accessed November 2010).
National Council of Churches of Christ. http://www.ncccusa.org (Accessed November 2010).
National Religious Partnership for the Environment. http://www.nrpe.org (Accessed November 2010).
U.S. Conference of Catholic Bishops. http://www.usccb.org (Accessed November 2010).

Patricia K. Townsend
Independent Scholar

Representations of Green Architecture

There is no widely accepted definition of what constitutes "green" in green architecture or of the other terms that are commonly used interchangeably with it, like "sustainable" or "environmental." Despite this vagueness, green is a commonly used term, understood to refer to an architecture that fulfills expectations of energy conservation (sometimes of energy generation) and has a minimal impact on the environment. The earlier uses of the term represented an understanding of architects' responsibility toward Earth's energy supplies within the framework of global environmental ethics. Recent uses of the term, however, represent a movement of the focus of architectural discourse from tectonic to thermodynamic and biotechnical conceptions of architecture.

There were series of events and publications that affected the conception of green architecture: biologist Rachel Carson's seminal book, *Silent Spring* (1962), was one of the first significant warnings that showed how, without consideration of the ecological balance of Earth, people could turn science into a weapon that destroys all forms of life. Her book was viewed as a wake-up call for many, including architects and planners. After nearly five decades, it continues to be one of the introductory readings on sustainability. Other seemingly less relevant events, like the growing U.S. involvement in the Vietnam War after 1965, contributed to a further development of "a crisis of confidence in science and technology." Another equally important event was the arrival of the visual images illustrating the already known fact that the planet Earth was a finite material entity floating in its orbit in space. The photos sent from the National Aeronautics and Space Administration (NASA) 1967 Apollo 8 mission transformed people's perception of the planet. The new perception of Earth seen from space as a relatively small and beautiful blue planet was further reinforced in a short film demonstrating the "relative size of things in the universe" by architect/designer couple Charles and Ray Eames in *Powers of Ten* in 1968.

The 1960s witnessed the publication of many important books on architecture and environmental conditions. Among them, Victor Olgyay's book *Design With Climate* (1963) was one of the first to be completely devoted to the effects of climate on the built environment. At that time, architects' role in changing Earth's climate was not even considered; however, books like Olgyay's were influential in fostering an awareness of the complex interactions between the natural and the built environments. Later, two very influential books, *Design With Nature* by Ian McHarg (1967) and *The Architecture of the Well-Tempered Environment* by Reyner Banham (1969), were published on both sides of the Atlantic. While the planner and landscape architect McHarg described "the place of nature in man's world," architectural critic Banham criticized the "narrow-eyed aesthetic look" in architecture and declared environmental issues as an integral concern in architecture.

The 1970s energy crisis brought the international recognition needed for the environmental cause. A 1985 report published in the scientific journal *Nature* showed that the hole in the ozone layer required immediate attention. In 1987, the report of the United Nations World Commission on Environment and Development (the Brundtland Report) was first to define sustainability as "development that meets the needs of the present without compromising the ability of future generations to meet their own needs." This was followed by the Montreal Protocol in 1989, and by the UN Conference on the Environment and Development, that is, the Earth Summit, in Rio de Janeiro in 1992. Not independent from these developments was the formation of the U.S. Green Building Council (USGBC) in 1993, which later led to the development of the green building rating system called Leadership in Energy and Environmental Design (LEED) in 2000. In the meantime, studies like Ed Mazria's well-known article "It's the Architecture, Stupid!" (1993) showed that the building industry and architecture consumed approximately half of all the energy produced in the United States and was responsible for nearly half of the carbon dioxide (CO_2) emissions annually. This was unequaled in any other sector.

The development of the internationally recognized green building rating system LEED showed that the green approach to design was becoming mainstream. Currently, LEED certification (basic, silver, gold, or platinum) is given to a project that acquires enough points based on evaluation of issues like energy and water use, CO_2 emissions, indoor environmental quality, materials and resources, and environmental impacts. However, the question of which materials would qualify as green, the lack of any scale of importance on its point grading system, and the criticisms of the understanding of green architecture independent of design qualities continue to be among the controversial issues in architectural discourse.

The aesthetics of green architecture, or the lack thereof, which was secondary to any discussion of global environmental ethics, was, however, primary from an architectural point of view and received more attention from architects as green architecture became more popular. A majority of green projects from the 1960s and 1970s were either derivatives of the green houses (solar houses) with mono-pitch roofs and glass-covered walls and roofs or were forms inspired by nature, built of natural materials like wood, stone, and earth. These projects were followed by forms reminiscent of traditional architecture and low-tech buildings, adopting passive heating and cooling strategies. In opposition to this approach were high-rise, high-tech green buildings. Despite the variety, it is hard to claim that either approach has moved away from existing architectural norms, but only "subtracted" (low-tech) or "added" (high-tech) layers to conventional buildings.

The 2009 special issue of *Harvard Design Magazine* titled "(Sustainability)+Pleasure Vol. I: Culture and Architecture" charted the new directions that green architecture has taken since the late 1990s. Articles by Inaki Abalos and Christopher Hight especially showed that this new direction treats environmental conditions like temperature as a transformative, space-defining force and moves away from the previous additive and subtractive formats. As a result, it cannot be evaluated according to conventional building standards, for it requires new ways of understanding space. This paradigm shift from tectonic to bioclimatic will require adaptations of architectural pedagogy, theory, and practice in the near future.

See Also: Global Warming and Culture; Greenwashing; Home; Space/Place/Geography and Materialism; Sprawl/Suburbs/Exurbs; Toxic/Hazardous Waste; Work.

Further Readings

Abalos, I. "Beauty From Sustainability." *Harvard Design Magazine* (Spring–Summer 2009).

Banham, R. *Architecture of the Well-Tempered Environment*. Chicago, IL: University of Chicago Press, 1984.

Carson, R. *Silent Spring*. Bristol, UK: Penguin Books, 1965.

Farmer, J. *Green Shift: Towards a Green Sensibility in Architecture*. Oxford, UK: Architectural Press & WWF-UK, 1997.

Gauzin-Muller, D. *Sustainable Architecture and Urbanism: Design, Construction, Examples.* Basel, Switzerland: Birkhauser, 2002.
Hight, C. "The New Somatic Architecture." *Harvard Design Magazine* (Spring–Summer 2009).
Koleeny, J. and C. Linn, eds. *Emerald Architecture: Case Studies in Green Building.* New York: McGraw-Hill, 2008.
Mazria, E. "It's the Architecture, Stupid!" *Solar Today* (May–June 2003).
McHarg, I. L. *Design With Nature.* Hoboken, NJ: John Wiley & Sons, 1992.
Olgyay, V. *Design With Climate: Bioclimatic Approach to Architectural Regionalism.* Princeton, NJ: Princeton University Press, 1963.

Sebnem Yucel Young
Izmir Institute of Technology

Russia

Russia is a leader in natural resource deposits, but its utilization of that energy is very inefficient. This copper melter of Karabash, Chelyabinsk Oblast, Russia, is one of the biggest pollutants of Russia.

Source: Wikimedia Commons

Russia faces a historical legacy of environmental problems from the Soviet era as well as modern environmental issues. While public environmental awareness has increased, lack of knowledge, local focus, and the ongoing economic difficulties associated with the breakup of the former Soviet Union have hindered the development of a public green culture in Russia. Although the atmosphere for environmentalism became more open beginning with the presidency of Mikhail Gorbachev in the late 1980s, environmental organizations active in Russia continue to face difficulties due to government and corporate resistance to environmental protection legislation.

With the breakup of the Soviet Union, many scientists and environmentalists as well as the general public saw Russia as an environmental disaster, highlighted by the national publicity received during the Chernobyl nuclear power plant disaster in 1986, which spewed lethal amounts of radiation into the air. In the 1960s, fertilizers, pesticides, and other chemicals had heavily polluted much of the arable land around the Aral Sea. Lake Baikal has been threatened by polluted waste runoffs from nearby industrial complexes. Soviet-era nuclear testing sites have produced dangerous radiation levels.

The Chernobyl disaster helped spur the coalition of individual environmental groups into organized unions. The last Soviet leader, Mikhail Gorbachev, was very supportive of environmental causes, realizing the significance of the country's environmental crisis.

He established a State Committee on Nature in 1988, whose duties were similar to the United States' Environmental Protection Agency (EPA). Internationally, environmental organizations such as Greenpeace and the World Wildlife Fund began publicizing Russia's pressing environmental issues while other green organizations worked to publicize such problems at the local level.

Gorbachev's policies of openness and restructuring enabled Russian environmentalists to spark nationwide protests against environmental problems. Thousands of people took to the streets of major Russian cities to demonstrate against nuclear power plants and industrial expansion. Environmental groups became critics of Soviet policies, while at the same time gaining widespread support for the cause of strengthening environmental protection and going "green." The collapse of the Communist government and the resulting economic turmoil disrupted any gains that the limited earlier Soviet green movement had accomplished. By that time, however, Russia's green movement had developed a stable organizational base.

The Russian government's and the Russian public's attitudes toward the environment have changed with the identification and growing public awareness of Russian environmental problems, resulting in a growing public discourse of the most prominent issues and their potential solutions. A 1993 poll showed that about 65 percent of Russians said that their local environmental quality was either fairly bad or very bad, and 85 percent rated their national environment as fairly bad or very bad. Russia is a leader in natural resource deposits, but it is still inefficient when it comes to utilizing energy. Russia uses about twice the level of energy that China does and about six times that of the United States. Nevertheless, Russia produces the same value of goods as those two countries. Still, Russia has promised to invest in energy-cutting technologies under President Dmitry Medvedev.

Much of the public's concern, however, is confined to the local level. There is less concern with national and global issues such as the loss of biodiversity, deforestation, acid rain, and potential climate change and global warming. Survey data among Russian people showed that many Russians were not willing to pay for new environmental policies. This may have to do with the economic instability that Russia faced in a post-Soviet world. There are also many people in Russia who are not knowledgeable about the environmental organizations in the country or who do not have a full understanding of the personal and human consequences of environmental degradation. While a significant green movement arose in Russia beginning in the late 1980s, parts of the green movement have also branched out into other areas of activism such as human rights and social justice, making it hard to identify a uniform green culture.

Green Organizations

Russian green culture and environmentalism are also well represented in society and politics. The number of environmental organizations in Russia rose in the 1990s. Scholars have identified three distinct types of green organization in the post-Soviet landscape. These are grassroots organizations, professionalized organizations, and government affiliates. The largest and strongest environmental organization within Russia is the Socio-Ecological Union. Other prominent organizations include the Russian Ecological Union, the Moscow Green Party, and the Moscow Ecological Foundation. The Socio-Ecological Union sets the tone for the green culture agenda in Russia. The largest international environmental organization in terms of size in Russia is Greenpeace, with over 4,000 members in 2000. Still, some of these members do not pay dues, and some are members of other

European countries. The green culture movement has had a mixed record of effectiveness in acting as intermediaries between the state and society.

In the 1990s, Russian experts identified the key threats to the environment within their country as well as global threats that will impact Russia. These threats included pollution from coal mining, radioactive contamination, soil erosion, desertification, chemical contamination, water pollution, and pollution from vehicle emissions. There are also oil spills, leaks in natural gas pipelines, and the flaring of natural gas. All of these environmental issues pose key threats to Russia's ecosystems and biodiversity.

Green organizations in Russia adhere to the belief that the government has neglected environmental protection in favor of economic development. Many green organizations have found it very difficult to influence policy making and enforcement. The Boris Yeltsin and Vladimir Putin administrations did not respond positively to green activism, but instead de-emphasized environmental protection.

In 1991, Boris Yeltsin signed Russia's first comprehensive environmental legislation into law, titled "On Environmental Protection." This legislation contained general statements about the environmental rights of citizens without setting any specific environmental goals. This law also defined the environmental functions for every level of government as well as Russian citizens. The law also specified the environmental regulation of every aspect of society in a general sense. Enforcing this law, however, was impractical, as Russia's judiciary had little practical experience in the realm of environmental law.

In 1993, Russia's total investment in environmental preservation was around $2.3 billion. This figure represented just a small fraction of the national budget that was earmarked for the government called "industrial construction," in which environmental expenditures were included. The structure of expenditures for environmental "construction" had remained basically unchanged since 1980. State budget deficits occurred in the 1990s, causing the amounts of funding for environmental and "green" costs from these sources to decline, in addition to the fact that there was already little government enthusiasm for the promotion of beneficial environmental projects. The only other sources of funding came from local budgets and private environmental foundations.

The Socio-Ecological Union argues for the purchase of cheap land that can then be utilized in an environmentally sound fashion. Still, this runs contrary to widespread traditional Russian cultural beliefs as to the proper use of land. Business interests also have their own thoughts on how that land should be used. Much of the green philosophy of Russian environmental organizations, such as the Socio-Ecological Union, is based on the plans and ideas of Western environmental organizations. Even though green organizations try to get their message out to the public, there is not yet enough support to achieve the widespread public pressure necessary to win the enactment of effective environmental measures.

The Russian Government

The Russian government has plans to enact new legislation requiring manufacturers to label the energy performance of household appliances. Other proposed new measures include requiring buildings to have meters installed to measure utility use, including gas, electricity, heating, and water. Some critics argue, however, that there is a wide gap between theory and the actual practice of green culture within Russian society, business, and government. Green parties and organizations face stiff opposition from hard-line former Soviet-era politicians and businessmen out to protect their own interests. For example, green activists claim that the national government and large corporations are ignoring positive environmental change in favor of immediate economic profits. Major Russian oil producers such as Rosneft, Lukoil, and TNK-BP admit that they are investing hundreds of millions of dollars to meet the government's target of increasing the utilization of flare gas (gas emitted as a by-product of oil drilling).

In the early 2000s, environmental pollution statistics were not publicly available, and government officials often accused researchers seeking such information of "spying" or other hidden agendas rather than pure scientific or environmental interests. Green World and Greenpeace in Russia, for example, have been subjected to tax audits and security investigations. Some green activists have been arrested and charged with treason for revealing the military's illegal dumping of nuclear waste. If there is compensation awarded to citizens because of environmental damage or disaster, local officials instead often simply pocket that money. In the spring of 2000, Russian President Vladimir Putin issued a decree dissolving the Environmental Protection Committee and the Forest Service. Environmental education has also been affected, with some key aspects of green culture not taught in the school system.

Environmentalists and green agencies have enjoyed some success at monitoring environmental degradations committed by private sector individuals and organizations. For instance, Green World of St. Petersburg protested the construction of a petroleum port in the Finnish Gulf. It also documented the lax safety practices at the Leningrad Atomic Agency Station and the Russian government's failure to clean up areas contaminated by radioactive waste around the city. In the early 1990s, powerful environmental commissions were formed in Moscow and in other cities throughout the country. These commissions blocked environmentally dubious projects proposed by the central government.

While some Russian proponents of green culture have always loved nature and sought its protection, others are new to the environmental movement. These newcomers proclaim that they gained their appreciation of nature and the environment much later in life, prompting them to join green culture organizations. Many of Russia's environmental leaders are scientists and engineers. Like other green culture advocates, Russian environmentalists support a reduction in society's consumption and the promotion of social justice in order to achieve a sustainable society. Specific public practices advocated by environmentalists include driving at reasonable speeds, and the use of electric vehicles to reduce dependence on gasoline and the pollution caused by vehicle exhaust fumes.

Most Russian green organizations remain quite small, with anywhere from 5 to 13 individuals who have active memberships. Many environmental organizations have a hard time finding new members, a time-consuming task that many environmental groups dread. With economic instability, most Russians do not want to pay the membership fees to become active members, even when such fees are fairly modest. Those who are active members cannot contribute that much financially, so many Russian environmental organizations are routinely short of funds. Environmental projects, therefore, cannot be carried out in an effective manner. Because of this, many green groups attempt to find funding outside the country or go without any financial assistance whatsoever.

At the beginning of the 21st century, Russian corporations and investors began to realize the profit potential of catering to the green culture movement. For example, Mikhail Prokhorov, Russian billionaire and new owner of the National Basketball Association (NBA) team the New Jersey Nets, is attempting to take Russia's car industry in a new, more environmentally friendly direction. He plans to invest over 100 million Euros in electric cars to be built at a new factory in St. Petersburg. Prokhorov faces stiff competition, however, as other financial stalwarts want to be the first to market a competitive ecofriendly vehicle.

Environmental activism has had a rocky relationship with the Russian government. Russia is committed to a certain extent to reducing the harmful emissions that cause air, water, and soil pollution and other environmental problems, but it is slow at providing energy-efficient technology and cutting pollution. There are some critics in Russia who claim that the new ecofriendly initiatives or legislation proposals from the national government are strictly public relations ploys. Green activists have also not been able to attain a solid footing with the general public. It remains to be seen whether Russia and green culture can create a relationship that will flourish and address the environmental issues plaguing the country in the 21st century.

See Also: Environmental Peacemaking; Human Geography; Organizations and Unions; Social Action, Global and Regional.

Further Readings

Bridges, Olga and Jim Bridges. *Losing Hope: The Environment and Health in Russia*. Avebury Series in Green Research. Aldershot, UK: Ashgate, 1996.

Curtis, Glenn E. *Russia: A Country Study*. Washington, DC: Federal Research Division, Library of Congress, 1998.

Davidson, Art. *Endangered Peoples*. San Francisco, CA: Sierra Club Books, 1993.

Edelstein, Michael R. and Maria Tysiachniouk. *Cultures of Contamination: Legacies of Pollution in Russia and the U.S.* Amsterdam: Elsevier JAI, 2007.

Feshbach, Murray. *Ecological Disaster: Cleaning Up the Hidden Legacy of the Soviet Regime*. New York: Twentieth Century Fund Press, 1995.

Feshbach, Murray. *Environmental and Health Atlas of Russia*. Moscow: Paims Publishing, 1995.

Goldman, Marshall I. *The Spoils of Progress: Environmental Pollution in the Soviet Union*. Cambridge, MA: MIT Press, 1972.

Hammond, Allen L. *Which World? Scenarios for the 21st Century*. Washington, DC: Island Press, 1998.

Henry, Laura. "Russia's Environmental Movement: Evaluating the Greens' Role in Civil Society." Paper presented at the annual meeting of the American Political Science Association, Chicago, IL, September 2, 2004. http://www.allacademic.com/meta/p_mla_apa_research_citation/0/5/9/7/4/p59741_index.html (Accessed September 2010).

Mills, Robin M. *The Myth of the Oil Crisis: Overcoming the Challenges of Depletion, Geopolitics, and Global Warming*. Westport, CT: Praeger, 2008.

Pryde, Philip R. *Conservation in the Soviet Union*. New York: Cambridge University Press, 1972.

Read, Piers Paul. *Ablaze: The Story of the Heroes and Victims of Chernobyl*. New York: Random House, 1993.

Smil, Vaclav. *Global Catastrophes and Trends: The Next 50 Years*. Cambridge, MA: MIT Press, 2008.

Wagner, Viqi. *Russia*. Farmington Hills, MI: Greenhaven Press, 2009.

Zakharov, Vladimir M. *The Health of the Environment*. Moscow: Center for Russian Environmental Policy, Center for the Health of the Environment, 2001.

Marcella Bush Trevino
Barry University

S

SCULPTURE

Sculpture, an art form that has existed for thousands of years, has gained new attention as a result of the sustainability movement. "Sculpture" refers to a three-dimensional artwork that is created through the shaping or combining of hard materials (such as stone, metal, glass, or wood) or plastic materials (such as clay, polymers, softer metals, or textiles). Sculpture can also consist of found objects, those that exist for another purpose but are later incorporated into an artwork. Green sculpture is that which promotes the key principles of sustainability, including ecology, grassroots advocacy, nonviolence, and social justice. As a result, green sculpture may embrace the use of materials that are sustainable or that cause the viewer to ponder environmental issues. Green sculpture may also include environmental works, which generally refers to those compositions that create or alter the environment for the viewer.

Traditional Materials

Sculpture has historically been a significant form of public art, used across cultures and time to produce works that memorialized individuals and events considered momentous and worth remembering. As a result, sculptors often chose materials that were as permanent as possible, seeking to create art that would last as long as feasible. Popular materials for sculpture have included bronze and stone, especially marble, limestone, porphyry, and granite. Wood and clay, which were less expensive, have also been popular media for sculpture. Occasionally, precious materials, including gold, silver, jade, and ivory, have been used, although much more rarely because of their cost. Although materials used traditionally reflected those readily accessible to the sculptor, this decision was dictated by availability more than any other reason. This resulted in sculptors in certain regions traditionally working with certain materials. With the advent of less expensive transportation and greater access to global markets, sculptors began using materials once considered exotic. As concern about environmental issues and sustainability grew, sculptors began to contemplate making choices that took into account green considerations or pondered using different materials altogether.

Sustainable Materials

By the 1960s, the modernist movement had revolutionized traditional, classical approaches to sculpture. Movements such as surrealism resulted in compositions being described as "sculpture" that would not have been possible using prior definitions, including such categories as "involuntary sculpture." Modernist sculpture embraced both abstract and figurative portrayals of persons, objects, and concepts. As the movement expanded, it came to include such movements as abstract expressionism, geometric abstraction, and minimalism. To assist in their work, modernist sculptors began experimenting with a wide array of new materials in addition to those traditionally used, such as stone, bronze, and wood. New materials used included epoxy resins, plastics, and concrete.

Environmental concerns also began to affect sculptors' choices beginning in the 1960s. Because sculpture traditionally has striven to create works that are permanent, there was not a great deal of concern about recycling the works themselves. Sculptors interested in their works' environmental impact, however, began to think about the consequences of using certain materials. Recycled materials were favored by some, as these would both reduce the strain placed on landfills and decrease the amount of energy expended on mining, refining, or transporting metals, stone, or other materials. Other sculptors began to emphasize the use of "found objects," which consists of using articles not normally considered art to compose a sculpture. One of the first examples of found objects was Marcel Duchamp's use of a urinal in his work *Fountain* in 1917. Using found objects minimizes the amount of trash placed into landfills but also causes the viewer to reconceptualize and ponder how society determines which items have use and which can be disposed.

Environmental Sculpture

The term *environmental sculpture* entails differing and sometimes conflicting definitions. Some believe that simply taking an extant work and placing it in an outdoor setting amounts to environmental sculpture. Others maintain that the term is better applied to sculptures that are designed for specific locations and that exploit and incorporate aspects of that site. Others also include within the definition "land art," which moves and maneuvers the site itself to create a new and altered reality for the viewer. Most environmental sculpture is crafted to a different scale so that humans are forced to interact with a work that is significantly smaller or larger than normal. Some environmental sculpture so alters the site that it encompasses the viewer, therefore verging on architecture.

Examples of site-created sculpture abound. Richard Serra, for example, is an American minimalist sculptor who works primarily with large-scale assemblies of sheet metal. Serra has placed works in both urban and rural settings, and many emphasize the weight and nature of the materials used. Other contemporary sculptors who specialize in site-created work include Beth Galston, Andrew Rogers, and Alan Sonfist.

Land art often requires the excavation of earth and stone and replacement of these materials in such a way as to create a new experience for the viewer. Robert Smithson, whose self-describing work *Spiral Jetty,* located at the Great Salt Lake in Utah, was a leading proponent of this genre. Construction of land art projects is often very intense, involving surveying locations, aerial photographs, planning, designing, excavation, construction, and maintenance. Some land art uses chiefly materials present at the site and focuses on rearranging them to form new shapes. Other land art freely introduces other materials, such as concrete, steel, asphalt, mineral pigments, or other resources to alter the natural landscape. Some view land art as a protest against the plastic aesthetics and ruthless commercialization of the art world, insofar that works are too permanent to be transported to a gallery, private home, or museum. Well-known practitioners of this movement include Walter de Maria, Jan Dibbets, Hans Haacke, Michael Heizer, Nancy Holt, Neil Jenney, Richard Long, Andrew Rogers, and Alan Sonfist. Some environmentalists criticize land art, however, as it is perceived as diminishing natural sites.

See Also: Art as Activism; Artists' Materials; Corporate Green Culture; Sustainable Art.

Further Readings

Baker, George, Bob Phillips, Ann Reynolds, and Lytle Shaw. *Robert Smithson: Spiral Jetty*. Berkeley, University of California Press, 2005.

Beardsley, John. *Earthworks and Beyond: Contemporary Art in the Landscape*. New York: Abbeville Press. 2006.

Busquets, J. *Olympic Sculpture Park for the Seattle Art Museum: The Ninth Veronica Rudge Green Prize in Urban Design*. Cambridge, MA: Harvard Graduate School of Design, 2008.

McShine, Kynaston, Lynne Cooke, John Rajchman, and Richard Serra. *Richard Serra Sculpture: Forty Years*. New York: The Museum of Modern Art, 2007.

Michel, K. *Green Guide for Artists: Nontoxic Recipes, Green Art Ideas & Resources for the Eco-Conscious Artist*. Beverly, MA: Quarry Books, 2009.

Stephen T. Schroth
Kylee M. Norville
Knox College

Shopping

The environmental impact of shopping is considerable, even when the focus is limited to shopping for reasonable essentials and not the accumulation of trinkets and "stuff for stuff's sake" that are often implied in discussions of "consumerism" and the treadmill of production. As stores have moved farther away from homes and vice versa—today typically stocking items transported long distances even if they had once been produced locally or could have been—and various resource-intensive practices have developed—from the use of hundreds of billions of disposable plastic bags for carrying purchases home to the wasteful packaging of so many products—the ordinary process of shopping for fundamentals like food, clothing, appliances, furniture, and other household goods has become one with significantly variable environmental impact.

To attempt to be greener shoppers, consumers can focus on products that are used or reusable, are energy-efficient, are recyclable or made with post-consumer recycled content, and are either free of packaging or use environmentally sound packaging or not too much of it. Green shopping is a class of moral purchasing, in which the choice between two or more options is informed by a moral preference. Boycotts are a well-known example of moral purchasing, but favoring brands of cosmetics that are "cruelty-free," or drinking Fair Trade coffee, or switching to paper products made with recycled paper are also examples, any of which could be considered green preferences. While some of these choices are made with little conscious consideration, the green shopper is generally one with well-defined preferences whose green shopping choices are part of an overall and ongoing concern for environmental issues. A number of green brands have developed to capitalize on this market niche: In the food-and-drink category alone, green brands rose from 5 to 328 products in the five years from 2002 to 2007, when the most rapid growth was seen.

Products that specifically market themselves as green can be expected to have more responsible packaging material, made of recycled or recyclable material. *Sustainable packaging* is a term that has developed to refer to the use of packaging material and processes in a sustainable manner, using life-cycle inventory and life-cycle assessments to make packaging design choices with smaller ecological footprints and environmental impacts. For instance, most plastic used to wrap packages, whether to contain the product (in the case of meat at the supermarket) or to prevent tampering (bottles of mouthwash) or simply to present an image of newness, is derived from petroleum. The U.S. Department of Agriculture's in-house research agency, the Agricultural Research Service, is working on a number of polymers to replace these plastic films, using casein and whey, proteins found in milk. Dairy-derived polymers would be biodegradable, would reduce the usage of petroleum products, and would be superior at keeping out oxygen compared with existing petroleum-based films.

Sustainable packaging uses renewable resources and biodegradable materials (when appropriate and compatible with the desire to recycle the packaging) and avoids practices that impact the atmosphere or the health of workers. Laws governing agricultural products sometimes interfere with the use of straw as a packing material, but if they can be resolved, it is a sustainable and environmentally sound packing material that has been used for thousands of years. A more recent development is green cell foam, a compostable material made from cornstarch. Bioplastics made from vegetable oil or pea starch are likely alternatives that will be relied on in the future and can achieve a carbon footprint one-third less than other plastic packaging materials. There is also opportunity for an industry to develop means to convert agricultural by-products that would commonly be disposed of as waste into bioplastics for packaging and other uses. The principal concern is that when bioplastics are added to the same recycling containers as conventional plastics, they can interfere with one another in the recycling process; this is likely solvable by minor alterations to the recycling infrastructure, perhaps as simple as adopting a word other than *plastic* for bioplastics and imprinting them with a recognizable logo indicating a separate recycling destination.

Furthermore, sustainable packaging should be minimal. "Packaging creep" in recent decades has led to more and more layers of packaging, sometimes entirely unnecessary (such as individually plastic-wrapped fruit and vegetables in supermarkets or the use of plastic-wrapped trays of fruit and vegetables rather than open bins for the customer to bag himself in order to assure customers that other customers have not been spreading germs on the produce), sometimes simply redundant (multiple layers of packaging) or excessively bulky. In this latter category, online vendor Amazon.com has a history of excessively bulky shipping materials, which may include using boxes large enough for appliances to ship a single book or compact disc. In 2008, Amazon.com partnered with several manufacturers to move away from packaging methods with severe environmental impacts or that are sources of consumer frustration (such as the sturdy "clamshell" packaging designed to make in-store shoplifting more difficult) and continues to work to improve its packaging practices.

Sustainable packaging also needs to be cost effective and logistics efficient or it is unlikely to be used: Amazon packages some of the items the way it does simply because boxes ship more easily than irregularly shaped parcels, for instance. This becomes redundant when manufacturers or third parties package a product in a box for the same reason—an electric razor in a plastic crèche in a cardboard box wrapped in plastic, for instance, which is then packed by Amazon in a cardboard box filled with packaging materials. The solution is finding the compromise between logistics efficiency and responsible use of packaging materials, and it will most likely involve further direct involvement between Amazon and other vendors and manufacturers.

One thing that has almost disappeared from the marketplace is reusable packaging. The days of the milkman and the seltzer man retrieving empty bottles from customers and replacing them with full ones are long gone, but packaging material reuse need not fade out with them. The DVD rental service Netflix has done a relatively good job of minimizing the packaging material it uses in shipping its DVDs in order to keep its own costs down: with the removal of a portion of the paper mailer the DVD is shipped in, the outgoing envelope becomes a return envelope. It is especially easy for Netflix to use such reusable packaging because of the nature of its service and the fact that customers need to return DVDs anyway; but more stores could begin providing reusable packaging of some kind, much the way deposit systems on bottles and cans are used to encourage consumers not to throw them away indiscriminately.

Other major elements of what constitutes a green brand include the production methods and the environmental and carbon impact of the product over its life cycle. The Federal Trade Commission issues Green Guides that cover the substantiation necessary to support products' environmental claims, for example, what must be true for something to be advertised as "recyclable," "recycled," or "biodegradable." But the guides are drastically out of date, having last been updated in 1998—lacking terms like *carbon neutrality, sustainable,* or *renewable.* This is especially problematic given the temptation of companies to greenwash their products, presenting them as more environmentally responsible than they are. Ecolabels, for instance, can provide the appearance of environmental responsibility—most consumers consider dolphin-safe tuna to be an environmentally conscious choice, despite the fact that sparing dolphins from being netted has nothing to do with sustainable fishing or canning practices, and other bycatch species like turtles suffer instead of the dolphins. This is not a matter of deliberate deception on the part of tuna packagers, but it does show the usefulness of ecolabels, many of which are easily exploited.

Online shopping may prove to be the best way to "go green" with one's shopping. Online shopping frees up land and energy resources that would otherwise be expended on large, inefficient retail spaces, even beyond the savings, both economic and environmental, when consumers do not travel to a store to do their shopping. Further, green claims can be immediately investigated online, and quality or price comparisons can be done quickly without needing to visit multiple locales. Online auction site eBay has created a green hub promoting sustainable and used products, and other sites exist enabling registered users to swap goods with each other, especially media like books, CDs, DVDs, and video games, which is a good way to avoid mindless accumulation or throwing away media that are no longer desired.

See Also: Advertising; Fair Trade; Green Consumerism; Greenwashing; Malls; Plastic Bags; Recycling.

Further Readings

Redclift, Michael. *Sustainable Development*. London: Routledge, 1987.
Rogers, Elizabeth and Thomas Kostigen. *The Green Book*. New York: Three Rivers Press, 2007.
Szasz, Andrew. *Shopping Our Way to Safety*. Minneapolis: University of Minnesota Press, 2009.
Torgerson, Douglas. *The Promise of Green Politics: Environmentalism and the Public Sphere*. Durham, NC: Duke University Press, 1999.

Bill Kte'pi
Independent Scholar

Sierra Club, Natural Resources Defense Council, and National Wildlife Federation

Photographer Ansel Adams attracted new members to the Sierra Club with his photographs of the Pacific coastal region. Pictured is his photo of Paradise Valley in California.

Source: Ansel Adams/The National Archives

Among the many non-profit, 501(c)(3) tax-exempt organizations working toward the protection of nature, animals, and the environment in the United States are the Sierra Club, the Natural Resources Defense Council, and the National Wildlife Federation. Each of these organizations has a lengthy history, focused areas of advocacy, and strategic campaigns to raise funds and attract membership.

Sierra Club

The Sierra Club claims to be the oldest and largest environmental organization in the United States. The organization was founded in May 1892 by famed American conservationist John Muir, whose sublime writings about nature, especially in the western United States, prompted the development of the U.S. National Park System. Muir was both an activist and an avid outdoorsman, and he often participated in communal outings that brought him into contact with like-minded individuals. His activities led him to work and hike with notables in the study and protection of nature, including science professors Joseph LaConte and Henry Senger, Stanford University president David Starr Jordan, and San Francisco attorney Warren Olney, the original founders of the Sierra Club. In the original articles of conservation, the organization was designed to promote conservation of the Pacific coastal regions, to provide education about the region, and to facilitate recreational use of the area. The new Sierra Club boasted 182 charter members, was headquartered in San Francisco, and elected Muir its first president. For Muir and many of the other activists, creating this organization was the first step in a strategic campaign to protect the great wild areas of the Pacific, including a focus on the Sierra Nevada.

Important early accomplishments of the Sierra Club under the leadership of John Muir included advocating for the establishment of the Sierra Forrest Reserve in 1893 and Mt. Rainier National Park in 1899. Early Sierra Club leadership believed that the key to preserving the great wild lands of the western United States lay in making the general public aware of its existence, and then instilling in that public a desire and a will to protect these lands through federal order of protection. Toward that end, the Sierra Club first published maps of the Yosemite Valley and the central Sierra Nevada to assist hikers and scientists interested in those areas. In 1901, the organization sponsored its first "outings": group hikes to pristine spots in nature organized and lead by Sierra Club members and designed to promote a passion for protecting these locations; these trips continue to this day. In 1906, the Sierra Club began a seven-year battle against the city of San Francisco, which requested water rights in an area of the Yosemite called Hetch Hetchy. The city intended to build a reservoir in the area, flooding a region that had been previously celebrated by members of the Sierra Club. While the Sierra Club launched a publicity campaign to prevent the destruction of the Hetch Hetchy environ, Congress passed the required legislation for the building of the reservoir, known as O'Shaughnessy Dam. In 1928, famed photographer Ansel Adams was hired as the club's official photographer, and some of his earliest work was initially published in the *Sierra Club Bulletin*, a publication for members. Adams was an important factor in attracting new members to the organization.

The current mission of the Sierra Club remains similar to its founding mission, and it focuses its resources to attacking six problems that it has identified as target projects. These six interrelated club goals include moving beyond coal, discovering and promoting clean energy solutions, protecting the various habitats of the United States, promoting and encouraging green transportation alternatives, limiting greenhouse gas emissions, and safeguarding communities threatened by environmental pollutants. Sierra Club activities in these areas have resulted in a "Climate Recovery Agenda," which details plans to reduce carbon emissions by 80 percent by the year 2050.

The Sierra Club continues to recruit new members and to attract attention to its activities through a complex public relations media presence. The club also continues to provide "outings" that draw attention and provide education for participants about the world's natural places. Since their instigation, these outings have developed into various types of trips, including the traditional national and international nature excursions, but also including local outings in various locations throughout the United States and inner-city outings that focus on the concerns facing our cities. Membership in the Sierra Club has remained steady at about 1.2 million for most of the 21st century. The organization maintains a sophisticated website for member information and networking.

Natural Resources Defense Council

The Natural Resources Defense Council (NRDC) was cofounded in 1970 by a group of concerned attorneys and law students, including John Adams and Gus Speth, in an effort to bring legal expertise to the fight to protect the environment. According to the organization's mission statement, its ultimate goal is to safeguard the people, animals, and life systems on Earth, which is enacted through a large network of concerned lawyers, activists, and members.

The NRDC focuses on environmental protection issues involving laws and policies in the United States through the courts and legislatures. Founding members of the NRDC established early patterns of environmental activism that included challenging the validity and enactment of laws and policy impacting both natural and built habitats as well as lobbying Congress and other elected officials to promote conservation legislation. Examples of early successes for the organization include the passage of the Clean Water Act in 1971, beginning legal action that led to the elimination of lead in gasoline in 1973, and helping to win federal protection for 100 million acres of wilderness in Alaska in 1980.

Over the course of its existence, the NRDC has been involved in two Supreme Court cases. The first of these cases, *Vermont Yankee Nuclear Power Corp. v. Natural Resources Defense Council, Inc., 435 U.S. 51*, centered on a license to operate a nuclear power plant that was granted by the Atomic Energy Commission to the Vermont Nuclear Power Corporation after extensive hearings before the Atomic Safety and Licensing Board. The NRDC challenged the granting of the license based on complex procedural issues related to an agency's right to impose its own rules regarding decision making. After the case reached the Supreme Court in 1978, the justices ruled against the NRDC, establishing a precedent that courts cannot impose rulemaking procedures on federal agencies.

The second historical legal case fought by the NRDC before the Supreme Court occurred in 1984; it was *Chevron U.S.A., Inc. v. Natural Resources Defense Council, Inc.* (*467 U.S. 837*). This case revolved around the redefinition of a term within policy managed by the Environmental Protection Agency during the Reagan administration. The proposed change allowed existing energy producers to construct new equipment that did not meet the standards of the Clean Air Act provided the overall pollution levels of the company did not change. The NRDC challenged a government organization's right to redefine procedures, and the Supreme Court again ruled against the NRDC, stating that the courts should defer to the administration agency's interpretations of terms and policy.

The NRDC continues to work for environmental protection, and 21st-century priorities include curbing global warming, planning for a clean energy future, saving endangered wildlife and wilderness areas, reviving the world's oceans, and limiting the use of toxic chemicals. The NRDC also has an international initiative aimed at the greening of China. The NRDC also operates two large programs, Partnership for the Earth and the NRDC Action Fund; both of these programs generate operating revenues for the organization to continue its work on preservation. The NRDC also maintains a series of websites to educate the public on its efforts, to promote its activities, and to recruit new membership.

National Wildlife Federation

The National Wildlife Federation (NWF) is the largest organization in the United States working toward conservation of natural environments and includes both national headquarters offices and smaller state affiliate organizations representing 47 states. The mission of the organization is directed toward preserving the natural environment for future generations.

The NWF was first imagined by cartoonist Ding Darling, who was inspired to create political cartoons with an ecological message after witnessing the environmental devastation caused by the unsustainable farming practices that resulted in the U.S. Dust Bowl in the 1930s. His cartoons attracted the attention of President Franklin D. Roosevelt, who appointed Darling to be head of the United States Biological Survey, which would later become the United States Fish and Wildlife Service in 1934. While holding that position, Darling drew the illustration that would become the first Federal Duck Stamp, a federal program that uses the profits generated by the sale of the stamps for waterfowl management. However, despite this work, Darling still felt that more could be done to protect wild lands, and so in 1936, he worked with President Roosevelt to invite 3,000 hunters, anglers, and conservationists to attend a meeting in Washington, D.C., called the North American Wildlife Conference. This group became the founding members of the General Wildlife Federation, which would later be renamed as the NWF, and its first conference would become an annual member meeting that continues to this day.

Because the early membership of the organization was composed of hunters and anglers, legislation related to wild lands and game animals was an early focus of this organization. Important early victories of the NWF include the passage of the Federal Aid in Wildlife Restoration Act, which applied levies to firearms sales to be used to restore and maintain habitat for wildlife and to conduct research on wildlife management, and which continues to generate income for wildlife protection. In 1966, the NWF was fundamental to the passage of the Endangered Species Preservation Act and the related 1969 Endangered Species Conservation Act, which provided for the acquisition and protection of lands home to nonhuman animal populations threatened with extinction. Later, the NWF also advocated for the passage of the Endangered Species Act in 1973, which actually created and defined the terms *endangered* and *threatened* as they apply to species protection. In 1985, the NWF turned its attention to the preservation and use of lands devoted to agriculture by supporting the Food Securities Act; by the 1990s, the organization had begun to promote the use of sustainable and clean energy by supporting the Energy Policy Act and other such legislation.

The NWF continues its work in the conservation and preservation of national wild areas and animals. Remaining true to its original goals, the NWF recognizes the different and sometimes divergent attitudes and opinions of those interested in conservation and preservation of wild lands, so while its initiatives and programs change over time, it remains committed to a model of conservation that focuses upon human stakeholders impacted by environmental issues. Some of the contemporary issues concerning the organization include global warming, sustainable energy, and nonhuman animal habitat protection. The organization publishes a magazine for members titled *National Wildlife* that publishes news and educational information, and it also maintains a website with similar content. The NWF also publishes a popular magazine for children titled *Ranger Rick* that is designed to educate children about their natural environment and to instill an ethic of preservation in its young readers. Current membership in the NWF is about 4 million.

The Big Three

Together, the Sierra Club, the NRDC, and the NWF represent the three largest and most influential of the 503(c)(3) nonprofit organizations working toward preservation and conservation of the natural world and its human and nonhuman inhabitants. While these organizations emerged during different historical eras in the United States, each shares an ethic committed to preserving the environment and each focuses on similar contemporary environmental issues, including clean energy, pollution, and habitat preservation and conservation. Despite their shared interests, the memberships of these organizations sometimes clash as a result of the different causes and philosophies important to each.

Furthermore, although these are the three largest organizations working toward these goals, there are many other national and international organizations striving to meet the same goals that boast large memberships and activist success, including Defenders of Wildlife, The Nature Conservancy, Greenpeace, and the Rainforest Action Network, among others. However, their combined membership, history, and past successes make these organizations primary in the battle for environmental protection.

See Also: Air Pollution; Environmental Communication, Public Participation; Grassroots Organizations; Social Action, National and Local; Werbach, Adam.

Further Readings

Brulle, B. *Agency, Democracy, and Nature: The U.S. Environmental Movement From a Critical Theory Perspective*. Cambridge, MA: MIT Press, 2001.

Galen Rowell's Sierra Nevada. San Francisco, CA: Sierra Club, 2010.

National Wildlife Federation. "Our History and Heritage." http://www.nwf.org/About/History-and-Heritage.aspx (Accessed September 2010).

Natural Resources Defense Council (NRDC). "NRDC: Who We Are." http://www.nrdc.org/about/who_we_are.asp (Accessed September 2010).

Schwarz, W. *Voices for the Wilderness: From the Sierra Club*. New York: Ballantine Books, 1950.

Jennifer Adams
DePauw University

Social Action, Global and Regional

Global and regional social activists for a greener culture advocate an energy revolution, climate legislation, and investment in renewable resources to make the planet more environmentally friendly. Social action has included protests, boycotts, marches, and Internet websites, among other measures to raise environmental awareness and promote sustainable living. Key socially based environmental groups with regional and/or global presences include the Save Our Environment Action Center; ISAR: Resources for Environmental Activists; the Climate Action Network Europe; Action for Solidarity, Equality, Environment, and Diversity Europe; Greenpeace; and the World Wildlife Fund.

Many social action groups are working both regionally and globally to secure local, national, regional, and global environmental policies to reduce various types of pollution, save threatened and endangered species and natural areas, and conserve natural resources. Social action groups also promote investment in renewable resources and the development of green products and practices as a key component in the solution for making the world a greener, more sustainable place to live. Renewable energy resources such as wind, solar, and geothermal can provide a large percentage of the average family's electricity and heating demands. Social activist groups maintain that renewable resources would also create jobs and reduce the dependence on foreign oil.

Social green culture initiatives will also protect the environment and health of communities and ensure that their children and all future generations would inherit a viable, sustainable world. Individuals and groups interested in promoting social action have also targeted the reach of the Internet to extend their presence regionally and globally, creating websites such as Blue Planet Green Living and the Take Part Online Social Action Network that encourage grassroots social action and offer advice on organizing and incorporating green culture into our everyday lives.

The Save Our Environment Action Center is a joint effort of the United States' most influential environmental advocacy agencies, many of which also have regional and global presences, to change that nation's environmental policies. They use the Internet to increase public awareness and activism on the most pressing environmental issues and to solicit mass citizen participation. Some of the member groups of this organization include American Rivers (protects and restores rivers), Defenders of Wildlife (saves endangered wildlife, habitat, and biodiversity), Earthjustice (protects people, wildlife, and natural resources by providing free legal representation to citizen groups to enforce environmental laws), Friends of the Earth (champions a healthier and more just world), Greenpeace, and the National Audubon Society (conserves and restores natural ecosystems, including birds and the habitats of other wildlife), and many others.

One group that is also a part of the Save Our Environment Action Center is the Union of Concerned Scientists (UCS), formed in 1969. This organization supports national, regional, and global environmental initiatives and calls on all scientists, health professionals, engineers, and economists to build a healthier environment and a safer world. The UCS delves into such areas as global warming, clean vehicles, clean energy, and nuclear power. It combines independent scientific research with citizen action to create innovative solutions and changes in government policy, corporate practices, and consumer choices. The UCS has had a long list of environmental victories dating from the 1970s. In the 1990s, the UCS led a delegation of American nonprofit organizations to the Kyoto negotiations that led to the Kyoto Protocol climate change treaty. Their lobbying also helped convince President Bill Clinton not to build a national missile defense system, which had potential negative environmental impacts, among other objections.

Scientists are not the only professional group to include a social component in order to advocate a better environment. Doctors founded Physicians for Social Responsibility (PSR) in 1961. This is a nonprofit advocacy organization that provides a medical and public health voice to social efforts to reverse global warming, prevent the toxic degradation of the environment, and prevent nuclear war and weapons proliferation. In the early 1990s, PSR mobilized the medical community on environmental health issues. This led to a human health and environmental collaboration between the Massachusetts Institute of Technology (MIT), the Harvard School of Public Health, Brown University, and PSR's Boston chapter. PSR's strategy is to educate the medical community and the public about its environmental concerns, which is accomplished through research, analysis, collaboration, and communications. The ultimate goal is positive government and societal change at all levels, from local to international.

Social Action Worldwide

Europe is another region with an active social action movement dedicated to environmental preservation and sustainability issues. Several organizations with regional presences have sought to foster grassroots social action through information and support. ISAR: Resources for Environmental Activists, founded in 1983 as the Institute for Soviet and American Relations, is dedicated in part to facilitating citizen diplomacy and involvement in environmental issues through social action and grassroots nongovernmental organizations (NGOs). European grassroots organizations and NGOs dedicated to encouraging and aiding social action on environmental issues include the Climate Action Network Europe (CAN-E) and Action for Solidarity, Equality, Environment, and Diversity Europe (A SEED Europe).

One of the world's main global social activist groups dedicated to environmental issues is Greenpeace, founded in 1971. Greenpeace maintains over 40 offices around the world. Its mission is to fight the threat of global warming, the destruction of the forests, and the destruction of the oceans, among numerous other issues. It is funded strictly through individual contributions. Greenpeace activists engage in a variety of global social actions, sometimes utilizing more radical tactics to achieve their goals. In 1975, they stopped France from carrying out atmospheric tests of nuclear explosions by sailing into test zones. In 1978, they halted the grey seal slaughter in the Orkney Islands of Scotland. In the early 1980s, they played a key role in the European Commission's decision to ban the import of seal pup pelts. Also in that decade, they pressured the International Whaling Commission to adopt a moratorium on commercial whaling by disrupting whale hunts at sea, work that is continued by Paul Watson with Sea Shepherd Conservation Society.

Greenpeace activists also successfully ran a campaign to have the continent of Antarctica preserved as a "World Park" in the early 1990s. Antarctica is considered a global commons, a territory with international access governed by international law established by the United Nations. The activists fighting for Antarctica were concerned with the environmental impact of the various national military bases established there under the Antarctica Treaty and staged peaceful protests in order to raise public awareness of environmental abuses. For example, the French government was forced to give up plans to construct an airstrip in the early 1990s when activists blocked the construction site. In the mid-1990s, activists also occupied Brent Spar, an oil storage facility in the North Sea owned by Shell Oil Company, in order to stop plans to sink the 14,500-ton installation because it was no longer useful. Shell eventually agreed to instead dismantle and recycle the installation on land.

The World Wildlife Fund (WWF) is a global social action group whose members are dedicated to the global protection of the future of nature and the environment. They attempt to work with world governments to enact positive policy changes toward a sustainable future. Teams of scientists, field staff, and policy makers work locally, regionally, and globally. The WWF also forms partnerships with key corporations, governments, marketers, and humanitarians who share its goals. The WWF and its partner companies have a vested interest in conserving nature and creating a healthier environment for future generations through the prevention of global climate changes that may signify species extinctions and irreparable damage to the world's ecosystems.

See Also: Corporate Social Responsibility; Demonstrations and Events; Environmental Law, U.S.; Grassroots Organizations; Human Geography; India; Institutional/Organizational Action; Nongovernmental Organizations; Organizations and Unions; Public Opinion; Russia; Social Action, National and Local.

Further Readings

Beckerman, Wilfred and Joanna Pasek. *Justice, Posterity, and the Environment*. New York: Oxford University Press, 2001.

Blue Planet Green Living. http://www.blueplanetgreenliving.com/about-us (Accessed September 2010).

Dasgupta, Partha. *Human Well-Being and the Natural Environment.* New York: Oxford University Press, 2001.
Dobson, Andrew. *Green Political Thought.* London: Routledge, 2007.
Easterbrook, Gregg. *A Moment on the Earth: The Coming Age of Environmental Optimism.* New York: Viking, 1995.
Faber, Daniel. *The Struggle for Ecological Democracy: Environmental Justice Movements in the United States.* New York: Guilford Press, 1998.
Freeze, R. Allan. *The Environmental Pendulum: A Quest for the Truth About Toxic Chemicals, Human Health, and Environmental Protection.* Berkeley: University of California Press, 2000.
Friedman, Thomas L. *Hot, Flat, and Crowded: Why We Need a Green Revolution—and How It Can Renew America.* New York: Farrar, Straus & Giroux, 2008.
Goldfarb, Theodore D. *Taking Sides: Clashing Views on Controversial Environmental Issues.* Guilford, CT: McGraw-Hill/Duskin, 2001.
Kennedy, Paul M. *Preparing for the Twenty-First Century.* New York: Random House, 1993.
Lytle, Mark H. *The Gentle Subversive: Rachel Carson, Silent Spring, and the Rise of the Environmental Movement.* New York: Oxford University Press, 2007.
Manes, Christopher. *Green Rage: Radical Environmentalism and the Unmaking of Civilization.* New York: Little, Brown, 1990.
Squatriti, Paolo. *Nature's Past: The Environment and Human History.* Ann Arbor: University of Michigan Press, 2007.
Steffen, Alex. *Worldchanging: A User's Guide for the 21st Century.* New York: Abrams, 2006.
Stevis, Dimitris and Valerie J. Assetto. *The International Political Economy of the Environment: Critical Perspectives.* Boulder, CO: Lynne Rienner, 2001.
Take Part Online Social Action Network. http://www.takepart.com (Accessed September 2010).
Warren, Karen and Nisvan Erkal. *Ecofeminism: Women, Culture, Nature.* Bloomington: Indiana University Press, 1997.

Marcella Bush Trevino
Barry University

Social Action, National and Local

The record of social change action dates back to the beginning of U.S. history, yet most observers point to the 1960s as the time when stirring concern for the environment pushed ecological interests to local and national prominence. Everyday conversations, collective citizen action, and government policy decisions were responding in robust fashion to the fusion of a wide range of environmental threats, including rampant air pollution, indiscriminate use of pesticides, marked decreases in water quality, and ever-increasing urban population growth. Much of the country's economic activity at the time—manufacturing, agriculture, communications, transportation, and energy—led to the production of untenable levels of carcinogenic waste products. At first, activists pressed for and achieved well-intended state and federal legislation.

More recently, the wave of social action has shifted to affect lifestyle changes in an effort to advance an understanding of interdependency of the Earth and its people. Most often communicated through grassroots-based collectives, social change action for a better environment is mobilizing people across America to instigate systemic changes and the transformation of attitudes, values, habits, and everyday actions for sustainability. These efforts join others around the world that, according to environmentalist Paul Hawken, represent the largest social movement ever, merging individuals and groups in the pursuit of social justice by giving due appreciation to indigenous cultures and communities. The study of social action for the environment thus provides a glimpse into how dedicated individuals, ridiculed by some and revered by others, have persevered to affect both small-scale changes and larger policy transformations.

Environmentalists recognize that many of our seemingly "natural" ecological disasters are oftentimes caused by or exacerbated by for-profit entities. Thus, they reason, strong citizen action is necessary to monitor and implement changes to avoid further catastrophes. From the efforts of one to the collective will of many, victories have been noted—from the closing of Love Canal chemical waste dump in 1978 to the public outcry surrounding the *Exxon Valdez* oil spill in 1989 and the outrage over the British Petroleum–Deepwater Horizon oil spill of 2010. Activists have focused their tactics and approach not just to seal off pollutants to preserve human life and mitigate disasters as was the directive of the 1970 Clean Water Act but also to halt the release of toxins altogether. Activists argue that weakened local and federally mandated regulations have failed to live up to the promise of protecting people and the environment. Levying fines in lieu of closing down toxic operations means that regulation has devolved into a nuisance that is now accepted as a reasonable cost of doing business.

From 1960s Outrage to 1970s Regulation

Rachel Carson warned in her book *Silent Spring* (1962) that poisons dumped into our environment at alarming levels were forever changing our soil and groundwater supplies for the worse. Her book is credited with fueling the environmental movement. This era's thrust toward multi-issue social change mirrored the actions of the country's earlier populist movement, which challenged the industrial enrichment of the few at the expense of harms to the many. At the same time, a steady transformation of corporate law undermined the fundamental principle that corporations were created to serve the public good. Instead, corporations were increasingly granted protections (e.g., 1976 *Buckley v. Valeo*) and, for the purposes of law, treated as people, with rights to free speech, due process, and compensation before being deprived of property. In doing so, the balance of power shifted in favor of corporations since they maintained the advantage of size and money to parlay against more loosely organized and underfunded social change activists. Despite efforts through the years by prominent activists, including consumer advocate Ralph Nader, progressive political commentator and author Thom Hartmann, and former U.S. Vice President Al Gore, Congress failed to effectively challenge corporate dominance.

Minimal barriers were set to prevent the wholesale purchase of public policy by wealthy corporations and individuals, but loopholes and industry-financed Political Action Committees ensured that the polluting industries maintained their ways by influencing who was elected and how they would vote. Coupled with the message and movement embodied soon thereafter by President Ronald Reagan's condemning tax-and-spend big government, regulatory agencies were allowed to be underfunded, understaffed, and led by corporate executives operating with conflicting interests.

Evolving Discourses

Social change efforts employ discourses that have the potential to affect widespread changes in citizen behavior in the public and private spheres, an important feature of healthy governance in democratic life. The most well-known U.S. activist, Ralph Nader, built his career on advocating consumer protection, including environmental safety through affordable, clean, and sustainable energy.

Protest actions, educational campaigns, informal networking, lobbying efforts, and entrance into formal political structures are among the many tools used by individuals and groups to advance the need for conservation, ecological preservation, and other protective measures to preserve the ecosystems under threat. All environmental actions, however, are not the same. Some have targeted changes requiring little more than greater awareness of how we can "save" as exemplified through recycling programs. Other environmental programs, including ones detailed by author and "Story of Stuff" originator Annie Leonard, argue that fundamental change is called for—the oil, coal, and nuclear industries, for instance, have maintained control over energy production by marshalling three-quarters of all federal incentives for their benefit. The solution to the energy crisis will thus come not in less oil use, but instead by a reinvestment of government dollars into renewable and efficient energy products and programs such as the use of rooftop solar power, the capturing of wind as energy, and the limiting of carbon emissions in order to reduce the dependency on diminishing fossil fuels.

Even in the 21st century, with a history of successes behind social movements to sustain a green culture, from antinuclear demonstrations to resource replenishment programs, social change advocates continue to face rhetorical and material challenges. In our automobile-focused society, for example, we face a paradox surrounding social action. Citizens and government leaders embrace the need to use less energy. However, the resources devoted to mass transportation are not keeping pace with the dollars spent on private automobile usage and roadway maintenance. For environmental activists, the struggle extends beyond funding buses or fixing highways. It is instead a question of ameliorating long-standing injustices to the environment with alternative positions that transform the culture in ways that protect the Earth while also providing the means by which all citizens can benefit from additional affordable, clean, and sustainable transportation options.

Despite the ever-increasing number of voices calling for sustainable practices, social change advocates remain ostracized by their opponents with calculated campaigns to discredit their efforts as simply popular fads, generally short-lived, frequently punctuated by media-mongering antics, and operating without deep thought or merit. The term *tree-huggers* is the most recognized caricature of environmental protectionists.

The environmental controversy shifted direction toward positive social change in the United States when former Vice President Al Gore released his film *An Inconvenient Truth* in 2006 with the scientific evidence detailing global warming's eventual trajectory absent necessary changes. Gore's work was not new or without precedent by other environmental action groups. Local and regional chapters of the Sierra Club, Greenpeace, and Earth First! as well as a host of other national and local collectives had long been publicizing similar concerns. However, the visibility and legitimacy of the former vice president's message was unparalleled in conveying the depth and breadth of the problem, garnering Gore and the Intergovernmental Panel on Climate Change the Nobel Peace Prize in 2007.

Advancing in the 21st Century

The vast majority of U.S. citizens today recognize that global warming from carbon emissions from our oil- and coal-dependent nation is causing catastrophic climate changes that endanger the entire planet. Moreover, citizens are aware of the depletion of and difficulty accessing remaining world oil reserves. In response, local movements for sustainable lifestyles are popping up in small and large communities, nationwide. The task for the future, they proclaim, requires more than recycling and a bit of bicycling.

Social change agents recognize the stakes are great and ask, have we already reached the "tipping points" of irreversible harm to the planet? The answer is unclear, but what is more concrete is the shift and necessity to link all environmental issues to the survival of life on the planet. That sense of urgency is fueling hope and action among a greater number of citizens, many for the first time identifying themselves as social change agents.

See Also: Environmental Justice Movements; Grassroots Organizations; Political Persuasion; Social Action, Global and Regional; Tree Planting Movement.

Further Readings

Carson, Rachel. *Silent Spring*. New York: Houghton Mifflin, 1962.
Hawken, Paul. *Blessed Unrest: How the Largest Social Movement in History Is Restoring Grace, Justice, and Beauty to the World*. New York: Penguin, 2007.
Leonard, Annie. *The Story of Stuff: How Our Obsession With Stuff Is Trashing the Planet, Our Communities, and Our Health—and a Vision for Change*. New York: Free Press, 2010.

Spoma Jovanovic
Lewis Pitts
Independent Scholars

Socialization, Early Childhood

To sociologists, socialization refers to the process of teaching people a culture, that is, the habits of mind, emotion, and action typical in a group. This teaching and learning occurs formally, informally, and even implicitly or via observation. Effective socialization functions as a form of social reproduction, though the learners are not passive in the process and it does not provide perfect or complete cultural transmission. Children in nations with an established "green culture" have varying degrees of contact with that culture and therefore experience varying degrees of "green socialization." Key agents relevant to green socialization include those typically associated with childhood socialization more generally, such as education, the media, and the family—but contact with nature itself also plays a role. Considered broadly, research finds that many children are eager and optimistic environmentalists, and various agents of green culture succeed in moving children in a

Studies show that early hands-on and residential experiences, like this camping and hiking trip on the Pocosin Lakes Interpretive Boardwalk, can impact children's appreciation for nature, knowledge of and concern for the environment, and attitudes about nature and society.

Source: Kendall Lipsey/USFWS

pro-environment direction, though of course, there are factors that inhibit environmentalism.

Green socialization impacts children's appreciation for nature, knowledge of and concern for nature and environmental problems, and attitudes about nature and society. Notably, children's knowledge of nature and understanding of nature and environmental problems is often sorely lacking, and while for some issues there is improvement with age, for others, confusion tends to persist (e.g., ozone depletion and global warming are conflated). Perhaps because children's lives are circumscribed by the power of adults, less literature focuses on socialization affecting children's behavior than on changes in subjectivity, but there is some research showing that lessons children learn about the environment can be transmitted to parents, that is, children become the agents of socialization. Some scholars writing about children and environmentalism point also to capacities like "biophilia," an innate interest in nonhuman life, or to typical aspects of human development like empathy. Factors like these may interact with green socialization to push children in the direction of nature appreciation and environmental concern.

Environmental education is encountered by children in traditional school curricula (as a secondary subject) and in nonschool settings such as parks and nature centers. As a field, environmental education is devoted to the practice of teaching about nature, the interrelation between societies and nature, and about environmental decision making. Studies of the effectiveness of environmental education do not often focus on the very young, but many do examine the impact of programming on elementary-age children. While long-term impact remains a concern, there is evidence that some programs are more effective than others. For example, residential programs are better than classroom programs, and programs that incorporate hands-on elements are better than those that do not. One important criticism of environmental education is that it emphasizes personal action and fails to direct young people to civic engagement in support of the environment.

Communication media contribute to green socialization as well. Messages can be implicit (e.g., nature imagery used in advertising) or explicit (e.g., news about an environmental problem). Limited evidence suggests that viewing nature programming may have pro-environment outcomes, and there is research showing that television is an important source of children's environmental knowledge. In terms of the content of children's media, the small body of research notably finds a decline over recent decades in the presence of nature in sources like children's picture books or elementary school textbooks. This evidence is countered by other aspects of green culture, such as the numerous nature- and environment-themed books published for children each year. As with environmental education, there is concern that explicit environmental messages in the media tend to overemphasize personal action.

When considering green socialization occurring in the family, attention must be given to direct social interaction and to the effect of the home environment. Considering the former, a sizable body of research indicates that family members who model nature appreciation are important to the early life experiences of those who later become environmental activists and professionals. Socialization promoting empathy and nurturing among girls has likewise been related to the adult tendency for women to have higher levels of environmental concern than men. In terms of the general household environment, there is some evidence suggesting the environmentalism of youth is predicted by "high" culture in the home (e.g., emphasis on books rather than television) because it fosters abstract rather than utilitarian thinking. Parental shopping habits must also be mentioned here as this is undoubtedly where many children first encounter green consumerism. Again, dulled civic and political engagement is a concern: green products assuage fears, making those who can afford them feel they are protected from environmental threats and therefore need not take further action.

Family also guides the degree and character of children's contact with nature itself. This is important because research shows that early experiences in natural settings predict adult environmentalism. This was first uncovered by qualitative research that asked environmental activists and professionals to reflect on the roots of their passion for nature and environmental protection, and some scholars questioned the value of this finding on methodological and other grounds. The critiques have, however, been dulled by replications in qualitative and quantitative studies. Notably, social and geographic changes in recent decades have increasingly insulated children from the natural world. Children's lives are more interior and focused on structured activities and engagement with electronic media, leading a number of scholars and writers to be concerned both about the future of environmentalism and about healthy childhood development.

See Also: Education (Climate Literacy); Green Consumerism; Home; Nature Experiences; Popular Green Culture; Religion; Shopping.

Further Readings

Chawla, Louise. "Significant Life Experiences Revisited: A Review of Research." *Journal of Environmental Education*, 29/3 (1998).

Chawla, Louise and Debra Flanders Cushing. "Education for Strategic Environmental Behavior." *Environmental Education Research*, 13/4 (2007).

King, Donna Lee. *Doing Their Share to Save the Planet: Children and the Environmental Crisis*. New Brunswick, NJ: Rutgers University Press, 1995.

Louv, Richard. *Last Child in the Woods: Saving Our Children From Nature-Deficit Disorder*. New York: Algonquin Books, 2005.

Rickinson, Mark. "Learners and Learning in Environmental Education: A Critical Review of the Evidence." *Environmental Education Research*, 7/3 (2001).

Szasz, Andrew. *Shopping Our Way to Safety: How We Changed From Protecting the Environment to Protecting Ourselves*. Minneapolis: University of Minnesota Press, 2009.

Christopher W. Podeschi
Bloomsburg University of Pennsylvania

Soil Pollution

Soil pollution is defined as the accumulation of chemicals, physical matter, or biological products in amounts or concentrations that are potentially toxic to other organisms or waterways. Physical matter can cause pollution by obstruction of soil activity, leaching, or littering. Chemicals may become pollutants in soils but this depends on their structure, concentrations, behavior, persistence, and interactions with other substances. Levels of pollution in soils and waterways are usually determined by national and international agencies based on scientific reports, probability of health impacts, and other factors. In the United States, for example, the Environmental Protection Agency (EPA) and the Natural Resources Conservation Agency (NRCA) are the organizations that manage the data and determine federal and state policies.

Common methods of soil pollution or contamination include excessive accumulation of trash or other physical or chemical man-made (xenobiotic) materials. For example, when runoff transporting metals, fertilizers, pesticides, or other products seep into the soil, increasing the concentrations of these substances above acceptable levels, the soil is considered polluted. Other examples of soil pollutants include excess of organic or inorganic fertilizers, heavy metals, plastic and its derivates, and medications. Industrial by-products, mining residues, nuclear waste, hydrocarbons or other oil derivates, and radioactive materials from various sources can also significantly alter soils. Chemicals mixing within soils can permit the formation of other highly toxic compounds as well.

Pesticides and antibiotics used in farming may also promote soil pollution by themselves or by their effects on soil species (e.g., altering species composition, increasing antibiotic-resistant strains in soils). Animal manure and human feces are also common sources of soil pollution. In addition, soils can be contaminated by living organisms and nonliving organic carbon-based forms. Among the living organisms, pathogens, parasites, protozoa, bacteria, and fungi that may be harmful to humans or wildlife can be found in soils. Nonliving organic forms include viruses, prions, and decomposing animals carrying diseases (e.g., chronic wasting disease [CWD]).

Agricultural, forestry, horticultural, and other management practices modify soils to a high degree. These practices, however, are usually perceived as beneficial to the receiving soils and are only considered a form of pollution when excess of nutrients, presence of pathogens, or other erosion-moving materials enrich unintended areas, for example, non-cultivable land (forest, wetlands, prairies, conservation areas), influencing species composition. Nutrient enrichment, pathogens, pesticides, and heavy metals may also affect human health by being present in recreational areas or house yards. For example, gardening soil mixes are highly enriched with nutrients and contain organisms and materials foreign to the receiving soils. This is, however, mostly considered as a form of soil improvement.

Eroded soils, highly depleted agricultural areas that had been poorly managed and no longer contain sufficient organic matter, are low in fertility and form clods or crust that restrict air, are also considered polluted soils. In summary, the main problems related to soil pollution include lack of fertility and sedimentation that decrease the available agricultural land and excess of fertilizer. In all these cases, the contaminated soils or sediment may be carried by runoff to streams or other water bodies. Soil may also become unproductive by being covered by soil sediments. Soils, sediments containing high amounts of metals, soils adjacent to mining sites, and soils containing higher levels of radioactive materials or hydrocarbons are also considered polluted. Soils transported by wind or water to other areas, causing landslides, sediment deposition, eutrophication, turbidity of water, pathogens, or impairing water for drinking or recreational purposes are considered causal agents of "nonpoint source pollution."

Soil contamination is not necessarily intentional or easily tracked to a point source. For example, soil sediments or nutrients carried by runoff arrive from other areas. Soil sediments from farms, constructions, roadsides, or landsides also contribute to polluting other soils and waterways. Wind may also disperse soil particles containing microbes or chemical contaminants in dust. Point source pollution is traceable to, for example, mining, nuclear waste, landfills, or industrial operations. In such cases, the cost of soil remediation is assigned to the polluter. The cost of nonpoint source pollution is assumed by local and federal governments.

Every year, billions of dollars are utilized to remediate soils and water contaminated by soil pollutants. In the United States alone, most of this funding is managed through NCRA, based on the Clean Water Act. Mining companies pay a high price to deal with the need for soil remediation. In this case, metals are either removed, treated in situ with addition or phytoremediation (plants uptaking the pollutant). In many cases such as land mining, mining, and nuclear waste, the cost of cleaning is so high that the soils are left polluted for decades.

The sources and causes for soil pollution are as complex as is its definition. Not all scientists agree regarding the definition of soil pollution. Some scientists have eliminated agricultural fertilizers and pesticides from soil contaminants. As these substances are responsible for impairing waterways, they are considered here as pollutants. Controversies also arise concerning chemicals or substances, which are naturally abundant in certain soils. For example, radon is naturally abundant in certain areas, yet because it causes illness to people, the soils in which these chemicals are present in amounts that represent a risk for health are considered polluted.

Soil pollution is a prevalent problem. Although most of the focus on this topic is on its consequences for water pollution or the loss of productive land, soil pollution has more dimensions. It affects society through economic and health costs and through lost or unusable spaces. Remediation and prevention of soil pollution are fundamental long-term investments for any society.

See Also: "Agri-Culture"; Environmental Law, U.S.; Organic Foods; Organics; Permaculture.

Further Readings

Mirsal, Ibrahim A. *Soil Pollution: Origin, Monitoring and Remediation.* New York: Springer, 2010.

Natural Resources Conservation Agency. http://www.nrcs.usda.gov/programs/ccpi/2010docs/ResourceConcernDefinitions.pdf (Accessed September 2010).

Shortle, James S. and Richard D. Horan. "The Economics of Nonpoint Pollution Control." *Journal of Economic Surveys*, 15/3 (2001).

U.S. Department of Agriculture, National Resources Conservation Service. "Welcome to the NRCS Soils Website." http://soils.usda.gov (Accessed September 2010).

van Straalen, N. M and D. Roelofs. "Genomics Technology for Assessing Soil Pollution." *Journal of Biology*, 7 (2008).

Mirna E. Santana
Independent Scholar

Space/Place/Geography and Materialism

Pittsburgh's East Liberty area was the subject of a massive urban renewal project, which generally tries to position a place as an intersection between nature and culture. Pictured is a shopping development in East Liberty.

Source: The Zach Morris Experience/Wikimedia Commons

The idea of place is a compelling one. Though it refers only, in the simplest sense, to a physical location, it takes on a richer and more resonant dimension. "Place identity" has become recognized as a matter of great significance, one taken into consideration in urban design, landscape architecture, and environmental psychology. A place has a meaning for its inhabitants or users that extends beyond its geographical location. Urban planners have in recent decades begun working with local communities in order to integrate new places into old ones, or transform existing places, in accordance with hyperlocal culture—a process called "placemaking."

Placemaking began in the 1970s with architects and city planners working together with public input in the design of spaces like parks, waterfronts, plazas, and city squares, and incorporated both building design and the surrounding landscape. In particular, placemaking takes into account what is going to happen in a place or what a place will encourage to happen: the social encounters that transpire in a place, the sights and sounds of being there, the wind currents and the path of the sun, positioning a place as an intersection between nature and culture. Placemaking followed close on the heels of the urban renewal projects that had begun as part of President Lyndon Johnson's Great Society programs in the mid-1960s. Urban spaces had grown with, in most cases, little long-term planning, adapting to the rapidly changing needs and character of the American people in the 19th and early 20th centuries, precipitating the desire for urban renewal in the form of land redevelopment.

Urban renewal was often enabled by the legal instrument of eminent domain, in order to purchase private property for redevelopment projects conducted by the city. This could involve relocating businesses in order to create better-defined neighborhoods (or to further variegate existing neighborhoods), relocating residential areas, and demolishing old buildings to make room for new ones or new roads in an effort to fight traffic congestion. Eminent domain is an important concept in the consideration of place, and the British name for it illustrates why: compulsory purchase. The Fifth Amendment to the U.S. Constitution limits eminent domain to instances where the forced purchase of private property is for public use, and it requires compensation, but modern Americans are still sometimes taken aback by the practice when it occurs. Typically, it is necessary for the sake of public utilities (such as building a damn or reservoir) or to build roads or railways, but the function of eminent domain depends on the underlying idea that while a citizen may own land, the government owns it more; that is, that the citizen's right of ownership may be superseded by the government's. The milder form of this principle is the government's right to condemn a property, but this does not take the title away from the owner.

In the early 1950s, the federal government acquired many of the poorer parts of cities, splitting the cost with city governments (the federal government paid two-thirds of the price) and letting private developers build new housing. The city of Pittsburgh sought to redefine itself thusly, destroying large portions of downtown in what was once a factory city in order to build office buildings, parks, and a sports arena. It was an unexpected success, but when similar actions were taken in other cities, they were sometimes criticized for the effect they had in relocating populations: novelist James Baldwin referred to urban renewal as "Negro Removal." But urban renewal continued to accelerate, until the end of the 1960s, when organized opposition developed. One of the issues in the civil rights movement had been the unofficial creation of segregated neighborhoods as a result of city and federal policies and the tendency of real estate agents to steer buyers to one neighborhood or another according to their ethnicity. Johnson's War on Poverty policies changed some of that in order to attempt to curtail the creation of new ghettos.

All of this speaks not only to the importance of a sense of place but to its intersection with materialism. In the broadest sense, materialism is a philosophy that states that nothing matters but matter: that all phenomena observe from interactions between physical things. This continues to be an important meaning, but the word is often used in connection with Karl Marx's usage of the phrase *the materialist conception of history*, usually referred to as historical materialism. In Marx's framework, the fundamental material of the world is the social interaction among people (particularly between different social classes, such as workers and employers), and civilization originates out of the social interactions undertaken in order to meet the material needs of food, shelter, clothes, and medicine. All political, religious, cultural, and legal institutions originate with the fact that every day, human beings are principally concerned with meeting their material needs. Places develop accordingly, as shelter to house people, as places of work where labor can be done, as places where transactions are carried out. Attachment to place develops because of the association between place and the meeting of these material needs.

The term *sense of place* is used by sociologists, anthropologists, and cultural geographers to refer to the strong sense of identity that a place can have. Most cultures have sacred spaces of one kind or another, from places of worship to graveyards. Classrooms, libraries, and courtrooms take on a character similar to sacred space, in that there is the compelling idea that they are places that require and invoke a particular standard of behavior on their users. Materialism may motivate a sense of place. Architecture can encourage it, as the interior of historic churches quickly shows, or that of a library built a century or more ago. Throughout the 20th century, more and more places developed that were criticized for being "placeless," lacking any sense of place: chain stores, gas stations, shopping malls, college dormitories, condominiums, housing developments, all invoking Gertrude Stein's complaint that "there's no there there." Common to these places is that they have no sense of locality: there is nothing to show the way they fit with their surroundings, the way they would be different were they located in Philadelphia instead of Dallas.

A remedy to placelessness and to environmental concerns has been the greater inclusion of green spaces in urban planning, which began as part of "beautification" efforts but is recognized to have significant cultural and psychological value, as well as restoring plant life to city spaces and improving the quality of the air. Green spaces include parks, but also the greenways of the United States and Canada, where long strips of grassy land are used for recreation and bicycle traffic. The shrubbery and small trees alongside sidewalks help to prevent soil erosion and improve air and water quality and prevent desertification, as well as restoring a sense of place to an urban or suburban area. Incorporating green spaces in land use planning with an emphasis on their role in an overall ecosystem is often called "green infrastructure," a term that also applies to storm water runoff management through natural systems instead of concrete, and man-made urban systems that mimic or rely on natural systems.

See Also: Alternative Communities; Alternative/Sustainable Energy; Art as Activism; Auto-Philia/Auto-Nomy; Environmental Justice Movements; Home; Malls; Sprawl/Suburbs/Exurbs; Theoretical Perspectives.

Further Readings

Canizaro, Vincent, ed. *Architectural Regionalism: Collected Writings on Place, Identity, Modernity, and Tradition*. Princeton, NJ: Princeton Architectural Press, 2007.
Hague, Cliff and Paul Jenkins, eds. *Place Identity, Participation, and Planning*. London: Routledge, 2005.
Redclift, Michael. *Sustainable Development*. London: Routledge, 1987.
Relph, Edward. *Place and Placelessness*. London: Pion, 1976.

Bill Kte'pi
Independent Scholar

Sprawl/Suburbs/Exurbs

Suburbs, suburbia, sprawl—these words often evoke strong opinions, consternation, disgust, and, for some, remembrance of childhood. Scholars and the public alike critique the suburbs as elitist, segregated, unurban, unecological, monotonous, and oppressive, yet suburbs became the dominant urban form in the second half of the 20th century. Cultural, economic, and political forces helped produce the suburbs. Some theories emphasize the cultural dimension by looking at changes in attitudes, values, and desires of the urban elite in their move from the central city to the outer city. Other theories examine broad changes in the economy, including changes in production, job markets, and real estate markets. Yet other theories examine the connection between political decision makers and real estate developers, emphasizing the political decisions of suburban development. Central to suburban development is nature, both in terms of the transformation of city and nature as well as cultural ideas of nature. Because suburbs are a major urban form, one of the most pressing dilemmas is to what extent are suburbs inherently ecologically unsustainable, or whether they can be transformed into sustainable living spaces.

Early Suburbanization: Fleeing the City, Returning to "Nature"

During the Industrial Revolution in both the United States and England, cities had developed a host of environmental and social problems. In the early 20th century, the first urban researchers documented massive influxes of immigrants, poverty, crime, and pollution in the city. Urban researchers, in what became known as the Chicago School, borrowed models of analysis from the new field of ecology and applied them to the city. For them, the city was like an ecosystem that changed and developed according to the laws of the system. The Chicago School authors developed a theoretical paradigm called "human ecology" that privileged the environment as a determining factor in social behavior. Therefore, such urban quantities and qualities as population density and population heterogeneity shaped the development of the city. In this analysis, the city is a product of the "ecological" conditions of competition and selection around areas such as access to resources, economic base, and technological infrastructure.

Ironically, with regard to the question of ecological sustainability, the impetus for early suburban development was a return to more natural and rural living. During this time of massive urbanization and concentration, the first substantial suburbanization took place. Wealthier urban citizens, perceiving the condition of the inner cities as problematic, began building homes on the edges of the existing cities. An imagined and created nature underpinned these places. Today, we might not think of these areas as completely suburban but at the time, they were on the edge between rural and natural spaces and the city. For example, in San Francisco, areas such as the Sunset District, the area abutting Golden Gate Park, developed as a residential neighborhood away from the central business district, adjacent to the newly created park. Often, these neighborhoods were connected to newly developed trolley car networks to ferry the wealthier citizens to their jobs in the core and back to the perceived idyllic neighborhoods at night. Even in Chicago, many miles of newer suburbs and edge cities now surround the first suburban areas.

To account for these changing urban patterns, Ernest Burgess of the Chicago School developed the "concentric ring model": the inner circle is the central business district, then comes the industrial zone, then a zone of transition, then the working class or poorer residential zone, and finally the outer wealthy residential zone. Because the wealthy tend to live farthest from the central business district, as the city expands, the model predicts, each zone will expand with it. Since the theory's debut in the first half of the 20th century, it has been critiqued for failing to accurately model cities of his own time as well as for being out of date in regard to cities of the late 20th and early 21st centuries, as we will explore in following sections. Nonetheless, early urban researchers were vigorously trying to document, theorize, and analyze the rapid and massive changes in the urban form and condition. Furthermore, the idea of nature has been intimately tied to the imagining and actual building of suburban neighborhoods. These early suburbs were imagined to be an escape from the vice, bustle, and grit of the inner city.

Post–World War II Suburbanization

The scale and scope of suburban development intensified after World War II. This occurred for several important reasons. Whereas earlier suburban development followed trolley lines or water and sewer services, post–World War II suburbanization developed in tandem with the increased use of automobiles, then freeways, during the 1950s. Another significant factor was the expanding middle class, boosted by the gains of organized labor in the 1930s, the GI Bill, the strong postwar economy, and the mass production of housing units, which lowered costs. Ideas of nature continued to play a central role in these developments. However, at this time, the idyllic suburban dream could be sought by a much larger and less wealthy class of urban citizens. The building of suburbs rearranged rural and natural landscapes and became places where white working- and middle-class families could relocate. Suburban racial segregation was built and maintained by practices such as redlining, racial covenants, and the general patterns of "white flight."

Postwar suburban development created both ecological and social consequences, particularly with regard to race and gender. Moreover, suburbanization has played a strong role in creating and perpetuating "environmental racism," the social outcome where people of color disproportionately live in or near areas of pollution or industrial contamination. On the ecological front, rapid horizontal low-density development transformed farmland or more natural areas into housing tracts. This was not so much the destruction of nature but rather the large-scale transformation of nature and the city at the same time, which in many cases produced the conditions for potential vulnerability to hazards and natural disasters for residents.

For example, in Los Angeles, wide-scale suburbanization placed new residents in direct danger of earthquakes and fires. In Seattle, postwar suburbanization exacerbated the destruction of salmon habitat and, in doing so, undermined one of the principal culturally and economically significant animals of Puget Sound. Differently, in New Orleans, expansion turned marshes and swamps into neighborhoods whose only protection against hurricanes was a system of human-made canals, levees, and pumping stations. The failure of the protection system and the incursion into marshland proved disastrous during Hurricane Katrina. More ominously, the transformation of the swamps and marshes into neighborhoods has only produced more vulnerability for New Orleans as a whole, not just for the outlying neighborhoods. In general, during these crises, poorer residents tend to be more heavily impacted than wealthy residents. Suburbs do not replace nature; they transform the relationship between nature and the city, not in a simple good or bad way, but in a complex way. Suburbanization helped create new vulnerabilities for urban citizens, plants, animals, and ecosystems in unequal ways.

Postwar suburban development, much like its earlier phase, was constituted by racial and gender inequality. Racial segregation amplified. It was maintained by racial covenants, redlining, block busting, and the general phenomenon of "white flight," the outflux of white residents from the inner city. As a consequence, the inner city, inhabited predominately by African Americans, Latinos, and immigrants, began to decay, gaining greater levels of poverty and crime. This only served to reinforce the idea that the inner city was dangerous, polluted, crime ridden, and contained all the bad elements of urban civilization, thus also reinforcing the idea of the idyllic suburb. During this time, the urban/suburban dichotomy was solidified in common discourse, where urban was bad, unclean, and unnatural; and suburban was good, clean, and more natural.

Another dichotomy crystallized during this time, this one centered on gender. Urban was idealized as masculine. The city was a man's place to work and spend time in, whereas the suburb was idealized as feminine. The suburb, the space for homes, was women's domain: soft, safe, and domestic. Feminist urban scholars have documented how the gendered city has in fact become real, as well as critiqued these idealizations as manifestations of a patriarchal society. Moreover, feminist urban scholars debate the extent to which the suburban form is not only a manifestation of postwar capitalist economy but constituted by the patriarchal single-family living style of late 20th-century United States. The postwar suburban boom had great ecological, racial, economic, and gender ramifications.

Growth Machine

Through the 1980s and 1990s, suburban development proliferated. Critics named these hyper suburbs "sprawl," conjuring the image of unplanned development as a plague sweeping farther and farther out from the central city. Researchers turned their attention to the economic and political structures that encouraged such growth. Two closely related theories emerged to understand suburban growth, and urban growth more broadly. Both approaches wanted to account for the political economy of growth, that is, the close connection between political decision makers and economic decision makers when building the city (e.g., real estate developers, land owners, and residential consumers).

One set of researchers theorized that out of the general economic crisis of the 1970s, urban space (property, buildings, undeveloped land) became increasingly commodified as economic elites attempted to find new means of capital accumulation. This is known as the "spatial fix," meaning that during times of economic crisis, money is directed toward the owning, building, and redeveloping of cities. Real estate markets, housing construction, and redevelopment became extremely profitable. Competition between cities increased as cities maneuvered to attract industrial and commercial development and capital for redevelopment projects. As this happened, city governments became entrepreneurial as they enacted market-based policies and programs to attract capital investments in the city. In some places, California, for example, suburban strip mall development became the primary means by which small municipalities and unincorporated areas could gain tax revenue. Overall, these theorists argue that after the 1970s, capitalist economies expanded commodity production from industrial and agricultural products to the city itself, and in doing so, fundamentally altered the way cities are built.

The other set of researchers, not satisfied with deterministic economic explanations, examined what was called the "growth machine": a coalition between city government and developers with the purpose of sparking and perpetuating urban growth. Growth coalitions typically unify otherwise competitive elites into productive alliances focused on growth projects. In order to build the city, elites tentatively coalesce around a shared commitment to growth. This commitment to growth pays back, so to speak, to members. For government decision makers, it might be in the form of votes or increased revenue for the city, whereas for developers, it might be in the form of increased property value or large contracts to build subdivisions. In addition, professionals tied to universities or the arts might have a commitment to growth as a means of expanding their institutions. Interestingly, labor union bureaucrats (and perhaps some rank-and-file members) sometimes join growth coalitions in an effort to increase jobs. Just as in the wider national and global economy, growth becomes sacrosanct, as both a means and an end in itself.

As suburban development intensified and spread, a backlash of reverse migration to the central city by mostly young professionals and upper-class residents began to occur. This return to the city coincided with economic and cultural changes. Namely, property owners and developers recycled and recommodified the dilapidated older sections of the city. At the same time, changing cultural practices, such as living closer to work, enjoying the diversity of experience the central city offers, and delayed child rearing by the professional class brought people back to live in the central city. These new urban forms were given names such as urban villages and lofts. Reverse migration often involved the process of "gentrification." Gentrification is the process by which lower-value or income neighborhoods or former industrial areas are bought and redeveloped into higher-value residential and commercial spaces. This typically involves the displacement of the current residents, often along race but almost always along class lines. The preexisting residents face a double effect: rising property values and rents as well as changes in the cultural or ethnic makeup of the neighborhood. As a result, the suburbs, which were historically thought of as white urban spaces, have increasingly seen the relocation of immigrants, including Eastern Europeans, Asians, and Latinos as well as African Americans. As mainstream society and scholars increasingly demonized suburbs and called for a return to the city as a means to ameliorate social and ecological problems, suburbs have remained a place that an increasingly racially, ethnically, and economically diverse population calls home.

Suburbs and Sustainability

As we have seen, nature has been a prominent component of suburban development throughout its history. We find nature in narratives of the suburban elites' imagined idyllic suburb as well as with the actual existing social and natural transformations of city and nature that take place when building suburbs. Along with the rising ecological consciousness of the past 30 years has come an understanding that suburban growth represents a unique social and ecological dilemma. On the one hand, suburbs are a major urban feature of cities in the United States. Thousands upon thousands of people reside there and call them home. They are the spaces of everyday life for these people. They are also highly commodified spaces, important for the industrial capitalist economy, real estate capital, jobs, and as places of residence. And, as the 2008 housing market crash indicated, they are part of an unstable and volatile market where livelihoods and dreams can be dashed quickly. On the other hand, rapid horizontal expansion transformed the more natural spaces surrounding cities into hybrid spaces of nature and city. In doing so, people, animals, and ecosystems were often made more vulnerable to crisis, and at many times thrown into socioecological crisis. Therefore, the dilemma is this: to what extent society and natural processes, including the lives of other animals, can be reconciled in the suburbs, and in the city more generally.

It is only in the past 30 or so years that more and more people are realizing the dilemma and proposing some solutions. In urban planning, the so-called New Urbanism movement has sought to rectify the wayward horizontal development of the past 60 years by imagining changes to existing suburbs, for example, deemphasizing strip malls and the heavy use of cars, and at the same time finding ways for more people to live in a redeveloped higher-density central city. So-called Green Urbanism has gone even further to argue for deeper ecological planning in all aspects of urban development. Proponents of Green Urbanism believe that the city, including suburban areas, can be transformed into a socioecological sustainable place. In both paradigms a central role is given to planning, evoking the need for urban municipalities to take a leading role in transformation of the city.

However, critics in the loose field of urban political ecology challenge the urban planning paradigms for inadequately accounting for the ways in which existing political, economic, and cultural arrangements constrain attempts to make cities more sustainable. Urban political ecology more specifically argues that political, economic, and social arrangements shape urban development, and more importantly, that the categories "urban" and "nature" cannot be separated. Urban political ecology must be distinguished from Chicago School human ecology discussed above. Urban political ecology does not follow a systems approach that uses the metaphor of natural systems to understand cities. Instead, researchers acknowledge that there is a distinct urban ecology and largely examine how the city and nature produce one another. That is, ecological change and social change shape one another. Cities, in this view, are combined socio-natural constructions that are historically produced, both in terms of social and physical–environmental qualities. They argue that the qualities of physical and environmental change and the resulting conditions are not free from the particular historical, social, cultural, political, or economic conditions and institutions that coexist with them. Consequently, urban political ecology charts the inequalities of socio-natural changes, in which socio-natural change might reduce the stability of one place, social group, or ecology, while enhancing the sustainability of another. We saw this clearly in the above discussion of white flight suburban development and in gentrification.

Urban political ecology makes its goal to elucidate who or what gains; who benefits from; and who suffers in what ways from particular manifestations of social and natural processes and conditions. Moreover, urban political ecologists argue that environmental transformations are never independent from class, gender, ethnic, or other power struggles. Ultimately, urban political ecologists understand the dilemma of urban sustainability as an inherently political question. They argue that socioecological sustainability can only be achieved by a democratically controlled and organized production of city and nature.

A major debate within the urban studies literature has been to what extent there are distinct urban and suburban cultures. As seen throughout this article, it is hard to parse out political and economic forces from cultural forces. As has been shown, in the past 100 years cultural forces have played a tremendous role in shaping the suburban form. From the original nature narrative of the idyllic countryside to notions of the family to consumerism and malls, culture has played a large role in urban development. This is one aspect of the debate. The other involves whether there are in fact distinct urban and suburban cultures, or whether they are manifestations of larger American and global economic and cultural formations. For example, proponents of the suburban culture thesis have linked suburbs to what is called "car culture." A major hallmark of post–World War II suburbanization has been the building of roadways and highways. Some have argued that suburbanites have developed a suburban way of life, a set of cultural practices based on consumption habits, gender norms, work preferences, and racial and ethnic exclusion. A major critique of suburban culture, according to these authors, is that it is ecologically unsustainable. Mass production and consumption of houses, roadways, and strip malls, they argue, are tied to a way of life that is inherently unsustainable. Within this framework, solutions to this dilemma posit that in order to green the suburbs, there will need to be major changes in the ecological consciousness as well as daily cultural practices of those who reside in the suburbs. The debate continues on the uniqueness of suburban culture, but nonetheless, attention is being paid to the development of sustainable and ecologically conscious daily cultural practices.

There are no easy answers to the dilemma of urban sustainability. The dilemma is one of the challenges of the 21st century for urban studies and planning, but more generally for urban citizens, city governments, and national governments. The suburbs are a central feature of this dilemma. Natural relations have been a constituting force as well as a narrative in the construction of the suburbs; they will undoubtedly continue to play a significant part in the 21st-century city.

See Also: Alternative Communities; Auto-Philia/Auto-Nomy; Commuting; Environmental Justice Movements; Gardening and Lawns; Home; Malls; Space/Place/Geography and Materialism; Work.

Further Readings

Harvey, David. *The Urbanization of Capital*. Baltimore, MD: Johns Hopkins University Press, 1985.
Heynen, Nik, Maria Kaika, and Erik Swyngedouw. *In the Nature of Cities: Urban Political Ecology and the Politics of Urban Metabolism*. London: Routledge, 2006.
Keil, Roger. "Urban Political Ecology." *Urban Geography*, 24 (2003).
Molotch, Harvey. "The City as a Growth Machine: Towards a Political Economy of Place." *American Journal of Sociology*, 82 (1976).
Saegert, Susan. "Masculine Cities and Feminine Suburbs: Polarized Ideas, Contradictory Realities." *Signs*, 5 (1980).
Wirth, Louis. "Urbanism as a Way of Life." *American Journal of Sociology*, 44 (1938).

Nik Janos
University of California, Santa Cruz

Sundance Channel

The Sundance Channel premiered in 1996 as a vehicle for televising independent films. In 2007, it ventured into the world of green programming. By 2008, it was reaching 30 million homes. Rainbow Media, a subsidiary of Cablevision, bought the Sundance Channel on June 18, 2008, for $496 million. The U.S. network is part of Robert Redford's Sundance Institute, a nonprofit center for independent filmmakers. Sundance Channel is independent of both the institute and the film festival. A Canadian Sundance Channel debuted on March 1, 2010. It is independent of Rainbow Media, but the developer, Corus Entertainment, holds licensing agreements for the name and programming.

The Sundance Channel airs world cinema, independent feature films, documentaries, short films, and original programs, including news about the annual Sundance Film Festival. Sundance programming is uncut and commercial free.

CBS Corporation's Showtime Networks, NBC Universal's Universal Studios, and creative director Robert Redford collaborated on the network, named for Redford's role in *Butch Cassidy and the Sundance Kid*. But the collaboration did not last. The first hint of a sale came in 2007, when neither NBC nor CBS was willing to sell its share to the other, and an outside financial services firm started negotiating with potential buyers.

Rainbow also owns several other cable channels. Sundance is still available from some providers, including DIRECTV, which includes it in the Showtime Unlimited package. When Cablevision subsidiary Rainbow Media purchased the Sundance Channel for $496 million in stock and cash, the 12-year-old network reached 30 million homes. Redford remained tied to the network but not in an ownership capacity. Cablevision promised not to combine Sundance with the Independent Film Channel (IFC), Rainbow's film channel since 1994. Cablevision also owned American Movie Classics (AMC), among other channels.

In February 2010, Sundance announced that it was beginning scripted programming, emulating its sister AMC network. Among the ventures was a three-part series about Carlos the Jackal, acquired from a French company. The goal was to revitalize Sundance as an arts and entertainment network featuring short programming as interesting as the films shown in the festival. Sundance was also developing a crime drama set in South America, a suspense thriller about a medical researcher on the boundaries of human testing, an adaptation of *Fear of Flying*, and a series about a globetrotting female photojournalist. Nonfiction series include one about women whose best friends are gay males and a couple about designers, one incorporating native elements.

Sundance not only broadcasts but also creates its own programs. For instance, *Anatomy of a Scene* analyzes a scene from various perspectives, including acting, writing, directing, costume design, and production design. Documentaries and other films comprise 70 percent of Sundance programs, with the original series filling the other 30 percent.

Sundance programming deals with environmental themes as well as the arts. In 2007, Sundance introduced Sundance Channel Green. The Green was a three-hour block of environmental programming that sought to bring the same sex appeal to eco-themed programming as the Sundance film festival did to independent films. The debut began with a 30-minute *Big Ideas for a Small Planet*, a slick and fast-paced program on environmental problem solving. Following that was a full-length documentary, the first feature on the significance of peak oil. Redford claimed that the country was just ending a period of disastrous political policy, with basic environmental programs rolling back even as the ecology became more damaged. He sought to portray the grassroots efforts that persisted even during the environmental dark ages, the green solutions for home furnishings, buildings, transportation, fashion, food, and farming. The Green was his way of getting the stories out.

The effort included three hours of hosted programs on environmental topics in prime time. Both original series and documentary premieres dealt with ecology, green living, and environmental stewardship. Sundance Channel Green made Sundance Channel the first U.S. television network with a regularly scheduled significant component dedicated to environmental matters. *Change Agents* was an original documentary series devoted to stories and characters on the leading edge of environmental action and creativity. Sundance also offered interstitial series on various issues with suggestions for creative methods of changing the viewer's life. The third component was feature-length documentaries and news specials on environmental topics.

The Green features were a departure from traditional environmental stories in that rather than emphasizing the disaster looming, as had been the case for at least 20 years, each one was focused on innovation and solutions, and profit-making ones at that. The term is *activist media*. The Green lineup includes such titles as *Big Ideas for a Small Planet*, *Carbon Cops*, *Eco Biz*, *Is Your House Killing You?*, and *Green Porno*. *Green Porno*, the comedy about the mating habits of various creatures, spawned *Seduce Me,* about the seduction rituals of wildlife. The star acts out the behaviors in a scientifically accurate and entertaining way.

Redford contended that environmentalism had reached the tipping point, and it was not a fad destined to fade. Even Wall Street was environmentally friendly. And the signs of environmental crisis were more apparent—gas shortages and higher prices or more intense weather or health deterioration from environmental problems. But to maintain the bandwagon required education and engagement, encouragement of innovation for new industries, and new jobs. Most important, the movement had to reach the young before they became jaded.

See Also: Environmental Media Association; Redford, Robert; Television, Cable Networks; Television Programming.

Further Readings

Courtenay, Erin. "Robert Redford Announces Green Sundance Channel." *Culture & Celebrity* (July 21, 2006). http://www.treehugger.com/files/2006/07/robert_redford.php (Accessed September 2010).
Little, Amanda. "Robert Redford Chats About the New Green Programming on the Sundance Channel." Grist.org. (April 16, 2007). http://www.grist.org/article/redford (Accessed September 2010).
Schults, Chris. "Sundance Channel Green." Grist.org (July 20, 2006). http://www.grist.org/article/sundance-channel-green (Accessed September 2010).
Stelter, Brian. "Cablevision Unit Buys Sundance Channel." *New York Times* (May 8, 2008). http://www.nytimes.com/2008/05/08/business/media/08sundance.html (Accessed September 2010).
Sundance Channel. http://www.sundancechannel.com (Accessed September 2010).
Szalai, George. "Sundance Channel Starts Scripted Programming." *The Hollywood Reporter* (February 9, 2010). http://www.hollywoodreporter.com/hr/content_display/television/news/e3i4fe3d67e44c8b3ad9c2bce2bb0400729 (Accessed September 2010).

John H. Barnhill
Independent Scholar

Sustainable Art

Sustainable art is an artistic genre interconnected with the analogous branches of green art, ecological art (also known as eco art), land art, art in nature, Earth art, bio art, crop art, and the like. The field is in harmony with the key principles of sustainability, including ecology, grassroots advocacy, nonviolence, and social justice. Beginning in the 1960s and 1970s, these contemporary green artistic movements rose to prominence, united in their substantial depiction of ecologically and environmentally driven content, while distinct in their awareness regarding the environmental impact and sustainability of the specific medium and materials used. Sustainable art may also be thought of as art produced with thought of the impact of the art and its reception with regard to its biophysical, cultural, economic, historical, and social environments.

Background

With roots dating back to the Industrial Revolution, both the process and the product within sustainable art seek to renegotiate the relationship between humans and the natural environment. Sustainable art's origins can be traced to the 1960s and 1970s, when conceptual art stressed dematerialization and questioned the art system's functioning. The renegotiation of the relationship between humans and the natural world is attempted through several generalized compositional mechanisms, such as paramount and monumental scale of object placement. This compositional decision has become a characteristic trait of certain sustainable, site-specific sculptural pieces that closely border on architecture. Another characteristic trait of environmentally focused artwork is the intentional use of recycled or

Solar trees, like this one in Gleisdorf, Austria, are a more aesthetically pleasing production of energy than wind turbines or solar farms, and are an example of a significant creative endeavor within the environmentalist movement.

Source: Anna Regelsberger/Wikimedia

freely found materials. While this feature may have evolved from the convenience of using free or found materials, some environmentalist artists have chosen to do so in a manner that has, in fact, made the decision significantly inconvenient.

An aspect of environmentally sustainable art that has made the genre difficult to classify is the definitional subjectivity of the word *environment* and whether its use in this context pertains exclusively to the natural biosphere, and whether the historical, political, cultural, and religious context is to be regarded as integral to the artistic content portrayed within this movement. Little consensus exists regarding what comprises sustainable art. Some believe a focus on art and sustainability better captures the movement than the term *sustainable art*. Still others explicitly refute attempts to label the movement, preferring instead to focus on artistic work that inspires thought about sustainability.

Sustainable Art in Practice

The sustainable art movement initially gained prominence in urban locations. The initial movement focused on artists seeking to bring natural elements back into the city sprawl in an act of reclamation. Sustainable art has since shifted to also incorporate exhibitions of artwork in rural and undeveloped landscapes. One genre within the realm of sustainable art is known as green sculpture or land art. Artists working within this field have produced often massive and monumental art. Sometimes, however, the focus has been upon minimalistic works characterized by a great reshaping of the physical terrain through use of heavy machinery such as bulldozers. Others have sought to perfect the landscaping practices of hedge-trimming and ice-carving sculptural forms, creating whimsical and thought-provoking works in unexpected locations.

The field is not without controversy. Some environmentalist artists have questioned whether the objective in producing sustainable artwork ought to focus strictly on the minimalization of environmental degradation. The result of this line of questioning is perhaps best exemplified by the installation of so-called Solar Trees in numerous cities across the globe. Solar Trees use metallic structures that mimic the dendritic shape of organic flora, supporting an array of photovoltaic cells for the solar production of electrical energy. These structures, while having required advanced scientific and technical engineering, do in fact constitute a significant creative endeavor within the environmentalist movement. Whereas other forms of sustainable energy production, such as contemporary wind turbines and solar farms, have attracted criticism for their brutal and aesthetically unappealing subjection of agrarian and uncultivated landscapes, these systems have now begun a process of adaptation so that they may not only incorporate the aesthetic element of sustainable art, but continue to serve as mechanisms for the production of energy from renewable resources.

Other Explorations

While the contemporary environmentalist shift in artistic production gained prominence throughout the 1960s and 1970s, the role of the natural environment is historically inseparable from the advent of artistic production, as evidenced by Paleolithic European cave paintings depicting horses and other native flora and fauna. East Asian watercolor scroll/brush paintings also depict organic formations of landscape, and botanical and wildlife illustrations are likely the oldest and most significant tradition in watercolor painting. German Renaissance artist Albrecht Dürer featured a wealth of environmentally thematic content within his body of work, ranging from the botanic to the bestial. Wildlife illustration reached its zenith in the 19th century with the advent of photography. Noteworthy pioneers of such watercolor illustration include John James Audubon and Paul Cézanne. More modern pioneers of sustainable art include Christo, Duane Hanson, Michael Heizer, Edward Kleinholz, George Segal, and Robert Smithson. As with the field itself, debate continues regarding precisely who is a practitioner of sustainable art.

Sustainable art continues to attract interest, both from within the visual art world and from environmental activists. International symposia devoted to sustainable art have been held at Central European University in Budapest, Hungary, and at the German Society for Political Culture at the Art Academy of Berlin. Although sustainable art has often focused on temporary installations of limited duration, recently, increased interest has resulted in more lasting, permanent works. Building on the work of site-specific artists, many real estate developers now incorporate landscaping and permanently sited sculpture. This type of sustainable art has proved popular with prospective tenants and has enjoyed the support of local zoning commissions, which frequently mandate developers spend a certain percentage of the overall cost of a development project on sustainable art.

The lines between environmental art, environmental sculpture, and sustainable art have become increasingly blurred. As a result, some believe that environmental sculpture, that which is significantly larger, or smaller, than human scale, is used to create a reaction on the part of the viewer. Others, however, maintain that such works have little to do with sustainability and thus should be considered a separate category. Continuing debate regarding this issue is anticipated.

See Also: Art as Activism; Artists' Materials; Music; Sculpture.

Further Readings

Barbero, Silvia, Brunella Cozzo, and Paola Tamborrini. *Eco-Design*. Potsdam, Germany: Ullmann, 2009.

Ecoartspace. http://www.ecoartspace.org (Accessed December 2010).

Smith, S. and V. Margolin. *Beyond Green: Toward a Sustainable Art*. Chicago, IL: Smart Museum of Art/University of Chicago Press, 2009.

Stephen T. Schroth
Daniel O. Gonshorek
Knox College

T

Technology and Daily Living

Over the past two decades, green culture has typically come to mean making a conscious effort to reduce negative environmental impact and move mankind toward a more sustainable future. This has resulted in companies making products that are energy efficient, biodegradable, and therefore more sustainable. Such products today are routinely used in daily activities of one's life such as work, leisure, cooking, commuting, and so forth. For example, a company in Canada has been reported to manufacture ecofriendly foam used in sofas and other furniture. Similarly, several manufacturers today provide solar-powered lamps and other similar electronic appliances. Some automobile manufacturers are also producing cars that are more energy efficient and made of biodegradable material. Thus, the use of green technology and daily living is primarily affected by people and organizations modifying their purchasing, manufacturing, investment, and consumption decisions through the use of technology. Organizations advocate the use of technology in our daily routine more frequently than a few decades ago. For example, electric light bulb manufacturers have recently shifted their product lines to more energy-efficient lighting solutions and have advocated the use of energy-efficient products to the general public. Several computer and battery manufacturers have also started recycling programs for consumers to return used batteries and parts. On the whole, it seems organizations and individuals are making a conscious decision to change or improve aspects of their daily lives to incorporate more sustainable elements.

Worldwide efforts have been made by companies, governments, and not-for-profit organizations to use technology that will help reduce the carbon footprint. *Carbon footprint* is a term that quantifies how much carbon emission a particular process or a person is responsible for: the less one's carbon footprint, the less the impact on the environment. Historically, carbon emissions from vehicles, household appliances, and coal-powered plants became the topic of examination and were seen as cause for environmental pollution, ozone depletion, and global warming. To tackle these issues, better technology was seen as one of the important drivers of change. Thus, over the past decade, technology has primarily been seen as a solution to reduce the impact of humans on planet Earth. *Sustainability* and *carbon footprint* are two terms that are driving much of our logic of using green technology in day-to-day activities.

Companies in the technology, architecture, and construction and building industries have been at the forefront of advocating and diffusing the use of green technologies.

Architecture, building, and construction industries today use roofing, electrical, and solar technology to reduce the overall carbon footprint of the house. For example, Chicago's McCormick Place has a 3.5-acre rooftop engineered for more than 40,000 plants. Another example is that of the Atlantic City Convention Center (ACCC). It is the largest rooftop installation in the United States, capable of producing 2.36-megawatts of solar power, and provides approximately 26 percent of the convention center's electrical power needs. On the whole, architecture and construction companies are paving the way for bringing these elements closer to our daily lives by putting emphasis on the use of organic and biodegradable materials in the construction of houses.

However, critics argue that technology alone should not be seen as a solution to sustainable living; instead, the focus should be on lowering nonessential use of products. They claim that in a consumerist society, we are prone to simply consume technology rather than critically assess its need in our daily lives. For example, a teenager might buy a biodegradable music player, CD player, phone, and a laptop without realizing that he or she is actually increasing his or her carbon footprint because if he or she critically assesses the situation, in theory he or she could do without one or two devices.

Overall, there is general consensus that collective action of organizations and individuals should benefit the environment. For a company, this means making a conscious effort to reduce overall negative impact of doing business on the environment while for the consumer, it is using products that harm the environment least. In either case, the primary focus has been the use of technology in routine activities by a company or a person to minimize environmental impact over the long term. Thus, it is in this context that that we see a move away from traditional ways of thinking of using technology. And the way technology affects how we live our daily lives, and changing the way we consume furniture, lighting, gardening, and housewares.

See Also: Green Consumerism; Institutional/Organizational Action; Political Persuasion; Shopping.

Further Readings

Abdul, Umair. "Eco-Friendly Foam for the Furniture Market." *Canadian Plastics*, 66 (2008).

Doyle, Arthur. "Taking Green Initiatives to the Top." *Successful Meetings*, 58/3 (2009).

Herndl, Carl and Stuart Brown. *Green Culture: Environmental Rhetoric in Contemporary America*. Madison: University of Wisconsin Press, 1996.

Stone, George, Cameron Montgomery, and Japhet Nkonge. "Do Consumers' Environmental Attitudes Translate Into Actions: A Five Nation Cross-Cultural Analysis." *Society for Marketing Advances Proceedings* (2008).

Kaustubh Nande
Independent Scholar

Television, Advertising

During the 2010 Super Bowl, Audi debuted its "green police" ad. The 60-second ad shows ordinary citizens being arrested for using plastic instead of paper, throwing away batteries,

Only 25 percent of consumers have bought a green product other than organic food and energy-efficient lighting, and poorly performing compact fluorescent light bulbs could reinforce the notion that energy-saving products are worse than standard products.

Source: Rutherford County, North Carolina (www.rutherfordcountync.gov)

not composting orange rinds, using incandescent light bulbs, and setting their hot tub thermostats too high. The action occurs while Cheap Trick's lead singer Robin Zander sings redone lyrics to the group's 1970s mega-hit "The Dream Police." Audi's "green police" Super Bowl ad, which promotes Audi's diesel A3 TDI (which claims gas mileage of 42 miles per gallon on the highway and a 30 percent reduction in greenhouse gas emissions) ranked sixth best in *USA Today*'s ad meter consumer evaluation of Super Bowl ads. The commercial was viewed by nearly 125 million Super Bowl fans in the United States alone. This ad is important because it signals how far television advertising tackling environmental themes—so-called green TV advertising—has come.

Advertising is a marketing tool that promotes a product, brand, or service to a target audience. Advertising is most effective when it communicates the best points of a particular product or service to a targeted audience. Television advertising is the most popular—and ubiquitous—ways of promoting any brand, service, or product. Because of its audiovisual nature, television ads (which are usually 30 to 60 seconds in length), are still the preferred method for advertisers to try to reach their target audiences. According to Forrester Research, advertisers in the United States were expected to spend $69.5 billion on television advertising in 2010.

While television advertising performs an important economic service for consumers, it is not without controversy. TV ads have been attacked for unfairly influencing children to consume unhealthy products, for promoting unrealistic portrayals of female body types, and for undercutting political discourse with the plethora of candidate "attack" ads. In the area of TV advertising of environmental issues, some advertisers have been accused of "greenwashing," the deceptive use of green advertising to promote a misleading perception that a company's policies, goods, or services are environmentally friendly. Dolls are often wrapped in more plastic, and can be sending a message to little girls that they should be plastic, too. For example, Mattel tried to market the Barbie BCause doll as "green" because it used a line of accessories made from repurposed excess fabrics and trimmings left over from other Barbie products. The doll itself, however, is still plastic and wrapped in plastic, not to mention that the dolls are assembled in China, where production processes and labor laws are far below U.S. standards.

Corporate social responsibility (CSR) has been defined as "a company's commitment to minimizing or eliminating any harmful effects and maximizing its long-run beneficial impact on society." Though not everyone believes that companies should act as good citizens (cf. Milton Friedman), the CSR concept has gained more support in recent years following corporate scandals at Enron, Tyco, and Lehman Brothers. In those scandals, the companies highly (and illegally) inflated the value of their companies, and then upon regulatory and financial market scrutiny, watched the value of their stock tumble into bankruptcy. As a result, leaders at Enron (such as Kenneth Lay) and Tyco (Dennis Kozlowski) were found guilty of defrauding their shareholders, while the corporate management at Lehman Brothers is under investigation.

One area of CSR is environmental marketing, also known as green marketing. The United States makes up only 5 percent of the world's population, yet its consumption patterns have caused a number of concerns for environmental researchers and critics alike. The average U.S. citizen consumes five times more than an average Mexican, 10 times as much as the average Chinese, and 30 times more than the average Indian. This consumption pattern is one of the reasons socially responsible companies are looking to implement green marketing strategies to support green consumption and behaviors. As Ben Cohen of Ben & Jerry's ice cream company has noted, "Corporations and advertisers must be part of the solution to environmental problems." Green marketers modify their products, develop new products, or utilize persuasive communication tactics to promote their products (and, by extension, the organization) as environmentally responsible in an effort to attract customers.

Since the 1990s, there has been an increasing demand for new products designed to protect the environment. Consumers have been increasingly motivated by environmental concerns and have engaged in conservation activities such as recycling. Public opinion polls consistently indicate that public concern for the environment continues to grow. Consequently, marketers have rushed to take advantage of this new trend.

"Green consumers" are those whose purchase behavior is influenced by environmental concern. Such consumers have concerns about the environment that go beyond purchase and consumption patterns to such things as production processes and product disposal issues. In other words, green consumers are concerned for the preservation of the environment and a noninvasive lifestyle.

Consumers' environmental concerns have influenced marketers and advertisers to create new green products, ecofriendly packaging, and new environmental campaigns, such as General Electric's "Eco-Imagination" campaign, which began in 2005. Green advertisers attempt to promote green consumerism by making consumers believe that by purchasing environmentally friendly products, they are contributing significantly to protecting the environment. Many consumers appear quite willing to adopt green attitudes and beliefs because it is socially acceptable and chic—a psychographic characteristic that advertisers are happy to tap into.

Examples of Green Television Marketing

For example, in General Electric's (GE's) "Eco-Imagination" campaign, one ad, "Cliff Diver," depicts what would happen to a cliff diver if they used traditional ways of landing a plane; that is, the plane would "step down" in terms of altitude. The diver reaches a certain level, then is suspended, until they are released to the next altitude step. With GE's new True Course Flight Management system, such a step-down process is unnecessary. It is also more fuel-efficient and reduces carbon emissions.

The ad is a reflection of GE's commitment to a cleaner environment, more efficient sources of energy, reduced emissions, and abundant sources of clean water. The company has devoted more resources to solar energy, hybrid locomotives, fuel cells, lower-emission aircraft engines, lighter and stronger materials, and efficient lighting and water purification technology. When GE announced the campaign, it pledged to improve its own energy efficiency and environmental performance.

In is noteworthy that the environment itself has become a "product," or something consumable that companies market and benefit from by developing strategies to satisfy consumers' needs. Traditionally, green marketing has been seen as a form of social marketing. Social marketing uses tools, techniques, and ideas derived from conventional marketing in order to fulfill societal goals with an underlying intention to influence public behavior. Social marketing has influenced a range of different fields, such as health, injury prevention, environment protection, and community service. These efforts have proved quite effective. "Social advertising," of which green advertising is a form, relies on appeals that communicate cost reductions to consumers in exchange for engagement in pro-social environmental behaviors; or, at least, the *intention* to act pro-environmentally.

Consumer behavior research suggests that it does not only take the willingness to get involved to change behavioral patterns, but that it also requires that consumers' needs and preferences for ecologically friendly products must be met in terms of the availability and affordability. More importantly, marketers and advertisers need to communicate with clarity the green product's environmental benefits, product performance, and other attributes. Beyond this, companies have to adhere to governmental policies (such as those of the Federal Trade Commission) regarding guidelines covering environmental marketing claims.

Scholarship on the efficacy of green advertising has existed since the 1990s, though research in the area has not been extensive. One area examined has been the believability or credibility of green advertising. Research has demonstrated that much of green advertising has suffered from low credibility scores. One reason for this is that green messages in advertising are perceived as vague and ambiguous. In addition, many consumers are skeptical about the motives behind green marketing activities. British Petroleum (BP) has tried to position itself as a green marketing company with its "Beyond Petroleum" campaign. Unfortunately, the April 2010 oil spill in the Gulf of Mexico has damaged the company's—and its campaign's—credibility. Because green consumers have an anti-corporate attitude, and distrust advertisers, their consumer behavior tends to be unpredictable, posing a major challenge for green marketers.

IBM and Green Advertising

Another company that has moved into the area of green television advertising is IBM. In September 2008, the company debuted a TV ad campaign that illustrated that being green was good business. For example, in "Tree Huggers," a female subordinate is explaining her corporate environment proposal to her superior. At first he is dismissive, saying that it would be good for the "tree huggers." She replies that the proposal will save the company millions of dollars. Up to this point in the commercial, the visuals are in black and white. Upon hearing that the proposal will save millions of dollars, colorful animated characters (including plants and animals) begin to appear.

In "Elevator," the same female in "Tree Huggers" gets on an elevator with a male coworker. She is accompanied by a number of animated characters. When the coworker asks what is going on, she says the company is going "green" with software that will save millions of dollars in energy costs. The male asks how much the company spent on energy the previous year. She replies, "millions." She then says, "Get used to it." The tagline for the ad is "Stop Talking. Start Saving."

The green TV ads are a reflection of the actual "green" movement occurring within IBM itself. In 2009, IBM created an industry alliance with key leaders in metering, monitoring, automation, data communications, and software to provide smart solutions for energy, water, waste, and greenhouse gas management. Charter members of the Green Sigma(TM) Coalition are Johnson Controls, Honeywell Building Solutions, ABB, Eaton, ESS, Cisco, Siemens Building Technologies Division, Schneider Electric, and SAP. The coalition members will work with IBM to integrate their products and services with IBM's Green Sigma solution. According to IBM, Green Sigma is an IBM solution that applies Lean Six Sigma principles and practices to energy, water, waste, and greenhouse gas (GHG) emissions throughout a company's operations, including transportation systems, data centers and IT systems, manufacturing and distribution centers, office facilities, retail space, and research and development sites. It combines real-time metering and monitoring with advanced analytics and dashboards that allow clients to make better decisions that improve efficiency, lower costs, and reduce environmental impact. In other words, the TV ads are not just rhetoric, but representative of IBM's corporate philosophy. For IBM, they have stopped talking and started saving.

Subaru and NBC's Green Week

One of the most extensive efforts at green advertising has been that of Subaru. For most consumers, the process of choosing among the numerous car brands and models and ultimately purchasing a car is a long one that entails searching for lots of information due to the high financial risk inherent in spending thousands on a car. Thus, although research shows that single car commercials are regarded as relatively weak in changing consumers' attitudes, the collective force of all the commercials created for each brand is more influential in helping shape the brand image in consumers' minds.

Consumer research shows that the average car consumer uses both cognitive and affective decision making when purchasing a car, with a consumer's cognitive processes ultimately playing a more dominant role. Therefore, the features that advertisers spotlight in car commercials can influence how consumers categorize different brands in their memories. This categorization process can then play a role in their decision-making process when the time comes for them to purchase a car.

Thus, the sponsorship of NBC's Green Week by car manufacturer Subaru has major implications for viewers of NBC prime time, as well as NBC Universal-owned cable networks CNBC, MSNBC, NBC News, NBC Sports, (now Syfy) Channel and Sundance Channel, Bravo, USA. Part of its "Green is Universal" program, NBC's Green Week is a two-week initiative that airs twice a year, during which green topics are integrated into the network's shows, such as discussing recycling on *The Today Show.* Though critics dismiss it as a marketing ploy, Green Week appears do a lot of good. The integration of "green" storylines and topics into most all of the network's programs—which represent a substantial group when all of the cable networks NBC owns are taken into account—at the very least raises awareness among TV viewers of "green" issues. As consumer behavior research reveals, the first step in changing attitudes is for consumers to pay attention to the advertising information, to motivate consumers, and to have the opportunity to change their attitudes.

Subaru's sponsorship of Green Week cost the company a reported $10 million. NBCU created the commercials for Subaru, which feature environmentally conscious Subaru owners, and seamlessly integrate the commercials into the beginning or end of the commercial breaks by somehow connecting the copy or dialogue used by the Subaru owners to the given NBC-owned network on which the commercial aired. Consumer behavior research demonstrates that such tactics can increase consumers' exposure to a commercial because consumers will be more likely to watch a commercial at the beginning or end of the commercial break. If a consumer thinks the commercial is part of the show, the commercial's effectiveness will also be enhanced.

Subaru also benefits by being the dominant car manufacturer sponsoring Green Week, thereby not allowing consumers to compare similarly environmentally friendly car models manufactured by competing brands. In these Green Week commercials, Subaru frames its brand of cars as "environmentally friendly," specifically around the goal-derived category held by a growing number of consumers of decreasing their ecological footprint. By priming the "green" attributes of its cars in these commercials, which are so ingeniously integrated into a number of NBC's extensive lineup of shows, Subaru frames its brand as being superior in that frame of mind (environmentally friendly). Though the Toyota Prius has developed such a strong brand image as an environmentally friendly car, as well as the Smart Car, Subaru clearly foresees a stronger brand image and ultimately increases in sales—legitimate ends to justify the means of spending millions on this sponsorship. Subaru's sponsorship of Green Week is an incredibly genius move by all those involved in its brand development and media planning.

As media planners are constantly being forced to craft new ways of reaching consumers through ever-evolving mediums, this sponsorship is so effective both in increased likelihood of exposure of consumers to Subaru's Green Week commercials and from the creation of a strong brand image in alignment with this green initiative. Consumers are often lazy and do not pay attention to details, so merely seeing the Subaru logo as they fast-forward through the commercials of a program during Green Week can create the association between Subaru and their goal-derived category of "environmentally friendly."

The important issue of responsibility herein is the agenda employed by Subaru of aligning itself with a green initiative to be associated with consumers' goal-derived category of "environmentally friendly" or similar associations in consumers' minds. The sponsorship of Green Week has, according to an NBC analysis of Nielsen IAG data, brought Subaru increased recall of commercials by consumers, speaking to the effectiveness of the advertisements run during Green Week.

However, the Subaru campaign has drawn critics, who charge that the company has a great responsibility to deliver on the brand image they are building so effectively through their Green Week sponsorship. The fear is that the selection of sponsors for Green Week was based more on financial offers to NBC than on truly being "green."

The 2010 Green Week sponsors all seem to practice what they preach in terms of employing green practices to be worthy of green images. Care needs to be taken, however, that future Green Weeks by NBCU or similar initiatives by other networks are careful to make sure that companies "walk the walk" and not just "talk the talk" of environmentalism in their advertising. This has major implications for consumers, who very well may make consumption decisions based on the associations they form about sponsors of initiatives like Green Week.

Consumer Reaction to Green Advertising

In terms of what elements in a green advertising message consumers find important in their purchase decision, one study divided consumers into high-involvement consumers versus low-involvement consumers. The researchers found significant differences between the two groups. The high-involvement consumers perceived the green ads to be believable, favorable, and good, while low-involvement consumers perceived the opposite. Low-involvement consumers also perceived the brand as less favorable. High-involvement consumers liked the following green message elements: product recycling symbols, promotion of the company's image, and describing environmental benefits. By contrast, low-involvement consumers preferred elements such as promoting a company's environmental claims.

Research has identified and categorized seven different kinds of ad appeals used by advertisers to meet their objectives. The first category, zeitgeist, consisted of appeals that simply reflected the mood of the times by implying a pro-environmental stance. For example, some ads promoting a green product contained such statements as "environmentally friendly," "recyclable," and "we care about the environment" without giving any further information. The second category was emotional appeals, where five types of emotional appeals were identified: fear, guilt, humor, self-esteem, and warmth. Rational appeals and appeals that emphasized the financial aspects of environmentalism were coded as a third category. Typical financial tactics were coupons, premiums, contests, or cause subsidies for buying green products. The fourth category was organic appeals highlighting the health aspects of environmentalism or emphasizing the goodness of "natural" products and ingredients. Corporate greenness was the fifth category; ads that emphasized the advertisers' commitment to social welfare were coded as demonstrating social responsibility, whereas ads that described the advertisers' environmental activities were coded as demonstrating green actions. The sixth category was testimonial appeals, whereby advertisers used a celebrity, an expert, or an everyday consumer to endorse the environmental benefits of the product/service. The seventh category was comparative benefit appeals. Some ads either directly or indirectly compared a green product or service with another product or service, and mentioned the benefits of the green one in comparison with others.

In terms of issues, among seven common issues used in green advertising appeals—concern for waste, concern for wildlife, concern for the biosphere, concern for popular issues, concern for health, energy awareness, and concern for environmental technology—college educated consumers (especially college women) responded particularly well to wildlife and waste appeals. Additionally, working adults preferred waste and energy appeals.

One key experiment tested the efficacy of the green appeals. The experiment used a two (issue proximity: high versus low) times two (guilt appeals: guilt appeals versus non-guilt appeals) factorial design conducted to explore green advertising effectiveness for consumers with different levels of environmental involvement. The results indicate that guilt appeals are more effective than non-guilt appeals when issue proximity (how psychologically close the issue is to the consumer) is low in advertising. On the contrary, non-guilt appeals are more effective than guilt appeals when issue proximity is high. The results show the reverse is true for consumers with low environmental involvement. There are two reasons that can explain the results. Some believe that these consumers tend to rely on peripheral cues, advertising external cues aim at increasing their motivation to attend to the provided information, and that they have poor attitudes toward green advertising.

Green advertising can also affect the perception of an organization. Research has shown that company image perceptions play a significant role in shaping consumer response to environmental ads. When consumers believed a company was environmentally concerned, consumer perceptions of advertiser image, product image, and purchase intent were more favorable.

Conclusion

Despite the efficacy of green marketing and green advertising, there is still a long way to go. Research shows that only 33 percent of consumers are ready to buy green products and only 25 percent have bought a green product other than organic food and energy-efficient lighting. Even green goods that have caught on have small market shares. Organic foods account for only 3 percent of all food sales, and green laundry detergent makes up only 2 percent of household cleaner sales. And, despite millions in advertising expenditures and government incentives, hybrid cars make up only 2 percent of the auto market.

Interestingly, consumers are inclined to act green: 67 percent of respondents in a recent survey say they prefer to do business with environmentally responsible companies. Yet, these same consumers expect companies to lead the way. Research found that 61 percent of consumers want corporations to take the lead in tackling environmental issues such as climate change. That's why General Electric's "Eco-Imagination" campaign has received such good reviews from consumers and advertisers alike.

In order for green marketing and green advertising to grow, businesses are going to need to make consumers aware of ecofriendly products. Green advertising needs to show how and why a particular product qualifies as being green, and merely using green packaging will not suffice. In addition, many green products have negative perceptions associated with them. A Roper Poll revealed that 61 percent of green products are believed to perform worse than conventional items. Consumer experience with poorly performing hybrid cars and compact fluorescent light bulbs only serves to reinforce that perception. The best advertising in the world will not overcome a poor product or service, even if it is environmentally friendly.

See Also: Advertising; Corporate Green Culture; Green Consumerism; Hybrid Cars; Internet, Advertising, and Marketing; Organic Foods; Popular Green Culture; Shopping.

Further Readings

Banerjee, Subhabrata, Charles S. Gulas, and Easwar Iyer. "Shades of Green: A Multidimensional Analysis of Environmental Advertising." *Journal of Advertising*, 21/32 (1995).

Hartenberger, Michael. *Green Advertising of Energy Companies: A Critical Look Beyond the Text*. Saarbrücken, Germany: VDM Verlag, 2010.

Meister, Mark and Phyllis M. Japp. *Enviropop: Studies in Environmental Rhetoric and Popular Culture*. Westport, CT: Praeger, 2002.

Rod Carveth
Fitchburg State University

Television, Cable Networks

Green cable has been talked about since the mid-1990s, but serious green cable is a recent development, and there are only two major players, one more major than the other, and their programming differs markedly. The first to commit was Sundance, but its commitment was for only three hours. Planet Green followed as a full-time ecolifestyle network.

Precursor Efforts

A market research study in 1993 reported that 53 percent of American adults were environmentally aware. A 1994 poll indicated that 20 percent of Americans were active either as environmentalists or as green consumers. Another 35 percent were amenable to buying green products if they were advertised as such. This appeared to be a market for cable to tap.

In February 1995, three around-the-clock cable environmental channels were in the works. Planet Central offered investigative magazine formats, nightly news, and original programming from street-smart filmmakers with a grassroots environmentalist element. It struggled to find a cable system willing to take a chance on it. Baltimore-based Ecology Channel offered a conservative alternative to the green activism of Planet Central. The Ecology Channel began in December 1994 with a lifestyle emphasis. Its lineup included children's shows, environmentally friendly infomercials, science fiction, and news. Six hours a day, it offered suburban and rural gardeners *The Home Gardening Club*, which included direct ordering of supplies and tools.

Queens-based Earth Television Network (ETN) was available to satellite subscribers only. Founder Chaz Scardino claimed that he did not need cable because he could get 400,000 viewers from the 8 million dishes in use in North America. He also expected to get advertising from the national network of holistic healers and health food stores. His plan was to begin with three hours of airtime a day, expanding to six when feasible. He was also considering the unproved technology of Direct Broadcast Satellite, which was to provide TV through a desktop antenna. ETN was to offer news, product information, and magazine-format environmental shows.

Another possible contender for the environmental viewer was the corporate-sponsored Outdoor Life Channel, a rod-and-gun network based on Times Mirror's *Outdoor Life* magazine. Times Mirror, owner of *Ski*, *Yachting*, and *Field and Stream*, acknowledged that its environmentalism was intended to preserve fish and game for sportsmen, but expected to draw environmentalists to the shows that did not show anything getting killed, the shows on rock climbing and snowshoeing and such. And its magazines had 24 million readers, while its cable system had 1.3 million customers. It was rumored to be negotiating the sale of its cable system, with final purchase contingent on the successful buyer offering *Outdoor Life* to a customer base of 3–4 million subscribers. For the previous two years, Times Mirror had provided programs such as *Saltwater Sportsman* to ESPN and *Ocean Planet* to the Discovery Channel and in 1995, was developing programs with CNBC and the SciFi Network. Outdoor Life Channel later became a Comcast property and had 55 million subscribers but little to no ecological content.

The proposed eco-channels struggled for access. Operators worried that viewers would be as likely to opt for an eco-channel, given the smorgasbord of alternatives on cable, or that it would be a PBS sort of situation—small audiences and struggles for funding.

Channel heads countered that they would keep their subscription prices low, work to find local corporate sponsors, and reinvest part of the subscription fee in the community's environmental organizations with the operator getting credit.

Transitions

Networks not otherwise committed to eco-emphasis still offered environmental series in the mid-1990s. NBC had *Earth 2*, set in a space station refuge from a future environmentally devastated world. ABC's Saturday morning cartoons included an animated *Free Willy*. On The Discovery Channel and CNBC, Ed Begley, Jr., hosted *Today's Environment*, which showcased environmentally friendly companies and new technology. In 2009, Fox's *24* claimed a carbon-neutral season thanks to a 2,179-ton carbon dioxide offset due to use of hybrid and biodiesel vehicles and the purchase of renewable energy and carbon offsets.

TBS had a vice president of environmental policy and environmental specials and series. It produced *Network Earth,* aired on CNN. The TBS Environmental Unit won an Emmy. TBS aired *Captain Planet and the Planeteers* from 1990 to 1996 before putting it into syndication. The Disney Channel offered *Ocean Girl* and *Danger Bay*. PBS continued to offer animal shows and environmental documentaries. *Nature*, in its 13th season in 1995, was the longest-running natural history show. *Operation Earth* began in 1990.

Animal Planet was a cooperative venture of BBC and Discovery Communications that debuted October 1, 1996. When Discovery bought channel space from WWOR-TV on January 1, 1997, it went national and eventually was available in more than 70 countries as well as the United States. It has country-specific versions in India, Japan, Taiwan, Canada, and elsewhere. In 2006, BBC sold its 20 percent of the U.S. channel to Discovery but retained half of Discovery Europe, Discovery Asia, and the Latin American versions, and minority shares of Discovery Japan and Discovery Canada. In February 2008, Animal Planet replaced its soft and furry image with more aggressive programming to appeal to human instincts. Rather than paternalistic and preachy, it was entertaining and edgy, designed for adults 25 to 49.

Animal Planet featured the unscripted ecodrama *Whale Wars* about the Sea Shepherd Conservation Society's efforts to thwart Japanese whalers, and National Geographic Channel was noted for its impressive array of nature documentaries. In April 2008, the National Geographic Channel presented a documentary, *Human Footprint,* with such information as the average person using 3,796 diapers as a baby, consuming 13,056 pints of milk in a lifetime, and washing over 500 pounds of laundry in a year, requiring seven washing machines per lifetime.

Environmentalist Networks

Channel Green on Sundance debuted in 2007. It was three hours of environmental programming, bringing the same sex appeal to ecoprogramming as the Sundance Festival brought to independent films. The debut included 30 minutes of *Big Ideas for a Small Planet*, followed by a full-length documentary. Redford wanted the Green Channel to be in the forefront of the activist media effort to undo decades of environmental ruin, but his approach was to emphasize innovative and profit-making solutions to environmental problems rather than the traditional approach that emphasized imminent disaster. Sundance's "hot" show in 2009 was *Green Porno*, which featured Isabella Rossellini acting out odd animal sex methods while wearing strange costumes.

Planet Green was still dominant with a game show, the other ecofriendly style shows, and "eco-tainment." Planet Green debuted in June 2008 with 250 hours of original green living programming and a potential audience of 50 million homes. The network sought to provide all age groups with relevant, accessible, and entertaining programs from various perspectives and incorporating a variety of ideas. The network sought to be a catalyst for individual activism and environmental conversation.

Begun in 1998, originally to rival HGTV, Discovery Home/Discovery Home and Leisure featured cooking, do-it-yourself, party planning, and interior/exterior design. It was relaunched in 2008 as Planet Green, the first channel devoted to ecoliving. It began with advertising commitments from General Motors (GM), Johnson & Johnson, Dow Chemical, Wachovia Bank, Clorox, Whirlpool, Frito-Lay, and Home Depot, among others. GM was the official automobile sponsor of the new channel, with Chevrolet sponsoring Greensburg. GM wanted to feature its hybrids and other environmentally friendly vehicles.

Discovery expected growth within three years from the initial 52 million homes to 70 million. Discovery's initial investment was $50 million, and it was spending more than that, expecting to recoup through advertising. The expected growth was to be due to wider availability of digital video, and viewers were expected to be the sorts who shopped at organic stores but still drove the big SUVs and wanted more information about how to be ecoconsumers.

Planet Green retained familiar Discovery Home cable stars, among them Leonardo DiCaprio, Tom Green, Adrian Grenier, and SuChin Pak. New Orleans chef Emeril Lagasse hosted Emeril Green. Ludacris and Tommy Lee of Mötley Crüe starred in the green-themed challenge show *Battleground Earth*. Among other programs were *Alter Ego, Blue Planet*, *Greenovate*, *Living With Ed*, and *Whale Wars*. *Dumped*, the recently terminated BBC reality series, tracked a group of "survivors" attempting to build a society in a large dump site; other offerings included *Ten Ways to Save the Planet*, and a weekly eco-news program with ABC's Bob Woodruff. Celebrities provided green tips during the "we'll be back/we now return to" breaks.

Planet Green was associated with PlanetGreen.com and TreeHugger.com, websites with information and tips, as well as fan sites for Planet Green media figures. The sites claimed over 2 million visits a month. Through the websites, Planet Green linked to nongovernmental organizations (NGOs) such as Ashoka, Earthwatch Institute, Earth Pledge, the Environmental Media Association, Global Green USA, Global Inheritance, the Greenbelt Movement International, the National Wildlife Federation, Ocean Conservancy, Oceana, and The Nature Conservancy.

See Also: Begley, Jr., Ed; Corporate Social Responsibility; DiCaprio, Leonardo; Environmental Communication, Public Participation; Environmental Media Association; Nongovernmental Organizations; Public Opinion; Redford, Robert; Sundance Channel; Television, Advertising; Television News; Television Programming.

Further Readings

"Comcast Names Gavin Harvey President of Outdoor Life Channel." Comcast.com (February 24, 2004). http://www.comcast.com/About/PressRelease/PressReleaseDetail .ashx?PRID=257 (Accessed September 2010).

Courtenay, Erin. "Robert Redford Announces Green Sundance Channel, Culture and Celebrity." (July 21, 2006). http://www.treehugger.com/files/2006/07/robert_redford.php (Accessed September 2010).

Deggans, Eric. "Top Five Eco-Friendly TV Shows or Channels." *The Feed* (April 22, 2009). http://blogs.tampabay.com/media/2009/04/top-five-ecofriendly-tv-shows-or-channels.html (Accessed September 2010).

"Discovery Network Launches 'Green' Cable Channel." *Arizona Daily Star* (April 10, 2008). http://www.redorbit.com/news/entertainment/1334858/discovery_network_launches_green_cable_channel (Accessed September 2010).

Goldin, Greg and Jim Motavalli. "Is TV Going Green?" *E: The Environmental Magazine* (February 1995). http://findarticles.com/p/articles/mi_m1594/is_n1_v6/ai_16551326/pg_7/?tag=content;col1 (Accessed September 2010).

Guidera, Mark. "Ecology Channel Airs on Top U.S. Cable System." *Baltimore Sun* (November 17, 1994). http://articles.baltimoresun.com/1994-11-17/news/1994321078_1_tci-ecology-cable-service (Accessed September 2010).

Johnson. Drew. "GM to Sponsor New Green Cable Network." LeftLane. http://www.leftlanenews.com/gm-to-sponsor-new-green-cable-network.html (Accessed September 2010).

Planet Green. "About Planet Green." http://planetgreen.discovery.com/about.html (Accessed September 2010).

Reuters. "Discovery Sees Planet Green in 70 Million Homes." Environmental News Network (November 28, 2007). http://www.enn.com/ecosystems/article/26063 (Accessed September 2010).

Schults, Chris. "Sundance Channel Green." (July 20, 2006). http://www.grist.org/article/sundance-channel-green (Accessed September 2010).

Sundance Channel."Green Porno." http://www.sundancechannel.com/greenporno (Accessed September 2010).

John H. Barnhill
Independent Scholar

Television News

Though the Internet has changed the shape of media intake, television—including streamed or downloaded television programs on the internet—is still the dominant information and news medium in the United States. With newspapers dwindling in popularity and magazines Balkanized into catering to specific interests and subcultures, television is a major source of daily news, shaper of viewpoints, and a significant influence on the public discussion of issues, as well as the forum in which much of that discussion transpires.

Television news comes in various forms, including news flashes of breaking news, local (affiliate-provided) and national (network-provided) news programs in the morning and evening, magazine programs like the Public Broadcasting System (PBS) show *Frontline* that provide feature-length coverage of a timely issue, panel shows like *Meet the Press*, which discuss recent events, and opinion shows, which revolve around a television personality, often interspersed with interviews or panel discussions. News has been a staple of television programming since the 1940s, and has a long and respectable history of in-depth issues coverage, the equal of the print media.

Perhaps the most significant amount of coverage relevant to environmental issues has been offered by PBS. While the network evening news covered the antecedents to the green movement in the 1950s, in the wake of Rachel Carson's *Silent Spring*, and through the 1960s as Americans became more concerned with pollution, PBS's magazine programs have gone in-depth on environmental issues since its flagship science program, *Nova*, began airing in 1974. The spinoff progam *NOVA scienceNOW*, introduced in 2005, is a news magazine program focusing on current events in science, with frequent discussions of climate change, green consumer culture, and the potential environmental impacts of biotech and nanotechnology. Furthermore, *Frontline* has in recent years covered topics such as the mounds of electronic waste dumped in the Third World by unscrupulous recycling companies, and the potential benefits of nuclear power.

Wars have been instrumental in shaping the history of television news: coverage of the Vietnam War in the 1960s was a major factor in expanding the standard evening news program from 15 minutes to 30. The war increased the need for "human interest" stories to fill time on nights with insufficient hard news, and have become a significant element of the television news genre. When the Gulf War broke out in 1991, it provided the Cable News Network (CNN) with its first sustained ratings boost, resulting in an expanded budget and sense of worth for the cable network and soon inspired imitators.

What initially set CNN apart was its 24-hour news coverage, though many of those hours were filled with repeated content from earlier in the day. It was founded in 1980, and soon joined by CNN2 (since renamed Headline News [HLN]), which was designed as a sort of "weather channel of news," featuring a loop of tightly-edited 30-minute newscasts. (Since 2005, as the cable news landscape has become more competitive, HLN's coverage has expanded to include more opinion shows and pop culture news.) CNN and CNN2 helped to popularize the "scrawl" which has become a staple of cable news programming: a constant ticker tape of brief headlines, breaking news updates, stock prices, sports scores, and weather updates, as appropriate, at the bottom of the screen during programming. By the 1990s, CNN had been joined by Fox News and MSNBC, the "big three" of cable news networks. Other cable news networks are either limited in availability (the ABC News Network) or more specifically focused in their coverage (ESPN News, CNBC, Bloomberg), while the big three provide similar mixes of general interest news programs. Politically, Fox News skews heavily toward the right; MSNBC skews heavily toward the left; and CNN enlists political contributors from both sides of the aisle while maintaining a more or less apolitical network identity.

The cable news channels, along with the Internet, have brought about the much-discussed "24-hours news cycle," as demand has increased for faster news production and cable news networks have had to compete with one another for audience share and advertising dollars. Attracting advertising has required building an audience—hence the development of each of the big three's network identities. Advertising dollars are spent more eagerly when the advertiser feels well-informed about the demographics of the viewing audience, and by the 21st century, a definite sense had developed about the differences between the audiences of Fox News, CNN, and MSNBC. Personality-driven programming, such as that built around Rachel Maddow on MSNBC, Wolf Blitzer on CNN, and a large stable of political commentators at Fox News, has been another key to generating ratings.

As personality has become so key to the news and the packaging of news programs has revolved in some part around argumentative, even aggressive, personalities in contrast with respected anchors like Walter Cronkite and David Brinkley, the role of the news as opinion-maker has been made more obvious. This role has not necessarily become stronger. Television news has always shaped opinion by choosing what to cover and what not to cover, and by making tonal choices such as how to handle the airing of emotionally triggering footage. Rather, news programs are now more likely to wear their hearts on their sleeves, to explicate rather than implicate, and to directly engage rivals—as in the ongoing feud between MSNBC's Olbermann and Fox News' Bill O'Reilly, until the former's tenure with the network ended in early 2011.

Climate change has become a politicized topic, and just as climate change deniers are given airtime on Fox News, environmental advocates are given time on MSNBC. Coverage of various issues may be skewed accordingly, which is not to say that the facts are misrepresented. A story about a new electric car, for instance, could focus on the advantages of lessening fossil fuel usage; on the question of how much total energy will be consumed in the manufacture and operation of the vehicle, and what the difference in carbon footprint is between an electric car and a traditional car; on the cost of the car or disadvantages like insufficient space for a large family; or on issues like whether or not the car will be manufactured with American labor and will create American jobs. Similarly, in covering the green movement, producers may choose to interview typical Americans modifying their homes to use less energy and water; dropouts from society with whom middle Americans cannot relate; scientists; policy makers; or advocates of obscure interpretations of the facts. The "spin" that news producers put on the story creates a vastly different story out of the same facts and leaves viewers with very different experiences.

See Also: Advertising; Blogs; Communication, National and Local; Media Greenwashing; Print Media, Magazines; Print Media, Newspapers; Public Opinion; Sundance Channel; Television, Advertising; Television, Cable Networks; Television Programming.

Further Readings

Cox, J. Robert. *Environmental Communication and the Public Sphere*. Thousand Oaks, CA: Sage, 2009.

Hansen, Anders. *Environment, Media and Communication*. London: Routledge, 2010.

Lester, Libby. *Media and Environment: Conflict, Politics and the News*. Queensland, AU: Polity Press, 2010.

Wyss, Robert L. *Covering the Environment: How Journalists Work the Green Beat*. London: Routledge, 2007.

Bill Kte'pi
Independent Scholar

Television Programming

Green television is relatively new, and the quality and quantity of green programming varies from network to network, with cable television maintaining an edge in both. As of 2007, television was in the process of shifting away from its traditional indifference to environmental concerns (e.g., SUV ads paid a lot of bills, and big screen TVs were not exactly energy conserving). Ad-funded television was better known for home improvement shows sponsored by Home Depot and Lowe's. Green options were sometimes featured (e.g., Energy Star appliances, hybrid vehicles, or renewable wood) but often the ecofriendly choice was to buy nothing at all, not a pleasing idea to advertisers.

But the climate has shifted as the concept of ecofriendliness became more pervasive and more linked to popular culture, including being touted by celebrities such as Leonardo DiCaprio. Networks were now interested in viewers who might purchase a Toyota Prius or organic products; generally, the young and wealthy.

Network Efforts

NBC, after data showed that consumers will spend more if a brand is ecofriendly, began Green Week in 2007 as part of its "Green Is Universal" corporate effort. In November 2007, during sweeps week, Bravo, SciFi (now Syfy), and other sister channels, along with parent channel NBC, offered a week of green-themed news and entertainment shows. The week generated $20 million in ad revenue from 20 sponsors. NBC also acquired new sponsors, including the maker of Soy Joy. In April 2008, the network added another week, including green logos and on-air promos.

NBC committed to incorporate 100 hours of programming that featured green products and services during the year. NBC also set aside a month to emphasize healthy eating and exercise. The dedication of a set period allowed green-desirous Walmart and other advertisers to run green-friendly ads, such as Walmart ads featuring locally grown produce. Behavior placement was also seen as a way to circumvent viewers from using their DVRs to skip past commercials.

CBS added "Going Green" as a component of its *Early Show,* and Fox promised to include the topic of climate change within the series *24*. NBC actually incorporated green plotlines, with *Law and Order* investigating a cash-for-clunkers bike ride, Al Gore appearing on *30 Rock*, a group bike ride on *Mercy*, and a recycling superhero on *The Office*. NBC Universal referred to the approach as "behavior placement"—a way of motivating viewers to adopt more socially responsible attitudes. Rather than taking on climate change and creating controversy, NBC espoused taking better care of the environment.

Green Programming and Cable Networks

Cable programming currently has the "hottest" satires on television, *The Daily Show* with Jon Stewart and the *Colbert Report* with Stephen Colbert, both of whom consistently spoof such ecotopics as mountaintop removal, the missing oil from the British Petroleum spill, and home and health.

In 2007, Ed Begley, Jr.'s show *Living With Ed* aired on HGTV, and the Sundance Channel introduced "The Green." MTV's *The Real World: Hollywood* was set in a "green" house. In October 2008, the hot 15 shows included *Big Ideas for a Small Planet*, a Sundance series that featured grassroots activists finding solutions to small matters with broader implications. Fine Living Network offered *It's Easy Being Green,* a news-format 30 minutes of ecofriendly products and practices as well as cutting-edge trends. National Geographic aired a special, *Six Degrees Could Change the World*, about global warming.

The Discovery conglomerate had two channels that devoted time to environmental programming. The Science Channel offered *Eco-Tech*, 30 minutes about scientists trying to fight global warming and other harmful conditions. Animal Planet offered a miniseries about the conflict between Japanese whalers and the Sea Shepherd Conservation Society.

Planet Green remains a major player in environmental programming. The channel debuted as Discovery Home and Leisure (then Discovery Home) on June 1, 1998, competing with Home and Garden TV (HGTV). On June 4, 2008, it became Planet Green, the first channel dedicated to ecofriendly living. The initial investment was 250 hours and $50 million of Discovery Communications funds. The network debuted with a commitment to 250 hours of original programming and a market of 50 million homes. Planet Green attracted major sponsors such as General Motors, S.C. Johnson, Frito-Lay, Home Depot, Whirlpool, and others.

Among the celebrities featured were Emeril Lagasse, Tom Brokaw, Adrian Grenier, Tom Bergeron, Tom Green, and Maria Menounos. The various programs dealt with green homes, design, food, technology, and the range of green lifestyle concerns. Adrian Grenier showed how being green can be hip. *Greenovate* showed simple green home improvements from a new light bulb to a "living roof." *Focus Earth* ranged from world events to climate change to politics to ecofriendly jobs. *Living With Ed* moved to the channel from HGTV. *Stuff Happens* included science, humor, and interesting stories of the everyday.

In 2006, the Public Broadcasting Service (PBS) created the *e² design* television series to enforce the fact that sustainable architecture and design solutions could help sustain the Earth for future generations. During its three seasons, the series presented an optimistic and informative style, and touched on topics from the Middle East to China to sustainable cities. It gave viewers the sense that sustainable design was feasible and that it could make a significant impact on the well-being of people and the planet.

See Also: *e² design,* TV Series; Green Consumerism; Sundance Channel; Television, Advertising; Television, Cable Networks.

Further Readings

Chozick, Amy. "What Your TV Is Telling You to Do: NBC Universal's Shows Are Sending Viewers Signals to Recycle, Exercise and Eat Right. Why?" *Wall Street Journal* (April 7, 2010). http://online.wsj.com/article/NA_WSJ_PUB:SB10001424052702304364904575166581279549318.html (Accessed September 2010).

Chua, Jasmin Malik. "Tune in to Planet Green on June 4." (April 8, 2008). http://planetgreen.discovery.com/work-connect/tune-in-to-planet-green.html (Accessed September 2010).

Colbert Nation. http://www.colbertnation.com/home (Accessed September 2010).

The Daily Show. http://www.thedailyshow.com (Accessed September 2010).

Poniewozik, James. "Green Screens." *Time* (August 16, 2007). http://www.time.com/time/magazine/article/0,9171,1653637,00.html (Accessed September 2010).

Tilden, Tommi Lewis. "The 15 Must-See Green Shows on TV Right Now. The Best Environmental News, Documentaries and Entertainment on the Tube." *The Daily Green* (October 24, 2008). http://www.thedailygreen.com/living-green/blogs/celebrities/best-green-television-shows-461008 (Accessed September 2010).

John H. Barnhill
Independent Scholar

Theater

There has always been an overlap between the arts and activism, and particularly between theater and social conscience. A recent development in this overlap is green theater, taking to heart much of the cradle-to-cradle design philosophy articulated by Michael Braungart and William McDonough in the early 21st century. Also called ecotheater, ecological theater, environmental theater, or sustainable theater, there is no single centralized movement, just a number of theaters and theater programs "going green." Some theaters use the stage as a way to educate the public and promote green thinking; others simply use greener buildings and practices.

One of the resource-intensive activities in many theaters is set construction, and sustainability-minded theaters and economically minded theaters alike find it beneficial to build sets that are designed for disassembly. While some sort of reusability has always been an aspect of set construction, with certain set elements that can be redecorated or slightly altered to be used in many different plays, designing for disassembly goes further than that, aiming for set elements that can be taken apart entirely after the production and restored to their constituent parts (i.e., more screws, less glue). Sets can also be made modular, consisting of parts that can easily be recombined for a drastically different set, and the season can be based on particular set elements. Productions can use less clothing dyes and scenic mud, and can instead rely on nontoxic dyes and paints when possible. Costumes may be purchased at used clothing stores, of course, an option that theaters have often relied on for financial reasons, and a substantial costume inventory can be maintained so that new clothes do not need to be purchased or made. When new costumes are acquired, they can be of sustainable fabrics like organic cotton or wool.

The most energy-intensive aspect of the theater is lighting. When possible, operating a theater in a LEED-certified building is the greenest option, but many theaters have little choice in this matter, particularly community theaters or those operated by universities. Still, even when the building itself cannot be changed, more energy-efficient options are available, such as the use of dimmable LED lights in place of conventional houselights. Compact fluorescent lights can be used for lobby lighting.

Many of the green options available are the same as for any small business: use sustainable soy inks in printers, and print on both sides of the paper for internal use; use paper products made from recycled content; turn off computers and lights when not in use; use green cleaning products; install environmentally friendly plumbing fixtures that reduce usage; and whenever possible, use e-mail instead of paper communications like memos and letters.

When a theater company has the ability to make changes to the building in which it operates, there are more options. In London, the Arcola Theater has advertised itself as the world's first carbon-neutral theater and plans significant upgrades to its facility in the coming years, including rooftop solar panels and a biomass heating system for the winter. In 2010, the Arcola debuted its HyLight fuel cell system, designed in-house in association with partners BOC (an industrial gas supplier) and White Light (a lighting equipment supplier). HyLight is a portable lighting and power supply that uses a hydrogen fuel cell generator without the noise or emissions of traditional diesel generators used by traveling theater companies. In Australia, actress Cate Blanchett has helped to fund the greening of the Sydney Theatre Company, installing a 2,000-panel rooftop solar array and a rainwater harvesting system, reducing the theater's energy consumption by 70 percent.

Seattle University in the 2010 season introduced its greenSquat theater program, consisting of consecutive productions that "squat" on the stage and reuse the established set and other materials.

One recent example of a production with a green message is *Fracturing*, an adaptation of Henrik Ibsen's *An Enemy of the People,* by Deanna Neil. The original Ibsen play dealt with the moral dilemmas surrounding water contamination in Norway in 1882; the adaptation, presented by Sweeter Theater Productions in summer 2010, dealt with hydraulic fracturing, an environmentally hazardous process for drilling for natural gas.

See Also: Art as Activism; Fashion; Film, Documentaries; Film, Drama/Fiction; Green Consumerism.

Further Readings

Cole, Toby and Helen Krich Chinoy. *Directors on Directing: A Source Book of the Modern Theatre*. Boston: Allyn & Bacon, 2009.

Fried, Larry K. and Theresa J. May. *Greening Up Our Houses: A Guide to More Ecologically Sound Theatre*. Hollywood, CA: Drama Publishers, 1994.

Johnstone, Bob. *Brilliant! Shuji Nakamura and the Revolution in Lighting Technology*. Amherst, NY: Prometheus Books, 2007.

Smith, Stephanie and Victor Margolin. *Beyond Green: Toward a Sustainable Art*. Chicago, IL: Smart Museum of Art, 2006.

Bill Kte'pi
Independent Scholar

Theoretical Perspectives

Theories aiming to explain the forces leading to modern environmental problems can roughly be situated in three different perspectives: human ecology, neo-Marxian political economy, and ecological modernization. Human ecology applies ecological principles to understanding societies and identifies basic material conditions such as demographic characteristics and geographic context as the forces shaping human interaction with the environment. Neo-Marxian political economy focuses on how the structure of the economic system and social inequalities contribute to environmental degradation and critiques the unsustainable character of capitalism. Ecological modernization, in contrast with neo-Marxian political economy, identifies modern institutions such as the market and techno-science as dynamic forces that can move societies toward environmental sustainability.

Human Ecology

The human ecology perspective aims to understand human societies using ecological principles. Human ecology recognizes that societies are embedded in the natural environment and are constrained by the same biophysical forces and natural limits that affect other animals. In particular, human ecology recognizes that all societies depend on finite resources and therefore cannot grow without limits. This leads to the recognition that demographic characteristics—including population size, growth, density, and age structure—are key forces influencing the degree to which societies degrade the environment, and that no societies, no matter how technologically "advanced," can fundamentally overcome ecological constraints. Human ecology also aims to understand how societies are shaped by their environments such as how factors like climate and biogeography influence social organization, technological development, and resource utilization. From this perspective, in order to solve environmental problems, cultures need to transform themselves so that the recognition of natural limits and an understanding of the dependence of societies on ecosystems are central to all social processes.

Neo-Marxian Political Economy

The neo-Marxian political economy perspective examines how economic forces and social inequalities, both within nations and among nations, influence the degree, kind, and location of environmental degradation that societies produce. One of the major foci of this perspective is on the growth dynamics of capitalist economies and how growth leads to environmental degradation. This position is exemplified by Allan Schnaiberg's "treadmill of production" model. The treadmill of production focuses on how the profit-seeking behavior of producers eliminates jobs by developing "labor-saving" technology, where mechanization makes it so that fewer workers are needed per unit of production. However, mechanization requires higher levels of natural resource consumption and leads to pollution, creating environmental degradation along with unemployment. Treadmill-dominated societies seek to solve these problems by further expanding production so as to create jobs and increase government revenue (via taxes on both workers and employers) for funding the cleanup of the environmental problems created by production. However, each round of expanding production generates another round of mechanization, unemployment, and environmental degradation. This is the heart of the treadmill analogy, where one keeps moving but stays in place. One important aspect of keeping the treadmill in motion is the manipulation by capitalists of people's identity, desires, and "needs" via marketing and other methods of ideological control, so that people identify with the interests of capitalists and accept their roles as workers and as consumers, focusing their lives on the acquisition of consumer goods. From the perspective of the treadmill of production, the only way to overcome environmental crisis is to restructure society so that it is not dominated by elite producers (capitalists) and is not focused on economic growth. Part of this transformation requires a rejection of consumer culture.

Complementary to the treadmill of production, metabolic rift is a neo-Marxian theory developed by John Bellamy Foster, who drew on the work of Karl Marx. Metabolic rift focuses on how the metabolic exchange between societies and nature is disrupted by modernization. Marx noted that urbanization, driven by rapid industrialization, separated people from the land, and he argued that this separation led to one of the major environmental crises of his time: the depletion of soil nutrients in farmland. In pre-industrial societies, agricultural products were consumed near where they were grown. Thus, the nutrients from the soil, in the form of agricultural and human waste, were recycled into the soil, sustaining agricultural production by maintaining soil fertility. However, as a greater share of the population was concentrated in cities, agricultural goods were shipped far away from where they were produced, and the nutrients that originated in the soil ended up in urban sewage systems, polluting cities, rivers, and the ocean. Thus, urbanization broke the nutrient cycle that predominated in pre-industrial societies, leading to a metabolic rift. Beyond Marx's original theory focusing on soil nutrients, metabolic rift analyses examine how social processes alter exchanges of material between society and nature, potentially creating a variety of metabolic rifts. The metabolic rift points to the need for developing place-based cultures that draw on local resources and maintain the quality of the local environment.

Another neo-Marxian approach, the world-systems perspective, examines how the structure of the global economy contributes to the production and distribution of environmental problems. The world-systems perspective is based on the recognition that all nations in the world are integrated into one global economy, and that what happens in one nation cannot fully be understood without reference to the context of the larger world-system. The modern world-system took form beginning in the 15th century with the expansion of European influence via global trade networks and colonization. European imperialism led to a world-system characterized by dramatic inequality, where a few "core" nations dominate the global economy, controlling trade and concentrating wealth in the hands of an elite class residing predominantly in North America, western Europe, and Japan. Most consumption of resources occurs in these core nations, which are thereby predominantly responsible for environmental problems around the world. The least powerful nations, those in the "periphery," are typically former colonies that achieved independence only recently, such as those in sub-Saharan Africa. Nations in the periphery typically are not heavily industrialized and are not well networked in the global economy and therefore are dependent on trade relationships with a few core nations. They typically have economies based on the extraction of natural resources (e.g., timber, minerals) for export. Due to this, problems like deforestation are often most intense in the global periphery. Semi-peripheral nations, such as Mexico and Malaysia, occupy a position of intermediary power in the world-system and are often the sites of rapid industrialization. As heavy industry is increasingly shifted from the core to the semi-periphery, pollution is often worse in semi-peripheral nations. Thus, the structure of the world-system makes it so that the forces driving environmental degradation (e.g., the resource consumption of the affluent) are spatially separated from where the degradation occurs (often in the periphery and semi-periphery).

Ecological Modernization

Ecological modernization theory counters neo-Marxian political economy, arguing that the forces of modernization—the processes connected to industrialization/postindustrialization, urbanization, and globalization, which are associated with science and technology, democracy, and the logic of the market—rather than inevitably leading to environmental degradation, can lead to environmental sustainability. In particular, ecological modernization focuses on the central role rationality plays in modernity. In earlier phases of modernization, economic rationality came to dominate social and political reasoning. Ecological modernization theorists suggest that in late modernity, rationality percolates into a growing number of social institutions, and that, as part of this, ecological rationality is gaining prominence, where ecological costs and benefits are carefully examined by both producers and consumers. It is supposed that ecological rationality will rise to be of equal importance with economic rationality, which will lead to a transformation of modern societies along lines of ecological sustainability. Part of this shift is the emergence of ecologically rational cultural norms, where people modify their lifestyles and consumer habits to minimize their environmental impacts.

The expectations of ecological modernization theory are similar to those behind the environmental Kuznets curve thesis in economics. The environmental Kuznets curve suggests that environmental degradation follows an inverted-U-shaped curve with respect to economic development (usually measured as per capita gross domestic product [GDP]). In this formulation, environmental degradation is expected to rise in the early stages of economic development, but as affluence grows, a turning point is reached, after which further economic development leads to declining environmental degradation. This change in affluent societies is supposed to arise due to greater investment in more efficient, "greener" technologies and changes in cultural values with respect to the environment.

See Also: "Agri-Culture"; Antiglobalization Movement; Corporate Social Responsibility; Environmental Justice Movements; International Law and Treaties; Kyoto Protocol.

Further Readings

Bell, Shannon Elizabeth and Richard York. "Community Economic Identity: The Coal Industry and Ideology Construction in West Virginia." *Rural Sociology*, 75/1 (2010).

Catton, William R., Jr. *Overshoot: The Ecological Basis of Revolutionary Change*. Chicago: University of Illinois Press, 1980.

Catton, William R., Jr., and Riley E. Dunlap. "Environmental Sociology: A New Paradigm." *The American Sociologist*, 13 (1978).

Clark, Brett and Richard York. "Carbon Metabolism: Global Capitalism, Climate Change, and the Biospheric Rift." *Theory and Society*, 34/4 (2005).

Diamond, Jared. *Guns, Germs, and Steel: The Fates of Human Societies*. New York: W. W. Norton, 1997.

Dinda, Soumyananda. "Environmental Kuznets Curve Hypothesis: A Survey." *Ecological Economics*, 49 (2004).

Foster, John Bellamy. *Marx's Ecology*. New York: Monthly Review Press, 2000.

Jorgenson, Andrew K. and Brett Clark. "The Economy, Military, and Ecologically Unequal Relationships in Comparative Perspective: A Panel Study of the Ecological Footprints of Nations, 1975–2000." *Social Problems*, 56 (2009).

Mol, Arthur P. J. *The Refinement of Production: Ecological Modernization Theory and the Chemical Industry*. Utrecht: Van Arkel, 1995.

Mol, Arthur P. J. and David A. Sonnenfeld. *Ecological Modernization Around the World: Perspectives and Critical Debates*. London: Frank Cass Publishers, 2000.

Moore, Jason W. "The Modern World-System as Environmental History? Ecology and the Rise of Capitalism." *Theory & Society*, 32 (2003).

Schnaiberg, Allan. *The Environment: From Surplus to Scarcity*. New York: Oxford University Press, 1980.

Schnaiberg, Allan and Kenneth A. Gould. *Environment and Society: The Enduring Conflict*. New York: St. Martin's Press, 1994.

York, Richard and Philip Mancus. "Critical Human Ecology: Historical Materialism and Natural Laws." *Sociological Theory*, 27/2 (2009).

York, Richard and Eugene A. Rosa. "Key Challenges to Ecological Modernization Theory: Institutional Efficacy, Case Study Evidence, Units of Analysis, and the Pace of Eco-Efficiency." *Organization & Environment*, 16/3 (2003).

Richard York
University of Oregon

Toxic/Hazardous Waste

The large amounts of toxic and hazardous waste produced globally have made its disposal and environmental and health consequences a serious issue. The green culture movement seeks to raise public awareness of the problems associated with toxic and hazardous wastes such as disposal, environmental degradation, and alternatives to their use. Green culture activists have played a key role in the implementation of federal legislation and alternative disposal practices to create safer and more effective ways of dealing with this environmental problem.

Toxic substances are defined as chemicals that are fatal to humans in low doses, and scientists and governments often set limits for each chemical that are deemed safe or unsafe to humans based on laboratory studies that often extrapolate information from animal testing. Hazardous substances are defined as those materials that can cause human injury or disease, economic loss, or environmental damage. Health problems associated with toxic and hazardous wastes include cancer and birth defects. They are especially dangerous to children, and scientific studies have revealed a tendency of such illnesses to occur in clusters that alert the public to a potential hidden toxic hazard.

Waste can come from a variety of sources, including chemical and industrial plants, commercial industries, and family households. Examples of hazardous waste include chemicals, metals, radioactive waste, and medical waste. Toxic and hazardous waste must be disposed of properly to avoid environmental contamination. Toxic chemicals can easily enter the air, water, and land. These waste products, if left undetected and unmonitored, will affect humans, animals, plants, and aquatic life, creating the potential for an environmental disaster.

Industrialized countries, countries with large populations, or countries with little environmental regulation or awareness and activism produce much of the world's toxic and hazardous waste. For example, the United States produces over 40 million tons of hazardous waste each year, and the former Soviet Union and Mexico became notorious for lax environmental oversight that allowed widespread cultures of environmental degradation to develop.

Beginning with the rise of the environmental movement in the 1960s and 1970s, new grassroots movements emerged as public awareness of environmental problems rose. The popularity of such issues grew with the publicity surrounding environmental disasters involving toxic and hazardous wastes, such as the revelations of toxic contamination in New York's Love Canal neighborhood and the accidents at the Union Carbide Plant in Bhopal, India, the Three Mile Island nuclear power plant in the United States, and the Chernobyl nuclear power plant in the Soviet Union.

One of the emphases of the green culture movement has been the roles individual people and families can play in proper disposal of toxic and hazardous household wastes or in encouraging the production and use of more environmentally friendly materials and production methods to create products. In short, people can reduce the amount of hazardous waste they use.

The green culture movement has also placed great emphasis on the development and use of green building technology, especially notable in urban areas. Green buildings utilize more efficient models of construction, renovation, maintenance, and demolition. The green building movement has also enjoyed the support of national government environmental organizations such as the U.S. Environmental Protection Agency (EPA), noting the statistics proving the effectiveness of green building technologies. In the United States, buildings account for 39 percent of total energy use, 12 percent of total water consumption, 68 percent of total electricity consumption, and 38 percent of carbon dioxide emissions.

Green building technologies can be utilized in any structure, even in retrofitting older preexisting structures, and cover a variety of environmental issues such as energy efficiency, ecofriendly cleaning products, and recycling. Green construction and building maintenance and energy systems are an important component in the reduction of the output of toxic and hazardous wastes associated with construction, maintenance, cleaning, energy, and waste.

Green chemistry, also called sustainable chemistry, is another avenue in reducing the environmental impact of toxic and hazardous wastes emphasized by the green culture movement. Green chemistry involves the development of chemically based products designed to reduce or eliminate the use of toxic and hazardous substances, making them more environmentally friendly. Results include reduced wastes, safer products, and reduced use of energy and other resources. Chemical processing allows materials to be changed into nontoxic substances. Bioremediation utilizes microorganisms to detoxify toxic compounds. Green engineering, defined as the design of products that are meant to reduce pollution and lower the risk of harm to humans and the environment, is closely related to green chemistry and is another tool promoted by green culture.

Corporations and Environmentalism

Companies around the world are making environmentalism a top priority, which includes the reduction of toxic and hazardous waste, as a result of public demand for green products. Resulting changes include the development and sale of new green products and services, among other measures. Public awareness has also led to improvements in worker safety and new regulations to protect workers who handle or are exposed to toxic and hazardous wastes on the job.

Proponents of green culture support government, corporate, and individual efforts to incorporate green purchasing and use programs that support these environmentally friendly and sustainable products and services. For example, the EPA's Environmentally Preferable Purchasing Program, established in 1993, helps federal government agencies comply with green purchasing requirements. This is extremely significant since the U.S. government is the world's largest consumer. Many educational institutions, private corporations, and individual households practice similar measures. Green purchasing programs target a number of areas with environmental consequences, such as the manufacturing process, product packaging, distribution methods, use of recycled materials in manufacture, and product reuse and recycling potential.

The storage and disposal of toxic and hazardous wastes are popular targets of the green culture movement. As a result, in many countries national regulatory agencies such as the EPA work closely with businesses and local authorities to ensure the proper handling and disposal of wastes. While most industrialized countries have banned open dumping, it is still a common method of waste disposal in developing countries, where illegal dumping is a major problem partly responsible for huge amounts of toxic materials contaminating the water. There are also millions of tons of waste dumped in the ocean. Until recent times, many cities and corporations dumped municipal and industrial waste and sewage into the world's oceans. Today, federal and international legislation, such as the United Nations' Law of the Sea, prohibits this type of dumping. There are people, however, who still argue that the ocean is the most inaccessible and least harmful area to dump wastes.

Permanent retrievable storage places waste storage containers in a secure building or another location, allowing for their regular inspection. Secure landfills are specially designed for the safe storage of toxic wastes. Disposal facilities such as landfills, surface impoundments, tanks, waste piles, land treatment units, and injection units are designed to permanently contain waste, but are unwelcome neighbors due to fears of possible environmental contamination. Environmental activists and neighborhood residents often lead campaigns against such facilities when the proposed location is nearby, giving rise to the phrase "not in my backyard" (NIMBY). Many people argue that rising land prices and shipping costs are making landfills an expensive waste disposal option.

Many industrialized countries ship their waste to other countries or from place to place internally, looking for a place that will accept toxic and hazardous waste. Some industrialized countries have agreed to stop shipping hazardous waste to less-developed countries, but the practice continues and is even expanding into new categories of toxic and hazardous waste, such as the e-waste generated from used computer and electronic components. India and China are among the largest acceptors of international e-waste. Environmental activists have pushed for the de-manufacturing of an item that has possible toxic waste as an alternative to the international shipment and disposal of e-waste. De-manufacturing is defined as the disassembly and recycling of obsolete consumer products, such as TV sets, personal computers (PCs), and air conditioners that would otherwise end up as hazardous waste. De-manufacturing allows for the reuse of still-valuable materials and disposal of toxic materials before they are released into the environment.

Green activism has raised public awareness of the contamination of various sites designated for the storage of toxic and hazardous wastes and the environmental consequences for neighboring populations. Problem sites include old industrial facilities, mining districts, or dumping grounds. Urban "brownfields" are properties known or suspected of hazardous material contamination. Key examples include the so-called Superfund sites in the United States, defined as uncontrolled or abandoned hazardous waste disposal sites.

The EPA alleges that there are 36,000 seriously contaminated sites in the United States that are either leaching contamination into the environment or have the serious potential to do so in the near future. The EPA concedes that toxic materials are known to have contaminated groundwater at 75 percent of Superfund sites. It has been estimated that 5 billion metric tons of highly poisonous chemicals were improperly disposed of in the United States between 1950 and 1975. The EPA has estimated the costs of hazardous waste cleanup in the United States at somewhere between $370 billion and $1.7 trillion. Cleaning such properties, however, allows for their reinvestment and property value increases in addition to their environmental benefits.

Green culture advocates note that many industrial wastes can be recycled safely and effectively. A hazardous waste can be recycled through a variety of processes, including recycling, reuse, and reclamation. Reclamation involves processing a material in order to reclaim a usable product. In 2007, approximately 1.8 million tons of hazardous waste was recycled. Hazardous waste treatment involves changing a material's physical, chemical, or biological nature to reduce its potential harm to the environment. The benefits of hazardous waste treatment include the neutralization of waste or reduction of its toxicity, the recovery of useful energy or material resources, and safer transportation, storage, and disposal. Combustion or incineration is one common way to reduce waste volume and destroy its hazardous organic components.

The Green Solution?

Green activists suggest that there are many options available for the proper removal of toxic and hazardous wastes. The safest and least costly alternative is to eliminate their production. The 3M Company, for instance, reformulated its products and redesigned its manufacturing processes to eliminate tons of waste and pollution from its factories, proving its cost effectiveness. Other viable alternatives include recycling, reuse, and incineration. Activists also note that many food and other consumer products contain excess packaging for marketing purposes, one of the greatest sources of unnecessary waste. The use of less packaging and biodegradable packaging, which can be broken down and decomposed by microorganisms over time, will reduce the need for waste disposal.

Green activists have also advocated waste incineration and resource recovery. The heat generated by waste incineration can be converted into energy. Some waste is sorted into burnable, unburnable, and recyclable before its incineration. Incinerated waste is known as refuse-derived fuel due to its higher energy content than that of raw trash. Green debates surrounding the effectiveness of waste incineration includes cost effectiveness, environmental cost/benefit analysis, and the potential of use fees.

The public has become increasingly aware of the problems associated with hazardous medical waste disposal through such highly publicized incidents as used hypodermic needles washing up on the shores of beaches, the result of medical wastes being illegally dumped into the world's oceans. Public outcry resulted in government intervention in countries such as the United States. For example, in 1998, the EPA and the American Hospital Association signed an agreement designed to eliminate pollution in America's hospitals. The agreement had three main goals: to eliminate mercury-containing waste from the healthcare waste stream, to reduce the volume of both regulated and unregulated waste, and to identify hazardous substances to prevent pollution and increase waste-reduction opportunities.

Recycling is one of the most popular waste disposal methods among green culture advocates, including community, institutional, and governmental campaigns; the use of recycling bins and containers; recycling competitions; and financial incentives such as bottle and can deposits. Many environmentalists maintain that recycling is a better alternative to burning or dumping based on such factors as cost effectiveness, energy use, land space needed, and potential pollution reduction. Problems associated with recycling include fluctuations in the market prices for recyclable commodities and the potential contamination of products due to recycled containers.

EPA estimates state that approximately 256 million tons of officially classified hazardous wastes are produced in the United States each year, not including toxic and hazardous waste produced by those industries that are not monitored by the EPA. The EPA maintains that the biggest sources of hazardous wastes in the United States are the chemical and petroleum industries. Green culture activists have targeted reductions in this area through the promotion of hybrid and electric vehicles, mass transit, and other alternative transportation methods that will also reduce dependence on fossil fuels.

See Also: Body Burden; Corporate Social Responsibility; Environmental "Bads"; Environmental Justice Movements; Garbage; Gardening and Lawns; Indoor Air Pollution; Love Canal and Lois Gibbs; Soil Pollution; United Farm Workers (UFW) and Antipesticide Activities; United Church of Christ; Water Pollution.

Further Readings

Brister, Jennifer. "Toxic Waste Disposal: How to Prevent Hazardous Waste." http://www.suite101.com/content/toxic-waste-disposal-a207945 (Accessed September 2010).
Brown, Michael Harold. *Laying Waste: The Poisoning of America by Toxic Chemicals.* New York: Pantheon Books, 1980.
Brown, Phil and Edwin J. Mikkelsen. *No Safe Place: Toxic Waste, Leukemia, and Community Action.* Berkeley: University of California Press, 1990.
Cozic, Charles P. *Garbage and Waste.* Farmington Hills, MI: Greenhaven, 1997.
Doherty, Brian. *Ideas and Actions in the Green Movement.* New York: Routledge, 2002.
Edelstein, Michael R. *Contaminated Communities: The Social and Psychological Impacts of Residential Toxic Exposure.* Boulder, CO: Westview Press, 1988.
Fawcett, Howard H. *Hazardous and Toxic Materials: Safe Handling and Disposal.* Hoboken, NJ: Wiley, 1984.
Freeze, R. Allan. *The Environmental Pendulum: A Quest for the Truth About Toxic Chemicals, Human Health, and Environmental Protection.* Berkeley: University of California Press, 2000.
Gerdes, Louise I. *Pollution: Opposing Viewpoints.* Detroit, MI: Greenhaven, 2006.
Grossman, Elizabeth. *High-Tech Trash: Digital Devices, Hidden Toxics, and Human Health.* Washington, DC: Island Press, 2006.
Harte, John. *Toxics A to Z: A Guide to Everyday Pollution Hazards.* Berkeley: University of California Press, 1991.
"Hazardous and Toxic Substances." U.S. Occupational and Safety Health Administration. http://www.osha.gov/SLTC/hazardoustoxicsubstances/index.html (Accessed September 2010).
Long, Robert Emmet. *The Problem of Waste Disposal.* New York: H. W. Wilson, 1989.
Maczulak, Anne. *Waste Treatment, Reducing Global Waste.* New York: Facts on File, 2009.
Markowitz, Gerald E. and David Rosner. *Deceit and Denial: The Deadly Politics of Industrial Pollution.* Berkeley: University of California Press, 2002.
McDonough, William and Michael Braungart. *Cradle to Cradle: Remaking the Way We Make Things.* New York: North Point Press, 2002.
Mitchell, Ralph. *Environmental Microbiology.* Hoboken, NJ: Wiley-Liss, 1992.
Norris, Ruth, Karim A. Ahmed, et al. *Pills, Pesticides and Profits: The International Trade in Toxic Substances.* Great Barrington, MA: North River Press, 1982.
Olson, Eric G. *Better Green Business: Handbook for Environmentally Responsible and Profitable Business Practices.* Philadelphia, PA: Wharton School Publishing, 2010.
Streissguth, Thomas. *Nuclear and Toxic Waste.* Farmington Hills, MI: Greenhaven, 2001.
Szasz, Andrew. *EcoPopulism.* Minneapolis: University of Minnesota Press, 1994.

"Toxic Waste: Man's Poisonous Byproducts." *National Geographic*. http://environment.nationalgeographic.com/environment/global-warming/toxic-waste-overview.html (Accessed September 2010).
Young, Mitchell. *Garbage and Recycling*. Farmington Hills, MI: Greenhaven, 2007.

Marcella Bush Trevino
Barry University

Tree Planting Movement

Tree-planting programs often take place at the local level, such as this one in Iowa. Here, the correct way to plant most trees is shown: placing the plant in a hole about twice as wide in diameter as its pot.

Source: Lynn Betts/Natural Resources Conservation Service

Environmentalists have long understood the importance of trees in promoting sustainability. In addition to contributing to biodiversity by providing natural habitats for birds, animals, and insects, trees perform an essential task in protecting the environment by absorbing carbon dioxide and releasing oxygen into the atmosphere. When trees are cut or burned down, pollutants remain in the air, threatening lives, health, and habitats. Countries around the world are reacting to this knowledge in various ways. In some countries, the tree planting movement is more than a century old; in others, it is a modern phenomenon. In the United States, the forestry movement began with the establishment of the American Forestry Association in 1876. Several decades later, President Franklin Roosevelt promoted tree planting as an important element in his New Deal program, establishing the Civilian Conservation Corps and sending young men around the country to plant 3 billion trees in the years between 1933 and 1942. Around the world, both government-sponsored and private tree-planting groups have been established to promote environmentalism and to repair existing environmental damage. Tree-planting activities have ranged from China's official Tree-Planting Day that occurs each spring to extensive efforts such as the one taking place in Kenya, where a major reforestation effort is under way.

While national governmental bodies often take an active role in promoting sustainability, tree-planting programs frequently take place at the local level. Under the leadership of mayors concerned with sustainability, several large American cities are currently vying for the status of America's greenest city. One of those is Chicago, Illinois, where Mayor

Richard M. Daley has launched a number of green initiatives, including one involving tree planting, with the ultimate goal of ensuring that 20 percent of all electricity used is derived from renewable sources. In Seattle, Washington, Mayor Greg Nickels issued a call to other American mayors by creating the Mayors' Climate Protection Agreement. Internally, Nickels pledged to slash greenhouse emissions by 170,000 tons each year.

In some countries, the planting of trees goes far beyond efforts to protect the environment. In Kenya, for instance, years of stripping away forest cover led to extensive crop failure and water shortages. The Nairobi-based Green Belt Movement (GBM) responded by working with communities, training volunteers through a 10-step process to engage in planting trees on public and private lands and in areas where forests had been degraded. GBM, which began with a group of women who were convinced that tree planting could change both the environment and the world, received international recognition in 2004 when founder Wangari Maathai was awarded the Nobel Peace Prize. In her acceptance speech, Maathai declared that world peace would only be possible if all nations agreed to promote sustainability. Since the late 1970s, more than 45 million trees have been planted in Kenya, concentrated in the five areas that furnish water to more than 90 percent of all Kenyans.

Local groups have been instrumental in promoting tree planting in a number of other countries also. In Uganda, the Uganda Women Tree Planting Movement was launched in 1985 to encourage women to become active environmentalists with the express purposes of protecting the environment and Ugandan natural resources as well as ensuring the supply of tree products. In Canada, local groups such as Ground Breakers Oakville, which has planted more than 8,000 native trees and shrubs at the urging of journalist Liz Benneian, are making a difference in how both private citizens and public officials engage in tree planting to reverse environmental damage. In Europe, the British Trust for Conservation Volunteers has been combining tree planting with exercise under the auspices of the "Green Gyms" movement. Since 1999, in more than 50 cities, volunteers plant trees and participate in other conservation activities as part of a regular exercise routine.

In many areas, tree planting has become a major part of both government and environmental planning. In a state ceremony held in Indonesia in 2010, President Susilo Bambang Yudhoyono distributed seedlings of trembesi trees, which are native to Latin America. According to the state environment minister, the trembesi were to be used to absorb 28 tons of greenhouse gas emissions each year. During a Christmas celebration that same year, coconut trees and rain tree seedlings were also handed out. Officials announced that a billion trees were to be planted over the course of the year. When it became necessary in 1998 to evacuate residents from a floodplain in Rahway, New Jersey, instead of leaving the empty land to remain useless, local officials and environmentalists worked together to turn the area into a natural habitat for a score of animals, birds, and insects by planting bushes, trees, and flowers and creating a river walk for local residents to enjoy nature. Flooding was circumvented by the use of basins designed to hold potential overflow whenever necessary.

One of the most important elements in the tree planting movement is planning for the future of the movement by teaching children about the importance of planting trees as a way to protect the environment as well as to provide beauty and shade.

See Also: Education (Climate Literacy); Environmental "Goods"; Maathai, Wangari; Social Action, Global and Regional; Social Action, National and Local.

Further Readings

Cahn, Robert and Patricia Cahn. "Did Earth Day Change the World?" *Environment*, 32/7 (September 1990).

Foster, John Bellamy. *The Ecological Revolution: Making Peace With the Planet*. New York: Monthly Review Press, 2009.

Goodspeed, Brianne. "It's Not Easy Being Greenest: 10 Cities to Watch." *E: The Environmental Magazine*, 17/4 (2006).

Green Belt Movement. "Tree-Planting Program." http://gbmna.org/w.php?id=13 (Accessed September 2010).

Hattam, Jennifer. "Whole-Earth Workout." *Sierra*, 91/1 (2006).

Los, Fraser. "Fighting the Good Fight." *Alternatives Journal*, 35/5 (2008).

MacGregor, Sherilyn. *Beyond Mothering Earth: Ecological Citizenship and the Politics of Care*. Vancouver: UBC Press, 2006.

Polacco, Elinore. "Yes, I Can Go Home Again." *Ecological Restoration*, 253 (2007).

Priesnitz, Wendy. "Helping Children Nurture Nature." *Natural Life*, 116 (2007).

"SBY Launches 'Trembesi' Tree Planting Movement." *Jakarta Post* (September 12, 2010). http://www.thejakartapost.com/news/2010/01/14/sby-launches-trembesi039-tree-planting-movement.html (Accessed September 2010).

Schor, Juliet B. and Betsy Taylor. *Sustainable Planet: Solutions for the Twenty-First Century*. Boston, MA: Beacon Press, 2002.

Steffen, Alex. *Worldchanging: A User's Guide for the 21st Century*. New York: Abrams, 2006.

Torgerson, Douglas. *The Promise of Green Politics: Environmentalism and the Public Sphere*. Durham, NC: Duke University Press, 1999.

Elizabeth Rholetter Purdy
Independent Scholar

United Church of Christ

The United Church of Christ (UCC) is an American religious denomination, the fifth-largest of the so-called mainline Protestant bodies following the United Methodist, Evangelical Lutheran, Presbyterian Church (USA), and the American Baptists. All of the mainline denominations have taken positions in support of many green issues and worked together as members of the Eco-Justice Working Group of the National Council of Churches of Christ. This group is one member of the National Religious Partnership for the Environment. Of all these faith communities, the UCC is noted because of its work for environmental justice.

When the story of the environmental justice movement is told, the tale usually begins in 1982 in rural Warren County, North Carolina. A landfill for disposing of contaminated soil containing polychlorinated biphenyl was proposed to be built in a poor African American community. Protests by the residents drew support from a wide variety of environmental and civil rights organizations. Among the protestors who came to the community were two individuals: a young Chinese American labor organizer, Charles Lee, and Benjamin Chavis, a civil rights activist then serving as the southern regional program director for the UCC Commission for Racial Justice (UCC-CRJ). Chavis regarded Warren County as an instance of *environmental racism*, a term coined to indicate the injustice of burdening communities of color with toxic wastes.

Chavis advanced to serve as executive director of the UCC-CRJ in 1985, and Charles Lee was hired to direct a special project on toxic injustice. Lee used publicly available data from the U.S. Census and the U.S. Environmental Protection Agency (EPA) to conduct a statistical study of the distribution of toxic waste sites in relation to communities of color in the United States. The study found that three of the largest five commercial hazardous waste sites in the United States were located in predominantly black or Hispanic communities. These three sites accounted for approximately 40 percent of the U.S. landfill capacity for hazardous waste. The study further concluded that uncontrolled waste sites, those that might ultimately require evaluation and cleanup through the Superfund process, were also disproportionately concentrated in black and Hispanic urban communities. The report made a series of recommendations to federal, state, and city governments to address the systematic inequities that it had revealed.

When the UCC-CRJ report, *Toxic Wastes and Race in the United States*, was published in 1987, it had powerful impact well beyond the religious community. The environmental justice movement that it promoted led to the establishment of the Office of Environmental Justice in the EPA in 1992 and the signing of Executive Order 12898 by President Clinton in 1994, requiring that federal agencies implement environmental justice policies.

Both Lee and Chavis had interesting careers after their work on the 1987 UCC-CRJ report. In the 1990s, Chavis went on to a controversial tenure as director of the NAACP. He subsequently lost his status as a UCC minister after he joined the Nation of Islam. Lee worked for the UCC-CRJ for 15 years on environmental injustice and helped to organize the First National People of Color Leadership Summit in Washington, D.C., in 1991. In 1999, Lee went to work for the EPA and in 2007 became director of its Office of Environmental Justice.

In 2007, the UCC-CRJ published a follow-up appraisal of the historic report "Toxic Wastes and Race." The strength of the environmental justice movement was reflected in the fact that they were able to commission authors who had already made respected academic careers in the study of environmental racism. The authors of the new report "Toxic Wastes and Race at Twenty" were Robert D. Bullard, Paul Mohai, Robin Saha, and Beverly Wright.

Growing out of its work on toxic wastes and race, the UCC came to specialize, among its interfaith partners, in concerns about toxic disasters. As a participant in Church World Service, it responded to the health issues of workers at the World Trade Center after 9/11. The UCC also was active in supporting the Native American community affected by exposure to heavy metals at the Tar Creek Superfund site in Oklahoma. The UCC created a print resource, "The Silent Disaster" (2000, revised 2008), to advise faith communities confronting disasters caused by technology.

The UCC originated with the English Puritan settlers of Massachusetts Bay Colony. New Englanders took their congregational form of government with them as they moved westward to the frontier. In this form of church polity, individual congregations have more independence than relatively more hierarchical denominations. This independence, allowing congregations that disagree with a denominational action to simply ignore it, is one factor that has allowed the Congregational churches to take progressive positions. This activism dates back to their work for the abolition of slavery in the 19th century and most recently to the acceptance of gay and lesbian clergy.

The UCC acquired its present name in 1957 when the Congregational churches merged with the Evangelical and Reform denominations, founded primarily by German and Swiss immigrants. The UCC membership is 1,111,691 (as reported in 2008, the most recent year for which data is available). It has its strongest presence in the northeast and its headquarters in Cleveland, Ohio. Although the UCC is predominantly a white denomination, its largest congregation is the 8,500-member Trinity UCC in Chicago, where President Barack Obama and his family are former parishioners.

See Also: Clinton, William J. (Executive Order 12898); Environmental Justice Movements; National People of Color Environmental Leadership Summit; Religious Partnership for the Environment.

Further Readings

Bullard, Robert D., Paul Mohai, Robin Saha, and Beverly Wright. "Toxic Wastes and Race and Toxic Wastes and Race at Twenty." United Church of Christ. http://www.ucc.org/environmental-ministries/environment/toxic-waste-20.html (Accessed November 2010).

Johnson, Daniel L. and Charles E. Hambrick-Stowe. *Theology and Identity: Traditions, Movements, and Polity in the United Church of Christ*. Cleveland, OH: United Church Press, 2008.

United Church of Christ. "The Silent Disaster." http://www.ucc.org/disaster/technology-disasters (Accessed November 2010).

Patricia K. Townsend
Independent Scholar

United Farm Workers (UFW) and Antipesticide Activities

César Chávez, cofounder of the United Farm Workers, utilized nonviolent protest methods to raise awareness of pesticide dangers. A major player in labor rights, Chávez is pictured here with the Labor Council for Latin America Advancement.

Source: U.S. Department of Labor

The United Farm Workers (UFW) is the largest and oldest union of agricultural laborers in the United States. It has affected much policy and agenda over the years in the field of environmental public policy, operations, and worker conditions, including the movement to protect farmworkers from the harmful effects of pesticides. Cofounder César Chávez utilized nonviolent protest methods to raise awareness of pesticide dangers in the 1960s through the 1980s. The UFW became the first labor union to insist on government protection for farmworkers from harmful and deadly pesticides, leading to local and national laws regulating pesticide use and exposure.

César E. Chávez and Dolores Huerta founded the UFW in 1962. The two shared a common vision and passion for organizing agricultural workers into a cohesive movement with an effective agenda for getting the best possible working conditions for them. Farmworkers wanted to secure worker control over their workplaces. The labor union was first called the National Farm Workers Association (NFWA) and then later changed its name to the UFW. From the beginning of the UFW, Chávez and Huerta had concerns about the effects that harmful pesticides had on farmworkers. In the 1960s, the UFW launched an antipesticide campaign.

Routes of harmful pesticides include inhalation, ingestion, and skin penetration. Inhalation occurs when a pesticide is sprayed into the air. Pesticides can also be ingested through contaminated water, residues on food, and hand-to-mouth activities such as when children put objects that have been on the floor in their mouths. For many farmworkers, however, skin penetration is the primary source of exposure. Harvesting exposes farmworkers to the pesticide residue left on plant foliage. There could also be several or numerous pesticides applied during one season, which further increases the chances of workers becoming affected via skin penetration.

The antipesticides campaigns of the late 1960s and early 1970s proved successful for the UFW. In the late 1960s, Chávez marched with a crowd of several hundred people to the national headquarters of the U.S. Food and Drug Administration (FDA) and demanded that the government increase its oversight of the pesticides used on food crops. Chávez's strategy was that of nonviolent resistance, including fasts, protests, and boycotts. In the 1970s, Chávez and the UFW were still fighting for government regulation to limit or prevent the harmful effects of pesticides on farmworkers. The UFW and its supporters demonstrated the effectiveness of collective action.

Huerta negotiated contracts with many California grape growers that required protective clothing for workers in fields that were sprayed with pesticides. The UFW also played a key role in the successful campaign to ban the use of DDT (dichlorodiphenyltrichloroethane) and other harmful pesticides. The UFW also established contracts with the growers that required a longer period of time between pesticide application and workers' reentry into those fields where pesticides had been applied. They also made it mandatory for farmworkers to be tested for pesticides on a regular basis. These key steps in the UFW's fight against pesticides were accomplished well before there were any similar government regulations enacted for the farmworkers.

The UFW also became the first union to require joint union–management committees to enforce safety regulations regarding the use of pesticides in vineyards. A Health and Safety Committee was established that was controlled by workers and worker representatives. The committee obtained information about pesticides in the fields, which, in turn, aided them in preventing and treating pesticide-related illnesses. The committee also approved the use of less-toxic pesticides. Farmworkers' rights were always at the forefront of these efforts. For example, a worker had a right not to work if he or she felt that his or her health or safety were endangered. The growers had to provide personal protective equipment for the workers. They also had to supply adequate toilet facilities, water, and rest periods.

Chávez and the UFW also sought the involvement of other individuals, groups, and organizations from within and without the farmworker industry in the antipesticide movement, pointing to its broader dangers outside the industry. He encouraged activists like Marion Moses to become doctors aimed at addressing the health dangers that farmworkers face by encountering pesticides. Moses was a former UFW nurse who became Chávez's personal physician and union researcher. Moses later created the Pesticide Education Center in San Francisco, California, with the goal of public education about the adverse health effects of exposure to pesticides at home, at work, and in the surrounding community. Chávez also encouraged farmworker Pablo Romero, who attended medical school and afterward formed a community task force that established new rules aimed at lessening the risk of accidental pesticide exposure.

The UFW's antipesticide campaigns continued into the 1980s. The UFW boycotted grape growers in 1984 after the discovery that residents in central California were suffering from pesticide-related illnesses and a high incidence of childhood cancers. Chávez urged

Americans to stop buying grapes until the grape industry stopped using the pesticides that caused these deadly effects. In the mid-1980s, near Salinas, California, there were two occasions in which workers were accidentally sprayed with pesticides. Following these events, the UFW pressed Monterey County to enact one of the toughest pesticide restriction laws in the country at that time. The state of California followed suit and changed its policies in favor of the farmworkers.

The UFW became the first labor union to insist on government protection for farmworkers from harmful and deadly pesticides. Chávez died in 1993, but the UFW continued its fight against pesticides. The labor union continued to push for more government protection of farmworkers and others who live and work around the agricultural fields. In the end, Chávez and the UFW undoubtedly raised U.S. awareness of the dangers of agricultural pesticides, succeeded in changing public policy, and set the farmworkers' agenda for the 21st century.

See Also: Body Burden; Chávez, César (and United Farm Workers); Huerta, Dolores; Organizations and Unions.

Further Readings

Barger, W. K. and Ernesto M. Reza. *The Farm Labor Movement in the Midwest: Social Change and Adaptation Among Migrant Farmworkers*. Austin: University of Texas Press, 1994.

Ferriss, Susan, Ricardo Sandoval, and Diana Hembree. *The Fight in the Fields: Cesar Chavez and the Farmworkers Movement*. New York: Harcourt Brace, 1997.

Mooney, Patrick H. and Theo J. Majka. *Farmers' and Farm Workers' Movements: Social Protest in American Agriculture*. New York: Twayne Publishers, 1995.

Pulido, Laura. *Environmentalism and Economic Justice: Two Chicano Struggles in the Southwest*. Tucson: University of Arizona Press, 1996.

Shaw, Randy. *Beyond the Fields: Cesar Chavez, the UFW, and the Struggle for Justice in the 21st Century*. Berkeley: University of California Press, 2008.

Thompson, Heather Ann. *Speaking Out: Activism and Protest in the 1960s and 1970s*. Upper Saddle River, NJ: Prentice Hall, 2010.

Marcella Bush Trevino
Barry University

V

Vacation

One popular trend is taking "climate tours" to environmentally threatened areas, like Warming Island. Near Greenland, the island was uncovered and revealed this strait in response to Greenland's melting ice sheet.

Source: John Rudolf/Wikimedia

While many vacations advertised as ecofriendly are legitimate, some unscrupulous companies or individuals use the trend toward ecotourism simply for profit. Well aware of potential profitability, many nations are now developing national plans for attracting ecotourists. Because hotel owners in some countries are more interested in making money than in protecting the environment, certification agencies have surfaced to protect consumers from scams. To prevent vacationers from becoming victims of scams, some 60 certification programs have been launched around the globe. While some programs use very loose criteria for being considered environmentally responsible, others such as the Sustainable Tourism Certification Network of the Americas, Sustainable Travel International, the Rainforest Alliance, and Green Global are generally considered reliable. There are also groups, such as the Ski Club of Great Britain, that provide travelers with advice on choosing green hotels. Using an online *Green Resort Guide*, this club provides would-be travelers with information on the environmental practices of particular hotels, including recycling efforts, the use of green power, traffic reduction, sewage management, and climate and green-building policies. The Green Hotel Association, which provides information on environmental practices, is also a good resource when planning trips.

Ecologically minded travelers need to be aware of all aspects of green travel, including the fact that methods of travel have a serious environmental impact. Climate Action

Network Europe suggests that traveling by plane should be considered irresponsible for distances less than 625 miles. The concern over carbon dioxide emissions suggests that buses and trains are preferable to traveling by private automobile. Traveling by motor coach may cut carbon dioxide emissions by 55 to 75 percent as opposed to other modes of travel, and bus travel is both environmentally friendly and low-cost, with some companies offering fares as low as $1 each way. Instead of traveling to exotic resorts, many ecoconscious individuals now use their vacation time to volunteer for projects that improve local environments, taking advantage of such opportunities as organic farming and rebuilding temples. Others opt for low-impact travel such as visiting nature reserves or biking through Europe or the Adirondacks.

Hotel owners often claim to be ecofriendly if they take such simple steps as using energy-saving light bulbs or encouraging vacationers to reuse towels to save energy. More responsible hotel owners take additional steps such as using solar or wind energy. The Post Ranch Inn in Big Sur, California, for instance, engages in extensive use of solar panels to generate its own energy and provides vacationers with the added benefits of watching deer and wild turkey graze underneath the massive solar panels. Inn Serendipity near Monroe, Wisconsin, uses a solar-powered water-heating system, and its bathroom tiles have been fashioned from recycled automobile windshields.

At the same time that the *New York Times* was chronicling the importance of ecotourism in popular culture, the *New Oxford American Dictionary* was identifying the term *carbon neutral* as the "word of the year" for 2006. As result of the consciousness-raising surrounding carbon neutrality, many vacationers have begun to take advantage of carbon-offset programs, and a number of websites, including Sustainable Travel International (www.sustainabletravelinternational.org), have started offering calculators to teach prospective travelers how to accomplish this.

There are other motives for vacations besides ecotourism, of course, and ways to approach it while being environmentally conscious. Like the tradition of a work-free weekend, vacation has taken on a great level of importance in the labor-intensive American culture, despite or because of the fact that vacation time is so much shorter in the United States than in Europe and Australia (where workers typically enjoy over a month of vacation time, compared to the two weeks or less that is typical for American workers). Americans no longer attending school lack the tradition of a long summer break spent outside the city or in vacation homes, as practiced in much of western Europe, but the idea of a summer beach vacation, for instance, is still very evocative in the public imagination. Particularly in advertising, the idea is often conjured up of "getting away from it all," which can have almost ritualistic connotations: the beach is transformed into an oasis away from the hubbub and frenzy of it all, a retreat back to nature and away from civilization.

See Also: Corporate Social Responsibility; Ecotourism; Weekends; Work.

Further Readings

Anderson, Rachel E. "Going Green: Top Ten Rules of Eco-Travel." *E: The Environmental Magazine,* 17/4 (2006).

Binshtock, Avital. "Low-Carbon Cupid." *Sierra,* 95/1 (2010).

"Budget-Savvy Green Travel Guide." *Mother Earth News,* 235 (2009).

Foster, John Bellamy. *The Ecological Revolution: Making Peace With the Planet*. New York: Monthly Review Press, 2009.
Green Hotels Association. "Why Should Hotels Be Green?" http://greenhotels.com/index.php (Accessed September 2010).
Higgins, Michelle. "If It Worked for Costa Rica . . ." *New York Times* (January 22, 2006).
International Ecotourism Society. "What Is Ecotourism?" http://www.ecotourism.org/site/c.orLQKXPCLmF/b.4835303/k.BEB9/What_is_Ecotourism_The_International_Ecotourism_Society.htm (Accessed September 2010).
MacGregor, Sherilyn. *Beyond Mothering Earth: Ecological Citizenship and the Politics of Care*. Vancouver: UBC Press, 2006.
Priesnitz, Wendy. "Have a Sustainable Summer." *Natural Life*, 104 (2005).
Priesnitz, Wendy. "Planning Your Green Vacation." *Natural Life*, 115 (2007).
Schendler, Auden. "Ski Green." *Business and the Environment*, 17/1 (January 2006).
Schor, Juliet B. and Betsy Taylor. *Sustainable Planet: Solutions for the Twenty-First Century*. Boston, MA: Beacon Press, 2002.
Ski Club of Great Britain. "Green Resort Guide." http://www.skiclub.co.uk/skiclub/skiresorts/greenresorts/default.aspx (Accessed September 2010).

Bill Kte'pi
Independent Scholar

Vanity Fair Green Issue

In May 2006, 2007, and 2008, *Vanity Fair* published what editors nicknamed the "Green Issue" in an effort to raise awareness among its readers about matters affecting the environment. Each of these issues, devoted to exploring and analyzing topics ranging from climate change to saving polar bears, found popularity with readers. The covers portrayed celebrities caring about Earth, which encouraged readers to behave in a similar manner. Surprisingly, however, editors only released three official green issues, and *Vanity Fair*'s sudden termination of the tradition sparked debate about whether conversations about environmental concerns had become common and mainstream, rather than marginalized.

Known for its vibrant, insightful writing and capturing the pulse of trends in culture, *Vanity Fair* typically offers its readers tips on high fashion, reflections on politics, and interviews with writers, filmmakers, and movie stars. When *Vanity Fair* devoted an issue a year to both couture and the environment, being green became, in a word, fashionable. Green activists, stereotypically portrayed as tree-hugging hippies, now appeared in the forms of well-dressed, happy, successful celebrities. Indeed, the magazine recruited powerhouse stars for the covers of the Green Issues: Julia Roberts, George Clooney, Robert F. Kennedy, Jr., and Al Gore (2006); Leonardo DiCaprio (2007); and Madonna (2008). The rhetoric of the magazine covers also emphasized the power celebrities have, more so than ordinary citizens, to make the world a greener place. On the 2006 cover, Julia Roberts, wearing a silken green gown and a wreath of green leaves on her head, suggested that being green leads to a relaxed, luxurious life. The 2008 cover depicted Madonna as Atlas, holding the world behind her; dressed like a dominatrix, she appeared to be protecting the world fiercely (though her interview focused more on her latest film than what conscious choices she makes to be ecofriendly and environmentally aware).

Each of the Green Issues championed different environmental causes. In 2006, the magazine's cover headline announced that global warming should be considered more dangerous than terrorism. In 2008, topics like the status of farmers and climate control received attention on the cover. Between advertisements for jewelry, cologne, vacation hotspots, and skin creams, editors of the 2006 issue placed sporadic green-focused content like the column "Green Gifts" with the subtitle "Looking Smart and Doing Your Part." Although one of the major articles in the issue concerned mining in the Appalachian Mountains, the majority of the magazine's green content occurred late in the issue, beginning on page 169 with Al Gore's piece "The Future Is Green." The magazine included brief profiles of celebrities, researchers, scholars, and activists who make the environment a priority, including Edward Norton and Bette Midler. Mark Hertsgaard's article "While Washington Slept" raised awareness about global warming, and the insert "What You Can Do: 50 Ways to Help Save the Planet" offered suggestions like buying a hybrid vehicle or being sure to use both sides of the paper when printing.

The last Green Issue, published in 2008, provided readers with a wealth of material on being green, especially in comparison to the amount of green content in the first *Vanity Fair* issue devoted to the environment. The 2008 issue included many more green advertisements than in 2006; the last Green Issue also included a score of lengthy, well-researched articles like Matt Tyrnauer's "Industrial Revolution, Take Two" (which focused on ecofriendly buildings), a piece about perils faced by farmers in "Monsanto's Harvest of Fear," and other articles about the Arctic oil rush, a green museum, and the plight of polar bears. With each issue, editors devoted more investigative content to green topics.

Rather suddenly, in 2009, *Vanity Fair* editors announced that they had canceled the Green Issue. In "It's So Last Year: *Vanity Fair* Abandons the 'Green Issue,'" Rachel Shields discussed the controversial decision. Editors explained that *Vanity Fair* would cover topics relating to the environment in each issue. Shields wrote that publishers of the magazine claimed that environmental issues were now "integral" to the media, so a Green Issue was no longer necessary. Publishers also claimed that concerns like the failing economy moved to the forefront as the issue most important to *Vanity Fair*'s readers. Shields reported that during 2008 the discussion of green issues plummeted by over 25 percent in comparison to the previous year. While some readers and critics did not fault *Vanity Fair* for focusing attention on the most prominent societal concerns like the troubled economy, green activists felt troubled. Shields referred to Colin Butfield of the World Wildlife Fund, an activist unsettled by *Vanity Fair*'s decision. Butfield argued that caring about environmental issues should not be presented as a fad; similarly, the publication's sudden termination of the Green Issue suggested to some readers that green issues are now passé.

Ultimately, *Vanity Fair*'s Green Issues educated and cultivated awareness in readers, with the help of both celebrity power and investigative reporting. The magazine's reputation for covering topics that are timely and fashionable, however, remains a cause of concern now that the publication no longer devotes an entire issue to the environment.

See Also: DiCaprio, Leonardo; Gore, Jr., Al; Green Consumerism; Media Greenwashing; Popular Green Culture; Print Media, Magazines.

Further Readings

"The Green Issue." *Vanity Fair*, 549 (May 2006).
"The Green Issue." *Vanity Fair*, 561 (May 2007).
"The Green Issue." *Vanity Fair*, 573 (May 2008).
Shields, Rachel. "It's So Last Year: *Vanity Fair* Abandons the 'Green Issue.'" *The Independent*. http://www.independent.co.uk/news/media/its-so-last-year-vanity-fair-abandons-the-green-issue-1662661.html (Accessed September 2010).

Karley Adney
Independent Scholar

Veganism/Vegetarianism

Vegans and vegetarians cite not only ethical but also environmental reasons to not eat meat: the livestock sector is responsible for 18 percent of the world's greenhouse gas emissions and 37 percent of all human-induced methane.

Source: Philip Fischer/USAID

The ongoing paradigm shift in culture and lifestyles toward promoting a sustainable environment is changing culture and the way we eat. Food is front and center of this change with a public becoming increasingly aware of organic foods, the slow food movement, buying produce locally, and the health benefits of a plant-based diet. An increasing number of people are eating locally produced or organic food, with plants as a major component of their diet. This "green enlightenment" is part of the ongoing paradigm shift in culture and lifestyle empowering the individual and society as well as promoting a more sustainable environment.

The word *vegan* was devised in 1944 by Donald Watson, who founded the Vegan Society. Author Joanne Stepaniak (2000) states that Watson combined the first three and the last two letters of "vegetarian" to form the word *vegan*, which he perceived as the beginning and end to the word *vegetarian*. Watson, along with several other members of the Vegetarian Society, wanted to form an alliance of nondairy vegetarians as a subgroup of the society. The Vegan Society in 1951 officially defined veganism based on the principle that humans should live without exploiting animals. In this context, veganism is not in itself a set of actions or practices but instead is a set of guiding principles on how one should conduct a life.

The Vegan Society has also used other definitions. Primarily, this philosophy places more emphasis on veganism as a practice and a lifestyle. It is a philosophy and a way of living that seeks to exclude all forms of exploitation of and cruelty to animals for food, clothing, or any other purposes. Veganism promotes the development and use of animal-free alternatives for the benefit of humans, animals, and the environment. In terms of diet, it promotes the practice of eliminating all products wholly or partly derived from animals. An animal product is defined as any product derived from animals, including meat, poultry, seafood, eggs, dairy products, honey, fur, leather, wool, and silk. Statistics vary but the current data points to the fact that approximately 0.5 percent of Americans identify themselves as vegans and 4 percent as vegetarians.

A number of the philosophical foundations that veganism was founded on include the ethical question related to whether it is right for humans to use and kill animals. The argument has been made by a number of scholars that animals are entities that possess an inherent value and therefore have basic moral rights. The principal moral right they possess is the right to be civilly treated. Additionally, the argument has been made that animal agriculture is unfair even when animals are raised humanely. Various scholars have contended that the use of animals for food becomes open to discussion, particularly when animal products are a luxury rather than a necessity. This philosophy entails that killing animals for food is always wrong unless it is needed for survival. Veganism still today describes a lifestyle and belief system that revolves around a respect for life.

In 1960, the American Vegan Society was founded by Jay Dinshah, advocating a strictly plant-based diet and a lifestyle free of animal products. This society championed the philosophy of *ahimsa*, a Sanskrit word interpreted as "dynamic harmlessness," along with encouraging service to humanity, nature, and creation. Accordingly, in order to practice veganism, it is not sufficient to simply avoid specific foods and products; it is necessary to actively participate in beneficial selfless actions as well. Veganism is not merely passive resistance; it compels practitioners to make deliberate and dynamic choices about each and every activity in their lives and to consider the impact and benefit of their actions.

Transition to Veganism

Becoming vegan is a process. Someone does not convert to complete veganism overnight; more typically people transition to a vegan lifestyle. This begins by altering their diet and then gradually replacing their clothing and other necessities with other options. Veganism encourages the individual to embark on a personal journey. It encourages the individual to strive to do more; understand more clearly the world around him; and to be more loving and humble. Author Joanne Stepaniak (2000) asserts that the gift of veganism is a guide for compassionate living, a path to honoring our roots, our planet, all life, and ourselves.

The practice of veganism entails abstaining from the use of animal products in every aspect of daily living as much as is possible and practical. The American Vegan Society espouses the following six pillars of a compassionate way of living based on the first letter of each pillar, which spells out the word *ahimsa*, the philosophy of nonviolence practiced and promoted by Mahatma Gandhi:

1. Abstinence from animal products
2. Harmlessness with reverence for life
3. Integrity of thought, word, and deed

4. Mastery over oneself
5. Service to humanity, nature, and creation
6. Advancement of understanding and truth

Many individuals become vegans for the health benefits. The American Dietetic Association annually publishes its position on vegan and vegetarians diets. It states that vegetarian and vegan diets are healthful, nutritionally adequate, and provide health benefits in the prevention and treatment of certain diseases. Many medical practitioners believe that animal fat and protein diets such as the standard American diet are detrimental to health. They also state that a lifestyle change incorporating a low-fat vegetarian or vegan diet could not only prevent various diseases, including artery disease, but actually reverse them. The American Dietetic Association concludes that diets that avoid meat tend to have lower levels of saturated fat, cholesterol, and animal protein and higher levels of carbohydrates, fiber, magnesium, potassium, folate, antioxidants such as vitamins C and E, and phytochemicals.

Veganism and the Environment

Those who adopt veganism for environmental reasons believe that veganism consumes fewer resources and causes less environmental damage than does an animal-based diet. Animal agriculture is linked to climate change, water pollution, land degradation, and a decline in biodiversity. An animal-based diet uses more land, water, and energy than a vegan diet. The United Nations Food and Agriculture Organization's (2006) report "Livestock's Long Shadow" concludes that the livestock sector—cows, chickens, and pigs—emerges as one of the top two or three most significant contributors to our most serious environmental problems at every scale from local to global. The authors of this report conclude that the livestock sector is responsible for one of the largest sources of greenhouse gases, is responsible for 18 percent of the world's greenhouse gas emissions as measured in carbon dioxide (CO_2) equivalents, and produces 65 percent of human-related nitrous oxide and 37 percent of all human-induced methane. By comparison, all of the world's transportation (including cars, trucks, buses, trains, ships, and planes) emits 13.5 percent of the CO_2. Geophysics researchers Gidon Eshel and Pamela Martin at the University of Chicago in a 2006 study concluded that a person switching from the average American diet to a vegan diet would reduce CO_2 emissions by 1,485 kilograms per year.

Likewise, a vegetarian diet historically was closely related to the idea of nonviolence toward animals and was promoted by religious groups and philosophers. The earliest records of vegetarianism come from ancient India and ancient Greece in the 6th century B.C.E. Following the Christianization of the Roman Empire, vegetarianism practically disappeared from Europe. Orders of monks in medieval Europe restricted or banned the consumption of meat for ascetic reasons, but none of them avoided fish. Vegetarianism reemerged during the Renaissance, becoming more widespread in the 19th and 20th centuries. In 1847, the first Vegetarian Society was founded in England, Germany, the Netherlands, and other countries followed. The International Vegetarian Union, a union of the national societies, was founded in 1908. In the Western world, the popularity of vegetarianism grew during the 20th century as a result of nutritional, ethical, environmental, and economic concerns.

There are a number of types of vegetarianism depending on the use of certain of foods. The following is a listing of some of the variations in a vegetarian diet inclusive of vegans:

- Ovo-lacto vegetarianism includes animal products such as eggs, milk, and honey
- Lacto vegetarianism includes eggs but not milk
- Ovo vegetarianism includes eggs but not milk
- Veganism excludes all animal flesh and animal products, including milk, honey, and eggs
- Raw veganism includes only fresh and uncooked fruit, nuts, seeds, and vegetables
- Fruitarianism permits only fruit, nuts, seeds, and other plant matter that can be gathered without harming the plant
- Su vegetarianism excludes all animal products as well as vegetables in the allium family: onions, garlic, scallions, leeks, or shallots
- Macrobiotic diets consist mostly of whole grains and beans

Strict vegetarians avoid products that may use animal ingredients on their labels or that use animal products in their manufacturing. For example, cheese that uses animal rennet (enzymes from animal stomach linings), gelatin from animal skin, bones, and connective tissue; some sugars that are whitened with bone char (e.g., cane sugar, but not beet sugar); and alcohol clarified with gelatin or crushed shellfish and sturgeon. Some vegetarians will describe themselves as vegetarian while practicing a semi-vegetarian diet. Some people call themselves flexitarians, in this manner reducing their consumption of meat as a way of transitioning to a vegetarian diet for health, environmental, or other reasons. There is another term called *semi-vegetarian* that includes fish and sometimes other seafood.

Health Benefits

The health benefits of a vegetarian diet are similar to those of a vegan diet. Estimates vary but about one in 20 Americans is vegetarian. The American Dietetic Association states that a vegetarian diet is healthy, nutritionally adequate, and provides health benefits in the prevention and treatment of certain diseases. Essential nutrients, proteins, and amino acids for the body's sustenance can be found in vegetables, grains, nuts, soymilk, eggs, and dairy. Vegetarian diets offer lower levels of saturated fat, cholesterol, and animal protein, and higher levels of carbohydrates, fiber, magnesium, potassium, folate, antioxidants such as vitamin C and E, and phytochemicals. Vegetarians tend to have lower body mass index, lower levels of cholesterol, lower blood pressure, and less incidence of heart disease, hypertension, and type 2 diabetes.

Many vegetarians are guided by the practice of economic vegetarianism. Their philosophical viewpoint encompasses the belief that the consumption of meat is economically and ecologically unsound. Their belief is grounded in the assumption that massive reduction in meat consumption in industrial nations will ease their healthcare burden and improve public health. Declining livestock herds will take pressure off rangelands and grain lands, allowing agriculture resource bases to rejuvenate. As the world populations grow, lowering meat consumption will allow more efficient use of land and water, while at the same time making grain more affordable to the world's growing population.

People may choose to be vegans or vegetarians for many reasons. They may have been raised in a household that embraces this form of eating, or have a partner, family member, or friend who believes in a plant-based diet. A great influence is the ramification of the current sustainability and environmental movement and the issues related to global climate change. Animals fed on grain and those that rely on grazing require more water than grain crops. According to a U.S. Department of Agriculture 2001 report, growing the crops needed to feed farmed animals requires nearly half of the U.S. water supply and 80 percent of its agricultural land. Animals raised for food in the United States consume 90 percent of the agricultural land that is available. These animals raised for food consume 90 percent of the soy crop, 80 percent of the corn crop, and 70 percent of the grains. It requires 10 times as many crops to feed animals being bred for meat production as it would to feed the same number of people on a vegetarian diet. Many proponents of a plant-based diet believe that it is ecologically irresponsible to consume meat. To continue to produce animal-based food is less efficient than the harvesting of grains, vegetables, legumes, seeds, and fruits.

See Also: Diet/Nutrition; Global Warming and Culture; Individual Action, National and Local; Organic Foods; Organics; Social Action, National and Local; Veganism/Vegetarianism as Social Action.

Further Readings

Koerner, Brendan I. "Vegans vs. Vegetarians: What Kind of Diet Is Best for the Environment?" *Slate* (October 23, 2007). http://www.slate.com/id/2176420 (Accessed August 2010).

Stepaniak, Joanne. *Being Vegan: Living With Conscience, Conviction, and Compassion.* Los Angeles: Lowell House, 2000.

"Why Eating Less Meat Could Cut Global Warming." *The Guardian* (November 11, 2007). http://www.guardian.co.uk/environment/2007/nov/11/food.climatechange (Accessed August 2010).

Carl A. Salsedo
University of Connecticut

Veganism/Vegetarianism as Social Action

At the heart of the green culture movement lies a conflict over the relationship between humans and the rest of nature, and this conflict is nowhere more personal than in choices about food—in particular, the choice to abstain from consuming animal products. Ethical veganism/vegetarianism is a grassroots social, cultural, and sometimes political movement that incorporates elements of postindustrial and new social movements. While participants vary in their ideological and tactical stances toward the broad issue of speciesism (preference for human interests in relations between species), the ethical vegetarian (veg) movement at its core is related to the animal rights movement, with the goal of countering the idea that animals should be used as human food or raw materials. Like many other recent social movements, the veg movement is loosely defined, and participation is often motivated by diverse concerns, such as animal rights, environmentalism and environmental justice, anti-capitalism, pacifism, population health, or radical feminism. The recent broadening in motivations for ethical veganism contributes to a rising profile and legitimacy for the movement even as it complicates what it means to be vegan or vegetarian.

Social activism can be defined as intentional action in solidarity to bring about change or support an argument, and might consist of protest, dissent, or formulation of an alternative movement. In that, the motivation for ethical vegetarianism is not typically based on human material welfare; the veg movement is a New Social Movement, focusing on how individual lifestyle and identity contribute to collective action. Like other New Social Movements such as environmentalism and feminism, the veg movement is broad and informal, mainly composed of "supporters" rather than "members." Where political movements aim to shift the balance of power and resources within a social system, so the tactics employed—such as protests—tend to emphasize conflict; cultural movements often use cultural products such as art and literature to shift the social framing of an issue, shifting values in order to spur collective action. In this case, the primary tactic of the core of the ethical veg movement is social and cultural: to frame treatment of animals as a civil rights issue, such that animals and humans are no different in that both have rights to humane treatment. Segments of the ethical veg movement outside the animal rights core may offer alternate social frames for the decision to abstain from animal products, such as environmentalists who advocate the health of ecosystems and future generations of humans, but still tend to emphasize social and cultural tactics.

Ethical Reasons

Ethical reasons to avoid animal consumption are more diverse and controversial, even among those committed to ethical vegetarianism. Because the focus of the veg movement is on individual action, it can be characterized as a transformational cultural movement: the goal is to change oneself and inspire change in others. Although vegetarianism has a history dating back to ancient Greek philosophers and Buddhist teachers, the Western veg movement took root in the United Kingdom in the early 1800s. An outgrowth of progressive reform movements sometimes linked to the temperance movement, health fads, and Protestant religious movements, early vegetarians often focused on emotional disgust at the sight of slaughterhouses or spiritual and physical purity. The contemporary veg movement is sometimes dated to the founding of the American Vegan Society (AVS) by Jay Dinshah in 1960. The AVS, along with the North American Vegetarian Society and the Vegetarian Union of North America, which promote veg culture through publications and a network of local organizations, relies on Eastern religious traditions of nonviolence, the relatedness of all living beings, and benevolent concern for the results of individual action for the collective good. Likewise, a variety of organizations (such as Vegetarian Resource Group and Vegan Outreach) focus primarily on providing information to individuals interested in changing their diets. Farm Sanctuary raises awareness about the treatment of animals on factory farms by engaging in rescues. However, some branches of the movement do focus on political aims, mostly centered on fostering an institutional environment that produces fewer obstacles to a plant-based diet. For instance, the Physicians Committee for Responsible Medicine (which promotes veganism through the promulgation of scientific evidence for the health benefits of veganism) is involved in a political campaign to increase availability of vegetarian options in school lunches. Still, the highest-profile veg organization is almost certainly PETA, People for the Ethical Treatment of Animals, which primarily uses cultural tactics, including shocking street theater using the human body to bring attention to the bodies of animals, but also media campaigns to raise awareness of vegan actresses and athletes.

The 1975 publication of *Animal Liberation* by Peter Singer reintroduced a strong philosophical foundation for the veg movement. Rather than using the concept of animal rights, he promotes a utilitarian framework in the idea that animals have interests in that they are capable of suffering, and that their suffering deserves equal consideration with that of humans. This argument borrows language from the familiar and more socially accepted civil rights movements against racism and sexism to frame animal consumption as an example of speciesism. By contrast, Tom Regan rejects the balance of human and animal needs or focus on achieving incremental improvements in animal welfare by taking a moral absolutist stance for animal abolition.

While utilitarian, absolutist, and religious arguments focus on an external moral imperative for animal welfare, postmodern arguments focus on the role of humans in the social construction of the relationship between humans and (other) animals. For instance, antiglobalization activists who are concerned with negative externalities of capitalism may see the commodification of animals as connected with the commodification of labor and other trends to focus on seeming improvements in the efficiency of production at the expense of quality of life. Ecofeminist participants in the veg movement, such as Carol Adams and Marjorie Spiegel, abstain from animal products because they see connections between human domination of animals and nature and other forms of oppression, especially racial and gender oppression. Some pacifists, harkening back to Tolstoy's writings, build on this framework by arguing that violence against animals makes violence against other humans more culturally acceptable, a view that dovetails well with recent criminological research.

The postmodern shift in ideology is also reflected in a shift from religion to science as the source of knowledge about the natural world. Environmental motivations for a plant-based diet have been gaining media attention, starting with Frances Moore Lappé's 1971 *Diet for a Small Planet*. Research, including work done by the United Nations, shows that consumption of animal products (particularly by factory farming) is a key contributor to climate change and pollution. For instance, EarthSave International, which has a broader environmental agenda, advocates for veganism rather than sustainable animal farming. Despite the rising global popularity of the Western diet, recent medical research suggests that vegetarians have lower rates of obesity, diabetes, heart disease, cancer, and overall mortality, and that reducing meat consumption can lower risk and alleviate severity of disease. Concerns about food safety and welfare of farm and slaughterhouse workers may also result in boycotts. Because animals raised on grain diets consume far more calories than they can produce, animal production also tends to result in higher global prices, which disproportionately hurt the poor in developing countries. Environmental justice advocates may also join in the ethical veg movement based on concerns that disadvantaged groups within regions may disproportionately bear the burdens of industrial agriculture, such as pollution and reduced property values, without receiving fair compensation. As developing countries adapt to a Western diet and to Western agricultural practices, and as agricultural production continues to be controlled by ever-fewer firms, these concerns are likely only to grow worse globally.

Many Motivations

While concerns about the environment, population health, and poverty contribute to the rising popularity of the ethical veg movement, this diversity of motivations also complicates analysis of the movement. For "green" ethical vegetarians, the emphasis is on reduction and reform—rather than elimination—of animal consumption, and the focus is often on the well-being of humans and ecosystems rather than of animals. Environmentalists focus on broader issues affecting species and habitats, aiming more toward ecological balance rather than focusing on treatment of individual animals. Even within the core of the ethical veg movement, which can still be seen as connected with the animal rights movement, there is broad agreement with the idea of balancing the interests of species, but conflicts emerge with respect to tactics. While some advocate gradual animal welfarist approaches such as advocating for better treatment for farm animals, others see small improvements as inimical to the main goal of complete animal liberation. This variety reflects a broader lack of agreement within the veg movement over the appropriate relations between species. Animal welfarists are not against using animals for food, research, hunting, or other purposes if animal suffering is minimized. By contrast, animal liberationists advocate the abolition of all animal exploitation.

Although promoting veganism is a common side tactic of the animal rights movement and increasingly of environmentalists, there are also organizations that exist primarily for the purpose of promoting veganism. However, as a cultural and transformational movement, success for the movement is measured in terms of inspirational ability to persuade individuals to adopt veg principles. Because some organizations view the terms *vegan* and *vegetarian* as politically loaded and unacceptable to many members of the public, some activists have shifted from an all-or-nothing attempt to convert individuals to veganism in favor of a gradualist encouragement to shift to a plant-based diet, a more flexible and politically neutral term. This strategic language is related to a tactical shift to focusing on the emerging research on health and ecological benefits of a plant-based diet as being much more acceptable to a broad audience than ethical or emotional arguments would be.

Like many other cultural movements, some people active in promoting veg values may not belong to any veg organization or even adhere to basic vegetarian principles. Where stereotypes from the 1970s held that vegan food would necessarily be unpalatable and vegans emaciated, recently, cultural trendsetters have set out to capitalize on and promote a positive view of a vegan diet. Food writer Mark Bittman at the *New York Times* promotes a plant-based diet that may include some meat. Chloe Coscarelli, a vegan celebrity chef, recently won *The Food Network's Cupcake Wars* contest against conventional chefs. Michael Pollan's best-selling books such as *The Omnivore's Dilemma* urge readers to ponder the ethics of animal consumption and welfare, although he ultimately remains an omnivore. The film *Food, Inc.* also popularized discussion of factory-farming practices. A character in the film *Scott Pilgrim* derives superpowers from his vegan diet. While these cultural products encourage consideration of the core principles of the ethical veg movement, they often do so outside the animal rights framework and may result in a dilution of the meaning of the terms *vegetarian* and *vegan*. When surveyed, many people who consider themselves vegan sometimes consume meat, not to mention using other animal products. In this way, the popularity of veg ethics can be problematic as the terminology and boundaries of the movement are diluted. On the one hand, given that green culture and ethical vegetarianism are seen as "cool," veg movement activists have achieved their goals. On the other hand, the boundaries of the movement become blurry.

See Also: "Agri-Culture"; Animals (Confinement/Cruelty); Cooking; Diet/Nutrition; EcoPopulism; Green Consumerism; Obesity and Health; Popular Green Culture; Social Action, Global and Regional; Social Action, National and Local; Veganism/ Vegetarianism.

Further Readings

Cherry, Elizabeth. "Veganism as a Cultural Movement: A Relational Approach." *Social Movement Studies*, 5/2 (2006).

Maurer, Donna. *Vegetarianism: Movement or Moment?* Philadelphia, PA: Temple University Press, 2002.

Phelps, Norm. *The Longest Struggle: Animal Advocacy From Pythagoras to PETA*. New York: Lantern Books, 2007.

Singer, Peter. *Animal Liberation: A New Ethics for Our Treatment of Animals*. New York: New York Review/Random House, 1975.

Katherine King
University of Michigan

W

Water (Bottled/Tap)

Water purification has been around since as early as 2000 B.C.E. as a symbolic act—pure water is sacred—and a practical one. This bottle of freshly purified water was made in a water treatment facility at Camp Liberty, Iraq.

Source: Pfc. Samantha Schutz/U.S. Department of Defense

Although water has always been essential to human life, the way that humans maintain and perceive a supply of safe drinking water has shifted over time. Water is both a physiological necessity and a culturally bound phenomenon to which humans attach meaning and value. This is reflected in the notion that water distribution and consumption occurs within a sociohistorical context imbued with meaning. Accordingly, shifts in the way in which water is obtained, the perceptions and maintenance of pure water, and the situation of water within the market structure of society evidence the culture surrounding drinking water.

Water Distribution

Throughout the history of human societies, access to a water supply became a hallmark of whether an area could be settled and developed. Water transportation systems increased the ability of civilizations to geographically move out from a source of water as well as allowed a water supply to sustain a larger population. One of the most advanced systems of water transport in ancient times was the aqueducts of the Roman Empire. This system provided relatively clean water to cities throughout the empire.

The current water distribution system, which enables tap water in homes, began in the early 19th century in Europe. Much of the drinking water infrastructure in the United States was built in the 30 years following World War II, a time when the population in the United States was growing. The ability to have drinkable water by turning on a tap relies on a complex infrastructure, which involves water treatment plants, sewage lines, drinking water distribution lines, and storage facilities. Additionally, the functioning of this system impacts the growth and development of industrial, commercial, and residential centers.

The Symbolic Value of Pure Water

Societies not only require water, but demand water that is pure. However, the way in which a society defines pure water and seeks to obtain this level of purity is culturally conditioned. First, the purity of water has symbolic value. Within many societies, water is sacred or represents a sacred act. For example, the Balinese construct water temples as an act of worship for the supply of water that sustains their rice crops. Water is used in many religions to symbolize a spiritual cleansing.

However, it was acknowledged relatively early in history that not all water sources were as pure as they could be. Historical records indicate an early recognition that various treatment methods could improve water quality. Sanskrit writings from as early as 2000 B.C.E. detail methods to clean water. In addition, Egyptian tomb paintings from as early as the 15th and 13th centuries B.C.E. portray water purification systems.

The use of early filtration and purification systems seems to be linked to improving the aesthetic characteristics of water (i.e., taste, odor, hardness). However, there is evidence to suggest an awareness that when techniques were used to "clean" the water, the health of the population improved. For example, Hippocrates's treatise on public hygiene included a water purification system. Similarly, Mesopotamia in 200 B.C.E. had sanitation laws that required cisterns and wells to be a specified distance from "unclean" areas such as cemeteries and slaughterhouses. Additionally, at various points in history, bottled water was perceived to have medicinal benefits. This water was valued as a healing element.

With the development of microbiology, a clear connection was made between water quality and human health. During this time, pure water was no longer defined simply in terms of aesthetic qualities or religious symbolism. Rather, pure water was free from contaminants that were harmful to human health. This resulted in an awareness of water treatment as an essential component of protecting human health and recognition of the need for a consistent method to improve water quality. Early treatment of water was done through the use of mechanical filtration, a process that often used sand. The first municipal water authority, which began in Scotland in the 19th century, used this sand filtration method. Today, sand filtration is only one component of water treatment. Scientific advancements, most notably the disinfection of water with chlorination, have further improved water quality and decreased morbidity and mortality from waterborne pathogens.

Purity and Bottled Versus Tap Water

In part, the bottled water versus tap water debate reflects broader cultural concepts of "purity" and society's response to water pollution. Bottling water is not a new phenomenon. It started in 1820 in the United States and was quite popular throughout the 19th century. During this time period, bottled water was considered safer to drink than tap water because it was free from harmful contaminants. The practice and popularity of bottling water declined when safe/clean drinking water could be obtained from the tap through disinfecting water with chlorination.

However, an increased awareness and public response to water pollution during the modern environmental movement challenged the purity of drinking water obtained from the tap. During the modern environmental movement of the 1960s and 1970s, citizens responded to water pollution through collective action in the form of social movements. The public response and increased demand for clean water, in part, influenced legislation to maintain a standard of water quality. In relation to drinking water, the Safe Drinking Water Act was passed in the United States. This act set a standard of quality and ensured the water quality was met. If one's tap water comes from a community water system, which is a public water distributor in the United States, then the quality of water is regulated by the Environmental Protection Agency (EPA).

In the late 1980s, public distrust in the safety of tap water, coupled with marketing claims that bottled water was superior in terms of purity, led to an unprecedented rise in sales and consumption of bottled water. In contrast to the environmental social movements of the 1960s and 1970s, the rise in bottled water consumption reflects an individual-level response to a perceived collective threat. It is suggested that advanced societies have shifted priorities in relation to environmental issues such as water pollution from protection of the environment to protection of individuals from environmental hazards. Purchasing bottled water, in part, reflects a cultural shift to buy environmental protection through products that reduce exposure to harmful substances.

Water as a Commodity

Human consumption of water also occurs within changing political and economic conditions. Therefore, obtaining pure water not only reflects consumer demands, but also occurs within the broader capitalist structure of society. As a commodity, drinking water is a time-bound status symbol, with the type of drinking water representing a position within society. Additionally, bottled water has represented a more pure version of drinking water. Today, one could pay up to 10 times more for water in a glass or plastic bottle than for water from the tap.

The consumer market for bottled water was fueled in part by the notion that the purity of water could be transformed into a commodified good, and perceptions of pure water could be influenced through marketing techniques. During the early 1900s, bottled water was used mostly in factories among lower- and working-class groups. However, in the late 1970s and 1980s, the Perrier company led the marketing campaign for bottled water, and through this campaign, bottled water consumption was attached to an elite consumer who could afford higher levels of purity. It was during this time that drinking bottled water became a fad among the elite upper class.

Drinking water as a consumer good, in part, reflects consumer trends. As a backlash to the bottled water trend, green marketing now encourages the use of personal home filtration systems and reusable water bottles. The use of disposable plastic water bottles, in part, symbolizes increased waste production by users. Reusable water bottles are growing in popularity, particularly those made of products other than plastic.

Another issue related to the commodification of water is whether water is a public good, or whether water distribution systems should be privatized. As a public good, water would be provided and regulated by the government. The highway system in the United States is an example of a public good. On the other hand, as a private good, distribution of water would occur through various companies. These companies would obtain ownership of a source of water, distribute and maintain the water, and charge a fee for service. The privatization of water would mean that water for human consumption is in part subject to market fluctuations.

See Also: Greenwashing; Media Greenwashing; Popular Green Culture; Recycling; Shopping; Water (Bottled/Tap); Water Pollution.

Further Readings

American Water Works Association (AWWA). *Water Quality and Treatment: A Handbook of Public Water Supply*. Denver, CO: AWWA, 1971.
Baker, M. N. and Michael J. Taras. *The Quest for Pure Water: The History of the Twentieth Century*, Volume 1 and 2. Denver, CO: AWWA, 1981.
Chapelle, Francis. *A Natural History of Bottled Spring Waters*. New Brunswick, NJ: Rutgers University Press, 2005.
Szasz, A. *Shopping Our Way to Safety: How We Changed From Protecting the Environment to Protecting Ourselves*. Minneapolis: University of Minnesota Press, 2007.
U.S. Environmental Protection Agency. "Water Infrastructure" (2010). http://www.epa.gov/waterinfrastructure (Accessed June 2010).
U.S. Environmental Protection Agency, Office of Water. "Water on Tap: What You Need to Know" (2009). http://www.epa.gov/safewater (Accessed June 2010).

Anjel Stough-Hunter
The Ohio State University

Water Pollution

Only 1 percent of the world's water supply is freshwater that is suitable for drinking. Because this relatively small amount of drinkable water is essential for human life, pollution of this water supply has detrimental impacts. Every year, more people die from unsafe water than from all forms of violence, including war. At its most basic level, pollution of water occurs when any substance (gas, liquid, or solid) changes the quality of a body of water. These changes in water quality negatively impact living organisms. Humans rely on water, while simultaneously polluting the supply on which they depend. The way in which water is polluted and the management of this pollution is reflective of the culture and dominant activities of a society.

Water Pollution as a Result of the Dominant Activities of Society

Human activities as diverse as industrial processing, agriculture, and maintenance of one's lawn can contribute to water pollution. The list of substances that are considered water pollutants is lengthy. However, most types of water pollution fall into one of six major categories: sediment, oxygen-depleting nutrients, organic chemicals, inorganic compounds, thermal pollutants, and biological pollutants.

The first category of water pollutants is sediment, which is simply the dirt in the stream bed or lake bottom. In correct amounts, sediment is not problematic. However, increases in the amount of sediment in a body of water can increase the temperature of the water, impair photosynthesis, reduce visibility, and clog the gills of fish. Increases in sediment often come from erosion of soil next to the stream or lake bank. This soil erosion is often linked to agricultural practices that focus only on increasing crop production or construction projects.

The second category of water pollutants is referred to as oxygen-depleting nutrients, because they decrease the level of oxygen in a body of water. This type of water pollution includes substances such as household cleaners, fertilizers, faulty septic systems, and runoff from agricultural activity. Phosphorus and nitrogen are the two most common of these nutrients. Excess nutrients cause eutrophication or increased plant growth called algae blooms in streams and lakes. The decomposition of these plants after death depletes the oxygen supply needed to support aquatic life. This results in the formation of "dead zones," or hypoxic water. These are areas where the concentration of dissolved oxygen in the water is so low that little life can survive. The intensification of agricultural and improper disposal of human and animal waste are key contributors to this type of pollution. According to the United Nations, eutrophication from excess nutrients is the most prevalent water-quality issue in the world.

The next category of water pollutants is organic chemicals. Organic chemicals are substances that contain a carbon chain. Two common organic chemical pollutants are PCBs (polychlorinated biphenyls) and DDT (a pesticide used to kill mosquitoes). Discharge of industrial waste and pesticide use are some of the key contributors to this type of water pollution. Water pollution from organic chemicals is problematic for several reasons. First, because they do not dissolve in water, they stay in the water for long periods of time. Thus, they affect the health of many organisms. Second, they bioaccumulate throughout the food chain. This is the process in which chemicals collect as one progresses up the food chain. Consuming organisms that are contaminated with organic chemicals can result in severe health problems.

Similar to organic chemicals, but lacking the carbon chain, are inorganic substances. This category refers to metals (e.g., arsenic, mercury, lead, and cadmium) and acids (e.g., sulfuric and nitric acids). Inorganic substances can enter the stream through activities such as improper disposal of waste, industry discharges, mining, and electrical generation. Many inorganic substances are harmful to human health and can bioaccumulate, like organic chemicals.

A thermal pollutant is one that changes the temperature of the water. While not specifically a substance, an increase in water temperature can harm or kill various forms of aquatic life. Thermal pollutants often result from large discharges of hot water into surface water at factory outfalls. However, climate change and habitat alteration, such as loss of coverage from foliage, could also lead to increases in water temperatures and therefore an alteration in water quality.

The final major category of water pollutants is biological pollutants. This category includes bacteria, viruses, and parasites. Animal and human feces are the most common source of biological pollution. Ingestion of biological pollutants by humans can cause a number of health problems that range in severity.

The Culture of Water Pollution Management

While water pollution is a measurable reality within societies, it is cultural differences rooted in the history of societies that influence how water pollution is perceived and managed. First, human societies have always depended on a supply of clean drinking water. Once a direct link was established between poor water quality and human health, societies focused on the advancement of water treatment technology to maintain this clean drinking supply. Through the Industrial Revolution, water pollution management revolved around the public health need of a safe source of drinking water. With the improvement of water quality through water treatment processes, morbidity and mortality rates in societies have declined.

Even within advanced societies, the purity of drinking water has remained a salient issue. Within these societies it is not just measurably safe water that underlies perceptions of water quality, but also culturally formed notions of "pure" and "pristine." This shift in perception can be seen in the drastic increase in bottled water consumption, which claimed a superior level of purity over tap water in the late 1980s and 1990s.

Increased attention to pollution of natural bodies of water from human activity occurred during the modern environmental movement of the 1960s and 1970s. This movement, influence by writings such as Rachel Carson's *Silent Spring* in 1962, focused public attention on how humans were causing harm to the environment and the impacts of this harm. At this time, humans began to demand clean water, and social movements surrounding environmental issues reflected this demand. This movement, in part, resulted in the formation of the federal Environmental Protection Agency (EPA) in 1970 and the passing of a series of legislation to protect natural bodies of water. This legislation was known as the Clean Water Act (CWA).

The CWA (passed in 1972 and amended in 1977) gave precedence to water pollution that resulted from industrial activities, which was reflective of the social movements and dominant modes of production of the time. This type of pollution falls into the broader category of point source pollution. Point source pollution is specifically defined by the EPA as pollution that enters a water supply from one known point. Point source pollution includes discharges from pipes, drains, channels, and so forth. The defining characteristics of this pollution are that the source of the contamination is readily identified and directly deposited into a body of water.

Point source pollution in the United States was drastically reduced after the passing of the CWA. In its expanded form, the CWA established the regulation of pollutant discharges into surface water and the standards of water quality for surface water. Specifically, this federal act made it unlawful to discharge any waste directly into a surface body of water without obtaining a National Pollutant Discharge Elimination System (NPDES) permit from the EPA. By implementing the requirement of a permit, the EPA was given authority to regulate and enforce standards on the substances directly discharged into surface water.

Since the 1970s, the response to water pollution has involved assessing water quality and enforcing policies to regulate point source pollution. In general, water quality is determined by comparing the physical and chemical characteristics of a sample of water to a set of water quality standards. Water quality guidelines for human usage have been successful in maintaining safe and clean water for humans. These water quality standards vary based on how the water supply will be used. However, water quality guidelines that ensure the viability of aquatic life have been difficult to set and maintain.

In the United States, water quality standards are set by the EPA. These standards involve assigning and maintaining designated uses of bodies of water as well as establishing specific drinking water standards. In addition, the EPA monitors whether a body of water is meeting the total maximum daily load requirements (TMDL). The TMDL is the level of a pollutant that a body of water can receive and still meet water quality standards. If a body of water is not able to meet its designated use and/or is above the TMDL requirement, it is classified as impaired.

A Societal Shift in Water Pollution Management

Although policy has focused on point source pollution, this is not the only way that water pollution occurs. Water pollutants can be transported to a body of water from a number of distant places. This is called nonpoint source pollution. This form of pollution does not enter a body of water at one specific location or discharge, but rather enters at diffuse locations often far downstream from a source of pollution. Nonpoint source pollution is a result of runoff carrying pollutants into surface bodies of water. Runoff is the excess storm water that flows over land and picks up contaminants as it drains into surface water. Nonpoint source pollution can also enter groundwater through a process called leaching. Leaching is the process of pollutants dissipating through land into water below. Nonpoint source pollution can originate from a large variety of activities such as excess application of fertilizers, oil and gas runoff from transportation corridors, excess nutrients from livestock waste, or faulty septic systems. The unifying characteristic of these nonpoint source pollutants is that they do not directly discharge into a body of water, but rather are transported to surface water through runoff or by leaching into groundwater.

Today, nonpoint source pollution is the leading cause of water-quality issues. Regulation of nonpoint source is more difficult than regulation of point source pollution because by definition the source of this form of pollution is not readily identifiable. By contrast, a wide variety of human activities contribute to nonpoint source pollution. Currently, the most prevalent nonpoint source pollution is inadequately treated human and animal waste. Failed on-lot septic systems and agricultural runoff are the leading contributors of this waste. Every individual is responsible for preventing nonpoint source pollution.

In response to the problem of nonpoint source pollution, the improvement of water quality has been approached from a watershed perspective. This approach looks not only at improving the water quality of a specific body of water, such as a single lake or stream, but also at the holistic ecosystem improvement of a watershed. A watershed is simply an area of land where all water that is under it or drains off of it goes into the same place. Watershed boundaries occur along ridges that divide the direction water flows. The concept behind a watershed approach is that what is done upstream affects everything else downstream.

At the policy level, programs are being developed that address nonpoint source pollution, most notably, water-quality trading programs. Rooted in the neoliberal market culture of society, these programs offer economic incentives for implementing management practices that reduce nonpoint source pollution.

Similarly, the focus of environmental social movements today reflects this shift in perceptions of water pollution. Rather than industrial discharge, there is an increased effort to raise awareness of how consumers' decisions can influence environmental outcomes and improve water quality. Industries face pressure to make products that are environmentally safe. Additionally, consumer pressure, through organic and local food movements, challenges the practices of large agricultural operations.

See Also: Environmental "Bads"; Environmental Protection Agency; Fish; Water (Bottled/Tap).

Further Readings

Chewonki. "Clean Water: Our Precious Resource." http://www.chewonki.org/cleanwater/water_pollution.asp (Accessed June 2010).

Liquid Assets: The Story of Our Water Infrastructure. Documentary film. University Park: Penn State Public Broadcasting, 2006. http://liquidassets.psu.edu (Accessed June 2010).

United Nations International Decade for Action: Water for Life (2006). http://www.un.org/waterforlifedecade/quality.html (Accessed June 2010).

U.S. Geological Survey. "Eutophication" (2008). http://toxics.usgs.gov/definitions/eutrophication.html (Accessed June 2010).

Water Encyclopedia. "Drinking Water and Society." http://www.waterencyclopedia.com/Da-En/Drinking-Water-and-Society.html (Accessed June 2010).

Anjel Stough-Hunter
The Ohio State University

WEEKENDS

Weekends are seen as a time for rejuvenation and escape; many people, like this couple, use the time for exploring area parks or nature trails.

Source: Sandra Friend/USDA Forest Service, Ocala National Forest

The weekend has been a cultural institution in the West for hundreds of years; the current institution of a two-day weekend and five-day workweek was first instituted at a New England cotton mill in order to accommodate both Jewish workers who recognized the Sabbath on Saturday and Christian workers who recognized it on Sunday. Labor unions spread the practice in the 1920s as the only one that was fair to workers without attempting to run businesses staffed only by Jews on Sundays and only by Christians on Saturdays. As Americans get older and their workweek takes more and more of their attention and dictates their schedule, the weekend becomes an important time for activity that feels more personal and expressive, a time not only for family and leisure but for spiritual development, and perhaps especially for spiritual development *through* family and leisure. The weekend is important in the "work–life balance" that psychologists have discussed as key to avoiding burnout, undue stress, and the stereotypical midlife crisis.

Common to many Americans' weekend experiences is the idea of "getting away from it all," a phrase that conjures up an escape not only from the headspace and labor of the workplace, but the fluorescent-lit, cubicle-bound, urban/suburban environment of it. One "gets away from it all" by escaping to the green areas of the public sphere, especially parks, lakeside and seaside areas, the countryside, or nature trails. There can even be a ritualistic feel to such weekends, conjuring up the idea of a periodic renewal through contact with nature.

Weekends, then, become a time when green consumers have the opportunity to disengage from the workweek in a clean, environmentally friendly way. There are a wide variety of green weekend opportunities both at and away from home, organized or informal. The weekend of course lends itself well to home improvement projects such as the installation of solar panels, a wind turbine, or an energy-efficient air conditioning system; it is also an ideal time for gardening or for doing home cooking, preparing lunches that can be brought to work in order to avoid a reliance on restaurants.

Many hiking tours are offered on the weekends, either self-guided or with a guide, and conservation areas and parks often look for volunteers to help with various simple on-site work. Green trips are a somewhat questionable proposition: Few forms of travel are truly green, although the per-passenger environmental impact of train and plane travel is comparatively reasonable if there are a higher number of passengers. The benefit of train travel is the scenic nature of the travel itself, in many parts of the country, although flying leaves more time for weekend activities, especially if taking a direct flight to the destination.

Bicycling tours are perfect for weekends, and there are more and more organized bicycling events, some designed for local residents, some as destination events for tourists. Bicycle tours are an excellent green way to explore a city and provide a fuller view than what is seen from the back seat of a taxi while waiting to arrive at one's destination. Responsible camping, swimming in natural bodies of water, kayaking, canoeing, fishing, bird-watching, and foraging for local wild foods (guides are available in much of the country) are all excellent green leisure activities.

See Also: Auto-Philia/Auto-Nomy; Ecotourism; Gardening and Lawns; Leisure; Nature Experiences.

Further Readings

Morgan, Erinn. *Picture Yourself Going Green: Step-by-Step Instruction for Living a Budget-Conscious, Earth-Friendly Lifestyle in Eight Weeks or Less*. Florence, KY: Course Technology PTR, 2009.

Redclift, Michael. *Sustainable Development*. London: Routledge, 1987.

Rogers, Elizabeth and Thomas Kostigen. *The Green Book*. New York: Three Rivers Press, 2007.

Bill Kte'pi
Independent Scholar

Werbach, Adam

Adam Werbach, founder of Act Now, an environmental consulting company, is one of the most widely recognized conservationists of his generation. He was also, at the age of 23, the youngest person to ever become the president of the Sierra Club, the oldest and largest environmental organization in the United States.

Born in Tarzana, California, in 1972, he took his first major environmental stance at the age of 8, when he circulated a petition among his second-grade classmates to dismiss Ronald Reagan's then-Secretary of the Interior, James Watt. While a student at Brown University pursuing a double major in political science and modern culture and media, Werbach founded the 30,000-member Sierra Student Coalition before becoming president of the national organization itself.

In 1997, two years after becoming president of the Sierra Club, he wrote *Act Now, Apologize Later*, a memoir intended to convince more people to participate in the quest for a safer and healthier environment. One year after the publication of this work, he founded Act Now, an organization ostensibly designed to engage the media and corporate worlds to get involved in economic, cultural, social, and environmental change. Ten years later, Act Now merged with global advertising giant Saatchi & Saatchi to form Saatchi & Saatchi S, the world's largest sustainability agency.

Seven years after founding Act Now, Werbach sent shockwaves through the environmental community when he delivered a talk in San Francisco titled "Is Environmentalism Dead?" (2004). In this speech, he declared the environmental movement dead and announced that he would no longer call himself an environmentalist. Saying that the movement was unprepared to reach out to the general public with an effective message to solve the underlying social and economic issues surrounding climate change and other environmental dilemmas, he threw down a challenge to the green movement to align its goals with broader social and economic initiatives in order to gain wider appeal for its cause.

In 2006, acting upon his belief that films are far better at bringing people together and spurring social change than the average political campaign, Werbach's agency Act Now launched Ironweed, a San Francisco–based company to produce media, help nonprofit organizations communicate their missions, and support independent filmmakers and movies that focus on subjects such as the environment, politics, and social responsibility.

Eighteen months after founding Ironweed, Werbach created more controversy when he accepted Walmart's offer to spearhead Walmart's sustainability efforts through its Personal Sustainability Project (PSP) initiative, a voluntary program designed for Walmart associates to integrate sustainability into their own lives by making small changes in their everyday routines. Many members of the Sierra Club (which funds the watchdog Walmart Watch) pleaded with him to reconsider joining forces with a corporation that he had previously criticized quite harshly for its abominable business practices and negative impacts on the environment. Opponents to this change in Werbach's direction argued that the world's largest retailer was engaging in consumer manipulation and greenwashing by hiring the former Sierra Club president and his agency. Others, however, responded that Walmart was not going to disappear, so it was pragmatic on Werbach's part to work with the company in order to change the status quo and push the company to make some positive changes in its operating procedures and behaviors.

In April 2008, Werbach returned to the Commonwealth Club to receive its 21st Century Visionary Award and delivered a follow-up speech to his 2004 declaration of the death of environmentalism in an address titled "The Birth of Blue." In this speech, he articulated a vision to broaden the green movement to what he terms a "blue" one, what he calls a billion-person-strong, consumer-based movement for sustainability with goals that reach far beyond those of the green movement. Elaborating on these ideas in greater detail, Werbach then published a book in 2009 titled *Strategy for Sustainability: A Business Manifesto*, a work in which he offers advice to businesses on how to incorporate policies and systems into their organizations for business longevity, profitability, and relevance. Striving to resolve the argument between sustainability activists and thinkers with diverging opinions about how to move from discussions concerning sustainability to its actual implementation in people's everyday lives, Werbach suggests that sustainability advocates need to engage people at the consumer level, a view that has also created a lot of controversy and disagreement within activist circles.

Werbach has appeared many times on television shows such as *Charlie Rose* and *Politically Incorrect* and was also chosen to host the environmental newsmagazine *Thin Green Line* on the Outdoor Life Network.

See Also: California AB 32; Communication, Global and Regional; Communication, National and Local; Corporate Green Culture; Corporate Social Responsibility; Environmental Communication, Public Participation; Environmental Media Association; Global Warming and Culture; Gore, Jr., Al; Green Consumerism; Green Jobs; Greenwashing; Media Greenwashing; Popular Green Culture; Print Media, Advertising; Public Opinion; Recycling; Sierra Club, Natural Resources Defense Council, and National Wildlife Federation

Further Readings

"Ironweed Film Club." http://www.ironweedfilms.com/about (Accessed September 2010).
Werbach, Adam. *Act Now, Apologize Later*. New York: Cliff Street Books, 1997.
Werbach, Adam. *Strategy for Sustainability: A Business Manifesto*. Boston, MA: Harvard Business Press, 2009.

Danielle Roth-Johnson
University of Nevada, Las Vegas

Work

The United Nations Environment Programme (UNEP) defines green jobs as those involving "work in agricultural, manufacturing, research and development, administrative, and service activities that contribute substantially to preserving or restoring environmental quality. Specifically, but not exclusively, this includes jobs that help protect ecosystems and biodiversity; reduce energy, materials, and water consumption through high-efficiency strategies; decarbonize the economy; and minimize or altogether avoid generation of all forms of waste or pollution." Green work is work that carries out not only formal environmental regulations that all industries are bound by, but with a deeper understanding of sustainability, energy efficiency, and the impact of human activity on the environment. Green jobs include environmental or sustainability consultants working with firms to advise them on sustainability issues; architects for green buildings; engineers to design green vehicles; alternative energy engineers; environmental lawyers; ecology educators; and sustainable farmers. But it also relies heavily on unskilled labor or vocational workers to implement the plans devised in those white-collar green jobs: the plumbers and electricians who do the actual work to make a building more efficient, the construction workers

"Green jobs" in the environmental sector, such as a technician for a solar panel array grid, focus on sustainability, energy efficiency, and the human impact on the environment.

Source: iStockphoto

who erect green buildings, wind farms, and hydroelectric generators and rebuild existing structures to make them more energy efficient. Though the term *green collar* has been in use since the environmental movement in the 1970s, the term *green jobs* has been more common in the 21st century, in part because of the dual usage of "green collar" (which can also refer to agricultural workers, regardless of the sustainability of their work). Since 2007, green job creation has been an increasingly prominent element in U.S. energy and economic policy.

The Green Jobs Initiative began in 2007 as a joint initiative by UNEP, the International Labor Organization, the International Trade Union Confederation, and the International Employers Organization (which joined in 2008) to promote the creation of green jobs and to encourage efforts by governments, employers, and trade unions to do likewise as a remedy for the challenges of climate change and environmental danger. As of 2010, the first phase of the Green Jobs Initiative was still under way: preparing a major comprehensive study of the emerging green economy and identifying green job opportunities and the green job-creation strategies that have worked best. The second phase is in the planning stages, when analysis of the information from the first phase will inform policy advice.

One of the issues concerning the initiative is the notion of "shades" of green. Jobs green in principle—even in intent—may not be green in practice if policies and procedures have not been correctly evaluated and put into place. Furthermore, the initiative is deeply concerned with ensuring that there are green jobs that constitute what it calls "decent work," jobs that people want to have, jobs that can compete for the best employees. Many waste management and recycling jobs as well as construction, biomass energy, and hazardous materials jobs do not so qualify. The creation of decent green jobs would be a step toward meeting two of the eight Millennium Development Goals (MDGs) adopted by the United Nations' member-states for 2015: poverty reduction (MDG 1) and protecting the environment (MDG 7). As conflicts between developed and developing nations have arisen in environmental challenges, and the consequences of climate change have been unfairly distributed to disproportionately affect the poor both on the local scale (poor neighborhoods) and the global scale (the global south and developing nations), this would be especially productive.

In the United States, the nongovernmental organization Green For All advocates socially responsible green investments and a transition to a "clean-energy economy" with green jobs as a solution to both environmental and economic malaise. Like many green movements and organizations, Green For All sees environmental and economic problems and disparities as having common solutions. Green For All programs include leadership development, local clean energy advocacy in target cities, "Green the Block" (a national campaign attempting to mobilize low-income communities of color in the call for clean energy), and green jobs projects. Green For All was one of the contributors helping the 2007 Green Jobs Act get passed and has been at work since advocating that the secretaries of labor and energy work together to create middle-class green construction jobs accessible to those in low-income communities.

The Green Jobs Act was introduced by Representatives Hilda Solis (D-CA) and John Tierney (D-MA) and authorized $125 million in funding to establish state and national green job training programs administered by the Department of Labor in an attempt both to create jobs as an economic remedy and to help move the country toward renewable power, biofuels development, and energy-efficient construction. The subsequent American Recovery and Reinvestment Act of 2009 allocated more money for green jobs in the energy, construction, and manufacturing sectors.

Anthony "Van" Jones, the founder of Green For All, was appointed by President Barack Obama as the first Special Adviser for Green Jobs, Enterprise, and Innovation in March 2009, but he resigned six months later after political skirmishes over his public criticisms of congressional Republicans and his apparent association with a Marxist group in the 1990s. There has been no replacement yet appointed as of this writing.

All jobs, green or otherwise, have some environmental impact. In addition to the resource consumption of the workplace and of particular business practices, there is the impact of commuting, particularly as suburban sprawl places many workers' homes farther and farther from their place of business. Office buildings, factories, retail stores, and other buildings devoted to work are frequently responsible for considerable resource usage, and may offer particular environmental challenges, whether it is their impact on the natural environment around them, or their impact on workers' health, because of the nature of the work conducted, or through issues like the spread of germs among personnel by recirculated air.

See Also: Body Burden; Commuting; Green Jobs; Leisure.

Further Readings

Cassio, Jim and Alice Rush. *Green Careers: Choosing Work for a Sustainable Future.* Gabriola Island, British Columbia, Canada: New Society Publishers, 2009.

Deitche, Scott M. *Green Collar Jobs: Environmental Careers for the 21st Century.* Westport, CT: Praeger, 2010.

Jones, Van with Ariane Conrad. *The Green Collar Economy.* New York: HarperOne, 2008.

Bill Kte'pi
Independent Scholar

Green Culture Glossary

A

Air Pollution: Contaminants or substances in the air that interfere with human health or produce other harmful environmental effects.

Albedo: The ratio of light reflected by a surface to the light falling on it.

Alternative Energy: Usually environmentally friendly, this is energy from uncommon sources such as wind power or solar energy, not fossil fuels.

Anthropocentric: Literally "human-centered." Anthropocentrism is the belief, taken for granted in most cultures through most of human history and argued more explicitly by some schools of philosophy today, that humans are the figurative "center of the universe," and that ethical systems should thus be principally concerned with human benefit.

Anthropogenic: Man-made; used especially to underscore the human origins of a substance or phenomenon, as in "anthropogenic climate change" or "anthropogenic toxic compounds."

B

Behavioral Change: As it affects energy efficiency, behavioral change is a change in energy-consuming activity originated by, and under control of, a person or organization. An example of behavioral change is adjusting a thermostat setting or changing driving habits.

Best Available Control Measures (BACM): A term used to refer to the most effective measures (according to Environmental Protection Agency guidance) for controlling small or dispersed particulates and other emissions from sources such as roadway dust, soot and ash from woodstoves, and open burning of rush, timber, grasslands, or trash.

Biodiversity: The total variety of life on Earth. Modern science considers biodiversity to be an inherently good thing for the ecosystem, and for the loss of species and of species diversity to be an alarming consequence of environmental damage. From an evolutionary standpoint, genetic diversity—the diversity of genes within a species—is also especially important.

Biomass: Any organic matter that is available on a renewable basis, including agricultural crops and agricultural wastes and residues, wood and wood wastes and residues, animal wastes, municipal wastes, and aquatic plants.

Brownfields: Abandoned, idled, or under-used industrial and commercial facilities/sites where expansion or redevelopment is complicated by real or perceived environmental contamination. They can be in urban, suburban, or rural areas. The Environmental Protection Agency's Brownfields Initiative helps communities mitigate potential health risks and restore the economic viability of such areas or properties.

C

Carbon Footprint: A popular term describing the impact a particular activity has on the environment in terms of the amount of climate-changing carbon dioxide and other greenhouse gases it produces. A person's carbon footprint is the amount of greenhouse gases that his or her way of life produces overall. It is also a colloquialism for the sum total of all environmental harm an individual or group causes over their lifetime. People, families, communities, nations, companies, and other organizations all leave a carbon footprint.

Carbon Offsets: Financial instruments, expressed in metric tons of carbon dioxide equivalent, which represent the reduction of carbon dioxide or an equivalent greenhouse gas. Carbon offsets allow corporations and other entities to comply with caps on their emissions by purchasing offsets to bring their totals down to acceptable levels. The smaller voluntary market for carbon offsets exists for individuals and companies that purchase offsets in order to mitigate their emissions by choice. There is a great deal of controversy over the efficacy and truthfulness of the offsets market, which is new enough that, in a best-case scenario, the kinks have not yet been worked out, while in the worst, it will turn out to be a dead end in the history of environmental reform.

Certified Organic: Food products that meet or exceed standards set forth by the U. S. Department of Agriculture National Organic Program (NOP). Products "made with organic ingredients" include 70 percent organic ingredients and cannot contain the organic label. "Organic" products must have at least 95 percent organic ingredients and may feature the USDA organic seal. "100% Organic" is the most stringent, but does not count water or salt.

Climate Change: A term used to describe short- and long-term effects on Earth's climate as a result of human activities such as fossil fuel combustion and vegetation clearing and burning.

Closed-Loop Recycling: The system of recycling in which a used product is broken down and remanufactured into a very similar, if not exactly the same, product (e.g., recycling a plastic soda bottle into a plastic water bottle).

Compost: A process whereby organic wastes, including food wastes, paper, and yard wastes, decompose naturally, resulting in a product rich in minerals and ideal for gardening and farming as a soil conditioner, mulch, resurfacing material, or landfill cover. Consumers can make their own compost by collecting yard trimmings and vegetable scraps.

Conscientious Consumption: An ethic that acknowledges the power of consumer activism in the movement toward sustainability.

Conservation: Preserving and renewing, when possible, human and natural resources.

D

Dioxin: Any of a family of compounds known chemically as dibenzo-p-dioxins. Concern about them has arisen from their potential toxicity as contaminants in commercial products. Tests on laboratory animals indicate that it is one of the more toxic anthropogenic (man-made) compounds.

E

Energy: The capability of doing work; different forms of energy can be converted to other forms, but the total amount of energy remains the same.

Energy Star: A joint program formed between the Environmental Protection Agency and the Department of Energy to identify and label high-efficiency building products.

Entropy: A measure of the unavailable or unusable energy in a system.

Environmental Equity/Justice: Equal protection from environmental hazards for individuals, groups, or communities regardless of race, ethnicity, or economic status. This applies to the development, implementation, and enforcement of environmental laws, regulations, and policies, and implies that no population of people should be forced to shoulder a disproportionate share of negative environmental impacts of pollution or environmental hazard due to a lack of political or economic strength.

Epidemiology: Study of the distribution of disease, or other health-related states and events in human populations, as related to age, sex, occupation, ethnicity, and economic status in order to identify and alleviate health problems and promote better health.

Ethics: The study of moral questions. Ethics can refer to specific types of ethics (such as applied ethics or medical ethics) or to specific systems of ethics (such as Catholic ethics or Marxist ethics). Though the religions of the world always include an ethical dimension to their belief systems, ethics and religion are not coequal, and the term "secular ethics" is sometimes used to describe systems of ethics that derive their conclusions from logic or moral intuition rather than from religious teachings or revealed truths. Secular ethics and religious ethics can and often do reach the same conclusions, and may do so by the same means; there are both secular and religious articulations of utilitarianism, for instance. Major types of ethics include descriptive ethics (which describes the values people live by in practice), moral psychology (the study of how moral thinking develops in the human species), and applied ethics (addressing the ethical concerns of specific real-life situations and putting ethics into practice).

Exposure: The amount of radiation or pollutant present in a given environment that represents a potential health threat to living organisms.

F

Fair Trade: A certification scheme that evaluates the economic, social, and environmental impacts of the production and trade of agricultural products, in particular coffee, sugar, tea, chocolate, and others. Fair trade principles include fair prices, fair labor conditions, direct trade, democratic and transparent organizations, community development, and environmental sustainability.

Farmers Market: A farmers market is a place where local farmers gather to sell their produce or specialty goods in a specific area at a designated time. All food bought at a farmers market is probably not produced using green or organic practices, but in general, the selection of organic food is broader than at a supermarket.

FSC: The Forest Stewardship Council (FSC) is an international nonprofit organization promoting responsible stewardship of the global forests. FSC certifies forests and forest products that fulfill their requirements for responsible forest stewardship.

Fugitive Emissions: Emissions not caught by a capture system.

G

Geothermal Energy: Any and all energy produced by the internal heat of the earth.

Greenhouse Effect: The warming of Earth's atmosphere attributed to a buildup of carbon dioxide or other gases. Some scientists think that this buildup allows the sun's rays to heat the Earth, while making the infrared radiation atmosphere opaque to infrared radiation, thereby preventing a counterbalancing loss of heat.

Greenhouse Gas Emissions: Any emissions that are released by humans (though naturally occurring in the environment), mainly through the combustion of fossil fuels, and have a warming potential as they persist in the atmosphere, contributing to the greenhouse effect.

Green Purchasing: The practice of selecting products and services that minimize the ecological impact of an individual or organization's day-to-day activities. Many organizations implement a green purchasing policy with guidelines for purchasing agents to select the "greenest" products and services available.

Greenwashing: A marketing ploy for businesses to jump onto the green movement bandwagon. They are not genuinely interested in sustainability, but are simply trying to improve their standing with the public by paying lip service. A company interested in "going green" for public relations reasons is greenwashing.

H

Hybrid Vehicle: Vehicles that use both a combustible form of fuel (gasoline, ethanol, etc.) and an electric motor to power them. Hybrid vehicles use less gasoline than a traditional combustion engine, and some even have an electric plug-in to charge the battery.

I

Irradiation: Exposure to radiation of wavelengths shorter than those of visible light (gamma, x-ray, or ultraviolet), for medical purposes, to sterilize milk or other foodstuffs, or to induce polymerization of monomers or vulcanization of rubber.

L

LD 50/Lethal Dose: The dose of a toxicant or microbe that will kill 50 percent of the test organisms within a designated period. The lower the LD 50, the more toxic the compound.

LEED (Leadership in Energy and Environmental Design): An organization that created the Green Building Rating System, which encourages and accelerates global adoption of sustainable green building and development practices through the creation and implementation of universally understood and accepted tools and performance criteria.

Life Cycle of a Product: All stages of a product's development, from extraction of fuel for power to production, marketing, use, and disposal.

Lifetime Exposure: Total amount of exposure to a substance that a human would receive in a lifetime (usually assumed to be 70 years).

Lowest Acceptable Daily Dose: The largest quantity of a chemical that will not cause a toxic effect, as determined by animal studies.

M

Megawatt: One thousand kilowatts, or 1 million watts. It is the standard measure of electric power plant generating capacity.

Moral Relativism: The acknowledgment that different cultures have different moral standards. There are various levels of moral relativism, from the weak descriptivist articulation that simply acknowledges and describes those differences, to the normative position which says that there is no universal moral standard, only culturally derived morals. Moderate positions often propose that there are certain key moral standards that form a universal ethical core, such as taboos on murder, incest, or parental neglect. The question of which moral standards are universal becomes important when cultures deal with one another and when international bodies mediate between them. Most differences are not about matters as obvious or seemingly clear-cut as murder, but may instead bear on matters of justice or on the distribution of responsibility. The questions of who has the responsibility to do something about climate change, or of the ethical importance of avoiding polluting behaviors, vary widely around the world. The opposite of relativism is universalism.

N

Net Metering: A method of crediting customers for electricity that they generate on-site in excess of their purchased electricity consumption. Customers with their own generation offset the electricity they would have purchased from their utility. If such customers generate more than they use in a billing period, their electric meter turns backward to indicate their net excess generation. Depending on individual state or utility rules, the net excess generation may be credited to the customer's account (in many cases at the retail price), carried over to a future billing period, or ignored.

Net-Zero Energy: Characteristic of a building that produces as much energy as it consumes on an annual basis, usually through incorporation of energy production from renewable sources such as wind or solar.

NGO: A nongovernmental organization operating independently from government that does not function as a private business. Also known as civil society organizations, these groups typically act in the public interest or at some broader political, cultural, or social goals.

NIMBY: An acronym for "not in my backyard" that identifies the tendency for individuals and communities to oppose noxious or hazardous materials and activities in their vicinity. It implies a limited or parochial political vision of environmental justice.

Normative: Describing how a thing ought to be. A normative statement describes what *should* be done, regardless of what *is* done. Normative ethics are concerned with how people ought to behave and what actions they ought to take. "Alcohol intake impairs judgment" is a descriptive statement; "Drunk driving is wrong" is a normative statement.

North–South: A model of the world that contrasts the industrialized, developed, wealthy countries of the global north with the developing, poorer countries of the global south. Geography applying to this model is partially figurative, with Australia and New Zealand included in the global north, and a number of African, Middle Eastern, and Asian nations in the northern hemisphere included in the global south. The term became popular in the wake of the Cold War, when a new way of distinguishing between the developed (first and second worlds) and developing (third world) was desired. However, while there are many political and cultural ties between the nations of the global north, the global south—much like the third world—is varied enough to invite criticism of the model's accuracy and usefulness.

P

Pacific Gyre: Otherwise known as the Great Pacific Garbage Patch, it is a gyre of small bits of marine garbage, including chemical sludge and pelagic plastic, thought to be larger than the state of Texas.

Persistent Toxic Chemicals, Persistent Pollutants: Detrimental materials, like Styrofoam or DDT, which remain active for a long time after their application and can be found in the environment years, and sometimes decades, after they were used.

Photochemical Smog: Air pollution caused by chemical reactions of various pollutants emitted from different sources.

Planned Obsolescence: The art of making a product break/fail after a certain amount of time. The failure of the product does not occur in a period of time that you will blame the manufacturer, but soon enough for you to buy another one and make more profit for the manufacturer.

Political Ecology: A field of research concerned with the relationship of systems of social and economic power to environmental conditions, natural resources, and conservation.

Pollution: Generally, the presence of a substance in the environment that, because of its chemical composition or quantity, prevents the functioning of natural processes and produces undesirable environmental and health effects. Under the Clean Water Act, for example, the term has been defined as the man-made or man-induced alteration of the physical, biological, chemical, and radiological integrity of water and other media.

Pollution Prevention: Identifying areas, processes, and activities that create excessive waste products or pollutants in order to reduce or prevent them through alteration, or eliminating a process. Such activities, consistent with the Pollution Prevention Act of 1990, are conducted across all Environmental Protection Agency programs and can involve cooperative efforts with such agencies as the Departments of Agriculture and Energy.

Polychlorinated Biphenyls: A group of toxic, persistent chemicals used in electrical transformers and capacitors for insulating purposes, and in gas pipeline systems as lubricant. The sale and new use of these chemicals, also known as PCBs, were banned by law in 1979.

Postconsumer Waste: In the recycling business, material that has already been used and discarded by consumers, as opposed to manufacturing waste. Using products with "postconsumer" recycled content actually keeps waste out of landfills and incinerators, unlike "postindustrial" recycled content, most of which would get recycled anyway.

Power: Energy that is capable or available for doing work; the time rate at which work is performed.

Precautionary Principle: A philosophy which states that policy makers should not wait for scientific proof of harmful effects before taking steps to limit harmful environmental and human health impacts from new products or activities. Specific areas of application include genetically modified food products and chemicals that may have harmful developmental effects in low doses.

R

Radioactive Waste: Any waste that emits energy as rays, waves, streams, or energetic particles. Radioactive materials are often mixed with hazardous waste, from nuclear reactors, research institutions, or hospitals.

Recycling: The process by which materials that would otherwise become solid waste are collected, separated or processed, and reused in the form of raw materials or finished goods.

Risk: A measure of the probability that damage to life, health, property, and/or the environment will occur as a result of a given hazard.

Risk Assessment: Qualitative and quantitative evaluation of the risk posed to human health and/or the environment by the actual or potential presence and/or use of specific pollutants.

S

Semiconductor: Any material that has a limited capacity for conducting an electric current.

Smog: Air pollution typically associated with oxidants. The word is a portmanteau of "smoke" and "fog."

Sustainability: To give support to, relief to, to carry, withstand, or to meet the needs of the present without compromising the ability of future generations to meet their needs.

Sustainable Seafood: The act of not overfishing, which causes the possibility of extinction or adverse effects on a habitat.

T

Toxicity: The degree to which a substance or mixture of substances can harm humans or animals.

Turbine: A device for converting the flow of a fluid (air, steam, water, or hot gases) into mechanical motion.

U

U.S. Department of Agriculture (USDA): Established by President Lincoln in 1862, the USDA is an umbrella organization encompassing all aspects of farming production that has executive and legislative authority to assure food safety and protect national resources. Active operating units include the National Organic Program, Agricultural Resource Service, Food Safety and Inspection Service, Risk Management Agency, and Animal and Plant Health Inspection Service.

V

VOCs (Volatile Organic Compounds): Gases emitted from liquid or solid substances that may cause short-term and long-term harmful health effects. Examples of products containing VOCs include paints and lacquers, paint strippers, cleaning supplies, pesticides, building materials and furnishings, office equipment such as copiers and printers, correction fluids and carbonless copy paper, graphics and craft materials including glues and adhesives, permanent markers, and photographic solutions.

W

Water Pollution: Includes chemicals and debris that render water unusable for natural habitat, human consumption, and recreation.

Watershed Approach: A coordinated framework for environmental management that focuses public and private efforts on the highest-priority problems within hydrologically defined geographic areas taking into consideration both ground and surface water flow.

Wildlife Refuge: An area designated for the protection of wild animals, within which hunting and fishing are either prohibited or strictly controlled.

Dustin Mulvaney
University of California, Berkeley

Sources: U.S. Environmental Protection Agency (http://www.epa.gov/OCEPAterms), U.S. Energy Information Administration (http://www.eia.doe.gov/tools/glossary)

Green Culture Resource Guide

Books

Adams, Anthony. *Your Energy-Efficient House: Building and Remodeling Ideas.* Charlotte, VT: Garden Way Publishing, 1975.
Albright, Horace M. and Marian Albright Schenck. *Creating the National Park Service: The Missing Years.* Norman: University of Oklahoma Press, 1999.
Ali, Saleem H., ed. *Peace Parks: Conservation and Conflict Resolution.* Cambridge, MA: MIT Press, 2007.
Beder, Sharon. *Global Spin: The Corporate Assault on Environmentalism.* London: Green Books, 2002.
Bekoff, Marc. *Minding Animals: Awareness, Emotion and Heart.* Oxford, UK: Oxford University Press, 2002.
Belasco, Warren. *Appetite for Change: How the Counterculture Took on the Food Industry.* New York: Pantheon Books, 1989.
Brooks, Karl B. *Before Earth Day: The Origins of American Environmental Law, 1945–1970.* Lawrence: University of Kansas Press, 2009.
Canavari, Maurizio and Kent D. Olson. *Organic Food: Consumer's Choices and Farmer's Opportunities.* New York: Springer, 2007.
Carson, Rachel. *Silent Spring.* Boston, MA: Houghton Mifflin, 1962.
Chiras, Daniel. *The Solar House.* White River Junction, VT: Chelsea Green Publishing, 2002.
Chryssavgis, John, ed. *Cosmic Grace, Humble Prayer: The Ecological Vision of the Green Patriarch Bartholomew.* Grand Rapids, MI: Wm. B. Eerdmans, 2009.
Connell, John. *Homing Instinct: Using Your Lifestyle to Design and Build Your Home.* New York: Warner Books, 1993.
Connelly, Matthew. *Fatal Misconception: The Struggle to Control World Population.* Cambridge, MA: Harvard University Press, 2008.
Cox, Robert J. *Environmental Communication and the Public Sphere.* Thousand Oaks, CA: Sage, 2009.
Critser, Greg. *Fat Land.* Boston, MA: Houghton Mifflin, 2003.
Dawkins, Marion. *Animal Suffering: The Science of Animal Welfare.* London: Chapman & Hall, 1980.

Deffeyes, Kenneth S. *Beyond Oil: The View From Hubbert's Peak*. New York: Hill & Wang, 2005.

DeLuca, Kevin Michael. *Image Politics: The New Rhetoric of Environmental Activism*. New York: Guilford Press, 1999.

Dessler, Andrew E. and Edward A. Parson. *The Science and Politics of Global Climate Change: A Guide to the Debate*. New York: Cambridge University Press, 2006.

Dunlap, Riley E. and William Michelson. *Handbook on Environmental Sociology*. Westport, CT: Greenwood Press, 2003.

Duram, Leslie A. *Good Growing: Why Organic Farming Works*. Lincoln: University of Nebraska Press, 2005.

Ehrlich, Paul. *The Population Bomb*. New York: Ballantine Books, 1968.

Epstein, Barbara. *Political Protest and Cultural Revolution*. Berkeley: University of California Press, 1991.

Esty, Daniel and Andrew Winston. *From Green to Gold: How Smart Companies Use Environmental Strategy to Innovate, Create Value, and Build Competitive Advantage*. New Haven, CT: Yale University Press, 2006.

Fischer, Frank. *Citizens, Experts, and the Environment: The Politics of Local Knowledge*. Durham, NC: Duke University Press, 2000.

Fisher, Dana R. *National Governance and the Global Climate Change Regime*. Lanham, MD: Rowman & Littlefield, 2004.

Fletcher, K. *Sustainable Fashion and Textiles: Design Journeys*. London: Earthscan, 2008.

Fromartz, Samuel. *Organic, Inc.: Natural Foods and How They Grew*. New York: Harcourt, 2006.

Frome, Michael. *Whose Woods These Are: The Story of the National Forests*. New York: Doubleday, 1962.

Gibbs, Lois. *Love Canal: My Story*. Albany: State University of New York Press, 1982.

Gould, K. A. and T. L. Lewis, eds. *Twenty Lessons in Environmental Sociology*. New York: Oxford University Press, 2009.

Guthman, Julie. *Agrarian Dreams: The Paradox of Organic Farming in California*. Berkeley, CA: University of California Press, 2004.

Hawken, P. *The Ecology of Commerce: A Declaration of Sustainability*. New York: HarperBusiness, 1993.

Hays, Samuel P. *Conservation and the Gospel of Efficiency: The Progressive Conservation Movement*. New York: Atheneum, 1980.

Hays, Samuel P. *Environmental Politics Since 1945*. Pittsburgh, PA: University of Pittsburgh Press, 2000.

Hemenway, Toby. *Gaia's Garden—A Guide to Home-Scale Permaculture*. White River Junction, VT: Chelsea Green Publishing, 2001.

Henderson, H. and W. Woolner, eds. *FDR and the Environment*. New York: Palgrave Macmillan, 2005.

Hethorn, J. and C. Ulasewicz, eds. *Sustainable Fashion: Why Now?* New York: Fairchild, 2008.

Hird, John A. *Superfund: The Political Economy of Environmental Risk*. Baltimore, MD: Johns Hopkins University Press, 1994.

Holmgren, David. *Permaculture: Principles and Pathways Beyond Sustainability*. Hepburn, Australia: Holmgren Design Services, 2002.

Jaffee, Daniel. *Brewing Justice: Fair Trade Coffee, Sustainability, and Survival.* Berkeley: University of California Press, 2007.
Jones, Van. *The Green-Collar Economy: How One Solution Can Fix Our Two Biggest Problems.* New York: HarperOne, 2008.
Juris, Jeffrey S. *Networking Futures: The Movements Against Corporate Globalization.* Durham, NC: Duke University Press, 2008.
Katsifiacas, George. *The Subversion of Politics.* Atlantic Highlands, NJ: Humanities Press, 1997.
Klare, Michael T. *Resource Wars: The New Landscape of Global Conflict.* New York: Owl Books, 2002.
Kraft, Michael E. *Environmental Policy and Politics.* Upper Saddle River, NJ: Pearson Longman, 2007.
Landy, Marc A., Marc J. Roberts, and Stephen R. Thomas. *Environmental Protection Agency: Asking the Wrong Questions, From Nixon to Clinton.* New York: Oxford University Press, 1994.
Lappé, Frances Moore. *Diet for a Small Planet.* New York: Ballantine Books, 1971.
Latouche, Serge. *Farewell to Growth.* Cambridge, UK: Polity Press, 2009.
Lazarus, Richard J. *The Making of Environmental Law.* Chicago: University of Chicago Press, 2006.
Levenstein, Harvey. *Paradox of Plenty: A Social History of Eating in Modern America.* Berkeley: University of California Press, 2003.
Mazur, Alan. *A Hazardous Inquiry: The Rashomon Effect at Love Canal.* Cambridge, MA: Harvard University Press, 1998.
McMichael, P. *Development and Social Change: A Global Perspective.* Thousands Oaks, CA: Pine Forge Press, 2008.
Meikle, Grahame. *Future Active: Media Activism and the Internet.* New York: Routledge, 2002.
Nelson, Willie and Turk Pipkin. *The Tao of Willie: A Guide to the Happiness in Your Heart.* New York: Gotham Books, 2006.
Nestle, Marion. *Food Politics.* Berkeley: University of California Press, 2002.
Newell, Peter. *Climate for Change: Non-State Actors and the Global Politics of the Greenhouse.* New York: Cambridge University Press, 2000.
Palmer, A. and H. Clark, eds. *Old Clothes, New Looks: Second Hand Fashion.* Oxford, UK: Berg, 2005.
Podobnik, Bruce. *Global Energy Shifts: Fostering Sustainability in a Turbulent Age.* Philadelphia, PA: Temple University Press, 2006.
Pollan, Michael. *Omnivore's Dilemma.* New York: Penguin, 2006.
Pollan, Michael. *Second Nature: A Gardener's Education.* New York: Dell, 1991.
Purchase, Graham. *Anarchism and Ecology.* Montreal: Black Rose Books, 1997.
Purchase, Graham. *Anarchism and Environmental Survival.* Tucson, AZ: Sharp Press, 1994.
Runte, Alfred. *National Parks, the American Experience.* Lincoln: University of Nebraska Press, 1987.
Schlosser, Eric. *Fast Food Nation.* New York: Houghton Mifflin, 2001.
Sellars, Richard West. *Preserving Nature in the National Parks: A History.* New Haven, CT: Yale University Press, 1997.
Shiva, Vandana. *Water Wars: Privatization, Pollution, and Profit.* Cambridge, MA: South End Press, 2002.

Singer, Peter. *Animal Liberation*. New York: HarperCollins, 2001.
Singer, Peter, ed. *In Defense of Animals: The Second Wave*. Oxford, UK: Wiley-Blackwell, 2005.
Smil, Vaclav. *Energy in Nature and Society: General Energetics of Complex Systems*. Cambridge, MA: MIT Press, 2008.
Szasz, Andrew. *EcoPopulism: Toxic Waste and the Movement for Environmental Justice*. Minneapolis: University of Minnesota Press, 1994.
Tallamy, Douglas W. *Bringing Nature Home: How You Can Sustain Wildlife With Native Plants*. Portland: Timber Press, 2006.
Ward, Colin. *Anarchism: A Very Short Introduction*. New York: Oxford University Press, 2004.
Wark, Kenneth, Cecil F. Warner, and Wayne T. Davis. *Air Pollution: Its Origin and Control*. Menlo Park, CA: Addison-Wesley, 1998.
Zimring, Carl A. *Cash for Your Trash: Scrap Recycling in America*. New Brunswick, NJ: Rutgers University Press, 2005.

Journals

Agriculture and Human Values
Audobon

Earth First! Journal
ecohome Magazine
The Ecologist
Environment & Behavior
E – The Environmental Magazine

Forest Products Journal

Green Source

International Journal of Environment and Health
International Journal of Environment and Sustainable Development

Journal for Nature Conservation
Journal of Environmental Biology
Journal of Environmental Education
Journal of Environmental Science & Health
Journal of Forestry
Journal of International Wildlife Law and Policy
Journal of Popular Culture
Journal of the History of Ideas

Mother Earth News

Natural Life

One Planet
Organic Family

Sierra Magazine

Urban Farm

Vegetarian Times

Whole Living

Websites

Best Green Blogs
www.bestgreenblogs.com

Brighter Planet Blog
numbers.brighterplanet.com

CorpWatch
www.corpwatch.org

The Daily Green
www.thedailygreen.com

Earth Charter
www.earthcharter.org.au

Environmental Justice
www.epa.gov/oecaerth/environmentaljustice

Environmental Media Association
www.ema-online.org

Ethical Weddings
www.ethicalweddings.com/blog

Farm to School
www.farmtoschool.org

Forest History
www.foresthistory.org

Global Greens
www.globalgreens.org

Green Blog
www.green-blog.org

Green Highways Partnership
www.greenhighways.org

Green Lashes and Fashion
www.greenlashesandfashion.blogspot.com

The Green Parent
www.thegreenparent.com

The Greenwashing Index
www.greenwashingindex.com

Grist: A Beacon in the Smog
www.grist.org

Little Green Stilettos
www.littlegreenstilettos.com

Mother Nature Network
www.mnn.com

National Geographic's Green Guide
www.thegreenguide.com

The Natural Nursery Blog
www.naturalnurseryblog.co.uk

No Impact Man
www.noimpactman.typepad.com

The Oil Drum
www.theoildrum.com

Pew Center on Global Climate Change
www.pewclimate.org

Practice Greenhealth
www.practicegreenhealth.org

Sierra Club
www.sierraclub.org

Smart Growth
www.smartgrowth.org

Style Saves the World
www.stylesavestheworld.blogspot.com

TreeHugger
www.treehugger.com

United Nations Environment Programme
www.unep.org

United States Department of Agriculture
www.usda.gov

The Vegetarian Resource Group
www.vrg.org

The Wilderness Society
www.wilderness.org

Green Culture Appendix

CorpWatch: Holding Corporations Accountable

www.corpwatch.org

The official website of CorpWatch, a nonprofit research and journalism organization based in San Francisco that investigates corporate behavior and advocates for corporate accountability and transparency. Information on the website is organized by industry and by issue, in either case providing a short summary of issues and links to news articles and other sources of information. Industries covered are chemicals, construction, energy, food and agriculture, manufacturing, media and entertainment, natural resources, pharmaceuticals, retail and mega-stores, technology and telecommunications, tobacco, tourism and real estate, transportation and war, and disaster profiteering. Issues covered are consumerism, corruption, environment, executive compensation, globalization, health, human rights, labor, money and politics, privatization, regulation, trade justice, and world financial institutions. The CorpWatch Hands-On Research Guide offers detailed, step-by-step suggestions for researching corporations (for instance, to support lawsuits, activist campaigns, and investigative articles) and is hosted on this website, as is the CorpWatch blog.

Environmental Justice

www.epa.gov/oecaerth/environmentaljustice

The website run by the U.S. Environmental Protection Agency (EPA), which explains the concept of environmental justice, reviews the history of the environmental justice movement, including the work of the National Environmental Justice Advisory Council (created in 1992), and provides information about environmental justice resources, including local contacts for different regions of the country. It includes information about grants and programs administered by the EPA, information about the Federal Interagency Working Group (established in 1994), and the Environmental Justice Achievement Awards. Multimedia resources, including podcasts, relating to environmental justice may be downloaded from the site, and there are many print information resources available for reference or download from the website, including policy documents, fact sheets, and planning documents, as well as a searchable bibliography (EJBib online) that currently contains references to about 4,000 documents.

Fair-Trade Labeling Organizations International: Tackling Poverty and Empowering Producers Through Trade

www.fairtrade.net

The website of an international coalition of 24 organizations from three continents headquartered in Bonn, Germany. The coalition set international standards for fair trade and supports fair-trade producers. This organization controls the FAIR TRADE certification mark, which is now used in over 50 countries in the world to signify products that meet fair-trade standards. The website includes basic information about fair trade and its history; annual reports of Fair-Trade Labeling Organization (FLO) International; statistics about fair trade; the certification process; standards for different products; multimedia information including films, brochures, and publications; news items about fair trade; and a searchable database of minimum prices and premiums for different products.

The Greenwashing Index

www.greenwashingindex.com

This website, run by Enviromedia Social Marketing (a social marketing agency working in the field of sustainability consulting) and the University of Oregon School of Journalism and Communication, collects information about greenwashing, a practice in which a company or organization claims to be following environmental practices but in fact does so in a minor way for the purposes of advertising (therefore analogous to whitewashing, but in a "green" or environmental sense). The website allows users to post ads that make environmental claims and they are then rated by other users on a scale of 1 (authentic) to 5 (bogus), and previously posted ads are also available for view (with filters by date and subject, such as agriculture, financial, or retail). The site also explains basic concepts relating to greenwashing, posts expert commentary on greenwashing issues, and collects links to news items about greenwashing.

Smart Growth Online

www.smartgrowth.org/default.asp

A website supported by the U.S. Environmental Protection Agency and maintained by the National Center for Appropriate Technology. It provides an overview of the concept of smart growth, focused on maintaining older towns and suburbs, preserving green space, developing mixed land use that combines housing, commercial, and retail uses, and developing mass transit and pedestrian options for transportation. Information is organized by issue including community quality of life, design, economics, environment, health, housing, and transportation. The website has information on the Smart Growth Network (founded in 1996) and the Smart Growth Speaker Series, including podcasts and videocasts of previous talks. The website also includes an index of smart growth news (by date and state) and a number of online tools, reports, and case studies relevant to smart growth.

The Vegetarian Resource Group

www.vrg.org

This website, run by the Baltimore-based Vegetarian Resource Group (VRG), offers two types of information: that provided from the VRG and links to information from other organizations such as the American Dietetic Association. Categories include nutrition, recipes, information for vegans, foodservice, restaurants, the environment, vegetarian business, and teens, family, and kids. The website includes data from several years of polls conducted by Harris Interactive, a market research firm, on the number of adult and teenage vegetarians in the United States and links to polls and studies conducted from other organizations addressing topics such as trends in marketing organic foods. Many guides, articles, and handouts (some in Spanish) from the VRG, covering topics like vegetarian fast food items and vegan nutrition for kids, are available for reading and/or download from the website. Press releases from the VRG are available from the website, and the *Vegetarian Journal* (published by the VRG) may be read online.

Worldchanging

www.worldchanging.com

The website of the nonprofit media organization Worldchanging.com, a global network of independent journalists founded in 2003 and headquartered in Seattle, Washington. News reports and analysis are provided by correspondents around the world, and the editorial staff is led by Executive Editor Alex Steffen and Managing Editor Amanda Reed. Besides a "front page" of recent news reports, information is organized into seven categories: (1) Stuff (including collaboration, emerging technologies, food and farming, purchasing green, and sustainable design); (2) Shelter (including green building, energy, water, and refugees and relief); (3) Cities (including leapfrogging, megacities, transportation, and urban planning and design); (4) Community (including arts, education, empowering women, health, philanthropy, and social entrepreneurship); (5) Business (including branding and marketing, green economy, socially responsible investment, and transforming business); (6) Politics (including communications and networking, media, transparency, human rights, and activism); and (7) Planet (including collaboration, emerging technologies, food and farming, purchasing green, and sustainable design). The website also includes a calendar of Worldchanging events and links to three local blogs (for Seattle, Denver, and Canada).

Sarah Boslaugh
Washington University in St. Louis

Index

Article titles and their page numbers are in **bold.**